D1519197

LINEAR WATER WAVES

The book gives a self-contained and up-to-date account of mathematical results in the linear theory of water waves. The study of these waves has many applications, including naval archtecture, ocean engineering, and geophysical hydrodynamics. The book is divided into three sections that cover the linear boundary value problems serving as the approximate mathematical models for time-harmonic waves, ship waves on calm water, and unsteady waves, respectively. These problems are derived from physical assumptions set forth in the introductory chapter, in which the linearization procedure is also described for the nonlinear boundary conditions on the free surface. In the rest of the book, a plethora of mathematical techniques is applied for investigation of the problems. In particular, the reader will find integral equations based on Green's functions, various inequalities involving the kinetic and potential energy, and integral identities. These tools are applied for establishing conditions that provide the existence and uniqueness of solutions, and their asymptotic behavior at infinity and near singularities of the boundary of the water domain. Examples of nonuniqueness usually referred to as "trapped modes," are constructed with the help of the so-called inverse procedure. For time-dependent problems with rapidly stabilizing and high-frequency boundary data, the perturbation method is used for obtaining the asymptotic behavior as the perturbation parameter tends to a limiting value.

Linear Water Waves will serve as an ideal reference for those working in fluid mechanics and engineering, as well as a source of new applications for those interested in partial differential equations of mathematical physics.

Nikolay Kuznetsov heads the Laboratory for Mathematical Modelling of Wave Phenomena at the Institute of Problems in Mechanical Engineering, Russian Academy of Sciences, St. Petersburg, where he has worked since 1986. Before that, Dr. Kuznetsov taught Mathematics at the Leningrad Shipbuilding Institute for nearly 15 years. The mathematical theory of water waves is the main topic of his research.

Vladimir Maz'ya is a professor of Applied Mathematics at the University of Linköping, Sweden, where he has worked since 1990. He is the author of 15 books and about 400 papers on various problems of applied mathematics, differential equations, functional analysis, spectral theory, asymptotic analysis, and the history of mathematics.

Boris Vainberg is a professor of Mathematics at the University of North Carolina at Charlotte. Before coming to the United States, Dr. Vainberg taught Mathematics at the Moscow State University for nearly 30 years. He is the author of two monographs and numerous papers on differential equations, scattering theory, and spectral theory.

LINEAR WATER WAVES

A Mathematical Approach

N. KUZNETSOV
Russian Academy of Sciences, St. Petersburg

V. MAZ'YA
University of Linköping, Sweden

B. VAINBERG
University of North Carolina at Charlotte

PUBLISHED BY THE PRESS SYNDICATE OF THE UNIVERSITY OF CAMBRIDGE
The Pitt Building, Trumpington Street, Cambridge, United Kingdom

CAMBRIDGE UNIVERSITY PRESS
The Edinburgh Building, Cambridge CB2 2RU, UK
40 West 20th Street, New York, NY 10011-4211, USA
477 Williamstown Road, Port Melbourne, VIC 3207, Australia
Ruiz de Alarcón 13, 28014 Madrid, Spain
Dock House, The Waterfront, Cape Town 8001, South Africa

http://www.cambridge.org

First published 2002

Printed in the United Kingdom at the University Press, Cambridge

Typeface Times Roman 10.25/13 pt. *System* LaTeX 2ε [TB]

A catalog record for this book is available from the British Library

Library of Congress Cataloging in Publication Data

ISBN 0 521 80853 7 hardback

To our wives
Natasha, Tatyana, and Anastasia

Contents

Preface *page* xi

Introduction: Basic Theory of Surface Waves 1
Mathematical Formulation 1
Linearized Unsteady Problem 5
Linear Time-Harmonic Waves (the Water-Wave Problem) 10
Linear Ship Waves on Calm Water (the Neumann–Kelvin Problem) 15

1. Time-Harmonic Waves 19
1. Green's Functions 21
1.1. Three-Dimensional Problems of Point Sources 21
1.2. Two-Dimensional and Ring Green's Functions 37
1.3. Green's Representation of a Velocity Potential 42
1.4. Bibliographical Notes 48
2. Submerged Obstacles 50
2.1. Method of Integral Equations and Kochin's Theorem 50
2.2. Conditions of Uniqueness for All Frequencies 67
2.3. Unique Solvability Theorems 87
2.4. Bibliographical Notes 96
3. Semisubmerged Bodies, I 99
3.1. Integral Equations for Surface-Piercing Bodies 99
3.2. John's Theorem on the Unique Solvability and Other Related Theorems 116
3.3. Bibliographical Notes 140
4. Semisubmerged Bodies, II 142
4.1. Trapped Waves 143

4.2. Uniqueness Theorems 164
4.3. Bibliographical Notes 212
5. Horizontally Periodic Trapped Waves 214
5.1. Two Types of Trapped Modes 215
5.2. Edge Waves 219
5.3. Trapped Modes Above Submerged Obstacles 229
5.4. Waves in the Presence of Surface-Piercing Structures 237
5.5. Vertical Cylinders in Channels 254

2. Ship Waves on Calm Water 263
6. Green's Functions 265
6.1. Three-Dimensional Problem of a Point Source in Deep Water 265
6.2. Far-Field Behavior of the Three-Dimensional Green's Function 283
6.3. Two-Dimensional Problems of Line Sources 305
6.4. Bibliographical Notes 315
7. The Neumann–Kelvin Problem for a Submerged Body 318
7.1. Cylinder in Deep Water 319
7.2. Cylinder in Shallow Water 341
7.3. Wave Resistance 349
7.4. Three-Dimensional Body in Deep Water 353
7.5. Bibliographical Notes 359
8. Two-Dimensional Problem for a Surface-Piercing Body 361
8.1. General Linear Supplementary Conditions at the Bow and Stern Points 362
8.2. Total Resistance to the Forward Motion 393
8.3. Other Supplementary Conditions 396
8.4. Trapped Modes 405
8.5. Cylinder in the Supercritical Stream 411
8.6. Bibliographical Notes 415

3. Unsteady Waves 419
9. Submerged Obstacles: Existence and Properties of Velocity Potentials 421
9.1. The Initial-Boundary Value Problem and an Auxiliary Steady-State Problem 421
9.2. Operator Equation for the Unsteady Problem 425
9.3. Main Results 427
9.4. Bibliographical Notes 432

10. Waves Caused by Rapidly Stabilizing and High-Frequency Disturbances 435
10.1. Rapidly Stabilizing Surface Disturbances 435
10.2. Rapidly Stabilizing Underwater Disturbances 453
10.3. High-Frequency Surface Pressure 460
10.4. High-Frequency Underwater Disturbances 474
10.5. Bibliographical Notes 484

Bibliography 485
Name Index 505
Subject Index 509

Preface

> Now, the next waves of interest, that are easily seen by everyone and which are usually used as an example of waves in elementary courses, are water waves. As we shall soon see, they are the worst possible example, because they are in no respect like sound and light; they have all the complications that waves can have.
>
> —*The Feynman Lectures on Physics*, Vol. 1, Section 51.4 (86)

The aim of the present book is to give a self-contained and up-to-date account of mathematical results in the linear theory of water waves. The study of different kinds of waves is of importance for various applications. For example, it is required for predicting the behavior of floating structures (immersed totally or partially) such as ships, submarines, and tension-leg platforms and for describing flows over bottom topography. Furthermore, the investigation of wave patterns of ships and other vehicles in forward motion is closely related to the calculation of the wave-making resistance and other hydrodynamic characteristics that are used in marine design. Another area of application is the mathematical modeling of unsteady waves resulting from such phenomena as underwater earthquakes, blasts, and the like.

The history of water wave theory is almost as old as that of partial differential equations. Their founding fathers are the same: Euler, Lagrange, Cauchy, Poisson. Further contributions were made by Stokes, Lord Kelvin, Kirchhoff, and Lamb, who constructed a number of explicit solutions. In the 20th century, Havelock, Kochin, Sretensky, Stoker, John, and others applied the Fredholm theory of boundary integral equations to the field of water waves.

There are several general expositions of the classical theory by Crapper [42], Lamb [179], Lighthill [201], Sretensky [310], Stoker [312], Wehausen and Laitone [354], and Whitham [359]. Various aspects of the linear theory of water waves were considered in works of Havelock and Ursell and can be found in their collected papers (see [111] and [342], respectively). Other works are focused on various applied aspects of the theory. In particular, Haskind [106], Mei [242], Newman [262, 263], and Wehausen [352] consider the wave–body interaction. Also, there is the very recent monograph by Linton and McIver [208] on the mathematical methods used in the theory of such interactions, but it mainly discusses mathematical techniques from the point of view of their applications in ocean engineering. Problems in the theory of ship waves and wave resistance are considered by Kostyukov [147], Bhattacharya [26], Timman, Hermans, and Hsiao [318], and Wehausen [353], but like [208] these works illuminate those problems in a way more appropriate for applied research. There are books by Debnath [46] and Ovsyannikov et al. [273] concerned with nonlinear waves. However, there is no monograph on the progress achieved in the more mathematical approach to the linear water-wave theory during the last few decades.

Although the decades after World War II have brought a renewed interest in both mathematical and applied aspects of the theory, some fundamental questions still remained open. A number of (at the time) unsolved problems were listed by Ursell in 1992 [341]. Since then, substantial progress has been achieved. The new results and methods developed for obtaining them together with those dating from the 1970s and 1980s form the core of this book. We give an account of the state of the art in the field providing the reader with modern tools for further research. It is worth mentioning that these tools are not only applicable to problems of water waves but also have a much wider range of usage. Integral identities and energy inequalities for proving uniqueness theorems, the inverse procedure for constructing non-uniqueness examples, various versions of the integral equations method for solving boundary value problems, and asymptotic expansions for both transient and steady-state problems represent several of the techniques used in the book, and the list can be continued.

The book is arranged in three parts, each treating one of the main themes, which are, respectively, as follows: time-harmonic waves, waves caused by the uniform forward motion of a body on calm water, and unsteady waves. Also, there is an introductory chapter preceding Part 1 that is concerned with governing equations obtained on the basis of general dynamics of an inviscid incompressible fluid (water is the standard example of such a fluid). Linearized problems are derived there as well.

Part 1 is devoted to waves arising, in particular, in two closely related phenomena, which are radiation of waves by oscillating immersed bodies and scattering of incoming progressive waves by an obstacle (a floating body or variable bottom topography). Mathematically these phenomena give rise to a boundary value problem that is usually referred to as the water-wave problem. The difficulty of this problem stems from several facts. First, it is essential that the water domain is infinite. Second, there is a spectral parameter (it is related to the radian frequency of waves) in a boundary condition on a semi-infinite part of the boundary (referred to as the free surface of water). Above all, the free surface may consist of more than one component as occurs for a surface-piercing toroidal body. Thus the questions of solvability and uniqueness are far from being solved because usual tools applicable to other problems of mathematical physics fail in this case. The problem of uniqueness is particularly difficult, and it was placed first in Ursell's list of unsolved problems mentioned above. Different cases are possible, and we demonstrate in Part 1 that, for some geometries of the water domain, the so-called trapped modes (that is, nontrivial solutions of the homogeneous problem leading to non-uniqueness in the inhomogeneous problem) do exist for certain values of the spectral parameter whereas other geometries provide uniqueness for all frequencies.

Part 1 is divided into five chapters. In Chapter 1, we give an account of Green's functions in three and two dimensions. This material is frequently used in the sequel because, first of all, Green's function gives a key for proving the solvability theorem by reducing the water-wave problem to an integral equation on the wetted surface (contour) of an immersed body, or of a bottom obstruction (see Chapters 2 and 3). Second, Green's function is the tool that is applied in Chapter 4 for the construction of trapped waves, in other words, for examples of non-uniqueness in the water-wave problem.

Chapter 2 is concerned with those cases in which the free surface coincides with the whole horizontal plane. The application of the integral equation technique to the problem of a submerged body is developed in Section 2.1. It provides the solvability of the water-wave problem for all frequencies except possibly for a finite number of values. In Sections 2.2 and 2.3, sufficient conditions on the body shape and bottom profile are established that guarantee the unique solvability for all frequencies. Moreover, a certain auxiliary integral identity is derived for proving one of the uniqueness theorems. This identity finds further applications in Chapters 3 and 5.

In Chapter 3, semisubmerged bodies are allowed in the way that leaves no bounded components of the free surface. As in Chapter 2, we first apply the method of integral equations. However, the integral equation based on the source distribution over the wetted rigid surface gives rise to so-called irregular frequencies, that is, the frequencies at which the integral equation is not solvable for an arbitrary right-hand-side term. These values are not related to the water-wave problem and arise from the fact that a certain boundary value problem in the domain between the body surface and the free-surface plane has these values as eigenvalues. There are different ways that lead to other integral equations without irregular frequencies. We consider one of them in detail and give a survey of the others in Section 3.1. In Section 3.2, we present uniqueness theorems related to geometries under consideration. We begin with John's theorem and then consider extensions of John's method.

Chapter 4 deals with the case in which isolated portions of the free surface are present. This case is distinguished from the situations presented in Chapters 2 and 3, because examples of trapped waves involving such geometries have been constructed. In Section 4.1, we give two-dimensional examples as well as axisymmetric ones. They show that the exceptional values of frequency when the water-wave problem is not uniquely solvable do exist, at least for special geometries obtained by means of the so-called inverse procedure. We begin Section 4.2 with a number of geometric conditions providing uniqueness in the two-dimensional problem when either two bodies are symmetric about a vertical axis or the water domain has no mirror symmetry. Section 4.2 also deals with the uniqueness in the water-wave problem for a toroidal body. It occurs that for an axisymmetric toroid (similarly to the case of two symmetric cylinders), intervals of uniqueness alternate with intervals of possible non-uniqueness on thc frequency half-axis. However, if more restrictions are imposed on the geometry, then it is possible to prove that some intervals of possible non-uniqueness are free of it.

A survey of results obtained in the extensive field of trapped waves periodic in a horizontal direction is given in Chapter 5. A short Section 5.1 contains a classification of such trapped waves. Edge waves are treated in Section 5.2. We present results on trapped modes above submerged cylinders and bottom protrusions in Section 5.3. Modes trapped by surface-piercing structures are considered in Section 5.4. The last section, Section 5.5, is concerned with trapped modes near vertical cylinders in channels.

Part 2 is concerned with waves caused by the uniform forward motion of a body on calm water, and these waves are usually referred to as ship waves. They are familiar to everybody because of their typical V pattern. The first mathematical explanation of this pattern appeared in 1887, when Lord Kelvin applied for this purpose his method of the stationary phase. Thus a clear evidence was given that *the linear theory* explains ship waves at least qualitatively. The boundary value problem describing ship waves is known as the Neumann–Kelvin problem, and as in the case of the water-wave problem the corresponding water domain is infinite, and there is a spectral parameter (related to the forward velocity) in the boundary condition on the free surface. The two problems are distinguished in both the free surface boundary conditions and conditions at infinity. The latter are unsymmetric and axially symmetric in the Neumann–Kelvin and water-wave problems, respectively.

Part 2 consists of three chapters, and as in Part 1 we begin with the three- and two-dimensional Green's functions for the Neumann–Kelvin problem (Chapter 6). It is worth mentioning that in Section 6.2 we give an asymptotic formula that describes the behavior of waves generated by Kelvin's source uniformly in all horizontal directions and with respect to depth.

The next two chapters, Chapters 7 and 8, are mainly concerned with the simpler two-dimensional Neumann–Kelvin problems for totally submerged and surface-piercing bodies, respectively. For the former case, necessary and sufficient conditions of the unique solvability are given for both infinite (Section 7.1) and finite (Section 7.2) depth of water. It is shown that these conditions hold for a circular cylinder in deep water, which is the only geometry when the problem is known to be uniquely solvable for all values of the forward velocity.

In the case of a surface-piercing cylinder, two supplementary conditions must be imposed and several sets of such conditions are possible. For one set of supplementary conditions considered in Section 8.1, the analogues of necessary and sufficient conditions from Chapter 7 are obtained, and they guarantee the unique solvability of the problem for surface-piercing cylinders. Other supplementary conditions are treated in Section 8.3, and some of them lead to the existence of trapped modes having finite energy. Examples of trapped modes are constructed in Section 8.4. In Section 8.5, we show that supplementary conditions of the first type guarantee that the unique solvability theorem holds for supercritical values of the forward velocity (that is, values exceeding a certain critical number depending on the water depth). Formulae for the total resistance of surface-piercing cylinders to the forward motion are derived in Section 8.2 for deep and shallow water, and these formulae generalize those obtained in Section 7.3 and expressing the wave-making resistance of totally submerged cylinders.

Section 7.4 deals with the three-dimensional Neumann–Kelvin problem for a totally submerged body, and it is established that the problem is solvable for all values of the spectral parameter with a possible exception for a finite number of values. We note that less is known about the three-dimensional Neumann–Kelvin problem than is known about the two-dimensional one. For instance, there is no example of a totally submerged body for which the problem is uniquely solvable for all values of the forward velocity. One of the difficulties in this direction arises from the fact that the uniqueness of a solution having finite energy does not imply the uniqueness of an arbitrary solution, as the case is in the water-wave problem. Another important unsolved question in three dimensions is how to impose a supplementary condition on the contour, where a surface-piercing body intersects the free surface.

In Part 3, which consists of two chapters, we investigate unsteady wave motions that develop in time under various disturbances applied either to the free surface or beneath it. In addition, certain initial conditions must be given at the time moment $t = 0$. Such problems arise in oceanography (for example, when describing generation of tsunamis), as well as in ship research (in particular, in the theory of wave-making resistance). All unsteady problems may be divided into two large classes. One of them consists of problems describing waves on the surface of an unsteady flow, whereas problems in the second class deal with waves arising from disturbances that are motionless relative water, and that depend on time only.

We begin Part 3 with results on the uniqueness, existence, and smoothness of solution. They are presented in Chapter 9 and hold for both classes of problems mentioned above. It should be noted that these results are obtained under the essential restriction that the free surface coincides with the whole horizontal plane, and the rigid boundaries of the water domain are placed at a finite distance from the free surface. The case of rigid boundaries intersecting the free surface is still an open question. In the next chapter, Chapter 10, we are concerned with problems describing waves caused by rapidly stabilizing and high-frequency disturbances that are motionless relative water. For both cases we give an asymptotic analysis based on a two-scale expansion for the velocity potential, and this allows us to describe principal terms in asymptotics of hydrodynamic characteristics such as the free surface elevation, the force acting on submerged bodies, the energy of waves, and so on.

In the Bibliography, we tried to list as many works that were published after 1960 and that treat the mathematical aspects of water waves as we could. An extensive lists of papers published up to 1960 are given by Stoker [312] and Wehausen and Laitone [354], and an additional bibliography can be found in the survey papers published by Newman [263] and Wehausen [353] during the 1970s. The papers listed in our Bibliography are mostly described briefly in Bibliographical Notes (almost every chapter has such a title for its last section), but a few are not. Of course, despite our efforts, there are omissions in the Bibliography (this is inevitable when one is dealing with several hundreds of works published over several decades).

To complete the description of the book, we mention that parts are divided into chapters, which consist of sections that are divided into subsections (and some subsections are divided into subsubsections). The titles of chapters and sections are given on the top of even and odd pages respectively. The titles of sections and subsections are given as bold headlines and numbered by two and three numbers, respectively; for example, 4.2 is Section 2 in Chapter 4, and 2.4.2 is Subsection 2 in Section 2.4. The titles of subsubsections are numbered by four numbers and are not bold. Every chapter has independent numbering of formulae and figures; for example, (2.36) denotes the 36th formula in Chapter 2, and Fig. 2.3(a) refers to part (a) of the third figure in Chapter 2. Most of the references are collected in Bibliographical Notes, but this does not apply to review chapters and sections.

A substantial part of the book is based on authors' contributions to the theory. The presentation of material is mathematically rigorous, despite the fact that we usually avoid the lemma–theorem style. Instead, we adopt a more or less informal style, formulating, nevertheless, all proved assertions in italics.

The prerequisite for reading the book is a course in Mathematical Analysis, and a familiarity with Bessel functions and the Fourier transform. We assume also that the reader is aware of the elements of functional analysis (for example, the Fredholm alternative is widely used in the book).

The book is supposed to be a research monograph in applied mathematics. Some of its topics might be of interest to mathematicians who specialize in partial differential equations and spectral operator theory. We also hope it could be used as a reference book by experts in ocean engineering as well as an advanced text for applied and engineering mathematics graduate students.

Acknowledgments

We have had the good fortune to benefit from many discussions and collaborations with a number of colleagues in England and in St. Petersburg (formerly Leningrad, Russian Federation). We have profited not only from the generous sharing of their ideas, insights, and enthusiasm, but also from their friendship, support, and encouragement. We feel deeply indebted to D. V. Evans, C. M. Linton, M. McIver, P. McIver, R. Porter, M. J. Simon, and F. Ursell. Special thanks are due to O. V. Motygin, who, above all, generously helped with advanced work in TEX-setting the book files (producing and/or accommodating the graphic files and creating the indices). In the course of our work on the book, we visited each other and the hospitality of the University of Linköping and the University of North Carolina at Charlotte is thankfully acknowledged.

Introduction: Basic Theory of Surface Waves

Here we give a brief account of physical assumptions (first section) and the mathematical approximation (second section) used for developing a mathematical model of water waves. The resulting linear boundary value problems are formulated in the third and fourth sections for the wave–body interaction and ship waves, respectively.

Mathematical Formulation

Conventions

Water waves (the terms *surface waves* and *gravity waves* are also in use) are created normally by a gravitational force in the presence of a free surface along which the pressure is constant. There are two ways to describe these waves mathematically. It is possible to trace the paths of individual particles (a Lagrangian description), but in this book an alternative form of equations (usually referred to as Eulerian) is adopted. The motion is determined by the velocity field in the domain occupied by water at every moment of the time t.

Water is assumed to occupy a certain domain W bounded by one or more moving or fixed surfaces that separate water from some other medium. Actually we consider boundaries of two types: the above-mentioned free surface separating water from the atmosphere, and rigid surfaces including the bottom and surfaces of bodies floating in and/or beneath the free surface.

It is convenient to use rectangular coordinates (x_1, x_2, y) with origin in the free surface at rest (which usually coincides with the mean free surface), and with the y axis directed opposite to the acceleration caused by gravity. For the sake of brevity we will write x instead of (x_1, x_2). This has the obvious advantage that two- and three-dimensional problems can be treated simultaneously, where it is possible. *Two-dimensional problems* form an important class of problems considering water motions that are the same in every plane orthogonal to a certain direction. Subscripts will be used to denote (partial) derivatives, for example:

$$u_t = \frac{\partial u}{\partial t}, \quad u_y = \frac{\partial u}{\partial y}, \quad u_{x_i} = \frac{\partial u}{\partial x_i}, \quad i = 1, 2.$$

When this notation is inconvenient, we will apply the following one:

$$\partial_t u, \partial_y u, \partial_{x_i} u, \ldots .$$

As usual, $\nabla u = (u_{x_1}, u_{x_2}, u_y)$, and the horizontal component of ∇ will be denoted by ∇_x, that is, $\nabla_x u = (u_{x_1}, u_{x_2}, 0)$. Clearly, $\nabla u = (u_x, u_y)$ and $\nabla_x u = (u_x, 0)$ in two-dimensional problems.

In several chapters, in particular in those concerned with the forward motion of a body, we use (x, z) instead of (x_1, x_2).

Equations of Motion and Boundary Conditions

In the Eulerian formulation one seeks the velocity vector $\mathbf{v}$, the pressure p, and the fluid density ρ as functions of $(x, y) \in \bar{W}$ and $t \geq t_0$, where t_0 denotes a certain initial moment. Assuming the fluid to be inviscid without surface tension, one obtains the equations of motion from conservation laws (for details see, for example, books by Lamb [179], Le Méhauté [186], and Stoker [312]).

The conservation of mass implies the continuity equation

$$\rho_t + \nabla \cdot (\rho \mathbf{v}) = 0 \quad \text{in } W.$$

Under the assumption that the fluid is incompressible (which is usual in the water-wave theory), the last equation becomes

$$\nabla \cdot \mathbf{v} = 0 \quad \text{in } W. \tag{I.1}$$

The conservation of momentum in inviscid fluid leads to the so-called Euler equations. Taking into account the gravity force, one can write these three (or two) equations in the following vector form:

$$\mathbf{v}_t + \mathbf{v} \cdot \nabla \mathbf{v} = -\rho^{-1} \nabla p + \mathbf{g}. \tag{I.2}$$

Here $\mathbf{g}$ is the vector of the gravity force having zero horizontal components and the vertical one equal to $-g$, where g denotes the acceleration caused by gravity.

An irrotational character of motion is another usual assumption in the water-wave theory; that is,

$$\nabla \times \mathbf{v} = 0 \quad \text{in } W.$$

Note that one can prove that the motion is irrotational if it has this property at the initial moment (see, for example, books by Lamb [179] and Stoker [312] for the proof of this assertion known as the Helmholtz theorem). The last equation guarantees the existence of a velocity potential ϕ, so that

$$\mathbf{v} = \nabla\phi \quad \text{in } \bar{W}. \tag{I.3}$$

This is obvious for simply connected domains; otherwise (for example, when one considers a two-dimensional problem for a totally immersed body), the so-called no flow condition, see (I.8) below, should be taken into account.

From (I.1) and (I.3) one obtains the Laplace equation

$$\nabla^2\phi = 0 \quad \text{in } W. \tag{I.4}$$

This greatly facilitates the theory but, in general, solutions of (I.4) do not manifest wave character. Waves are created by the boundary conditions on the free surface.

Let $y = \eta(x, t)$ be the equation of the free surface valid for $x \in F$, where F is a union of some domains (generally depending on t) in $\mathbb{R}^n$, with $n = 1, 2$. The pressure is prescribed to be equal to the constant atmospheric pressure p_0 on $y = \eta(x, t)$, and the surface tension is neglected. From (I.2) and (I.3) one immediately obtains Bernoulli's equation,

$$\phi_t + |\nabla\phi|^2/2 = -\rho^{-1}p - gy + C \quad \text{in } \bar{W}, \tag{I.5}$$

where C is a function of t alone. Indeed, applying ∇ to both sides in (I.5) and using (I.2) and (I.3), one obtains $\nabla C = 0$. Then, by changing ϕ by a suitable additive function of t, one can convert C into a constant having, for example, the value

$$C = \rho^{-1}p_0. \tag{I.6}$$

Now (I.5) gives the *dynamic boundary condition* on the free surface:

$$g\eta + \phi_t + |\nabla\phi|^2/2 = 0 \quad \text{for } y = \eta(x, t),\ x \in F. \tag{I.7}$$

Another boundary condition holds on every "physical" surface $\mathcal{S}$ bounding the fluid domain W and expresses the kinematic property that there is no transfer of matter across $\mathcal{S}$. Let $s(x, y, t) = 0$ be the equation of $\mathcal{S}$; then

$$\mathrm{d}s/\mathrm{d}t = \mathbf{v}\cdot\nabla s + s_t = 0 \quad \text{on } \mathcal{S}. \tag{I.8}$$

Under assumption (I.3) this takes the form of

$$\frac{\partial\phi}{\partial n} = -\frac{s_t}{|\nabla s|} = v_n \quad \text{on } \mathcal{S}, \tag{I.9}$$

where v_n denotes the *normal velocity* of $\mathcal{S}$. Thus the *kinematic boundary condition* (I.9) means that the normal velocity of particles is continuous across a physical boundary.

On the fixed part of $\mathcal{S}$, (I.9) takes the form of

$$\partial\phi/\partial n = 0. \tag{I.10}$$

On the free surface, condition (I.8), written as follows,

$$\eta_t + \nabla_x\phi \cdot \nabla_x\eta - \phi_y = 0 \quad \text{for } y = \eta(x, t), x \in F, \tag{I.11}$$

complements the dynamic condition (I.7). Thus, in the present approach, two nonlinear conditions (I.7) and (I.11) on the unknown boundary are responsible for waves, which constitutes the main characteristic feature of water–surface wave theory.

This brief account of governing equations can be summarized as follows. *In the water-wave problem one seeks the velocity potential $\phi(x, y, t)$ and the free surface elevation $\eta(x, t)$ satisfying* (I.4), (I.7), (I.9), *and* (I.11). *The initial values of ϕ and η should also be prescribed, as well as the conditions at infinity (for unbounded W) to complete the problem, which is known as the Cauchy–Poisson problem.*

Energy and Its Flow

Let W_0 be a subdomain of W bounded by a "geometric" surface ∂W_0 that may not be related to physical obstacles and that is permitted to vary in time independently of moving water unlike "physical" surfaces described below. Let $s_0(x, y, t) = 0$ be the equation of ∂W_0. The total energy contained in W_0 consists of kinetic and potential components and is given by

$$E = \rho \int_{W_0} [gy + |\nabla\phi|^2/2] \, \mathrm{d}x\mathrm{d}y. \tag{I.12}$$

The first term related to the vertical displacement of a water particle corresponds to the potential energy, whereas the second one gives the kinetic energy that is proportional to the velocity squared. Using (I.5) and (I.6), one can write this in the form of

$$E = -\rho \int_{W_0} (\rho\phi_t + p - p_0) \, \mathrm{d}x\mathrm{d}y.$$

Differentiating (I.12) with respect to t we get

$$\frac{\mathrm{d}E}{\mathrm{d}t} = \rho \int_{W_0} \nabla\phi \cdot \nabla\phi_t \, \mathrm{d}x\mathrm{d}y + \int_{\partial W_0} \frac{s_{0t}}{|\nabla s_0|} (\rho\phi_t + p - p_0) \, \mathrm{d}S.$$

Green's theorem applied to the first integral here leads to

$$\frac{\mathrm{d}E}{\mathrm{d}t} = \int_{\partial W_0} \left[\rho\phi_t \left(\frac{\partial \phi}{\partial n} - v_n \right) - (p - p_0)v_n \right] \mathrm{d}S, \qquad \text{(I.13)}$$

where (I.4) is taken into account and v_n denotes the normal velocity of ∂W_0. Hence the integrand in (I.13) is the rate of energy flow from W_0 through ∂W_0 taken per units of time and area. The velocity of energy propagation is known as the group velocity. However, it does not play any significant role in considerations presented in this book, and we restrict ourselves to references to works of Stoker [312] and Wehausen and Laitone [354], where further details can be found.

If a portion of ∂W_0 is a fixed geometric surface, then $v_n = 0$ on this portion; the rate of energy flow is given by $-\rho\phi_t(\partial\phi/\partial n)$.

If a portion of ∂W_0 is a "physical" boundary that is not penetrable by water particles, then (I.9) shows that the integrand in (I.13) is equal to $(p_0 - p)v_n$. Therefore, there is no energy flow through this portion of ∂W_0 if either of two factors vanishes. In particular, this is true for the free surface ($p = p_0$) and for the bottom ($v_n = 0$).

Linearized Unsteady Problem

Linearization: Its Applicability and Justification

About 50 years ago, John [125] assessed the problem formulated at the end of the subsection on equations of motion and boundary conditions as follows:

> In this generality little can be done either toward a discussion of the motion or toward an explicit solution of the equations. The difficulties arising from the fact that ϕ is a solution of the potential equation determined by non-linear boundary conditions on a variable boundary are considerable, and have only been overcome in the special cases of permanent waves treated by Levi-Civita and Struik.

Here works [194, 314] by Levi-Civita and Struik, respectively, are cited (see also Nekrasov's work [261]).

Since then a large number of papers has been published and great progress has been achieved in the mathematical treatment of nonlinear water-wave problems (we list only a few works: Debnath [46], Kirchgässner [138], Olver [272] and Ovsyannikov et al. [273], where further references can be found). However, all rigorous results in this direction are concerned with water waves in the absence of floating bodies, and the present state of the art for the *nonlinear problem for floating bodies* is the same as 50 years ago. Of course, a substantial body of numerical results treating different aspects of the nonlinear problem has emerged during the past three decades, but this approach is beyond our scope.

To be in a position to describe water waves in the presence of bodies, the equations should be approximated by more tractable ones. The usual and rather reasonable simplification consists of a *linearization* of the problem under certain assumptions concerning the motion of a floating body. An example of such assumptions (there are other ones leading to the same conclusions) suggests that a body's motion near the equilibrium position is so small that it produces only waves having a small amplitude and a small wavelength. There are three characteristic geometric parameters:

1. A typical value of the wave height H.
2. A typical wavelength L.
3. The water depth D.

They give three characteristic quotients: H/L, H/D, and L/D. The relative importance of these quotients is different in different situations. Nevertheless, it was found (see, for example, Le Méhauté [186], Sections 15-2 and 15-3) that if

$$\frac{H}{D} \ll 1 \quad \text{and} \quad \frac{H}{L}\left(\frac{L}{D}\right)^3 \ll 1,$$

then the linearization can be justified by some heuristic considerations. The last parameter $(H/L)(L/D)^3 = (H/D)(L/D)^2$ is usually referred to as Ursell's number. Its role in a classification of water waves is presented in detail by Le Méhauté [186], Section 15-2. Further results treating the problem of linearization can be found in the paper [22] by Beale, Hou, and Lowengrub.

The linearized theory leads to results that are in a rather good agreement with experiments and observations. During the 1940s and 1950s, a substantial work in this direction was carried out by Ursell and his coauthors. Thus Barber and Ursell [19] discovered a good agreement between predictions of the linear theory for group velocity and values resulting from observations, and Ursell [325] demonstrated the same for frequencies. Some experiments were carried out by Dean, Ursell, and Yu [45] and by Ursell and Yu [343], and in a certain range of wave steepnesses a very close agreement was obtained between the measured wave amplitude (up to some corrections inevitable in an experiment) and theoretical predictions made on the base of the linear problem.

Furthermore, there is mathematical evidence that the linearized problem provides an approximation to the nonlinear one. For the Cauchy–Poisson problem describing waves in a water layer caused by prescribed initial conditions, the linear approximation is justified rigorously by Nalimov (see the book by Ovsyannikov et al. [273], Chapter 3). More precisely, under the assumption that the undisturbed water occupies a layer of constant depth, the following are proved:

1. *The nonlinear problem is solvable for sufficiently small values of the linearization parameter.*
2. *As this parameter tends to zero, solutions of the nonlinear problem do converge to the solution of the linearized problem in the norm of some suitable function space.*

Equations for Small Amplitude Waves

A formal perturbation procedure leading to a sequence of linear problems can be developed as follows. Let us assume that the velocity potential ϕ and the free surface elevation η admit expansions with respect to a certain small parameter ϵ:

$$\phi(x, y, t) = \epsilon\phi^{(1)}(x, y, t) + \epsilon^2\phi^{(2)}(x, y, t) + \epsilon^3\phi^{(3)}(x, y, t) + \cdots, \tag{I.14}$$

$$\eta(x, t) = \eta^{(0)}(x, t) + \epsilon\eta^{(1)}(x, t) + \epsilon^2\eta^{(2)}(x, t) + \cdots, \tag{I.15}$$

where $\phi^{(1)}$, $\phi^{(2)}, \ldots, \eta^{(0)}$, $\eta^{(1)}, \ldots,$ and all their derivatives are bounded. Consequently, the velocities of water particles are supposed to be small (proportional to ϵ), and $\epsilon = 0$ corresponds to water permanently at rest.

Substituting (I.14) into (I.4) gives

$$\nabla^2\phi^{(k)} = 0 \quad \text{in } W, \quad k = 1, 2, \ldots. \tag{I.16}$$

Furthermore, $\eta^{(0)}$ describing the free surface at rest cannot depend on t. When the expansions for ϕ and η are substituted into the Bernoulli boundary condition (I.7) and grouped according to powers of ϵ, one obtains

$$\eta^{(0)} \equiv 0 \quad \text{for } x \in F.$$

This and Taylor's expansion of $\phi[x, \eta(x, t), t]$ in powers of ϵ yield the following for orders higher than zero:

$$\phi_t^{(1)} + g\eta^{(1)} = 0 \quad \text{for } y = 0, x \in F, \tag{I.17}$$

$$\phi_t^{(2)} + g\eta^{(2)} = -\eta^{(1)}\phi_{ty}^{(1)} - \left|\nabla\phi^{(1)}\right|^2/2 \quad \text{for } y = 0,\ x \in F, \tag{I.18}$$

and so on; that is, all these conditions hold on the mean position of the free surface at rest.

Similarly, the kinematic condition (I.11) leads to

$$\phi_y^{(1)} - \eta_t^{(1)} = 0 \quad \text{for } y = 0, x \in F, \tag{I.19}$$

$$\phi_y^{(2)} - \eta_t^{(2)} = -\eta^{(1)}\phi_{yy}^{(1)} + \nabla_x\phi^{(1)} \cdot \nabla_x\eta^{(1)} \quad \text{for } y = 0, x \in F, \tag{I.20}$$

and so on. Eliminating $\eta^{(1)}$ between (I.17) and (I.19), one finds the classical first-order linear free-surface condition:

$$\phi_{tt}^{(1)} + g\phi_y^{(1)} = 0 \quad \text{for } y = 0, x \in F. \tag{I.21}$$

In the same way, one obtains from (I.18) and (I.20) the following:

$$\phi_{tt}^{(2)} + g\phi_y^{(2)} = -\phi_t^{(1)}\nabla_x^2\phi^{(1)} - \frac{1}{g^2}\left[\phi_t^{(1)}\phi_{ttt}^{(1)} + \left|\nabla_x\phi^{(1)}\right|^2\right]_t \quad \text{for } y = 0, x \in F.$$

Further free-surface conditions can be obtained for terms in (I.14) having higher orders in ϵ. All these conditions have the same operator in the left-hand side, and the right-hand term depends nonlinearly on terms of smaller orders. It is worth mentioning that all of the high-order problems are formulated in the same domain W occupied by water at rest. In particular, the free-surface boundary conditions are imposed at $\{y = 0, x \in F\}$.

Boundary Condition on an Immersed Rigid Surface

First, we note that the homogeneous Neumann condition (I.10) is linear on fixed surfaces. Hence, this condition is true for $\phi^{(k)}$, $k = 1, 2, \ldots$. The situation reverses for the inhomogeneous Neumann condition (I.9) on a moving surface $\mathcal{S}$, which can be subjected, for example, to a prescribed motion or freely floating. The problem of a body freely floating near its equilibrium position will not be treated in the book (for linearization of this problem see John's paper [125]). We restrict ourselves to the linearization of (I.9) for $\mathcal{S} = S(t, \epsilon)$ undergoing a given small amplitude motion near an equilibrium position S, that is, when $S(t, \epsilon)$ tends to S as $\epsilon \to 0$.

It is convenient to carry out the linearization locally. Let us consider a neighborhood of $(x^{(0)}, y^{(0)}) \in S$, where the surface is given explicitly in local Cartesian coordinates (ξ, ζ), where in the three-dimensional case $\xi = (\xi_1, \xi_2)$, having an origin at $(x^{(0)}, y^{(0)})$ and the ζ axis directed into water normally to S. Let $\zeta = \zeta^{(0)}(\xi)$ be the equation of S, and $S(t, \epsilon)$ be given by $\zeta = \zeta(\xi, t, \epsilon)$, where

$$\zeta(\xi, t, \epsilon) = \zeta^{(0)}(\xi) + \epsilon\zeta^{(1)}(\xi, t) + \epsilon^2\zeta^{(2)}(\xi, t) + \cdots. \tag{I.22}$$

After substituting (I.14) and $s = \zeta - \zeta(\xi, t, \epsilon)$ into (I.8), we use (I.3), (I.22), and Taylor's expansion in the same way as in the subsection on equations for small amplitude waves. This gives the following first-order equation:

$$\phi_\zeta^{(1)}(\xi, \zeta^{(0)}, t) - \nabla_\xi \phi^{(1)}(\xi, \zeta^{(0)}, t) \cdot \nabla_\xi \zeta^{(0)}(\xi) = \zeta_t^{(1)}(\xi, t),$$

which implies the linearized boundary condition:

$$\partial \phi^{(1)}/\partial n = v_n^{(1)} \quad \text{on } S, \tag{I.23}$$

where

$$v_n^{(1)} = \zeta_t^{(1)}/\left[1 + \left|\nabla_\xi \zeta^{(0)}\right|^2\right]^{1/2}$$

is the first-order approximation of the normal velocity of $S(t, \epsilon)$.

The second-order boundary condition on S has the form

$$\frac{\partial \phi^{(2)}}{\partial n} = \frac{\zeta_t^{(2)}}{\left[1 + \left|\nabla_\xi \zeta^{(0)}\right|^2\right]^{1/2}} - \zeta^{(1)} \frac{\partial^2 \phi^{(1)}}{\partial n^2} - \left[\frac{1 + \left|\nabla_\xi \zeta^{(1)}\right|^2}{1 + \left|\nabla_\xi \zeta^{(0)}\right|^2}\right]^{1/2} \frac{\partial \phi^{(1)}}{\partial n^{(1)}},$$

where $\partial \phi^{(1)}/\partial n^{(1)}$ is the derivative in the direction of normal to $\zeta = \zeta^{(1)}(\xi, t)$ calculated on S. In addition, further conditions on S of the Neumann type can be obtained for terms of higher order in ϵ.

Thus, all $\phi^{(k)}$ satisfy the same linear boundary value problem with different right-hand-side terms in conditions on the free surface at rest and on the equilibrium surfaces of immersed bodies. These right-hand-side terms depend on solutions obtained on previous steps. Solving these problems successively, beginning with problems (I.16), (I.21), and (I.23) complemented by some initial conditions, one can, generally speaking, find a solution to the nonlinear problem in the form of (I.14) and (I.15). However, this procedure is not justified mathematically up to the present time. Therefore, in this book we restrict ourselves to the first-order approximation, which in its own right gives rise to an extensive mathematical theory. Investigations in this field are far from being exhausted.

We conclude this subsection by summarizing the boundary value problem for *the first-order velocity potential* $\phi^{(1)}(x, y, t)$. *It is defined in* W *occupied by water at rest with a boundary consisting of the free surface* F, *the bottom* B, *and the wetted surface of immersed bodies* S, *and it must satisfy*

$$\nabla^2 \phi^{(1)} = 0 \quad \text{in } W, \tag{I.24}$$

$$\phi_{tt}^{(1)} + g\phi^{(1)} = 0 \quad \text{for } y = 0, x \in F, \tag{I.25}$$

$$\partial \phi^{(1)}/\partial n = v_n^{(1)} \quad \text{on } S, \tag{I.26}$$

$$\partial \phi^{(1)}/\partial n = 0 \quad \text{on } B, \tag{I.27}$$

$$\phi^{(1)}(x, 0, 0) = \phi_0(x) \quad \text{and } \phi_t^{(1)}(x, 0, 0) = -g\eta_0(x), \tag{I.28}$$

where ϕ_0, $v_n^{(1)}$, *and* η_0 *are given functions, and* $\eta_0(x) = \eta^{(1)}(x, 0)$*; see* (I.17). *Then*

$$\eta^{(1)}(x, t) = -g^{-1}\phi_t^{(1)}(x, 0, t)$$

gives the first-order approximation for the elevation of the free surface.

In conclusion of the section, it should be mentioned that for the case of a rigid body freely floating near an equilibrium position, a linearized system of coupled equations was proposed by John [125]. This system was investigated by John [126], Beale [21], and Licht [197, 200]. Another coupled initial-boundary value problem dealing with a fixed elastic body immersed in water was considered by Licht [198, 199].

Linear Time-Harmonic Waves (the Water-Wave Problem)

Separation of the t variable

We pointed out in the preface that this book is concerned with the steady-state problem of radiation and scattering of water waves by bodies floating in and/or beneath the free surface, assuming all motions to be simple harmonic in the time. The corresponding radian frequency is denoted by ω. Thus, the right-hand-side term in (I.23) is

$$v_n^{(1)} = \text{Re}\{e^{-i\omega t} f\} \quad \text{on } S, \tag{I.29}$$

where f is a complex function independent of t, and the first-order velocity potential $\phi^{(1)}$ can then be written in the form

$$\phi^{(1)}(x, y, t) = \text{Re}\{e^{-i\omega t} u(x, y)\}. \tag{I.30}$$

The latter assumption is justified by the so-called limiting amplitude principle, which is concerned with the large-time behavior of a solution to the initial-boundary value problem having (I.29) as the right-hand-side term. According to this principle, such a solution tends to the potential (I.30) as $t \to \infty$, and u satisfies a steady-state problem. The limiting amplitude principle has general applicability in the theory of wave motions, and its particular form for water waves was proved by Vullierme-Ledard [349]. Thus the problem of our interest describes waves developing at large time from time-periodic disturbances.

A complex function u in (I.30) is also referred to as velocity potential (this does not lead to confusion, because it will always be clear what kind of time dependence is considered in one part of the book or another). We recall that u is defined in the fixed domain W occupied by water at rest outside any bodies present. The boundary ∂W consists of three disjoint sets: (i) S, which is the union of the wetted surfaces of bodies in equilibrium; (ii) F, denoting the free surface at rest that is the part of $y = 0$ outside all the bodies; and (iii) B, which denotes the bottom positioned below $F \cup S$. Sometimes we will consider W unbounded below and corresponding to infinitely deep water. In this case $\partial W = F \cup S$.

Substituting (I.29) and (I.30) into (I.24)–(I.27) gives the boundary value problem for u:

$$\nabla^2 u = 0 \quad \text{in } W, \tag{I.31}$$

$$u_y - \nu u = 0 \quad \text{on } F, \tag{I.32}$$

$$\partial u / \partial n = f \quad \text{on } S, \tag{I.33}$$

$$\partial u / \partial n = 0 \quad \text{on } B, \tag{I.34}$$

where $\nu = \omega^2 / g$. Throughout the book a normal n to a surface always directs *into* the water domain W.

For deep water ($B = \emptyset$), condition (I.34) should be replaced by the following one:

$$\sup_{(x,y) \in W} |u(x, y)| \leq \text{const} < \infty. \tag{I.35}$$

Despite the fact that this condition has no direct hydrodynamic meaning, we impose it because it is essential for certain proofs in what follows. Besides, (I.35) implies the following natural behavior of the velocity filed (see Subsection 1.1.1.1 for the proof):

$$|\nabla u| \to 0 \quad \text{as } y \to -\infty; \tag{I.36}$$

that is, the water motion decays with depth. Conditions at infinity that are similar to the last two conditions are usually imposed in the boundary value problems for the Laplacian in domains exterior to a compact set in $\mathbb{R}^2$ and $\mathbb{R}^3$. A natural requirement that a solution to (I.31)–(I.35) should be unique also imposes a certain restriction on the behavior of u as $|x| \to \infty$. We discuss conditions providing uniqueness in the subsection after the following one.

Examples

Let us consider some simple examples of waves existing in the absence of bodies. The corresponding potentials can be easily obtained by separation of variables.

For a layer W of constant depth d, $F = \{x \in \mathbb{R}^2, y = 0\}$ and $B = \{x \in \mathbb{R}^2, y = -d\}$ are the free surface and bottom, respectively. *A plane progressive wave* propagating in the direction of a wave vector $\mathbf{k} = (k_1, k_2)$ has the following velocity potential:

$$\mathrm{Re}\{A \exp[i(\mathbf{k} \cdot x - \omega t)]\} \cosh k_0(y + d). \tag{I.37}$$

Here A is an arbitrary complex constant, $k_0 = |\mathbf{k}|$, and the following relationship,

$$\nu = \omega^2/g = k_0 \tanh k_0 d, \tag{I.38}$$

holds between ω and k_0. Tending d to infinity, we note that k_0 becomes equal to ν and instead of (I.37) we have

$$\mathrm{Re}\{A \exp[i(\mathbf{k} \cdot x - \omega t)]\} e^{\nu y}$$

for the velocity potential of plane progressive wave in deep water.

A sum of two potentials (I.37) corresponding to identical progressive waves propagating in opposite directions gives a standing wave. Putting $\exp \nu y$ instead of $\cosh k_0(y + d)$ in (I.37) and omitting $\tanh k_0 d$ in (I.38), one gets the potential of a progressive wave in deep water.

A standing cylindrical wave in a water layer of depth d has the following potential:

$$w_{\mathrm{st}}(x, y) \cos \omega t, \quad \text{where } w_{\mathrm{st}}(x, y) = C_1 \cosh k_0(y + d) J_0(k_0|x|),$$

where k_0 is defined by (I.38), C_1 is a real constant, and J_0 denotes the Bessel function of order zero. The same manipulation as above gives the standing wave in deep water.

A cylindrical wave having an arbitrary phase at infinity may be obtained as a combination of w_{st} and a similar potential with J_0 replaced by Y_0, which is another solution of Bessel's equation. This allows one to construct a potential of outgoing wave as follows:

$$\mathrm{Re}\{e^{-i\omega t} w_{\mathrm{out}}(x, y)\}, \quad \text{where } w_{\mathrm{out}}(x, y) = C_2 \cosh k_0(y + d) H_0^{(1)}(k_0|x|),$$

where k_0 is defined by (I.38), C_2 is a complex constant, and $H_0^{(1)}$ denotes the first Hankel function of order zero. Outgoing behavior of this wave becomes clear from the asymptotic formula (see handbooks by Abramowitz & Stegun [1], and Gradshteyn and Ryzhik [96]):

$$H_0^{(1)}(k_0|x|) = \left(\frac{2}{\pi k_0 |x|} \right)^{1/2} e^{i(k_0|x| - \pi/4)} [1 + O(|x|^{-1})] \quad \text{as } |x| \to \infty.$$

Therefore, wave w_{out} behaves at large distances like a radially outgoing progressive wave, but it is singular on the axis $|x| = 0$. This is natural from a physical point of view, because outgoing waves should be radiated by a certain disturbance. In the case under consideration, the wave is produced by sources distributed with a suitable density over $\{|x| = 0, -d < y < 0\}$. Replacing $H_0^{(1)}$ in w_{out} by the second Hankel function, $H_0^{(2)}$, one obtains an incoming wave.

Radiation Conditions

Examples in the previous subsection demonstrate that problem (I.31)–(I.34) should be complemented by an appropriate condition as $|x| \to \infty$ to avoid non-uniqueness of the solution, which follows from the fact that there are infinitely many solutions of the form of (I.37). On the other hand, the energy dissipates when waves are radiated or scattered; that is, there exists a flow of energy to infinity. On the contrary, there is no such a flow for standing waves and no net flow for progressive waves. Since we are going to describe radiation and scattering phenomena, a condition should be introduced for eliminating waves having no flow of energy to infinity. For this purpose a mathematical expression is used known as a *radiation condition*. To formulate this condition we have to specify the geometry of the water domain at infinity.

Let W be an $(m+1)$-dimensional domain $(m = 1, 2)$, which at infinity coincides with the layer $\{x \in \mathbb{R}^m, -d < y < 0\}$, where $0 < d \leq \infty$. We say that u satisfies the radiation condition of the Sommerfeld type if

$$u_{|x|} - ik_0 u = \sigma(y) o\left[|x|^{(1-m)/2}\right] \quad \text{as } |x| \to \infty \text{ uniformly in } y, \theta. \qquad \text{(I.39)}$$

Here $\sigma(y) = (1 + |y|)^{-m}$ if $d = \infty$, $\sigma(y) = 1$ if $d < \infty$, k_0 is defined by (I.38) for $d < \infty$, and $k_0 = \nu$ for $d = \infty$, and $\theta \in [0, 2\pi)$ is polar angle in the plane $\{y = 0\}$. Uniformity in θ should be imposed only for the three-dimensional problem $(m = 2)$.

Let us show that (I.39) guarantees dissipation of energy. For the sake of simplicity we assume that $d < \infty$. By C_r we denote a cylindrical surface $W \cap \{|x| = r\}$ contained inside W. By (I.13) the average energy flow to infinity through C_r over one period of oscillations is equal to

$$\mathcal{F}_r = -\frac{\rho\omega}{2\pi} \int_0^{2\pi/\omega} \mathrm{d}t \int_{C_r} \frac{\partial \phi}{\partial t} \frac{\partial \phi}{\partial |x|} \, \mathrm{d}S.$$

Substituting (I.30) and taking into account that

$$\int_0^{2\pi/\omega} e^{\pm 2i\omega t} \, \mathrm{d}t = 0,$$

one finds that

$$\mathcal{F}_r = -\frac{\rho\omega^2}{8\pi} \int_0^{2\pi/\omega} \mathrm{d}t \int_{C_r} (ie^{i\omega t}\bar{u} - ie^{-i\omega t}u)(e^{-i\omega t}u_{|x|} + e^{i\omega t}\bar{u}_{|x|}) \, \mathrm{d}S$$

$$= -\frac{\rho\omega}{4\pi} \int_{C_r} \left(i\bar{u}u_{|x|} - iu\bar{u}_{|x|}\right) \mathrm{d}S = \frac{\rho\omega}{2} \operatorname{Im} \int_{C_r} \bar{u}\, u_{|x|} \, \mathrm{d}S. \qquad \text{(I.40)}$$

This can be written as follows:

$$\mathcal{F}_r = \frac{\rho\omega}{4k_0} \left\{ \int_{C_r} \left(\left| u_{|x|} \right|^2 + k_0^2 |u|^2 \right) \mathrm{d}S - \int_{C_r} \left| u_{|x|} - i k_0 u \right|^2 \mathrm{d}S \right\}. \quad \text{(I.41)}$$

Moreover, $\mathcal{F}_r$ does not depend on r when the obstacle surface S lies inside the cylinder $\{|x| = r\}$, which can be proved as follows.

By W_r and F_r we denote $W \cap \{|x| < r\}$ and $F \cap \{|x| < r\}$, respectively. Let us multiply (I.31) by $\bar{u}$ and integrate the result over W_r. Then applying the divergence theorem we obtain

$$\int_{W_r} |\nabla u|^2 \,\mathrm{d}x\mathrm{d}y = -\int_{\partial W_r} \bar{u} \,\frac{\partial u}{\partial n} \,\mathrm{d}S,$$

where n is directed into W_r. Using (I.32) and (I.34) we get

$$\int_{W_r} |\nabla u|^2 \,\mathrm{d}x\mathrm{d}y = \nu \int_{F_r} |u|^2 \,\mathrm{d}x + \int_{C_r} \bar{u}\, u_{|x|} \,\mathrm{d}S - \int_S \bar{u} \,\frac{\partial u}{\partial n} \,\mathrm{d}S.$$

Comparing this with (I.40) we find that

$$\mathcal{F}_r = \frac{\rho\omega}{2} \operatorname{Im} \int_S \bar{u} \,\frac{\partial u}{\partial n} \,\mathrm{d}S$$

is independent of r.

This fact yields that $\mathcal{F}_r \geq 0$ because (I.39) implies that the last integral in (I.41) tends to zero as $r \to \infty$.

The crucial point in the proof that $\mathcal{F}_r \geq 0$ is equality (I.41). It suggests that (I.39) can be replaced by a "weaker" radiation condition of the Rellich type,

$$\int_{C_r} \left| u_{|x|} - i k_0 u \right|^2 \mathrm{d}S = o(1) \quad \text{as } r \to \infty. \quad \text{(I.42)}$$

Actually, (I.39) and (I.42) are equivalent (see the Subsection 1.3.2).

So, in what follows we consider problem (I.31)–(I.34) complemented by either (I.39) or (I.42). In various papers this problem appears under different names: the floating-body problem, the sea-keeping problem, the wave–body interaction problem, the water-wave radiation (scattering) problem, and so on. In what follows we use the simplest name: the water-wave problem.

Other Time-Harmonic Problems

In conclusion of the present section, we mention some boundary value problems that couple time-harmonic water waves with oscillations in other media. Hazard and Lenoir [113] considered scattering of an incident water wave by an elastic body immersed in water (the corresponding initial-boundary value problem was treated by Licht [199]). A linearized model of water-wave motion in a porous structure was proposed by Sollitt and Cross [308] for describing the interaction of water waves with rubble-mound breakwaters. This model was investigated by McIver [237], where further references are given. The most recent coupled problem was advanced by Pinkster [289] and investigated by Newman [267]. It is concerned with acoustic waves in a bounded air chamber placed on the free surface of water and open from below for interaction with water waves.

Linear Ship Waves on Calm Water (the Neumann–Kelvin Problem)

Separation of the t variable

Here we turn to waves created by a rigid body moving uniformly with constant velocity U on a calm water of constant depth d. It is convenient to denote the horizontal coordinates by (x, z) instead of (x_1, x_2). We assume (without loss of generality) that the motion is along the x axis of a fixed coordinate system. Moreover, we suppose waves to be steady with respect to a moving coordinate system attached to the body, or, in other words, one may speak about a uniform running flow about the body. The flow carries steady waves downstream (from the body to $x = -\infty$), so we set the following in (I.24)–(I.27):

$$\phi^{(1)}(x, y, z, t) = u(x - Ut, y, z), \tag{I.43}$$

where the (x, y, z)-coordinate system is fixed. Using the same notation (x, y, z) for the system attached to the body (since we use only these coordinates in what follows, this does not lead to any confusion), we see that the velocity potential $u(x, y, z)$ is defined in a fixed domain W occupied by water at rest outside the body's surface S. Since the water depth is constant, W is bounded below by $y = -d$ ($d \in (0, +\infty]$ and $d = +\infty$ for deep water), and we assume that S has no common points with this plane when $d < +\infty$. As in the third major section (the water-wave problem), we denote by F the free surface at rest that is the part of $y = 0$ outside the body.

Substituting (I.43) into (I.24)–(I.27), one obtains the following for u:

$$\nabla^2 u = 0 \quad \text{in } W, \tag{I.44}$$

$$u_{xx} + \nu u_y = 0 \quad \text{on } F, \tag{I.45}$$

$$\partial u/\partial n = f \quad \text{on } S, \tag{I.46}$$

$$u_y = 0 \quad \text{when } y = -d, \tag{I.47}$$

where $\nu = g/U^2$ in (I.45), and $f = U\mathbf{n} \cdot \mathbf{x}$ in (I.46) (by $\mathbf{n}$ and $\mathbf{x}$ we denote the unit normal to S directed into W and the unit vector directed along the x axis, respectively). However, in our considerations of problem (I.44)–(I.47) we do not use the specific form of the right-hand-side term in (I.46), and we use an arbitrary function as f. Also, for deep water, condition (I.47) should be replaced by the following one:

$$|\nabla u| \to 0 \quad \text{as } y \to -\infty.$$

That is, the water motion decays with depth.

In addition, we note that in the two-dimensional problem $u(x, y)$ could be a multiple-valued function if W is a doubly connected domain, that is, S is a totally submerged contour. In this case a velocity circulation should be prescribed on S. For the sake of simplicity, we assume this circulation to be equal to zero, and so $u(x, y)$ is a single-valued function in W.

Conditions at Infinity Upstream and Downstream

When we consider the forward motion of a body, different horizontal directions are not equivalent and the radiation condition (see the subsection on radiation conditions) is not appropriate for describing the behavior of the velocity potential as $x^2 + z^2 \to \infty$ (or $|x| \to \infty$ in the two-dimensional case). Since the x direction is chosen as the direction of the body motion in an infinite ocean undisturbed except for the body, the reasonable condition is that the water motion vanishes far ahead of the body; that is,

$$|\nabla u(x, y, z)| \to 0 \quad \text{as } x \to +\infty \tag{I.48}$$

(z should be omitted here for the two-dimensional problem). It is also obvious that in the two-dimensional problem

$$|\nabla u(x, y)| = O(1) \quad \text{as } |x| \to \infty. \tag{I.49}$$

In three dimensions a similar condition is, of course, true, but too rough to provide any uniqueness result. It is possible to impose a more precise condition instead of (I.49). However, as often occurs in the theory of water waves, the formulation of this condition cannot be completed until the problem is partly solved (see Section 7.4). Moreover, certain supplementary conditions should be imposed when the body is surface piercing. Various versions of such supplementary conditions for the two-dimensional problem are given in Chapter 8.

Finally, it is important to note the obvious fact that u satisfying problem (I.44)–(I.47) and (I.48)–(I.49) is defined up to an arbitrary constant term.

As for the water-wave problem, there are different names for the formulated problem, and in what follows it is referred to as the Neumann–Kelvin problem. Presumably, this name was first used by Brard [32] in 1972.

Other Problems for Ship Waves

In conclusion of the present section, we mention other statements describing the forward motion of a body in water. Peters and Stoker [286] (see also Stoker [312], Chapter 9) developed a mathematical approach describing the motion of a ship under the most general conditions compatible with a linearized theory and the assumption of an infinite ocean. For this purpose they applied a special formal perturbation procedure allowing the coupled pitching, surging, and heaving motions in a seaway consisting of a wave train having crests orthogonal to the course of ship's forward motion. The latter is assumed to be at a constant speed and along a straight line. In particular, the following boundary condition,

$$U^2 u_{xx} - 2i\omega U u_x - \omega^2 u + g u_y = 0,$$

arises on the free surface in the case when the uniform forward motion of a body at the speed U is coupled with the time-harmonic motion having the radian frequency ω.

An initial-boundary value problem describing unsteady waves produced by a rigid body in the uniform forward motion is formulated, for example, in Newman's survey paper [263]. In this case the free surface boundary condition takes the form

$$\phi_{tt} - 2U\phi_{xt} + U^2\phi_{xx} + g\phi_x = 0.$$

A mathematical treatment of this problem is given by Hamdache [103]. In [263], one can also find a statement unifying the simple harmonic time dependence of waves with the uniform forward motion of a rigid body immersed in water.

Part 1

Time-Harmonic Waves

1

Green's Functions

The simplest "obstacle" to be placed into water is a point source. The corresponding velocity potential (up to a time-periodic factor) is usually referred to as the Green's function. This notion is crucial for the theory we are going to present in this book, since a wide class of time-harmonic velocity potentials (in particular, solutions to the water-wave problem) admit representations based on Green's function (see Section 1.3).

Potentials constructed by using Green's functions form the basis for such different topics as proving solvability theorems (see Chapters 2 and 3) and constructing examples of trapped waves (nontrivial solutions to homogeneous boundary value problems given in Chapter 4).

The plan of this chapter is as follows. Beginning with Green's functions of point sources in water of infinite (Subsection 1.1.1) and finite (Subsection 1.1.2) depths, we proceed with straight line sources and ring sources (Section 1.2) arising in two-dimensional problems and problems with axial symmetry, respectively. Green's representation of velocity potentials and related questions are given in Section 1.3. Bibliographical notes (Section 1.4) contain references to original papers treating the material of this chapter as well as other related works.

1.1. Three-Dimensional Problems of Point Sources

1.1.1. Point Source in Deep Water

In the present subsection, we consider in detail Green's function describing the point source in deep water. In Subsection 1.1.1.1, we define it as a solution to the water-wave problem having Dirac's measure as the right-hand-side term in the equation. Also, a number of equivalent explicit representations of this function are given. In Subsection 1.1.1.2 we derive one of them by using the Fourier transform; in Subsection 1.1.1.3 we are concerned with the asymptotic behavior of Green's function at infinity; and in

Subsection 1.1.1.4 we describe its behavior as the source point approaches the free surface. The theorem establishing the uniqueness of Green's function is proved in Subsection 1.1.1.5. In Subsection 1.1.1.6, we obtain the asymptotic behavior of Green's function and its derivatives for large and small values of ν.

1.1.1.1. Green's Function: The Boundary Value Problem and Explicit Representations

Let a pulsating source of radian frequency ω be placed at a point (ξ, η), $\xi \in \mathbb{R}^2$, $\eta < 0$, beneath the free surface of water, occupying the lower half-space $\mathbb{R}^3_- = \{x \in \mathbb{R}^2, y < 0\}$. Representing the corresponding velocity potential in the form

$$\mathrm{Re}\{G(x, y; \xi, \eta) \exp(-i\omega t)\}, \quad \eta < 0,$$

we say that $G(x, y; \xi, \eta)$ is Green's function. It must satisfy the following boundary value problem:

$$\nabla^2_{(x,y)} G = -4\pi \delta_{(\xi,\eta)}(x, y) \quad \text{in } \mathbb{R}^3_-, \tag{1.1}$$

$$G_y - \nu G = 0 \quad \text{when } y = 0, \tag{1.2}$$

$$\int_{C_r} \left| G_{|x|} - i\nu G \right|^2 \mathrm{d}S = o(1) \quad \text{as } r \to \infty, \tag{1.3}$$

$$\sup\{|G(x, y; \xi, \eta) - R^{-1}| : (x, y) \in \mathbb{R}^3_-\} < \infty. \tag{1.4}$$

Here $R^2 = |x - \xi|^2 + (y - \eta)^2$, $C_r = \{|x| = r, y < 0\}$, and $\delta_{(\xi,\eta)}(x, y)$ is Dirac's measure at (ξ, η), that is, a linear functional acting on continuous functions as follows:

$$\left\langle \delta_{(\xi,\eta)}(x, y), \psi(x, y) \right\rangle = \psi(\xi, \eta).$$

It follows from (1.4) that *the velocity field defined by G tends to zero as* $y \to -\infty$; similarly, (I.36) follows from (I.35).

To prove this assertion we see from (1.1) that $H = G - R^{-1}$ is a harmonic function in $\mathbb{R}^3_-$ as well as its derivatives. Then the mean value theorem (see, for example, Courant and Hilbert [41], Chapter 4, Section 3.1) gives the following for $b < -\eta$:

$$\nabla H(x, y) = \frac{3}{4\pi b^3} \int_{R<b} \nabla H \, \mathrm{d}\xi \mathrm{d}\eta.$$

This and the divergence theorem lead to

$$\nabla H(x, y) = \frac{3}{4\pi b^3} \int_{R=b} H \mathbf{n} \, \mathrm{d}S,$$

where $\mathbf{n}$ is the unit normal to the sphere $R = b$ directed outward. Then

$$|\nabla H(x, y)| \leq 3b^{-1} \sup\{|H|\}.$$

From this and (1.4) we get

$$|\nabla\{G(x, y; \xi, \eta) - R^{-1}\}| = O(|y|^{-1}) \quad \text{as } y \to -\infty, \tag{1.5}$$

which proves the assertion.

The main aim of the present subsection is to prove the following result.

Problem (1.1)–(1.4) *has the unique solution*

$$G(x, y; \xi, \eta) = R^{-1} + R_0^{-1} + 2\pi i \nu e^{\nu(y+\eta)} J_0(\nu|x - \xi|)$$
$$+ 2 \int_0^\infty \frac{\nu}{k - \nu} e^{k(y+\eta)} J_0(k|x - \xi|) \, \mathrm{d}k. \tag{1.6}$$

Here $R_0^2 = |x - \xi|^2 + (y + \eta)^2$, *and the integral is understood as the Cauchy principal value.*

In Subsections 1.1.1.2–1.1.1.5 we derive (1.6), investigate properties of Green's function, and prove the uniqueness theorem for it. Here we begin with some other representations for G. They are equivalent, but for various purposes different representations are desirable.

Using the path of integration ℓ_- as shown in Fig. 1.1, we can write (1.6) as follows:

$$G(x, y; \xi, \eta) = R^{-1} + R_0^{-1} + 2 \int_{\ell_-} \frac{\nu}{k - \nu} e^{k(y+\eta)} J_0(k|x - \xi|) \, \mathrm{d}k. \tag{1.7}$$

Identity

$$(a^2 + b^2)^{-1/2} = \int_0^\infty e^{-kb} J_0(ka) \, \mathrm{d}k, \quad b > 0, \tag{1.8}$$

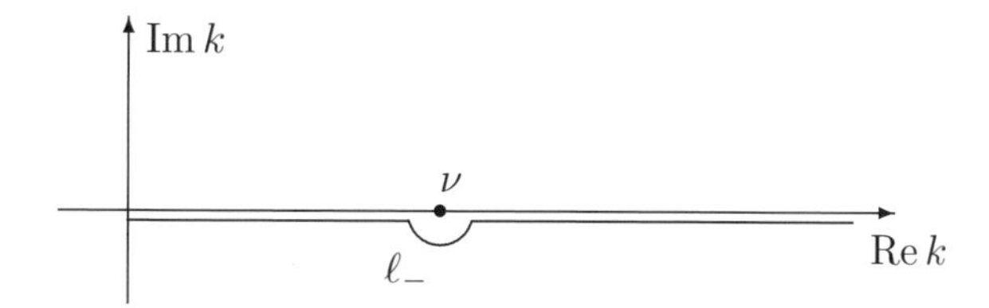

Figure 1.1.

(see 6.611.3 in Gradshteyn and Ryzhik [96]) leads to two other formulae:

$$G(x, y; \xi, \eta) = R^{-1} + \int_{\ell_-} \frac{k+\nu}{k-\nu} e^{k(y+\eta)} J_0(k|x-\xi|) \, dk, \tag{1.9}$$

$$G(x, y; \xi, \eta) = R^{-1} - R_0^{-1} + 2\int_{\ell_-} \frac{k}{k-\nu} e^{k(y+\eta)} J_0(k|x-\xi|) \, dk. \tag{1.10}$$

1.1.1.2. Derivation of (1.6)

We see from (1.1) and (1.4) that it is convenient to seek Green's function in the following form:

$$G(x, y; \xi, \eta) = R^{-1} + H(x, y; \xi, \eta), \tag{1.11}$$

where $H(x, y; \xi, \eta)$ is a bounded harmonic function in $\mathbb{R}^3_-$. To determine H we use the Fourier transform in x:

$$\hat{H}(\sigma, y; \xi, \eta) = \int_{\mathbb{R}^2} H(x, y; \xi, \eta) e^{-i\langle x, \sigma\rangle} \, dx,$$

where $\langle x, \sigma\rangle = x_1\sigma_1 + x_2\sigma_2$. The Laplace equation yields

$$\hat{H}_{yy} - |\sigma|^2 \hat{H} = 0 \quad \text{for } y < 0.$$

This has a solution bounded as $y \to -\infty$:

$$\hat{H} = A(\sigma; \xi, \eta) \, e^{|\sigma| y}.$$

For finding $A(\sigma; \xi, \eta)$ we apply the Fourier transform to (1.2) by using

$$\widehat{R^{-1}} = |\sigma|^{-1} \exp[-i\langle \xi, \sigma\rangle - |\sigma|(y-\eta)]. \tag{1.12}$$

Then we get

$$|\sigma|\{A - |\sigma|^{-1} \exp[|\sigma|\eta - i\langle \xi, \sigma\rangle]\} = \nu\{A + |\sigma|^{-1} \exp[|\sigma|\eta - i\langle \xi, \sigma\rangle]\},$$

and a simple manipulation gives

$$\hat{H}(\sigma, y; \xi, \eta) = |\sigma|^{-1} \frac{|\sigma|+\nu}{|\sigma|-\nu} \exp[|\sigma|(y+\eta) - i\langle \xi, \sigma\rangle].$$

Applying (1.12) again we write this in the following form:

$$\hat{H}(\sigma, y; \xi, \eta) = \widehat{R_0^{-1}} + \frac{2\nu}{|\sigma|(|\sigma|-\nu)} \exp[|\sigma|(y+\eta) - i\langle \xi, \sigma\rangle].$$

Formal application of the inverse Fourier transform yields

$$H(x, y; \xi, \eta) = \frac{1}{R_0} + \frac{\nu}{\pi} \int_{\mathbb{R}^2} \frac{\exp[|\sigma|(y+\eta) + i\langle x-\xi, \sigma\rangle]}{|\sigma|(|\sigma|-\nu)} \, d\sigma,$$

but we have to describe how the last integral should be understood.

Introducing two polar coordinate systems as follows,

$$x_1 - \xi_1 = |x - \xi| \cos\theta, \qquad x_2 - \xi_2 = |x - \xi| \sin\theta,$$
$$\sigma_1 = k\cos\varphi, \qquad \sigma_2 = k\sin\varphi, \quad \text{where } k > 0,$$

we get formally

$$\begin{aligned} &H(x, y;\xi, \eta) - R_0{}^{-1} \\ &= \frac{\nu}{\pi}\int_0^\infty \int_0^{2\pi} \frac{\exp k\,[(y+\eta) - i|x-\xi|\cos(\varphi - \theta)]}{k - \nu}\,\mathrm{d}\varphi \mathrm{d}k \\ &= \frac{4\nu}{\pi}\int_0^\infty \int_0^{\pi/2} \frac{\exp k(y+\eta)}{k-\nu} \cos(k|x-\xi|\cos\varphi)\,\mathrm{d}\varphi\mathrm{d}k \\ &= 2\int_0^\infty \frac{\nu}{k-\nu}\, e^{k(y+\eta)}\, J_0(k|x-\xi|)\,\mathrm{d}k. \end{aligned}$$

The last equality is based on the integral representation of the Bessel function J_0 (see 8.411.4 in Gradshteyn and Ryzhik [96]).

When ν in the formulae for $\hat{H}$ and H is replaced by a complex parameter λ having $\operatorname{Im}\lambda > 0$, the above calculations are rigorous, and function (1.11) satisfies (1.1), (1.4), and (1.2), where λ stands instead of ν. Moreover, as $\lambda \to \nu + i0$ the expression for H converges to

$$R_0^{-1} + 2\int_{\ell_-} \frac{\nu}{k-\nu} e^{k(y+\eta)}\, J_0(k|x-\xi|)\,\mathrm{d}k$$

uniformly in (x, y) belonging to an arbitrary bounded subset of $\mathbb{R}^3_-$. The same is true for all derivatives of H. Therefore, (1.7) satisfies (1.1), (1.2), and (1.4). We note that (1.6) is equivalent to (1.7), and in Subsection 1.1.1.3 we demonstrate that the radiation condition (1.3) holds for (1.7). So (1.6) is proved to be Green's function.

1.1.1.3. Asymptotics at Infinity

The behavior of $G(x, y; \xi, \eta)$ as R and R_0 tend to infinity depends on whether $|x - \xi|$ is bounded or not. First, we consider the case in which $|x - \xi| < |y + \eta|$, and therefore $|y + \eta| > R_0/2$. From (1.10)

$$\begin{aligned} G_1(x, y; \xi, \eta) &= G(x, y; \xi, \eta) - \left(R^{-1} - R_0^{-1}\right) \\ &= 2\int_{\ell_-} k(k-\nu)^{-1}\, e^{k(y+\eta)}\, J_0(k|x-\xi|)\,\mathrm{d}k. \end{aligned}$$

Let us split the last integral into a sum by dividing ℓ_- into two parts $(0, \tau)$ and $\ell_-(\tau)$, where $\tau \in (0, \nu)$. Then

$$|G_1(x, y; \xi, \eta)| \leq 2 \left| \int_0^\lambda k(k-\nu)^{-1} e^{k(y+\eta)} J_0(k|x-\xi|)\, \mathrm{d}k \right| + 2e^{\tau(y+\eta)} \left| \int_{\ell_-(\tau)} k(k-\nu)^{-1} e^{(k-\tau)(y+\eta)} J_0(k|x-\xi|)\, \mathrm{d}k \right|.$$

The first term on the right-hand side does not exceed

$$\text{const} \int_0^\tau k e^{k(y+\eta)}\, \mathrm{d}k = O(|y+\eta|^{-2}),$$

and the second one is $O[e^{\tau(y+\eta)}]$. The derivatives of G_1 can be estimated in the same way. Therefore, we arrive at the following conclusion:

$$|G_1| + |\nabla G_1| = O\big(R_0^{-2}\big) \quad \text{as } R, R_0 \to \infty \text{ and } |x-\xi| < |y+\eta|.$$

When $|x-\xi| > |y+\eta|$ we have $|x-\xi| > R_0/2$. To obtain the asymptotics of G under this assumption we need a slightly modified version of (1.10), which is derived now. First, we replace J_0 in the integrand in (1.6) by $\mathrm{Re}\, H_0^{(1)}$, and we transform (1.6) as follows:

$$\begin{aligned} G(x,y;\xi,\eta) - R^{-1} - R_0^{-1} &= 2\nu\, \mathrm{Re} \int_0^\infty \frac{e^{k(y+\eta)}}{k-\nu} H_0^{(1)}(k|x-\xi|)\, \mathrm{d}k + 2\pi i \nu e^{\nu(y+\eta)} J_0(\nu|x-\xi|) \\ &= 2\nu\, \mathrm{Re} \int_{\ell_+} \frac{e^{k(y+\eta)}}{k-\nu} H_0^{(1)}(k|x-\xi|)\, \mathrm{d}k \\ &\quad + 2\pi i \nu e^{\nu(y+\eta)} \left[J_0(\nu|x-\xi|) + i Y_0(\nu|x-\xi|) \right]. \end{aligned}$$

Here ℓ_+ is obtained by reflecting ℓ_- in the real axis, and Y_0 denotes the Bessel function of the second kind of order zero.

Since the integrand has no poles in the first quadrant, the path of integration ℓ_+ can be replaced by the path from zero to infinity along the positive imaginary axis. Then

$$G(x, y; \xi, \eta) = R^{-1} + R_0^{-1} + 2\pi i \nu\, e^{\nu(y+\eta)} H_0^{(1)}(\nu|x-\xi|) + 2\nu\, \mathrm{Re} \int_0^\infty \frac{i e^{ik(y+\eta)}}{ik-\nu} H_0^{(1)}(ik|x-\xi|)\, \mathrm{d}k. \qquad (1.13)$$

Using the integral representation of the modified Bessel function K_0 (see

8.432.9 in Gradshteyn and Ryzhik [96]), we get

$$
\begin{aligned}
\operatorname{Re}\int_0^\infty \frac{i e^{ik(y+\eta)}}{ik-\nu} H_0^{(1)}(ik|x-\xi|)\,\mathrm{d}k &= \frac{2}{\pi}\operatorname{Re}\int_0^\infty \frac{e^{ik(y+\eta)}}{ik-\nu} K_0(k|x-\xi|)\,\mathrm{d}k \\
&= \frac{2}{\pi}\operatorname{Re}\int_0^\infty \frac{e^{ik(y+\eta)}}{ik-\nu}\,\mathrm{d}k \int_1^\infty \frac{e^{-mk|x-\xi|}\,\mathrm{d}m}{(m^2-1)^{1/2}} \\
&= \frac{2}{\pi}\operatorname{Re}\int_0^\infty \frac{\mathrm{d}k}{ik-\nu} \\
&\quad\times \int_1^\infty \frac{\exp\{k[i(y+\eta)-m|x-\xi|]\}}{(m^2-1)^{1/2}}\,\mathrm{d}m.
\end{aligned}
$$

Integration by parts with respect to k gives

$$
\begin{aligned}
&\frac{2}{\pi\nu}\operatorname{Re}\int_1^\infty \frac{\mathrm{d}m}{[i(y+\eta)-m|x-\xi|]\,(m^2-1)^{1/2}} \\
&\quad+\frac{2}{\pi}\operatorname{Re}\int_0^\infty \frac{i\,\mathrm{d}k}{(ik-\nu)^2}\int_1^\infty \frac{\exp\{k[i(y+\eta)-m|x-\xi|]\}}{[i(y+\eta)-m|x-\xi|]\,(m^2-1)^{1/2}}\,\mathrm{d}m. \quad (1.14)
\end{aligned}
$$

The first term here is equal to

$$
-\frac{2}{\pi\nu}\int_1^\infty \frac{m|x-\xi|\,\mathrm{d}m}{[(y+\eta)^2+m^2|x-\xi|^2](m^2-1)^{1/2}},
$$

and calculating the integral we obtain $-(\nu R_0)^{-1}$. Substituting (1.14) into (1.13), we arrive at

$$
\begin{aligned}
&G(x,y;\xi,\eta) = R^{-1} - R_0^{-1} + 2\pi i\nu e^{\nu(y+\eta)} H_0^{(1)}(\nu|x-\xi|) \\
&\quad+\frac{4\nu}{\pi}\operatorname{Re}\int_0^\infty \frac{i\mathrm{d}k}{(ik-\nu)^2}\int_1^\infty \frac{\exp\{k[i(y+\eta)-m|x-\xi|]\}}{[i(y+\eta)-m|x-\xi|]\,(m^2-1)^{1/2}}\,\mathrm{d}m. \quad (1.15)
\end{aligned}
$$

The absolute value of the last integral in (1.15) does not exceed

$$
\begin{aligned}
&C\int_0^\infty \frac{\mathrm{d}k}{1+k^2}\int_1^\infty \frac{\exp\{-km|x-\xi|\}}{|x-\xi|m(m^2-1)^{1/2}}\,\mathrm{d}m \\
&= C\int_0^\infty \frac{e^{-k|x-\xi|}}{|x-\xi|(1+k^2)}\,\mathrm{d}k\int_1^\infty \frac{\mathrm{d}m}{m(m^2-1)^{1/2}} \leq \frac{C_1}{|x-\xi|^2} \leq \frac{4C_1}{R_0^2}.
\end{aligned}
$$

Similar estimate can be obtained for the gradient of the integral, but in the latter case one needs an extra integration by parts with respect to k.

Taking into account the asymptotic behavior of $H_0^{(1)}$ (see the Examples section in the Introduction) and the result obtained above for the case $|x-\xi| < |y+\eta|$, we formulate the following theorem.

Function (1.6) *has the following asymptotic behavior as* $R, R_0 \to \infty$: *(i) if* $|x - \xi| < |y + \eta|$, *then*

$$G(x, y; \xi, \eta) = R^{-1} - R_0^{-1} + G_1(x, y; \xi, \eta),$$

where $|G_1| + |\nabla G_1| = O(R_0^{-2})$; *(ii) if* $|x - \xi| > |y + \eta|$, *then*

$$G(x, y; \xi, \eta)$$
$$= \frac{1}{R} - \frac{1}{R_0} + 2\left(\frac{2\pi\nu}{|x-\xi|}\right)^{1/2} e^{\nu[y+\eta+i(|x-\xi|-\pi/4)]} + G_2(x, y; \xi, \eta),$$

where $|G_2| + |\nabla G_2| = O(R_0^{-2} + e^{\nu(y+\eta)}(1 + |x - \xi|)^{-3/2})$.

The last asymptotic formula implies that G satisfies the radiation condition (1.3), if $|\xi|^2 + \eta^2 <$ const. This completes the justification of (1.6) as a representation of Green's function.

It is also clear that G satisfies Sommerfeld's radiation condition (I.39) with $k_0 = \nu$ provided (ξ, η) belongs to a bounded region.

1.1.1.4. Asymptotic Behavior Near the Free Surface

Another important property of G is concerned with its behavior as $R \to 0$ and $y, \eta \to 0$ simultaneously. From (1.6) we see that $R^{-1} + R_0^{-1}$ gives the strongly singular part of G, and it is sufficient to verify that $G - R^{-1} - R_0^{-1}$ has a weaker singularity. In particular, *the following representation holds*:

$$G(x, y; \xi, \eta) = R^{-1} + R_0^{-1}$$
$$+ 2\nu e^{\nu(y+\eta)} \log(R_0 + |y + \eta|) + h(|x - \xi|, y + \eta), \qquad (1.16)$$

where

$$|h| + |\nabla h| \leq \text{const} \quad \text{as } R, y, \eta \to 0. \qquad (1.17)$$

In order to prove (1.16) and (1.17) it is sufficient to establish these formulae for a function which satisfies (1.1) and (1.2) and differs from Green's function by an infinitely smooth term. Therefore, we consider

$$G^*(x, y; \xi, \eta) = R^{-1} + R_0^{-1} + 2\nu \int_{-1}^{0} \frac{e^{-\nu\tau}}{R(\tau)} d\tau, \qquad (1.18)$$

where $R(\tau) = [|x - \xi|^2 + (y + \eta + \tau)^2]^{1/2}$.

If $\eta + \tau \leq 0$, then $[R(\tau)]^{-1}$ is harmonic in $\mathbb{R}^3_-$, and so we have

$$\nabla^2 G^* = -4\pi\delta_{(\xi,\eta)}(x, y) \quad \text{in } \mathbb{R}^3_- \text{ for } \eta < 0. \qquad (1.19)$$

A direct calculation gives that

$$\left(\frac{\partial}{\partial y} - \nu\right)\left(R^{-1} + R_0^{-1}\right) = -2\nu R_0^{-1} \quad \text{when } y = 0,$$

and

$$\frac{\partial}{\partial y}\int_{-1}^{0} \frac{e^{-\nu\tau}}{R(\tau)}\,\mathrm{d}\tau = \int_{-1}^{0} e^{-\nu\tau}\frac{\partial}{\partial\tau}[R(\tau)]^{-1}\,\mathrm{d}\tau$$
$$= \frac{1}{R(0)} - \frac{e^{\nu}}{R(1)} + \nu\int_{-1}^{0}\frac{e^{-\nu\tau}}{R(\tau)}\,\mathrm{d}\tau.$$

Therefore,

$$G_y^* - \nu G^* = \frac{-2\nu e^{\nu}}{[|x-\xi|^2 + (1+|\eta|)^2]^{1/2}} \quad \text{when } y = 0. \tag{1.20}$$

Let us demonstrate that $G - G^*$ is a smooth function and that (1.16) holds for G^*. From (1.1), (1.2), (1.19), and (1.20), we get

$$\nabla^2(G - G^*) = 0 \quad \text{in } \mathbb{R}^3_-, \quad (G - G^*)_y - \nu(G - G^*) = -f \quad \text{when } y = 0.$$

Here $\eta < 0$ and f denotes the right-hand side in (1.20). Since f is an infinitely differentiable function when $\eta \leq 0$, the smoothness of $G - G^*$ is a consequence of a priori estimates for the Laplacian. However, the application of a priori estimates requires that a certain (very weak) estimate must hold for $G - G^*$ an $\eta \to 0$. In order to avoid the latter estimate, another approach is developed below.

Let $v = (G - G^*)_y - \nu(G - G^*)$, where $\eta < 0$. Then we have that

$$\nabla^2 v = 0 \quad \text{in } \mathbb{R}^3_-, \quad v = -f \quad \text{when } y = 0. \tag{1.21}$$

Here f is an infinitely differentiable function when $\eta \leq 0$. Without loss of generality, we assume that $|\xi| \leq a < \infty$. By D we denote a bounded domain in $\mathbb{R}^3_-$ such that ∂D is a C^∞-surface and

$$\{|x| < 2a, \quad -b < y < 0\} \subset D, \quad \text{where } b > 0.$$

We get from (1.8) that

$$G - R^{-1} \in C^\infty \quad \text{when } y + \eta \leq -b < 0. \tag{1.22}$$

Formula (1.13) and the first equality in the formula following (1.13) imply that

$$G - R^{-1} \in C^\infty \quad \text{where } |x - \xi| \geq a > 0,\ y + \eta \leq 0. \tag{1.23}$$

It is clear that (1.22) and (1.23) hold also for $G^* - R^{-1}$, and so

$$G - G^* \in C^\infty \quad \text{when } y + \eta \le -b < 0 \\ \text{or } |x - \xi| \ge a > 0, \ y + \eta \le 0, \tag{1.24}$$

and (1.24) remains true for v. This fact and (1.21) yield

$$\nabla^2 v = 0 \quad \text{in } D, \quad v = v_0 \quad \text{on } \partial D,$$

where $\eta < 0$ and v_0 is an infinitely smooth function of all variables when $\eta \le 0$. Hence, there exists a limit of v as $\eta \to -0$ and also

$$v \in C^\infty \quad \text{when } (x, y) \in \bar{D}, |\xi| < a, \eta \le 0.$$

From the ordinary differential equation

$$(G - G^*)_y - \nu(G - G^*) = v, \qquad y + \eta \le 0,$$

where $v \in C^\infty$, and the fact that $G - G^* \in C^\infty$ when $y = -b$ and $\eta \le 0$ (see (1.24)), it follows that

$$G - G^* \in C^\infty \quad \text{when } y + \eta \le 0.$$

Now, it remains to show that (1.16) and (1.17) hold for G^* defined by (1.18). Taking into account that

$$\int_{-1}^{0} \frac{\mathrm{d}\tau}{R(\tau)} = \log(R_0 + |y + \eta + \tau|)\Big|_{\tau=-1}^{\tau=0} = \log(R_0 + |y + \eta|) + h_1,$$

where h_1 satisfies (1.17), we have to check that (1.17) holds for

$$h_2 = \int_{-1}^{0} \frac{e^{-\nu\tau} - e^{\nu(y+\eta)}}{R(\tau)} \mathrm{d}\tau \\ = -\int_{-1}^{0} \frac{\nu(y + \eta + \tau)}{R(\tau)} \mathrm{d}\tau + \int_{-1}^{0} \frac{e^{-\nu\tau} - e^{\nu(y+\eta)} + \nu(y + \eta + \tau)}{R(\tau)} \mathrm{d}\tau.$$

In the last integral, the integrand and its gradient are bounded, and so (1.17) is true for this integral. Since

$$-\int_{-1}^{0} \frac{\nu(y + \eta + \tau)}{R(\tau)} \mathrm{d}\tau = [|x - \xi|^2 + (y + \eta + \tau)^2]^{1/2}\Big|_{\tau=-1}^{\tau=0},$$

(1.17) holds for this integral as well. The proof is complete.

1.1.1.5. Uniqueness of Green's Function

In order to demonstrate that G given by equivalent formulae in Subsection 1.1.1.1 is the unique solution to (1.1)–(1.4) we have to prove the following proposition.

Only a trivial solution satisfies the following boundary value problem:

$$\nabla u = 0 \quad \text{in } \mathbb{R}^3_-,$$
$$u_y - \nu u = 0 \quad \text{when } y = 0,$$
$$\int_{C_r} \left| u_{|x|} - i\nu u \right|^2 \mathrm{d}S \to 0 \quad \text{as } r \to \infty,$$
$$\sup \{ |u(x, y)| : (x, y) \in \mathbb{R}^3_- \} < \infty.$$

First we note that the estimate

$$|\nabla u(x, y)| = O(|y|^{-1}) \tag{1.25}$$

holds because u is bounded (see Subsection 1.1.1.1 for the proof of a similar statement for $G - R^{-1}$). Then Green's identity over a semi-infinite domain $\mathbb{R}^3_- \cap \{|x| < r\}$ takes the following form:

$$\int_{\mathbb{R}^3_- \cap \{|x|<r\}} |\nabla u|^2 \,\mathrm{d}x\mathrm{d}y = \nu \int_{F_r} |u(x, 0)|^2 \,\mathrm{d}x + \int_{C_r} \bar{u}\, u_{|x|} \,\mathrm{d}S.$$

Here $F_r = \{|x| < r, y = 0\}$, $C_r = \{|x| = r, y < 0\}$, and the free-surface boundary condition is taken into account. From the latter equation it follows that

$$\operatorname{Im} \int_{C_r} \bar{u}\, u_{|x|} \,\mathrm{d}S = 0.$$

Since the left-hand side is equal to

$$\frac{1}{2\nu} \left\{ \int_{C_r} \left(\left| u_{|x|} \right|^2 + \nu^2 |u|^2 \right) \mathrm{d}S - \int_{C_r} \left| u_{|x|} - i\nu u \right|^2 \mathrm{d}S \right\},$$

the radiation condition implies that

$$\lim_{r\to\infty} \int_{C_r} \left| u_{|x|} \right|^2 \mathrm{d}S = \lim_{r\to\infty} \int_{C_r} |u|^2 \,\mathrm{d}S = 0. \tag{1.26}$$

The theorem in Subsection 1.1.1.3 implies that the integrals

$$\int_{C_r} |G(x, y; \xi, \eta)|^2 \,\mathrm{d}S_{(x,y)}, \qquad \int_{C_r} \left| G_{|\xi|}(x, y; \xi, \eta) \right|^2 \mathrm{d}S_{(x,y)} \tag{1.27}$$

are bounded when (ξ, η) is fixed and $r \to \infty$.

From (1.25), (1.5) and the free-surface boundary condition we obtain the following Green's representation:

$$u(\xi, \eta) = \frac{1}{4\pi} \int_{C_r} \left[G(x, y; \xi, \eta) u_{|x|}(x, y) - u(\xi, \eta) G_{|x|}(x, y; x, y) \right] \mathrm{d}S$$

for every $(\xi, \eta) \in \mathbb{R}^3_-$, if r is taken greater than $|\xi|$. Applying the Schwarz inequality to the absolute value of the right-hand side and taking into account

(1.26) and boundedness of the integrals (1.27), we get that $u(\xi, \eta) = 0$. The proof is complete.

1.1.1.6. Asymptotic Behavior for Small and Large Values of ν

It is natural to expect that the solution G of problem (1.1)–(1.4) converges to Green's functions of the Neumann and Dirichlet problems in a half-space as ν tends to zero and infinity, respectively. We recall that decaying at infinity Green's function of the Neumann problem in $\mathbb{R}^3_-$ is equal to $R^{-1} + R_0^{-1}$, whereas that of the Dirichlet problem is $R^{-1} - R_0^{-1}$. The aim of the present section is to establish the following assertion.

For any $\epsilon > 0$ *we have*

$$G(x, y; \xi, \eta) - R^{-1} - R_0^{-1} \to 0 \quad \text{as } \nu \to 0, \tag{1.28}$$

$$G(x, y; \xi, \eta) - R^{-1} + R_0^{-1} \to 0 \quad \text{as } \nu \to +\infty \tag{1.29}$$

uniformly with respect to $(x, y), (\xi, \eta) \in \mathbb{R}^3_-$ *such that* $y + \eta \le -\epsilon$.

For (1.28) to be proved the integral in (1.7) should be estimated. Changing k to νk as the variable of integration, we leave the same notation ℓ_- for the path of integration indented below at $k = 1$. Now ℓ_- is independent of ν. Because J_0 is a bounded function and $|e^{-\nu k \epsilon}| < (\nu|k|\epsilon)^{-1/2}$, the integral does not exceed

$$C\nu \int_{\ell_-} \left| \frac{e^{-\nu k \epsilon}}{k - 1} \right| \mathrm{d}|k| \le C\nu \int_{\ell_-} \frac{(\nu|k|\epsilon)^{-1/2}}{|k - 1|} \mathrm{d}|k| = C_1 \left(\frac{\nu}{\epsilon} \right)^{1/2},$$

where C and C_1 are certain constants. This proves (1.28).

Using (1.10) in the same way, we get

$$\begin{aligned} \left|G - R^{-1} + R_0^{-1}\right| &\le C\nu \int_{\ell_-} \left| \frac{k e^{-\nu k \epsilon}}{k - 1} \right| \mathrm{d}|k| \\ &\le C\nu \int_{\ell_-} \left| \frac{k}{k - 1} \right| (\nu|k|\epsilon)^{-3/2} \, \mathrm{d}|k| = \frac{C_1}{\epsilon \sqrt{\nu \epsilon}}. \end{aligned}$$

This means that (1.29) holds.

An extension of the last assertion is the following remark.

For functions (1.28) *and* (1.29) *all derivatives also converge to zero; that is,*

$$\left|\nabla \left[G(x, y; \xi, \eta) - R^{-1} - R_0^{-1}\right]\right| \to 0 \quad \text{as } \nu \to 0, \tag{1.30}$$

$$\left|\nabla \left[G(x, y; \xi, \eta) - R^{-1} + R_0^{-1}\right]\right| \to 0 \quad \text{as } \nu \to +\infty \tag{1.31}$$

uniformly with respect to $(x, y), (\xi, \eta) \in \mathbb{R}^3_-$ *such that* $y + \eta \le -\epsilon$.

We begin the proof by noting that $G - R^{-1} \mp R_0^{-1}$ are harmonic functions in $\mathbb{R}^3_-$. Therefore, the absolute values of their derivatives at any point (x, y) such that $y \leq -\epsilon$ can be estimated through the maximum values of functions themselves on the sphere of radius $\epsilon/2$ centered at (x, y). Hence, (1.28) and (1.29) imply the uniform convergence of derivatives when $y \leq -\epsilon$ and $\eta < 0$. For the uniform convergence of derivatives to be justified when $y + \eta \leq -\epsilon$, it remains to be noted that they depend only on $|x - \xi|$ and $y + \eta$; see (1.7) and (1.10).

We conclude this subsection with a demonstration that (1.17) allows (1.28) to be proved under a little bit weaker assumption. Instead of G_0 considered in Subsection 1.1.1.4 and depending on ν, we introduce $\tilde{G}_0$, which is obtained from G_0 by taking ν equal to one. Using k/ν as the integration variable in the integral in (1.13), we get

$$G_0(x, y; \xi, \eta) = \nu \tilde{G}_0(\nu x, \nu y; \nu \xi, \nu \eta).$$

This and (1.17) imply that for any $a > 0$ there exists a constant $C(a)$ such that we have

$$\left| G(x, y; \xi, \eta) - R^{-1} - R_0^{-1} \right| \leq C(a) \nu \log(\nu R), \qquad (1.32)$$

$$\left| \nabla \left\{ G(x, y; \xi, \eta) - R^{-1} - R_0^{-1} \right\} \right| \leq C(a) R_0^{-1}, \qquad (1.33)$$

when ν is small enough, and $|x| + |y| + |\xi| + |\eta| \leq a$.

1.1.2. Point Source in Water of Finite Depth

The plan of this subsection repeats that of the previous one. We begin with the definition of Green's function, explicit expressions for it, and its behavior at infinity (see Subsection 1.1.2.1). In Subsection 1.1.2.2 we consider some properties of Green's function.

1.1.2.1. Explicit Representations for Green's Function

Let a source be pulsating in a layer $L = \{x \in \mathbb{R}^2, -d < y < 0\}$ of a finite depth $d > 0$. The corresponding Green's function must satisfy the boundary value problem:

$$\nabla^2_{(x,y)} G = -4\pi \delta_{(\xi,\eta)}(x, y) \quad \text{in } L, \qquad (1.34)$$

$$G_y - \nu G = 0 \quad \text{when } y = 0, \qquad (1.35)$$

$$G_y = 0 \quad \text{when } y = -d, \qquad (1.36)$$

$$\int_{C_r} \left| G_{|x|} - i k_0 G \right|^2 \mathrm{d}S = o(1) \quad \text{as } r \to \infty. \qquad (1.37)$$

Here (ξ, η) is the source point, $C_r = \{|x| = r, -d < y < 0\}$, and k_0 denotes the unique positive root of $k_0 \tanh k_0 d = \nu$ (cf. the Examples section in the Introduction).

Here we prove two explicit formulae for G, and the main result can be stated as follows.

The unique solution to (1.34)–(1.37) *has the following form:*

$$\begin{aligned} G(x, y; \xi, \eta) = {} & R^{-1} + R_d^{-1} \\ & + 2 \int_{\ell_-} \frac{(\nu + k) e^{-kd} \cosh k(y + d) \cosh k(\eta + d)}{k \sinh kd - \nu \cosh kd} \\ & \times J_0(k|x - \xi|)\, \mathrm{d}k, \end{aligned} \tag{1.38}$$

where $R_d^2 = |x - \xi|^2 + (y + 2d + \eta)^2$, *and* ℓ_- *denotes the semiaxis* $k > 0$ *indented at* k_0 *in the same way as in* (1.7).

The uniqueness follows by the same method as in Subsection 1.1.1.5. Let us demonstrate that function (1.38) satisfies (1.34)–(1.37). We begin with an observation that the Laplace equation and the bottom boundary condition can be verified by a direct calculation.

Further, assuming that $-d < \eta < y \leq 0$ and taking into account (1.8), one gets

$$G(x, y; \xi, \eta) = \int_{\ell_-} q(k)\, J_0(k|x - \xi|)\, \mathrm{d}k,$$

$$q(k) = 2 \cosh k(\eta + d)\, \frac{k \cosh ky + \nu \sinh ky}{k \sinh kd - \nu \cosh kd}, \tag{1.39}$$

from which (1.35) follows.

For proving (1.37) we derive another representation [see (1.40) below]. Let q_0 denote the residue of $q(k)$ at the pole $k = k_0$; that is,

$$q_0 = \frac{2\nu(k_0 \cosh k_0 y + \nu \sinh k_0 y) \cosh k_0(\eta + d)}{\left(k_0^2 d - \nu^2 d + \nu\right) \sinh k_0 d}.$$

Now we note that (1.39) is valid for $-d < \eta \leq y \leq 0$ when $x \neq \xi$. So (1.39) is equivalent to

$$G(x, y; \xi, \eta) = \pi i q_0 J_0(k_0|x - \xi|) + \int_0^\infty q(k) J_0(k|x - \xi|)\, \mathrm{d}k,$$

where the integral is understood as the Cauchy principal value. Replacing J_0

by Re $H_0^{(1)}$, we get from here

$$
\begin{aligned}
G(x, y; \xi, \eta) &= \pi i q_0 J_0(k_0|x-\xi|) + \operatorname{Re}\int_0^\infty q(k) H_0^{(1)}(k|x-\xi|)\,\mathrm{d}k \\
&= \pi i q_0 J_0(k_0|x-\xi|) + \operatorname{Re}\int_{\ell_+} q(k) H_0^{(1)}(k|x-\xi|)\,\mathrm{d}k \\
&\quad + \operatorname{Re}\pi i q_0 H_0^{(1)}(k_0|x-\xi|) \\
&= \pi i q_0 H_0^{(1)}(k_0|x-\xi|) + \operatorname{Re}\int_{\ell_+} q(k) H_0^{(1)}(k|x-\xi|)\,\mathrm{d}k,
\end{aligned}
\tag{1.40}
$$

where ℓ_+ is the reflection of ℓ_- in the real axis.

Formula 8.472.3 in Gradshteyn and Ryzhik [96] gives

$$kH_0^{(1)}(kt) = \frac{1}{t}\frac{\mathrm{d}}{\mathrm{d}k}\big[kH_0^{(1)}(kt)\big].$$

Then the second term in (1.40) is equal to

$$\frac{-1}{|x-\xi|}\operatorname{Re}\int_{\ell_+}\left[\frac{q(k)}{k}\right]' kH_1^{(1)}(k|x-\xi|)\,\mathrm{d}k. \tag{1.41}$$

One readily verifies

$$\left|\left[\frac{q(k)}{k}\right]'\right| \le \frac{C}{1+|k|^2} \quad \text{when } k \in \ell_+,$$

where C is a constant independent of y and η. Taking into account

$$\big|H_1^{(1)}(z)\big| \le |z|^{-1/2}, \quad \operatorname{Im} z \ge 0,$$

we find that the absolute value of (1.41) does not exceed $C|x-\xi|^{-3/2}$. Derivatives of the second term in (1.40) can be estimated in the same way. From (1.40) and the asymptotics of $H_0^{(1)}$ (see the Examples section in the Introduction), one obtains

$$
\begin{aligned}
G(x, y; \xi, \eta) &= G_1(x, y; \xi, \eta) \\
&\quad + \frac{2\nu\cosh k_0(y+d)\cosh k_0(\eta+d)}{\nu d + \sinh^2 k_0 d}\left(\frac{2\pi}{k_0|x-\xi|}\right)^{1/2} \\
&\quad \times e^{i(k_0|x-\xi| - \pi/4)}, \\
|G_1| + |\nabla G_1| &= O\big(|x-\xi|^{-3/2}\big) \quad \text{as } |x-\xi| \to \infty.
\end{aligned}
\tag{1.42}
$$

This formula derived under assumption $-d < \eta \le y \le 0$ holds for $-d < y \le \eta \le 0$ as well. This becomes clear when one notes that Green's function is symmetric:

$$G(x, y; \xi, \eta) = G(\xi, \eta; x, y).$$

Thus (1.42) yields that G satisfies (1.37).

1.1.2.2. Properties of Green's Function

We conclude this section with more properties of Green's function. By (1.38) it has a cylindrical symmetry about the vertical axis through (ξ, η). We also give an expansion for G into a series. We note that the equation

$$k \sinh kd - \nu \cosh kd = 0$$

has a sequence of roots of the form

$$\pm ik_1, \pm ik_2, \dots, \pm ik_n, \dots,$$

apart from the unique positive root k_0. Here

$$0 < k_1 < k_2 < \cdots < k_n < \cdots.$$

The meromorphic function $q(k)$ is bounded for all complex k, if an ε neighborhood of the poles is excluded. So Mittag-Leffler's theorem on the decomposition of meromorphic functions into simple fractions (see, for example, Whittaker and Watson [360], Section 7.4) can be applied to $q(k)/(4k)$. Combining the contributions of $\pm ik_n$, we find that

$$\begin{aligned}\frac{q(k)}{4k} = {} & \frac{k_0^2 - \nu^2}{d\left(k_0^2 - \nu^2\right) + \nu} \, \frac{\cosh k_0(y+d) \cosh k_0(\eta + d)}{k^2 - k_0^2} \\ & + \sum_{n=1}^{\infty} \frac{k_n^2 + \nu^2}{d\left(k_n^2 + \nu^2\right) - \nu} \, \frac{\cos k_n(y+d) \cos k_n(\eta+d)}{k^2 + k_n^2}. \end{aligned} \tag{1.43}$$

On substituting this into (1.39), we can integrate term by term by using the following identity (see 6.532.4 in Gradshteyn and Ryzhik [96]):

$$\int_0^\infty \frac{k J_0(ak)}{k^2 + b^2} \, \mathrm{d}k = K_0(ab), \quad a > 0, \ \mathrm{Re}\, b > 0.$$

Since we integrate along ℓ_-, the last formula also applies to the first term in (1.43), but K_0 should be replaced by the first Hankel function of zero order.

Hence,

$$G(x, y; \xi, \eta) = \frac{2\pi i \left(k_0^2 - \nu^2\right) H_0^{(1)}(k_0|x - \xi|)}{d\left(k_0^2 - \nu^2\right) + \nu} \cosh k_0(y + d) \cosh k_0(\eta + d)$$
$$+ 4 \sum_{n=1}^{\infty} \frac{\left(k_n^2 + \nu^2\right) K_0(k_n|x - \xi|)}{d\left(k_n^2 + \nu^2\right) - \nu} \cos k_n(y + d) \cos k_n(\eta + d), \tag{1.44}$$

which gives an expansion of the source potential in terms of cylindrical waves.

Finally note that (1.16), (1.17), and their justification remain valid in the case of water having the finite depth.

1.2. Two-Dimensional and Ring Green's Functions

In the present section we give an account of results concerning two-dimensional and ring Green's functions. The former corresponds to a straight horizontal source-line in water of constant depth (finite or infinite) and is considered in Subsection 1.2.1. The latter describes (as its name points out) the velocity potential of a source having unit strength and positioned on a horizontal circle. In Subsection 1.2.2 we restrict ourselves only to the case of a ring Green's function for deep water.

1.2.1. Two-Dimensional Green's Functions

It was pointed out in the Conventions section of the Introduction that two-dimensional problems describe wave motions parallel to a certain plane, that is, invariant with respect to translation in the direction orthogonal to that plane. Therefore, two-dimensional Green's functions correspond to straight horizontal source-lines. Since methods applied in Section 1.1 work in two dimensions as well, we leave the majority of the details to the reader.

We begin with the case of finite depth. Let L denote a strip $-d < \operatorname{Im} z < 0$ [it is convenient to use complex notation $z\,(= x + iy)$ instead of (x, y)]. Green's function $G(z, \zeta)$ describing waves caused by a source placed at $\zeta = \xi + i\eta \in L$ must satisfy the following boundary value problem:

$$\nabla_z^2 G = -2\pi \delta_\zeta(z) \quad \text{in } L,$$
$$G_y - \nu G = 0 \quad \text{when } y = 0,$$
$$G_y = 0 \quad \text{when } y = -d,$$
$$G_{|x|} - ik_0 G = o(1) \quad \text{as } |x| \to \infty.$$

The unique solution of this problem is given by the following formula, which is similar to (1.38):

$$G(z,\zeta) = -\log|z-\zeta| - \log|z-\bar{\zeta}+2id| + 2\log d + 2\int_{\ell_-}\left[\frac{(\nu+k)\cos k(x-\xi)\cosh k(y+d)\cosh k(\eta+d)}{k\sinh kd - \nu\cosh kd} - 1\right] \times \frac{e^{-kd}\,\mathrm{d}k}{k}. \tag{1.45}$$

The path ℓ_- is the same as in Subsection 1.1.1.1 *(see Fig.* 1.1*), but indented at k_0 instead of ν.*

Using the well-known formulae

$$\int_0^\infty \frac{1-\cos ak}{k} e^{-kb}\,\mathrm{d}k = \log\frac{(a^2+b^2)^{1/2}}{b}, \quad \int_0^\infty \frac{e^{-kb}-e^{-kd}}{k}\,\mathrm{d}k = \log\frac{d}{b},$$

where $b > 0$, we get from (1.45)

$$G(z,\zeta) = \int_{\ell_-}\cos k(x-\xi)\frac{q(k)\,\mathrm{d}k}{k}. \tag{1.46}$$

Here either $-d < \eta < y \leq 0$ or $-d < \eta \leq y \leq 0$ and $x \neq \xi$, and $q(k)$ is the same function as in (1.39). Moreover, the representation analogous to (1.40),

$$G(z,\zeta) = \pi i q_0 k_0^{-1} e^{ik_0|x-\xi|} + \mathrm{Re}\int_{\ell_+} e^{ik|x-\xi|}\frac{q(k)}{k}\,\mathrm{d}k, \tag{1.47}$$

is true. As in Subsection 1.1.2.1, formulae (1.45), (1.46), and (1.47) allow us to verify that G satisfies the boundary value problem. Also, the uniqueness theorem can be proved in the same way as in Subsection 1.1.1.5.

Green's function for the two-dimensional problem in the lower half-plane can be obtained from (1.45) by letting $d \to \infty$:

$$G(z,\zeta) = -\log|z-\zeta| + \int_{\ell_-}\left[\frac{k+\nu}{k-\nu}e^{k(y+\eta)}\frac{\cos k(x-\xi)}{k} + \frac{e^{-k}}{k}\right]\mathrm{d}k.$$

Integral representations for the logarithm (see above) lead to other expressions:

$$G(z,\zeta) = -\log\left|\frac{z-\zeta}{z-\bar{\zeta}}\right| + 2\int_{\ell_-} e^{k(y+\eta)}\frac{\cos k(x-\xi)}{k-\nu}\,\mathrm{d}k \tag{1.48}$$

$$= -\log|z-\zeta| - \log|z-\bar{\zeta}| + 2\int_{\ell_-}\left[\frac{\nu}{k-\nu}e^{k(y+\eta)}\frac{\cos k(x-\xi)}{k} + \frac{e^{-k}}{k}\right]\mathrm{d}k. \tag{1.49}$$

In the same manner as in Subsection 1.1.1.4, one easily proves that the last integral is a bounded function of (x, y) and (ξ, η), and its gradient can be estimated by const $|\log|z-\bar{\zeta}||$ as $|x-\xi|, y, \eta \to 0$ simultaneously. Hence, (1.49) implies the proposition on the local behavior of Green's function near the free surface analogous to that in Subsection 1.1.1.4.

The contour integral in (1.48) is equal to

$$\pi i e^{\nu(y+\eta)} \cos \nu(x-\xi) + \int_0^\infty e^{k(y+\eta)} \frac{\cos k(x-\xi)}{k-\nu} \, dk,$$

where the integral is understood as the Cauchy principal value. Using results from Bochner's book [28], Sections 2, 5, and 8, one immediately finds the asymptotic behavior of the last Fourier type integral as $|z-\zeta| \to \infty$ and $|x-\xi| >$ const:

$$\int_0^\infty e^{k(y+\eta)} \frac{\cos k(x-\xi)}{k-\nu} \, dk = -\pi e^{\nu(y+\eta)} \sin \nu|x-\xi| + O\left(e^{\nu(y+\eta)}|x-\xi|^{-1}\right)$$

As $|z-\zeta| \to \infty$ but $|x-\xi| <$ const, the same reasoning as in Subsection 1.1.1.3 gives the asymptotics of the integral in (1.48). The following theorem summarizes the asymptotic behavior of G (cf. the theorem in Subsection 1.1.1.3).

Green's function (1.48) *has the following asymptotic behavior as* $|z-\zeta|$, $|z-\bar{\zeta}| \to \infty$: *(i) if* $|x-\xi| <$ const, *then*

$$G(z,\zeta) = -\log|z-\zeta| + \log|z-\bar{\zeta}| + G_1(z,\zeta),$$

where $|G_1| + |\nabla G_1| = O(|z-\bar{\zeta}|^{-1})$; *(ii) if* $|x-\xi| >$ const, *then*

$$G(z,\zeta) = -\log|z-\zeta| + \log|z-\bar{\zeta}| + 2\pi i e^{\nu(y+\eta+i|x-\xi|)} + G_2(z,\zeta),$$

where $|G_2| + |\nabla G_2| = O(e^{\nu(y+\eta)}|x-\xi|^{-1})$.

We note that integrals in (1.49) and (1.48) are similar to those in (1.7) and (1.10), respectively; the former have $\cos k(x-\xi)$ instead of $J_0(k|x-\xi|)$, which appears in the latter. Hence the results proved in Subsection 1.1.1.6 for the three-dimensional Green's function in the case of deep water remain valid for (1.49) and (1.48). So we give here only the formulation.

For any $\epsilon > 0$ *we have*

$$G(x,y;\xi,\eta) + \log|z-\zeta| + \log|z-\bar{\zeta}| \to 0 \quad \text{as } \nu \to 0,$$
$$G(x,y;\xi,\eta) + \log|z-\zeta| - \log|z-\bar{\zeta}| \to 0 \quad \text{as } \nu \to \infty$$

uniformly in $(x,y), (\xi,\eta) \in \mathbb{R}^2_-$ *such that* $y+\eta \leq -\epsilon$. *All derivatives of the functions in the above formulae also uniformly converge to zero.*

1.2.2. Ring Sources and Their Potentials

For geometries that have a vertical axis of symmetry (say the y axis), it is reasonable to introduce horizontal rings of sources. This leads to a sequence of ring-source potentials as follows.

Let (r, φ) and (ϱ, ϑ) denote the polar coordinates of x and ξ, respectively. By $G(x, y; \xi, \eta) = G(r, \varphi, y; \varrho, \vartheta, \eta)$ we denote Green's function of a point source at a distance $\varrho > 0$ from the y axis; see Subsection 1.1.1, (1.2.1) for expressions of this function in the case of infinite (finite) depth. We note that G can be expanded in the following form:

$$G(r, \varphi, y; \varrho, \vartheta, \eta) = \sum_{n=0}^{\infty} (2 - \delta_{0n})\, G^{(n)}(r, y; \varrho, \eta) \cos n(\varphi - \vartheta), \quad (1.50)$$

where δ_{kn} is the Kronecker delta. The product

$$G^{(n)}(r, y; \varrho, \eta) \cos n\varphi = \frac{1}{2\pi} \int_0^{2\pi} G(r, \varphi, y; \varrho, \vartheta, \eta) \cos n\vartheta \, \mathrm{d}\vartheta$$

will be referred to as the *ring-source potential* of order n, and $G^{(n)}$ will be called the *ring Green's function* of order n. The value of this potential at (r, φ, y) results from a distribution of sources having the density $(2\pi\varrho)^{-1}$ $\cos n\vartheta$ along a horizontal ring of radius ϱ about the y axis at the depth $-\eta$.

Let us give an expression for $G^{(n)}$ in deep water. From (1.8) and (1.9) we have

$$G(r, \varphi, y; \varrho, \vartheta, \eta) = \int_{\ell_-} \left[e^{-k|y-\eta|} + \frac{k + \nu}{k - \nu}\, e^{k(y+\eta)} \right] J_0(k|x - \xi|)\, \mathrm{d}k, \quad (1.51)$$

where

$$|x - \xi|^2 = r^2 - 2r\varrho \cos(\varphi - \vartheta) + \varrho^2.$$

Neumann's addition formula (see 8.531.1 in Gradshteyn and Ryzhik [96]) gives

$$J_0(k|x - \xi|) = \sum_{n=0}^{\infty} (2 - \delta_{0n})\, J_n(kr) J_n(k\varrho) \cos n(\varphi - \vartheta).$$

Substituting this into (1.51) and comparing the result with (1.50), we get

$$G^{(n)}(r, y; \varrho, \eta) = \int_{\ell_-} \left[e^{-k|y-\eta|} + \frac{k + \nu}{k - \nu}\, e^{k(y+\eta)} \right] J_n(kr)\, J_n(k\varrho)\, \mathrm{d}k. \quad (1.52)$$

Properties of the ring Green's function can be obtained from the corresponding properties of G or derived directly. It is obvious that

$$G^{(n)}(r, y; \varrho, \eta) = G^{(n)}(\varrho, \eta; r, y).$$

Similarly to the potential arising in the case of a straight infinite line of sources (see Subsection 1.2.1), $G^{(0)}$ must have a logarithmic singularity on the ring. However, (1.52) has the disadvantage that the singular nature of the integral for $n = 0$ is not readily apparent. Using formula 2.12.38.1 in Prudnikov et al. [293], we get

$$\int_0^\infty e^{-k|y-\eta|} J_0(kr)\, J_0(k\varrho)\, \mathrm{d}k = \frac{2K(k')}{\pi\,[(y-\eta)^2 + (r+\varrho)^2]^{1/2}}, \qquad (1.53)$$

where K denotes a complete elliptic integral of the first kind and

$$k'^2 = \frac{4r\varrho}{(y-\eta)^2 + (r+\varrho)^2}.$$

Since $k' \to 1-0$ as $y \to \eta$ and $r \to \varrho$ simultaneously, the asymptotic formula for (1.53) follows from 8.113.4 in Gradshteyn and Ryzhik [96]:

$$-\frac{1}{2\pi\varrho} \log[(y-\eta)^2 + (r-\varrho)^2] + O(1).$$

An equation for $G^{(n)}$ can be easily derived by separating variables in the three-dimensional Laplace equation. Furthermore, the construction of $G^{(n)}$ yields that it satisfies the free surface boundary condition and the radiation condition at infinity.

For water of finite depth, the nth-order ring Green's function can be found in the same manner from (1.39) and (1.43). The result is as follows:

$$\begin{aligned} G^{(n)}&(r, y; \varrho, \eta) \\ &= \frac{2\pi i \left(k_0^2 - \nu^2\right) J_n(k_0 r_<)\, H_n^{(1)}(k_0 r_>)}{d\left(k_0^2 - \nu^2\right) + \nu} \cosh k_0(y+d) \cosh k_0(\eta+d) \\ &\quad + 4 \sum_{m=1}^{\infty} \frac{\left(k_m^2 + \nu^2\right) I_n(k_m r_<)\, K_n(k_m r_>)}{d\left(k_m^2 + \nu^2\right) - \nu} \cos k_m(y+d) \cos k_m(\eta+d). \end{aligned} \qquad (1.54)$$

Here I_n and K_n denote modified Bessel functions; $r_< = \min\{r, \varrho\}$, and $r_> = \max\{r, \varrho\}$. Also, the following formulae (see Watson [350], p. 429),

$$\int_{\ell_-} \frac{k}{k^2 - k_0^2} J_n(kr)\, J_n(k\varrho)\, \mathrm{d}k = \frac{\pi i}{2} J_n(k_0 r_<)\, H_0^{(1)}(k_0 r_>),$$

$$\int_0^\infty \frac{k}{k^2 + k_m^2} J_n(kr)\, J_n(k\varrho)\, \mathrm{d}k = I_n(k_m r_<)\, K_n(k_m r_>),$$

are applied when (1.54) is derived.

Similar formulae allow us to transform (1.52) into

$$G^{(n)}(r, y; \varrho, \eta) = 2\pi i \nu e^{\nu(y+\eta)} J_n(\nu r_<) H_n^{(1)}(\nu r_>)$$
$$+ \frac{4}{\pi} \int_0^\infty (k \cos ky + \nu \sin ky)(k \cos k\eta + \nu \sin k\eta)$$
$$\times I_n(kr_<) K_n(kr_>) \frac{\mathrm{d}k}{k^2 + \nu^2}. \tag{1.55}$$

We may interpret this as a formal limit in (1.54) as $d \to \infty$ when the sequence $\{k_m\}$ becomes dense on $[0, \infty]$ and the series transforms to an integral. From (1.55), we see that the ring Green's function behaves at infinity like outgoing waves satisfying the radiation condition as a result of the first term on the right-hand side. The asymptotic behavior of K_n (see 8.451.6 in Gradshteyn and Ryzhik [96]) and Watson's lemma (see, for example, Chapter 3, Section 3 in Olver [271]) show that the integral in (1.56) decays like r^{-3} as $r \to \infty$.

1.3. Green's Representation of a Velocity Potential

It is well known that Green's function gives a representation of solutions to a partial differential equation in the integral form. However, it requires the solution to be smooth near the boundary. In Subsection 1.3.1 we consider Green's decomposition of a water-wave velocity potential in the form involving no information about smoothness. The equivalence of Sommerfeld's and Rellich's radiation conditions demonstrated in Subsection 1.3.2. In Subsection 1.3.3 we give a simple representation for bounded potentials given throughout a uniform layer.

1.3.1. Green's Decomposition

To be specific we consider a three-dimensional water domain of finite depth, but all results are also true for deep water and in two dimensions. More precisely, let a domain W be contained within the layer $L = \{x \in \mathbb{R}^2, -d < y < 0\}$ and coincide with it at infinity. Thus there may be some bounded bodies immersed in water totally or partially and local protrusions of the flat bottom. By S we denote the union of wetted surfaces of all bodies immersed in water; F is the free surface that is the part of $\{y = 0\}$ outside all bodies; B is the bottom, and $B_d = B \cap \{y = -d\}$ is the flat part of B having the same depth as at infinity.

In order to avoid superfluous assumptions on the smoothness of potential near S and $B \setminus B_d$, we need some auxiliary surfaces. Let $a > 0$ be such that $W \cap \{|x| > a_*\} = L \cap \{|x| > a_*\}$ for some $a_* < a$. We define a sequence of domains $\{W_a^{(m)}\}$ $(m = 1, 2, \ldots)$ so that they exhaust $W_a = W \cap \{|x| < a\}$ and $\partial W_a^{(m)}$ is the union of a truncated circular cylinder $\mathcal{C}_a = L \cap \{|x| = a\}$ and a surface $A_a^{(m)}$ that approximates $S \cup (B \cap \{|x| < a\})$ in such a way that $W_a^{(m)} \subset W_a^{(m+1)} \subset W_a$, and every point $(\xi, \eta) \in W_a$ belongs also to $W_a^{(m)}$ for sufficiently large values of m. Moreover, we assume $A_a^{(m)}$ to have the following properties: (*i*) it is such that $\partial W_a^{(m)}$ is a C^2-surface outside a finite number of edges and conic points; (*ii*) it coincides with $\{y = -d\}$ outside a certian neighbourhood of $S \cup (B \backslash B_d)$. The part of $A_a^{(m)}$ within that neighbourhood, we denote by A_m because it does not depend on a.

Let w be a harmonic function in W [and so $w \in C^2(W)$] belonging to $C^1(W \cup F \cup B_d)$ and satisfying

$$w_y - \nu w = 0 \quad \text{on } F, \qquad w_y = 0 \quad \text{on } B_d. \tag{1.56}$$

For an arbitrary point $(\xi, \eta) \in W$ we choose a and m so that $(\xi, \eta) \in W_a^{(m)}$. The standard application of Green's function $G(x, y; \xi, \eta)$ (this function is given in Subsection 1.1.2) for the finite domain of the depth d leads to the usual representation for w:

$$w(\xi, \eta) = \frac{1}{4\pi} \int_{\partial W_a^{(m)}} \left[w(x, y) \frac{\partial G}{\partial n_{(x,y)}} (x, y; \xi, \eta) - G(x, y; \xi, \eta) \frac{\partial w}{\partial n_{(x,y)}} \right] \mathrm{d}S.$$

Here $n_{(x,y)}$ is directed into $W_a^{(m)}$. The boundary conditions (1.56) and those valid for $G(x, y; \xi, \eta)$ reduce this representation to

$$\begin{aligned} w(\xi, \eta) = {} & \frac{1}{4\pi} \int_{\mathcal{C}_a} \left[G(x, y; \xi, \eta) w_{|x|}(x, y) - w(x, y) G_{|x|}(x, y; \xi, \eta) \right] \mathrm{d}S \\ & + \frac{1}{4\pi} \int_{A_m} \left[w(x, y) \frac{\partial G}{\partial n_{(x,y)}} (x, y; \xi, \eta) - G(x, y; \xi, \eta) \frac{\partial w}{\partial n_{(x,y)}} \right] \mathrm{d}S. \end{aligned} \tag{1.57}$$

Thus w is written as a sum that will be referred to as *Green's decomposition*. The first and second integrals will be denoted by u^L and u^W, respectively. We note that for a fixed value of m, functions w and u^W are independent of a whenever $(\xi, \eta) \in W_a^{(m)}$. From (1.57) we see that u^L is also independent of a. Thus it is a harmonic function in the whole layer L and the first term in (1.57) gives a representation of u^L for $(\xi, \eta) \in L \cap \{|x| < a\}$. Properties of Green's function guarantee that

$$u_y^L - \nu u^L = 0 \quad \text{when } y = 0, \qquad u_y^L = 0 \quad \text{when } y = -d. \tag{1.58}$$

Since w and u^L do not depend on m, (1.57) implies that u^W does not depend on m as well. It is clear that u^W satisfies the boundary conditions (1.56). Thus we arrive at the following assertion.

Every velocity potential w in W satisfying the boundary conditions (1.56) *can be decomposed into $u^L + u^W$, where u^L is a velocity potential defined throughout the layer L and that satisfies* (1.58), *and u^W is a velocity potential defined in W only and that satisfies* (1.56) *and the radiation condition* (I.42).

This assertion gives another confirmation (the first one was given in the Radiation Conditions section of the Introduction) of the fact that the boundary value problem (I.31)–(I.34) should be complemented by the radiation condition (I.42) in order to describe the interaction of water waves with obstacles. As we see from Green's decomposition, u^L and u^W correspond to incident and scattered waves, respectively, and the former is arbitrary whereas the latter depends on all the data of the problem.

We mentioned at the beginning of this subsection that *the last assertion is also true for W having the infinite depth.* Again the assertion is a consequence of formula (1.57). However, not only its derivation is a little more tedious in this case, but we need an extra assumption to be imposed on w. Namely, we suppose that w is bounded, which implies that $|\nabla w|$ decays as $y \to -\infty$; see derivation of (1.5). The amendments in the proof are as follows. We have to integrate over $\partial(W_a^{(m)} \cap \{y > -d\})$, where $0 < d < \infty$, in the formula preceding (1.57) because the standard integral representation is valid for bounded domains. Then we get (1.57) by letting $d \to \infty$, which is legitimate in view of the boundedness of w, the fact that $|\nabla w|$ decays as $y \to -\infty$, and the asymptotics of Green's function at infinity obtained in Subsection 1.1.1.3. The rest of the proof given above must be literally repeated.

Let us consider again the water domain W that is contained within the layer L and coincides with it for $|x| > a$. If w satisfies the boundary conditions (1.58) and the radiation condition (I.42), then the uniqueness theorem can be applied to u^L (in Subsection 1.1.1.5 such a theorem was proved for $\mathbb{R}^3_-$, but it is valid for L as well). Indeed, u^L is harmonic in L, satisfies (1.58) and (I.42) (the latter condition for u^L is a consequence of (1.57) and the fact that w and u^W do satisfy (I.42)), and so the uniqueness theorem gives that $u^L = 0$ in L. In addition, let ∂W consists of a finite number of surfaces that belong to the class C^2 outside a finite number of conic points, and let the adjacent surfaces form edges with nonzero dihedral angle along them. Then we can put $W_a^{(m)} = W_a$, and obtain the following assertion from (1.57).

Let W satisfy the just formulated conditions, and let (I.31)–(I.34) and (I.42) *hold for u. Then the following representation holds:*

$$u(\xi, \eta) = \frac{1}{4\pi} \int_S \left[u(x, y) \frac{\partial G}{\partial n_{(x,y)}}(x, y; \xi, \eta) - G(x, y; \xi, \eta) \frac{\partial u}{\partial n_{(x,y)}} \right] \mathrm{d}S$$
$$+ \frac{1}{4\pi} \int_{B \setminus B_d} u(x, y) \frac{\partial G}{\partial n_{(x,y)}}(x, y; \xi, \eta) \, \mathrm{d}S. \tag{1.59}$$

Similarly, we obtain the following assertion for deep water.

Let W be infinitely deep and satisfy the above smoothness conditions. If (I.31)–(I.33), (I.35), *and* (I.43) *hold for u, then the following representation holds:*

$$u(\xi,\eta) = \frac{1}{4\pi}\int_S \left[u(x,y)\frac{\partial G}{\partial n_{(x,y)}}(x,y;\xi,\eta) - G(x,y;\xi,\eta)\frac{\partial u}{\partial n_{(x,y)}} \right] \mathrm{d}S \tag{1.60}$$

1.3.2. Equivalence of Two Forms of the Radiation Condition

Two forms of the radiation condition were introduced in the Radiation Conditions section of the Introduction. Sommerfeld's condition (I.39) requires that as $|x| \to \infty$ the combination $u_{|x|} - ik_0 u$ must tend to zero uniformly with respect to y and polar angle θ in the (x_1, x_2) plane. According to Rellich's condition (I.42) only the integral over $\mathcal{C}_r$ of the above expression squared should tend to zero as $r \to \infty$. Thus (I.42) seems to be weaker than (I.39). It is obvious that (I.39) implies (I.42), but they are actually equivalent because *it follows from Green's formulae* (1.59) and (1.60) that (I.42) implies (I.39) as well.

To prove this assertion in the case when $W \subset L$ and coincides with L at infinity, we substitute asymptotics (1.42) in both integrals in (1.59). Changing the variables (ξ, η) to (x,y), we get

$$u = \beta(\theta)\cosh k_0(y+d)|x|^{-1/2}e^{ik_0|x|} + O\big(|x|^{-3/2}\big) \quad \text{as } |x| \to \infty. \tag{1.61}$$

Here $\beta(\theta)$ is a certain smooth function. The last formula means that Sommerfeld's radiation condition (I.39) holds for u.

If the water domain has the depth d at infinity but is not contained in L, then $W' = W \cap \{|x| > a\}$ must be considered instead of W. Since $W' \subset L$ when a is sufficiently large, W' can be considered as a water domain in which the cylinder $\{|x| < a\}$ is immersed. Therefore, (1.59) and (1.61) remain valid.

More information about u can be obtained by substituting (1.44) into both integrals in (1.59), which gives an expansion

$$u(\xi,\eta) = a_0(\xi)\cosh k_0(\eta+d) + \sum_{n=1}^{\infty} a_n(\xi)\cos k_n(\eta+d) \tag{1.62}$$

valid for $(\xi, \eta) \in W \cap \{|x| > a\}$. Here $k_0 > 0$ and $ik_1, ik_2, \ldots$ are the roots of $k \tanh kd = \nu$, and terms are referred to as "simple" waves of order 0, 1, ... Another way to prove this expansion is based on the completeness of the system

$$\cosh k_0(y+d), \cos k_1(y+d), \cos k_2(y+d), \ldots, \cos k_n(y+d)\ldots$$

in $L^2(-d, 0)$ demonstrated by Weinstein [356]. Expansion (1.62) shows that, generally speaking, u cannot be written simply as $a(\xi)\cosh k_0(\eta + d)$; that is, it is not equal to the principal term in (1.61).

Since each term in (1.44) satisfies the Laplace equation, the same must be true for each term in (1.62). Hence,

$$\nabla_x^2 a_n - (-1)^{\delta_{0n}} k_n^2 a_n = 0, \quad n = 0, 1, \ldots, \tag{1.63}$$

where δ_{mn} is the Kronecker delta. It follows from (1.44) that a_n decays exponentially at infinity for $n \geq 1$. The behavior of $a_0(x)$ as $|x| \to \infty$ is given by the first term in (1.61) divided by $\cosh k_0(y + d)$.

For the convenience of the reader, we list some other asymptotic formulae here. They can easily be derived from (1.57) in the same way as (1.61); that is by using the asymptotics of Green's function. For u satisfying the radiation condition in a three-dimensional domain of infinite depth, we have

$$u(x, y) = \beta(\theta)|x|^{-1/2} e^{\nu(y+i|x|)} + O\big((|x|^2 + y^2)^{-1} + e^{\nu y}(1 + |x|)^{-3/2}\big)$$

as $|x|^2 + y^2 \to \infty$. In two dimensions, we have

$$u(x, y) = A_\pm h(y) e^{\pm i K x} + \psi_\pm(x, y) \quad \text{for } \pm x > 0.$$

Here $K = \nu$, $h(y) = e^{\nu y}$ for water of the infinite depth and $K = k_0$, $h(y) = \cosh k_0(y + d)$ for water of finite depth. The remainder term has the following behavior:

$$|\psi_\pm| + |\nabla \psi_\pm| = O(|x|^{-1}) \quad \text{as } |x| \to \infty$$

when the depth is finite, and

$$|\psi_\pm| = O\big((x^2 + y^2)^{-1/2}\big), \qquad |\nabla \psi_\pm| = O((x^2 + y^2)^{-1}) \quad \text{as } x^2 + y^2 \to \infty$$

for deep water.

1.3.3. Representation of a Bounded Potential in a Layer

Unlike u^W, *the potential* u^L *is a simple wave of order zero, if it is bounded in* L; that is, this potential admits the representation

$$u^L(\xi, \eta) = A(\xi) \cosh k_0(\eta + d), \tag{1.64}$$

where $A(\xi)$ satisfies (1.63) for $n = 0$.

We begin the proof by noting that the boundary conditions (1.58) allow us to continue u^L analytically across $y = 0$ and $y = -d$. The second condition in (1.58) permits us to continue u^L as a harmonic function by means of the equation

$$u^L(x, y) = u^L(x, -2d - y).$$

The first condition in (1.58) permits us to continue $u^L_y - \nu u^L$ across $y = 0$ as an odd function of y that is harmonic. This leads to the functional equation

$$u^L(x, y) = u^L(x, -y) + 2\nu e^{\nu y} \int_{-y}^{0} e^{\nu\eta} u^L(x, \eta)\, \mathrm{d}\eta,$$

which continues u^L as a harmonic function across $y = 0$. Therefore $u^L(x, y)$ can be considered as a bounded harmonic function in the layer $\{x \in \mathbb{R}^2, -2d < y < +d\}$.

For any point $(x, y) \in \bar{L}$ the mean value theorem gives

$$\nabla u^L(x, y) = \frac{3}{4\pi d^3} \int_{R<d} \nabla u^L \, \mathrm{d}\xi \mathrm{d}\eta,$$

where $R^2 = |x - \xi|^2 + (y - \eta)^2$. The last equality implies (see Subsection 1.1.1.1) that the first derivatives of $u^L(x, y)$ are bounded:

$$|\nabla u^L(x, y)| \leq 3d^{-1} \sup\{|u^L|\}.$$

This and (1.44), where $K_0(k_n|x - \xi|)$ decays exponentially as $|x| \to \infty$, give

$$\lim_{r\to\infty} \left\{ \int_{C_r} \left[G u^L_{|x|} - u^L G_{|x|}\right] \mathrm{d}S - \int_{C_r} \left[G^{(0)} u^L_{|x|} - u^L G^{(0)}_{|x|}\right] \mathrm{d}S \right\} = 0,$$

where

$$G^{(0)}(x, y; \xi, \eta) = \frac{2\pi i \left(k_0^2 - \nu^2\right) H^{(1)}_0(k_0|x - \xi|)}{d\left(k_0^2 - \nu^2\right) + \nu} \times \cosh k_0(y + d) \cosh k_0(\eta + d).$$

Both integrals in the last limit are actually independent of r (see Subsection 1.3.1) and the equality obtained by dropping the limit sign is also correct. Hence

$$u^L(\xi, \eta) = \frac{1}{4\pi} \int_{C_a} \left[G^{(0)}(x, y; \xi, \eta) u^L_{|x|}(x, y) - u^L(x, y) G^{(0)}_{|x|}(x, y; \xi, \eta)\right] \mathrm{d}S.$$

From here (1.64) follows because $G^{(0)}$ depends on η only through the factor $\cosh k_0(\eta + d)$. Noting that the Laplace equation is valid for u^L, we conclude that $A(\xi)$ satisfies (1.63) for $n = 0$; note that the analogous fact holds for each term in (1.62).

1.4. Bibliographical Notes

There are numerous papers treating different Green's functions for time-periodic water waves. Here we mention a few works directly connected with our presentation.

1.1.1. The Fourier transform was applied by Wehausen and Laitone [354] to the derivation of Green's function G describing a pulsating source in deep water. Lenoir [187] used the same method in two dimensions for both infinite and finite depth. One more representation was found for Green's function by Haskind [104], who applied a different method than that given in Subsection 1.1.1.2. Deriving the asymptotic representation at infinity, we do not follow any published paper. An alternative method for this was developed by Maz'ya [222]. The method of investigation of the local behavior of G near the free surface was not published earlier. The proof of uniqueness theorem given in Subsection 1.1.1.5 is from John [126], who gives it for water of a finite depth, but it applies for water unbounded below as well (and holds in either two or three dimensions).

In addition to the papers just cited, one can find treatments of a point source in deep water in the works of Kochin [143], Haskind [106], and Havelock [109, 110]. Havelock gave a representation for G that was different from those presented in Section 1.1.1, and in [110] he developed wave potentials having higher-order singularities. Somewhat earlier these singularities had been given by Thorne [317], whose paper contains a rather complete census of the possible singular solutions for both two and three dimensions and for finite and infinite depths. Included are series expansions as well as integrals.

1.1.2. The widely known paper by John [126] contains, in particular, an account of properties of Green's function describing a point source in water of finite depth. The related results can also be found in the works of Haskind [106] and Wehausen and Laitone [354].

1.2. In connection with the two-dimensional Green's functions, the following papers should be listed: Haskind [106], John [126], Kochin [142], Thorne [317], and Wehausen and Laitone [354]. The ring Green's functions were developed by Hulme for infinite [118] and finite depth [119].

1.3. In [126], John obtained Green's representation under a superfluous assumption on the smoothness of a velocity potential. The definition of simple waves is also from John [126], Section 4, as well as the result given in Subsection 1.3.3. In two dimensions, formula (1.64) was proved by Weinstein [356].

Other results. In [67], Evans obtained a Green's function describing a source in water of finite depth containing a semi-infinite horizontal plate upon which a plane progressive wave is obliquely incident.

In addition to papers treating Green's functions in water of constant depth, we mention several papers in which pulsating sources are considered for domains of variable depth. In the two-dimensional case, a Green's function for a source above a slopping bottom is constructed by Sretensky [309] and Morris [250]. Both authors point out the existence of wave-free source positions, that is, such points in the water domain that no waves are radiated to infinity by a source placed at any of them. Moreover, in the second of her papers [250], Morris numerically investigated the geometric locus of wave-free sources.

Vainberg and Maz'ya [347] proved the existence of a Green's function (both in two and three dimensions) for a layer of variable depth, coinciding with a constant depth layer outside some vertical circular cylinder. They considered sources placed not only in the water domain but also in the free surface and on the bottom. These results are presented in Subsection 2.3.3.

2
Submerged Obstacles

It was pointed out in the Preface that methods of investigation of the uniqueness and solvability for the water-wave problem depend essentially on the type of obstacle in respect to its intersection with the free surface. Among various possibilities, the simplest one is the case in which the free surface coincides with the whole horizontal plane (and so rigid boundaries of the water domain are represented by totally submerged bodies and the bottom of variable topography); we restrict our attention to this case in the present chapter.

We begin with the method of integral equations (Section 2.1), which not only provides information about the unique solvability of the water-wave problem but also serves as one of the most frequently used tools for a numerical solution of the problem. In Section 2.2, various geometric criteria of uniqueness are obtained with the help of auxiliary integral identities. The uniqueness theorem established allows us to prove the solvability of the problem for various geometries of submerged obstacles in Section 2.3. The last section, Section 2.4, contains bibliographical notes.

2.1. Method of Integral Equations and Kochin's Theorem

When Green's function is constructed it is natural to solve the water-wave problem by applying integral equation techniques, which is a standard approach to boundary value problems. In doing so, a proof of the solvability theorem for an integral equation is usually based on Fredholm's alternative and the uniqueness of the solution to the boundary value problem.

When applying this approach to the problem of a totally submerged body, one finds both a similarity to, and a distinction from, the exterior Neumann problem for the Laplace equation. Because the kernels of the corresponding integral operators have the same singularity, Fredholm's alternative is applicable in both cases. Besides, the uniqueness theorem is available for the exterior Neumann problem, and it is an important ingredient in proving the solvability theorem for integral equations. However, the uniqueness theorem for the water-wave problem is far from obvious, and it has been proved for submerged obstacles only under certain geometric assumptions (see Section 2.2 in this chapter). Another distinction arises from the presence of the frequency parameter ν in kernels of integral equations for the water-wave problem.

In the present chapter, we deal with a body of an arbitrary geometry, and we involve no a priori knowledge of uniqueness. The main result here is a theorem, essentially from Kochin (1939–1940) [142, 143], which guarantees the unique solvability of the two- and three-dimensional problems for all $\nu > 0$, except possibly for a finite number of values. The proof relies upon the following two properties of integral operators: (i) they depend analytically on the parameter ν; and (ii) for sufficiently small (large) values of ν, they are close to the invertible integral operators, arising in the exterior Neumann problem obtained by even (odd) continuation across $\{y = 0\}$.

We also demonstrate that if the uniqueness theorem in the water-wave problem is true for a certain value of ν, then the standard argument shows that the integral equation is uniquely solvable for this value of ν.

The section is divided into four subsections. In Subsection 2.1.1, we reduce the water-wave problem to integral equations, and we investigate them in Subsections 2.1.2 and 2.1.3 for smooth and nonsmooth bodies, respectively. In the latter case, only two-dimensional bodies are considered, and special attention is paid to the behavior of solutions to integral equations near corner points.

2.1.1. Integral Equations of the Water-Wave Problem

Here we assume that a two- or three-dimensional (we shall treat both cases simultaneously) water domain W has a constant (possibly, infinite) depth, and that it contains one totally submerged body (this can be readily replaced by a finite number of such bodies). We suppose the body's boundary S to belong to the class $C^{1,\alpha}$; that is, a normal vector to S is Hölder continuous. In Subsection 2.1.1.1 we introduce a single-layer potential, state its properties, and use it for reducing the water-wave problem to a Fredholm integral equation. Another integral equation following from Green's formula is obtained in Subsection 2.1.1.2.

2.1.1.1. Single-Layer Potential and the Corresponding Integral Equation

We seek a solution to the water-wave problem in the form of a source distribution over S, in other words, as a *single-layer potential with unknown density* μ:

$$(V\mu)(x, y) = [(m-1)\pi]^{-1} \int_S \mu(\xi, \eta)\, G(x, y; \xi, \eta)\, \mathrm{d}S. \qquad (2.1)$$

This potential depends on $(x, y) \in L \setminus S$, where $L = \{x \in \mathbb{R}^{m-1},\ -d < y < 0\}$. Here $0 < d \leq \infty$ is the water depth, $m = 2, 3$ is the flow dimension, and when $m = 3$, $x = (x_1, x_2)$ and $\xi = (\xi_1, \xi_2)$ for the source point $(\xi, \eta) \in S$. Green's function G describing flow caused by a point source is investigated in Section 1.1 for $m = 3$ and in Subsection 1.2.1 for $m = 2$. The properties of G guarantee that if $\mu \in C(S)$, that is, μ is a continuous function on S, then the potential (2.1) has the following properties.

1. It is a harmonic function in W as well as in $L\setminus\bar{W}$.
2. It satisfies the free surface boundary condition.
3. It satisfies the homogeneous Neumann condition on $\{y = -d\}$ when the water depth is finite, or it tends to zero together with its gradient as $y \to -\infty$ in the case of infinite depth.
4. It satisfies Sommerfeld's radiation condition (I.39)

In what follows, we need the notion of a regular normal derivative on S. Let a differentiable function u be given on either side of S, and let $\mathbf{n}_x$ and $\mathbf{n}_y$ be the projections on the x plane and y axis, respectively, of the normal $\mathbf{n}_{(x,y)}$ to S at (x, y). We say that $\partial u/\partial n_\pm$ is the regular normal derivative, if uniformly in $(x, y) \in S$ we have

$$\frac{\partial u}{\partial n}(x + \tau\mathbf{n}_x, y + \tau\mathbf{n}_y) \to \frac{\partial u}{\partial n_\pm}(x, y) \quad \text{as } \tau \to \pm 0.$$

It is clear that the regular normal derivative belongs to $C(S)$.

When reducing the water-wave problem to a boundary integral equation and investigating the latter equation, we find that the following properties of $V\mu$ play an important role.

5. $V\mu$ is a continuous function in $\bar{L}$.
6. The integral operator

$$(T\mu)(x, y) = \frac{1}{(m-1)\pi} \int_S \mu(\xi, \eta) \frac{\partial G}{\partial n_{(x,y)}}(x, y; \xi, \eta)\, \mathrm{d}S \qquad (2.2)$$

   defines a continuous function of $(x, y) \in S$.
7. The regular normal derivative of $V\mu$ does exist on S and is equal to

$$\frac{\partial(V\mu)}{\partial n_\pm} = \mp\mu + T\mu, \qquad (2.3)$$

where the subscript $+$ $(-)$ denotes the derivative on the side directed to W $(L\setminus\bar{W})$.

8. An immediate consequence of (2.3) is the equality

$$\partial(V\mu)/\partial n_- - \partial(V\mu)/\partial n_+ = 2\mu. \tag{2.4}$$

Properties 5–8 immediately follow from the equality

$$G(x, y; \xi, \eta) = E(x, y; \xi, \eta) + H(x, y; \xi, \eta), \tag{2.5}$$

where $H(x, y; \xi, \eta)$ is a harmonic function in L in both (x, y) and (ξ, η),

$$E(x, y; \xi, \eta) = R^{-1} \quad \text{for } m = 3, \qquad E(x, y; \xi, \eta) = -\log R \quad \text{for } m = 2, \tag{2.6}$$

and R is the distance between (x, y) and (ξ, η). In fact, (2.5) means that properties 5–8 follow from the analogous properties of the potential

$$\int_S \mu(\xi, \eta)\, E(x, y; \xi, \eta)\, \mathrm{d}S,$$

which can be found in the textbooks by Mihlin [246], Chapter 4, Vladimirov [348], Chapter 5, and Kellogg [136], Chapter 6 (in the last book it is assumed that $m = 3$ and S belongs to class C^2).

From properties 1–4 we see that potential (2.1) satisfies all conditions of the water-wave problem except for the Neumann condition on S. By (2.3) *this potential is a solution of the problem if and only if*

$$-\mu(x, y) + (T\mu)(x, y) = f(x, y), \quad (x, y) \in S. \tag{2.7}$$

This is a Fredholm integral equation in $L^2(S)$, since T *is a compact operator in this space.* The latter assertion is a consequence of (2.5), which splits T defined in (2.2) into a sum of two operators. The second operator is compact because its kernel $\partial H/\partial n_{(x,y)}$ is a continuous function of variables $(x, y; \xi, \eta) \in S \times S$. The first operator is compact, as is demonstrated, for example, in the textbooks by Colton and Kress [40], Mihlin [246], and Vladimirov [348]. The proof uses the fact that the corresponding kernel has a weak singularity, that is,

$$\frac{\partial E}{\partial n_{(x,y)}} = O(R^{-m+1+\alpha}) \quad \text{as } R \to 0,$$

which is true since S belongs to $C^{1,\alpha}$.

The following assertion on continuity of solutions to (2.7) is also a corollary of this estimate (see the textbooks cited in the previous paragraph).

If $f \in C(S)$ [$C^{k,\alpha}(S)$, that is, f and its derivatives up to order k are Hölder continuous functions], *then every solution* $\mu \in L^2(S)$ *is continuous on* S [*belongs to* $C^{k,\alpha}(S)$].

2.1.1.2. Another Integral Equation

Let u satisfy the free surface and bottom boundary conditions

$$u_y - \nu u = 0 \quad \text{on } F, \qquad u_y = 0 \quad \text{on } B$$

(the last condition should be replaced by the requirement that $|u| \leq \text{const} < \infty$, when $B = \emptyset$). Moreover, let the radiation condition (I.42) hold for u. Then (1.59) and (1.60) and similar formulae in the two-dimensional case imply

$$u(\xi, \eta) = \frac{1}{2(m-1)\pi} \int_S \times \left[u(x, y) \frac{\partial G}{\partial n_{(x,y)}} (x, y; \xi, \eta) - G(x, y; \xi, \eta) \frac{\partial u}{\partial n_{(x,y)}} \right] \mathrm{d}S. \quad (2.8)$$

Here $m = 2, 3$ is the dimension of the water domain that allows (2.8) to hold in either case.

Now, assuming that the normal derivative of u is a prescribed continuous function f on S, we shall derive a boundary integral equation from (2.8). The right-hand side in (2.8) is a difference of two potentials whose properties are the same as those of the potentials having E instead of G (see, for example, the books cited in Subsection 2.1.1.1, and Maz'ya's survey [224], Section 1). The first potential is known as a double-layer potential, and the second one is the single-layer potential considered in Subsection 2.1.1.1. The latter is continuous in the whole layer L, whereas the former one has jumps on S. Thus letting $(\xi, \eta) \in W$ approach the surface S, we arrive at

$$-u(\xi, \eta) + \frac{1}{(m-1)\pi} \int_S u(x, y) \frac{\partial G}{\partial n_{(x,y)}} (x, y; \xi, \eta) \, \mathrm{d}S = \frac{1}{(m-1)\pi} \int_S f(x, y) G(x, y; \xi, \eta) \, \mathrm{d}S, \quad (\xi, \eta) \in S. \quad (2.9)$$

We note that the integral operator in this equation is adjoint to $\bar{T}$, where T appears in (2.7).

The method presented here and the corresponding boundary integral equation are sometimes referred to as *the direct method* and *the direct integral equation*, respectively (see, for example, Maz'ya [224]). The reason for this is the fact that the unknown function is the trace of u on S. Thus (2.9) has an advantage when it is sufficient to determine this trace for further calculations, as often occurs in applications. A drawback of this equation is a rather complicated form of the right-hand-side term.

2.1.2. Solvability Theorems

In Subsection 2.1.2.1, we prove the solvability theorem for the water-wave problem by using the boundary integral equation derived in Subsection 2.1.1.1, and imposing no a priori assumption about the uniqueness of solution. However, a finite number of exceptional values of ν could occur. So, in general, there is no solvability for the integral equation, and consequently no solution of the water-wave problem could be obtained by using this method for those values of ν. In Subsection 2.1.2.3, we show how the integral equation from Subsection 2.1.1.2 applies for establishing that the solution is unique for the nonexceptional values of ν. Simplifications arising from the assumption of uniqueness in the water-wave problem are shown in Subsection 2.1.2.2.

2.1.2.1. Kochin's Solvability Theorem

For the sake of simplicity, we consider the integral equation (2.7) only for water of infinite depth, but the same result is true for the finite depth case as well. We begin with the formulation of an auxiliary result from the linear operator theory, which will be applied when the integral equation is investigated.

Let X be a Hilbert space, and let T_ν be an operator function analytic in ν for ν in a (connected) domain D of the complex plane. If the operator T_ν is a compact operator in X for every $\nu \in D$, and if $I - T_\nu$ has a bounded inverse operator at least for one value of ν, then the operator $(I - T_\nu)^{-1}$ does exist and is bounded for all $\nu \in D$ with a possible exception for a set of isolated points.

If ν is an exceptional point, then there is a finite-dimensional subspace of solutions μ_0, satisfying $(I - T_\nu)\mu_0 = 0$.

Below this theorem will be referred to as the *invertibility theorem* (see, for example, the books by Gohberg and Krein [95] and Kozlov and Maz'ya [148] for the proof).

Let $E_0(x, y; \xi, \eta)$ [cf. (2.6)] denote R_0^{-1} for $m = 3$, and $-\log R_0$ for $m = 2$, where $R_0^2 = |x - \xi|^2 + (y + \eta)^2$. We need an auxiliary equation

$$-\mu(x, y) + (T_\infty \mu)(x, y) = f(x, y), \quad (x, y) \in S, \tag{2.10}$$

where $(T_\infty \mu)(x, y)$ is equal to

$$\frac{1}{(m-1)\pi} \int_S \mu(\xi, \eta) \frac{\partial}{\partial n_{(x,y)}} [E(x, y; \xi, \eta) - E_0(x, y; \xi, \eta)] \, \mathrm{d}S.$$

It is well known (see, for example, the books by Mihlin [246] and Vladimirov [348]) that

$$\frac{\partial E}{\partial n_{(x,y)}}(x, y; \xi, \eta) = \frac{\cos\left[\mathbf{n}_{(x,y)}, \mathbf{R}\right]}{R^{m-1}},$$

where $[\mathbf{n}_{(x,y)}, \mathbf{R}]$ denotes the angle between $\mathbf{n}_{(x,y)}$ and $\mathbf{R}$ directed from (x, y) to (ξ, η). A similar formula holds for the normal derivative of E_0. Then we can write

$$(T_\infty \mu)(x, y) = \frac{1}{(m-1)\pi} \int_{S \cup S_0} \mu(\xi, \eta) \frac{\partial E}{\partial n_{(x,y)}}(x, y; \xi, \eta)\, \mathrm{d}S,$$

where $S_0 = \{(x, y) : (x, -y) \in S\}$ is a mirror reflection of S by the free surface $\{y = 0\}$, and the density μ is extended to S_0 as an odd function with respect to y. If we extend f in (2.10) in the same manner, then (2.10) takes the form of the well-known equation arising when a single-layer potential is used for solving the Laplacian exterior Neumann problem in a domain outside $S \cup S_0$.

The extended equation (2.10) is shown to be uniquely solvable (see, for example, the books by Kellogg [136], Chapter 11, the $m = 3$ case only, Mihlin [246], Chapter 4, and Vladimirov [348], Chapter 5, where both cases $m = 2$ and $m = 3$ are considered). Its solution $\mu(x, y)$ must be odd in y, because $-\mu(x, -y)$ is a solution of the extended equation (2.10) with $-f(x, -y) = f(x, y)$, and the uniqueness theorem holds for the integral equation of the exterior Neumann problem. Hence, restricting $\mu(x, y)$ to S, one obtains the solution of (2.10). In other words, $I - T_\infty$ is an invertible operator.

The assertions given in Subsection 1.1.1.6 and at the end of Subsection 1.2.1 for $m = 3$ and $m = 2$, respectively, show that the kernel of $T - T_\infty$ tends to zero uniformly on $S \times S$ as $\nu \to \infty$. Hence the norm of $T - T_\infty$ can be made arbitrary small for a large enough ν. Then writing $I - T = I - T_\infty + (T_\infty - T)$, we see that $I - T$ is invertible for all sufficiently large values of ν as a sum of an invertible operator and an operator whose norm is small enough.

Moreover, from (1.9) for $m = 3$ and from (1.48) for $m = 2$, we find that T depends analytically on ν in a neighborhood of the positive real half-axis. Then the theorem on invertibility yields the following result.

The integral equation (2.7) *with an arbitrary* $f \in L^2(S)$ *has a unique solution for all* $\nu > 0$, *except possibly for a sequence of values tending to zero.*

If $f \in C(S)$, then the assertion on the continuity (see Subsection 2.1.1.1) implies that the corresponding solution μ to (2.7) is continuous as well. Hence property 7 applies to (2.1), and this property together with (2.7) give that $V\mu$ satisfies the Neumann condition on S. On account of properties 1–4, we see that $V\mu$ is a solution of the water-wave problem. Thus the following theorem is proved.

For all $\nu > 0$ *except possibly for a sequence tending to zero, and arbitrary* $f \in C(S)$, *the water-wave problem has a solution of the form* (2.1), *where* μ *satisfies* (2.7).

Two last assertions can be improved. Put

$$(T_0\mu)(x, y) = \frac{1}{(m-1)\pi}\int_S \mu(\xi, \eta)\frac{\partial}{\partial n_{(x,y)}}[E(x, y; \xi, \eta) + E_0(x, y; \xi, \eta)]\,\mathrm{d}S.$$

Extending μ to S_0 as an even function with respect to y, we get

$$(T_0\mu)(x, y) = \frac{1}{(m-1)\pi}\int_{S\cup S_0} \mu(\xi, \eta)\frac{\partial E}{\partial n_{(x,y)}}(x, y; \xi, \eta)\,\mathrm{d}S.$$

Then

$$-\mu(x, y) + (T_0\mu)(x, y) = f(x, y), \quad (x, y) \in S \cup S_0, \tag{2.11}$$

is an integral equation of the Neumann problem in the domain outside $S \cup S_0$. Here f is extended as an even in y function. Like the extended equation (2.10), the last equation is uniquely solvable. Moreover, its solution is even in y, which follows in the same way as the oddness of the solution satisfying the extended equation (2.10). Thus $I - T_0$ is also an invertible operator.

The assertions given in Subsection 1.1.1.6 and at the end of Subsection 1.2.1 for $m = 3$ and $m = 2$ respectively show that the kernel of $T - T_0$ tends to zero uniformly with respect to $(x, y; \xi, \eta) \in S \times S$ as $\nu \to 0$. Then the norm of $T - T_0$ tends to zero as $\nu \to 0$, and in the same way as above we demonstrate that (2.7) is uniquely solvable for all sufficiently small values of ν. Combining this with results already proved, we arrive at the following theorem.

For all $\nu > 0$, except possibly for a finite set of values, the water-wave problem with arbitrary $f \in C(S)$ has a solution of the form (2.1) *where μ satisfies* (2.7)*, which is a uniquely solvable equation for these values of ν.*

2.1.2.2. Solvability of the Integral Equation When the Uniqueness Holds for the Water-Wave Problem

In the introductory remarks to this chapter, we mentioned that the question of uniqueness is far from its final solution for the water-wave problem. Here we simply suppose that the problem has at most one solution. This facilitates the investigation of equation (2.7) as we demonstrate here.

Let $\mu_0 \in L^2(S)$ solve the homogeneous equation

$$-\mu_0 + T\mu_0 = 0. \tag{2.12}$$

The assertion on continuity (see Subsection 2.1.1.1) implies that $\mu_0 \in C(S)$.

Properties 1–4 and 7 guarantee that $V\mu_0$ satisfies the homogeneous water-wave problem. Then the uniqueness theorem for the water-wave problem implies that $V\mu_0$ vanishes in W. Therefore property 5 yields that $V\mu_0 = 0$ on S. Now by property 1, this potential is a solution of the homogeneous Dirichlet problem in $\mathbb{R}^m_- \setminus \bar{W}$. Consequently, $V\mu_0 = 0$ in the latter domain. Since $V\mu_0$ is equal to zero in both W and $\mathbb{R}^m_- \setminus \bar{W}$, the regular normal derivatives vanish on either side of S, and (2.4) implies that $\mu_0 = 0$.

Thus (2.12) has only a trivial solution, and Fredholm's alternative guarantees that (2.7) is uniquely solvable in $L^2(S)$ for an arbitrary right-hand term. If $f \in C(S)$, then the assertion on the continuity yields that the corresponding solution μ to (2.7) is continuous as well. Hence property 7 applies to (2.1), and together with (2.7) gives that $V\mu$ satisfies the Neumann condition on S. From properties 1–4 we see that $V\mu$ is a solution of the water-wave problem that is unique by assumption. Thus the following theorem is proved.

If the homogeneous water-wave problem has only a trivial solution, then the corresponding nonhomogeneous problem is uniquely solvable for an arbitrary continuous function f in the Neumann condition. The solution can be found in the form of (2.1)*, where μ should be determined from equation* (2.7)*, which is uniquely solvable as well.*

As was mentioned above, our considerations are valid when a finite number of bodies is contained in water. In addition, the assumption that water has an infinite depth was not used in the proof, and the theorem is true in the case of water bounded below by a horizontal bottom.

2.1.2.3. Unique Solvability of the Water-Wave Problem

First, let (2.7) be uniquely solvable for a certain ν. The last theorem in Subsection 2.1.2.1 implies that, on one hand, there is a solution u of the water-wave problem having the form (2.1) where μ satisfies (2.7). On the other hand, $u(\xi, \eta)$ satisfies the boundary integral equation (2.9) for $(\xi, \eta) \in S$, as was shown in Subsection 2.1.1.2. Our aim is to demonstrate that u is the unique solution of the problem.

Assuming that there are two solutions u_1 and u_2, we have from (2.9) that $u_0 = u_1 - u_2$ satisfies

$$-u_0(\xi, \eta) + (\bar{T}^* u_0)(\xi, \eta) = 0, \quad (\xi, \eta) \in S, \tag{2.13}$$

where T appears in (2.7). Since (2.7) is uniquely solvable, the homogeneous equation corresponding to (2.7), and consequently (2.13), has only a trivial solutions by the Fredholm theory. Thus $u_0 = 0$ on S, and also $\partial u_0/\partial n = 0$

on S because u_1 and u_2 satisfy the same Neumann condition there. Now the uniqueness theorem for the Cauchy problem for the Laplace equation (see, for example, Gilbarg and Trudinger's book [94], Chapter 2) implies that u_0 vanishes identically in W; that is, *the water-wave problem is uniquely solvable, if the boundary integral equation* (2.7) *is uniquely solvable.*

Moreover, it is shown in Subsection 2.1.2.2 that *the integral equation* (2.7) *is uniquely solvable [by the Fredholm theory this property holds simultaneously for* (2.9)*] if the water-wave problem is uniquely solvable.*

Thus, *the water-wave problem is uniquely solvable if and only if the boundary integral equations* (2.7) *and* (2.9) *are uniquely solvable.*

Let us turn to the case in which ν is an exceptional value. Let (2.12) have n linearly independent solutions $\mu_0^{(j)}$, $j = 1, \ldots, n$. Every solution is a continuous function on S by the assertion on continuity. Properties 1–4 and 7 guarantee that every potential $V\mu_0^{(j)}$ generated by a solution to (2.12) satisfies the homogeneous water-wave problem. Assuming that

$$\sum_{j=1}^{n} c_j V\mu_0^{(j)} = 0,$$

we get that $V\mu_0 = 0$ in W, where $\mu_0 = \sum_{j=1}^{n} c_j \mu_0^{(j)}$. Then $\mu_0 = 0$ as is shown in Subsection 2.1.2.2, and $c_j = 0$ because $\mu_0^{(j)}$ are linearly independent. So $V\mu_0^{(j)}$, $j = 1, \ldots, n$ are linearly independent.

Let us show that the subspace of solutions for the homogeneous water-wave problem is n dimensional. If there are more than n linearly independent solutions, then their traces on S give more than n linearly independent solutions of (2.13) (this follows from the above argument based on the uniqueness of the Cauchy problem), which is impossible since (2.13) and (2.12) have the same number of linearly independent solutions. Hence, the water-wave problem has n linearly independent solutions. Combining this with the previous result, we arrive at the following conclusion.

The water-wave problem for a totally submerged body is equivalent to the boundary integral equations (2.7) *and* (2.9), *which means that the following two statements hold:*

1. *The water-wave problem is uniquely solvable for a certain ν if and only if* (2.7) *and* (2.9) *are uniquely solvable for this ν.*
2. *For a certain ν the homogeneous water-wave problem has n linearly independent solutions if and only if* (2.12) *and* (2.13) *also have n linearly independent solutions each.*

2.1.3. Integral Equation for a Contour with Corner Points

Since interaction of water waves with bodies having edges is rather usual in applications, here we give an account of the corresponding mathematical approach. We present in detail the simplest case of a totally submerged two-dimensional body with corner points, leaving further results for Section 3.1, where the three-dimensional problem of a surface-piercing body is treated. In the present section, results analogous to those in the smooth case are obtained for solvability of the boundary integral equation and the water-wave problem. An analysis of singular behavior of solutions near corner points is also given.

2.1.3.1. Some Definitions and Auxiliary Results

Here we are concerned with the two-dimensional water-wave problem in a domain W having infinite depth and bounded internally by a simple closed piecewise smooth curve S. To simplify the presentation we assume that there are only two corner points P_- and P_+ on S, where two regular C^2 arcs make angles α_- and α_+ (either distinct from 0 and 2π) directed into W. Let the arc length s be measured on S clockwise from a point dividing one of the arcs into pieces of equal length. To be specific, let $0 < s_- < s_+ < |S|$, where $s = s_\pm$ corresponds to $P_\pm$, and $|S|$ is the total length of S.

To avoid strong singularities near $P_\pm$ we require that a solution u must satisfy the following inequality:

$$\int_{W_a} (|\nabla u|^2 + |u|^2)\,\mathrm{d}x\mathrm{d}y < \infty, \quad W_a = W \cap \{|x| < a\}, \tag{2.14}$$

where a is an arbitrary positive number; that is, the energy is locally finite. Despite this condition, seeking u in the form of (2.1), one cannot expect μ to be continuous everywhere on S. Thus we suppose that μ belongs to a Banach space $C_\kappa(S)$ $(0 < \kappa < 1)$ that consists of functions continuous on $S\backslash\{P_-, P_+\}$ and is supplied with a norm

$$\|\mu\|_\kappa = \max_\pm\{\sup\{|s - s_\pm|^{1-\kappa}|\mu(s)| : 0 < \pm(s - s_*) < s_*\}\}.$$

Here $s_* = (s_+ - s_-)/2$ is the middle of the second arc between P_- and P_+. So two points with $s = 0$ and $s = s_*$ divide the whole contour S into parts of equal length: $0 \leq s \leq s_*$ and $s_* \leq s \leq 2s_* = |S|$, on which weights depending on $s - s_-$ and $s - s_+$ respectively are given. Apart from functions that are continuous everywhere on S, the space $C_\kappa(S)$ contains functions having power growth as $|s - s_\pm| \to 0$. Moreover, $C_\kappa(S)$ is continuously embedded in $L^p(S)$ for $p < (1 - \kappa)^{-1}$.

It is obvious that the potential (2.1) with such μ satisfies the first four properties listed in Subsection 2.1.1.1. Furthermore, the direct value of the normal derivative of $V\mu$ is a continuous function of $(x, y) \in S\setminus\{P_-, P_+\}$ and formulae (2.3) and (2.4) are true there. The convergence of normal derivatives of $V\mu$ to the limit values (2.3) is uniform along normals erected to S on compact subsets of $S\setminus\{P_-, P_+\}$; that is, the regular normal derivatives do exist on these subsets. The continuity of $V\mu$ in $\overline{\mathbb{R}^2_-}$ can be established by virtue of general properties of integral operators with kernels of the potential type (see, for example, Kantorovich and Akilov's book [130], Chapter 11, Section 3, Theorems 6 and 7). We formulate here these results for further references.

Let D and D' be n- and m-dimensional piecewise smooth bounded manifolds in the Euclidean space whose points are denoted x and ξ. Let

$$(Kw)(x) = \int_D K(x, \xi)\, w(\xi)\, \mathrm{d}\xi, \quad x \in D'$$

have a kernel

$$K(x, \xi) = B(x, \xi)|x - \xi|^{-\lambda}, \quad \lambda > 0 \tag{2.15}$$

where $B(x, \xi)$ is a bounded function, which is continuous for $x \neq \xi$. If

$$q < \frac{mp}{n - (n - \lambda)p}, \quad n - (n - \lambda)p < m,$$

then K is a compact operator mapping $L^p(D)$ into $L^q(D')$, where $p, q > 1$.

If $(n - \lambda)p > n$, then K is a compact operator mapping $L^p(D)$ into $C(D')$.

To derive the continuity of $V\mu$ from this theorem, we note that for this potential $n = 1, m = 2, \mu \in L^p(S)$ for a certain $p > 1$, and the kernel can be written in the form of (2.15) where $\lambda > 0$ is arbitrarily small. It is clear that $(1 - \lambda)p > 1$ for sufficiently small λ, which allows us to apply the second assertion of the theorem.

Since derivatives of $E(x, y; \xi, \eta)$ $(m = 2)$ can be written in the form of (2.15) with $\lambda = 1$, the formulated theorem guarantees that (2.14) holds for $V\mu$. Thus for arbitrary $\mu \in C_\kappa(S)$ the potential (2.1) satisfies all the conditions of the water-wave problem except for the Neumann condition on S. The latter holds if and only if equation (2.7) is valid for $(x, y) \in S\setminus\{P_-, P_+\}$.

2.1.3.2. Fredholm's Aternative for an Integral Equation on a Piecewise Smooth Contour

Our aim is to show that the scheme applied in Subsection 2.1.2.1 works in the present situation, but corner points make considerations more complicated. First, we have to apply the invertibility theorem to $I - T$, which is an operator function depending analytically on a parameter ν and acting in the Banach space $C_\kappa(S)$ – not in the Hilbert space $L^2(S)$ as in Subsection 2.1.2.1. Second, T is not a compact operator because S has corner points. Fortunately, a more general form of the invertibility theorem exists (see Kozlov and Maz'ya's book [148], Section A.8):

Let X be a Banach space, and let B_ν be an operator function analytic in a domain D. If the operator B_ν is a Fredholm operator in X for every $\nu \in D$ and if B_ν has a bounded inverse operator at least for one value of ν, then the operator $(B_\nu)^{-1}$ does exist and is bounded for all $\nu \in D$ with a possible exception for a set of isolated points.

Thus this theorem can be used for $X = C_\kappa(S)$, and $B_\nu = I - T$, which is a Fredholm operator if $|T| < 1$ in $C_\kappa(S)$. Here $|T|$ is the essential norm of T, that is, $|T| = \inf \|T - K\|$, where the infimum is taken over all compact operators K on $C_\kappa(S)$. The last inequality guarantees that for a certain compact operator K the series

$$I + (T - K) + (T - K)^2 + \cdots$$

converges and defines the bounded inverse operator $[I - (T - K)]^{-1}$. Then (2.7) is equivalent to the following equation:

$$-\mu + [I - (T - K)]^{-1} K \mu = [I - (T - K)]^{-1} f,$$

where $[I - (T - K)]^{-1} K$ is a compact operator. Since the Fredholm alternative holds for the last equation, the same is true for (2.7). We estimate $|T|$ in the next subsection.

2.1.3.3. *Estimate for the Essential Norm*

For estimating $|T|$ we consider

$$(N\mu)(x, y) = -\frac{1}{\pi} \int_S \mu(\xi, \eta) \frac{\partial \log R}{\partial n_{(x,y)}} \, \mathrm{d}S, \quad (x, y) \in S \setminus \{P_-, P_+\}.$$

Note that replacing T by N in (2.7) leads to the equation corresponding to the exterior Neumann problem in a domain outside S. Since $G + \log R$ is a harmonic function in $\mathbb{R}^2_-$, see (1.48), the operator $T - N$ has a bounded kernel that can be written in the form of (2.15) with an arbitrarily small exponent λ and a continuous function B. Now we can apply the theorem formulated in Subsection 2.1.3.1, since $C_\kappa(S)$ is continuously embedded into $L^p(S)$ for $1 < p < (1 - \kappa)^{-1}$. Thus an appropriate choice of λ yields that $T - N$ is a compact operator mapping $C_\kappa(S)$ into $C(S)$. Hence, it is a compact operator mapping $C_\kappa(S)$ into itself. Then $|T| = |N|$, and the latter essential norm will be estimated now.

The kernel of N depends on (x, y) and (ξ, η) belonging to $S \setminus \{P_-, P_+\}$. Denoting by s and σ the values of arc length corresponding to (x, y) and (ξ, η), respectively, we have

$$-\frac{\partial \log R}{\partial n_{(x,y)}} = \frac{y'(s)\,[x(s) - \xi(\sigma)] - x'(s)\,[y(s) - \eta(\sigma)]}{[x(s) - \xi(\sigma)]^2 + [y(s) - \eta(\sigma)]^2}. \tag{2.16}$$

For $\sigma \neq s_\pm$ this is a continuous function of s except for $s = s_\pm$, where it has jumps. Let us investigate the local behavior of (2.16) near a corner point, say P_- with $s = s_-$. If $(s - s_-)(\sigma - s_-) > 0$, that is, both points belong to the same smooth arc, then (2.16) is a bounded continuous function (see, for example, Petrovskii [288], Section 35.2). Thus (2.16) contributes nothing to $|N|$ in this case.

When P_- separates (x, y) and (ξ, η), say $s - s_- > 0$ and $\sigma - s_- < 0$, we apply Taylor's formula with the remainder term depending on second derivatives. Then the right-hand side in (2.16) takes the form

$$\frac{[x'_-\eta'_- - \xi'_- y'_-](\sigma - s_-) + \Omega_2^{(-)}}{[x'_-(s - s_-) - \xi'_-(\sigma - s_-)]^2 + [y'_-(s - s_-) - \eta'_-(\sigma - s_-)]^2 + \Omega_3^{(-)}}$$

$$= \frac{(\sigma - s_-)\sin\alpha_- + \Omega_2^{(-)}}{(\sigma - s_-)^2 + (s - s_-)^2 - 2(\sigma - s_-)(s - s_-)\cos\alpha_- + \Omega_3^{(-)}}. \tag{2.17}$$

Here $(x'_-, y'_-) = (x'(s_-), y'(s_-))$ and $(\xi'_-, \eta'_-) = (\xi'(s_-), \eta'(s_-))$ are unilateral tangent vectors to the arcs making the angle α_- at P_-, and

$$\Omega_m^{(-)} = \Omega_m^{(-)}(s - s_-, \sigma - s_-)$$

is a homogeneous form of order $m = 2, 3$ with respect to $s - s_-$ and $\sigma - s_-$. The coefficients of $\Omega_m^{(-)}(s - s_-, \sigma - s_-)$ are bounded continuous functions of the same variables. From (2.16) and (2.17) we get

$$-\frac{\partial \log R}{\partial n_{(x,y)}} = \frac{(\sigma - s_-)\sin\alpha_-}{(\sigma - s_-)^2 + (s - s_-)^2 - 2(\sigma - s_-)(s - s_-)\cos\alpha_-} + M_-(s, \sigma),$$

where

$$M_-(s, \sigma) = \frac{\Delta_-\Omega_2^{(-)} - \Omega_3^{(-)}(\sigma - s_-)\sin\alpha_-}{\Delta_-\big[\Delta_- + \Omega_3^{(-)}\big]}$$

and $\Delta_- = (\sigma - s_-)^2 + (s - s_-)^2 - 2(\sigma - s_-)(s - s_-)\cos\alpha_-$. Since $\alpha_- \neq 0, 2\pi$, there exists a constant C_-, such that

$$\Delta_- > C_-[(\sigma - s_-)^2 + (s - s_-)^2] \quad \text{when } (\sigma - s_-)(s - s_-) < 0.$$

So $M_-(s,\sigma)$ is a bounded function in a neighborhood of $s, \sigma = s_-$, and it is continuous for $s \neq \sigma$. A similar calculation works when $s - s_- < 0$ and $\sigma - s_- > 0$. A neighborhood of P_+ can be treated in the same way. Thus we arrive at the following assertion.

If $(s - s_\pm)(\sigma - s_\pm) < 0$, then in a neighborhood of $s = \sigma = s_\pm$ we have

$$-\frac{\partial \log R}{\partial n_{(x,y)}} = \frac{|\sigma - s_\pm| \sin \alpha_\pm}{(\sigma - s_\pm)^2 + (s - s_\pm)^2 - 2|\sigma - s_\pm||s - s_\pm| \cos \alpha_\pm} + M_\pm(s, \sigma),$$

where $M_\pm(s,\sigma)$ is a bounded function that is continuous for $s \neq \sigma$.

Let δ be a positive number such that $\delta < \min\{s_-, s_*\}$; that is, δ is less than the half-length of either arc adjoining P_- because of the definition of s_* (see Subsection 2.1.3.1). In a domain defined by the inequalities

$$|s - s_\pm|, |\sigma - s_\pm| \leq \delta \quad \text{and } (s - s_\pm)(\sigma - s_\pm) \leq 0,$$

we put

$$g_\pm(s,\sigma) = \frac{|\sigma - s_\pm| \sin \alpha_\pm}{(\sigma - s_\pm)^2 + (s - s_\pm)^2 - 2|\sigma - s_\pm||s - s_\pm| \cos \alpha_\pm}.$$

Let $g_\pm(s,\sigma) = 0$ elsewhere in the square $0 \leq s, \sigma \leq |S|$. The choice of δ means that g_- and g_+ have disjoint supports. By $N_\pm$ we denote an integral operator with the kernel $\pi^{-1} g_\pm(s,\sigma)$. The theorem formulated in Subsection 2.1.3.1 and the assertion obtained above imply that $N - N_- - N_+$ is a compact operator in $C_\kappa(S)$. Then $|N|$ in $C_\kappa(S)$ does not exceed $\|N_- + N_+\|$ in this space. Let us estimate the last norm.

The definition of δ implies that $\|(N_- + N_+)\mu\|_\kappa$ is equal to

$$\max_\pm\{\sup\{|s - s_-|^{1-\kappa}|(N_-\mu)(s)| : |s - s_-| < \delta\},$$
$$\sup\{|s - s_+|^{1-\kappa}|(N_+\mu)(s)| : |s - s_+| < \delta\}\}.$$

If $s < s_-$, then $|s - s_-|^{1-\kappa}|(N_-\mu)(s)|$ is equal to

$$\frac{\sin|\pi - \alpha_-|}{\pi} \left| \int_{s_-}^{\delta + s_-} \frac{\mu(\sigma)|\sigma - s_-||s - s_-|^{1-\kappa} d\sigma}{(\sigma - s_-)^2 + (s - s_-)^2 + 2|\sigma - s_-||s - s_-| \cos(\pi - \alpha_-)} \right|.$$

Here the argument in trigonometric functions is changed in order to make it less than π in absolute value because this allows us to write that the last expression does not exceed

$$\|\mu\|_\kappa \frac{\sin|\pi - \alpha_-|}{\pi} \int_0^{\delta/|s - s_-|} \frac{\sigma^\kappa d\sigma}{1 + 2\sigma \cos(\pi - \alpha_-) + \sigma^2}.$$

Replacing $\delta/|s - s_-|$ by $+\infty$ in the upper limit and using formula 3.252.12

in Gradshteyn and Ryzhik [96], we arrive at the following estimate:

$$|s - s_-|^{1-\kappa} |(N_- \mu)(s)| \leq \|\mu\|_\kappa \frac{\sin \kappa |\pi - \alpha_-|}{\sin \kappa\pi},$$

which can be also obtained under the assumption that $s > s_-$. Similar estimation is applicable to $|s - s_+|^{1-\kappa} |(N_+ \mu)(s)|$. Thus, we arrive at the following theorem.

For the essential norm of T in the space $C_\kappa(S)$, $0 < \kappa < 1$, we have the estimate

$$|T| = |N| \leq \max_{\pm} \frac{\sin \kappa |\pi - \alpha_\pm|}{\sin \kappa\pi},$$

and $|T| < 1$ when

$$\kappa < \min_{\pm} [1 + |1 - \alpha_\pm/\pi|]^{-1}. \tag{2.18}$$

Estimate (2.18) for κ is obtained by solving the inequality

$$\max_{\pm} \frac{\sin \kappa |\pi - \alpha_\pm|}{\sin \kappa\pi} < 1.$$

2.1.3.4. Solvability of the Integral Equation on a Piecewise Smooth Contour

It follows from (2.18) that the Fredholm alternative holds for equations (2.7), (2.10), and (2.11) in $C_\kappa(S)$ for any $\kappa \in (0, 1/2]$. One can also take $\kappa > 1/2$, but then a choice of κ depends on $\alpha_\pm$. For smaller values of $|\alpha_\pm - \pi|$, larger values of κ are admissible, leading to μ with a weaker singularity near corner points.

The Fredholm alternative guarantees that (2.10) and (2.11) are uniquely solvable in $C_\kappa(S)$ as equations corresponding to the exterior Neumann problem in a domain outside $S \cup S_0$ (see Carleman [36], Chapter 1). As in Subsection 2.1.1.3, this fact and the invertibility theorem give that *for κ satisfying* (2.18), *equation* (2.7) *is uniquely solvable in $C_\kappa(S)$ for all $\nu > 0$, except possibly for a finite number of values.*

Hence, the theorem on solvability of the water-wave problem formulated at the end of Subsection 2.1.2.1 is valid for two-dimensional submerged obstacles with corner points. Moreover, theorems proved in Subsections 2.1.2.2 and 2.1.2.3 can be extended to this case as well. For this purpose, Carleman's results on an integral equation arising in the Dirichlet problem (see [36], Chapter 1) should be used.

2.1.3.5. Asymptotics of Solutions Near Corner Points

Here we give a brief extraction from the extensive theory describing behavior of solutions to elliptic boundary value problems in piecewise smooth domains. Our aim is to show how these results allow us to deduce exact asymptotics near a corner point for μ solving (2.7).

Since the velocity potential u satisfies the Neumann condition on S, we first write down the asymptotics of u near a corner point (they are justified in references listed in Section 2.4). For the sake of simplicity, we assume that W coincides with a sector in a vicinity of $P_\pm$. Taking P as the origin of local coordinates (x', y'), we represent this sector in polar coordinates $\rho e^{i\theta} = x' + iy'$ as follows:

$$\{0 < \rho < \rho_0, 0 < \theta < \alpha\}.$$

To simplify the notation we omit subscripts $\pm$ at P, α, and so on. Using a conformal mapping of W on a half-plane, we can readily verify that

$$u(\rho, \theta) = u(P) + O(\rho) \quad \text{when } 0 < \alpha < \pi, \tag{2.19}$$

$$u(\rho, \theta) \sim u(P) + c_1 \rho^{\pi/\alpha} \cos \frac{\pi\theta}{\alpha} \quad \text{when } \pi < \alpha < 2\pi, \tag{2.20}$$

as $\rho \to 0$.

We showed in previous subsections that u can be found in the form of (2.1), and the corresponding density μ, belonging to $C_\kappa(S)$, where κ satisfies (2.18), must be determined from (2.7). The asymptotics of μ can be written explicitly, giving the precise value of κ. Let v be the solution of the following auxiliary Dirichlet problem:

$$\nabla^2 v = 0 \text{ in } \mathbb{R}^2_- \backslash \bar{W}, \quad v = u \text{ on } S.$$

Properties 1 and 5 (see Subsection 2.1.1.1) show that v is given by the single-layer potential (2.1) in $\mathbb{R}^2_- \backslash W$. Then we get from (2.4) that

$$\mu = (\partial_n v - \partial_n u)/2, \tag{2.21}$$

where ∂_n denotes the normal derivative. This leads to the asymptotic formula for μ because formulae (2.19) and (2.20) can be differentiated and the same is true for the asymptotics of v, which is as follows. If $0 < \alpha < \pi$ (α is directed into W, which is the complementary domain to that where v is given), then

$$v(\rho, \theta) \sim u(P) + c_2 \rho^{\pi/(2\pi - \alpha)} \sin \frac{\pi(2\pi - \theta)}{2\pi - \alpha}, \quad \text{as } \rho \to 0,\ \alpha \le \theta \le 2\pi. \tag{2.22}$$

In the other case ($\pi < \alpha < 2\pi$) the asymptotics has the following form:

$$v(\rho,\theta) \sim u(P) - c_1 \rho^{\pi/\alpha} \cos \frac{\pi(\theta - \pi)}{\alpha} \sec \frac{\pi^2}{\alpha}, \quad \alpha \leq \theta \leq 2\pi. \quad (2.23)$$

Here c_1 is the constant from (2.20). We note (although we do not use it below), that c_2 and c_1 in (2.22) and (2.23), respectively, can be calculated by means of special solutions to auxiliary boundary value problems. These solutions are uniquely specified by prescribed singularities at P. For example,

$$c_2 = \frac{1}{\pi} \int_S [u - u(P)] \frac{\partial v_a}{\partial n} \, \mathrm{d}S,$$

where v_a is a harmonic function in $\mathbb{R}^2_- \backslash \bar{W}$, such that $v_a = 0$ on $S \backslash \{P\}$ and

$$v_a \sim \rho^{-\pi/(2\pi - \alpha)} \sin \frac{\pi(2\pi - \theta)}{2\pi - \alpha} \quad \text{as } \rho \to 0.$$

Substituting (2.19) and (2.22) into (2.21), we find that for $\alpha \in (0, \pi)$ the asymptotics of μ has the following form:

$$\mu(\rho) \sim \frac{\pi c_2}{2(\alpha - 2\pi)} \rho^{\frac{\alpha - \pi}{2\pi - \alpha}} \quad \text{as } \rho \to 0. \quad (2.24)$$

For $\pi < \alpha < 2\pi$, we similarly get the following from (2.20) and (2.23):

$$\mu(\rho) \sim \pm \frac{\pi c_1}{2\alpha} \rho^{(\pi - \alpha)/\alpha} \tan \frac{\pi^2}{\alpha} \quad \text{as } \rho \to 0, \quad (2.25)$$

where $+$ ($-$) corresponds to the ray $\theta = 0$ ($\theta = \alpha$). It should be noted that according to (2.24) and (2.25), the solution $\mu(\rho)$ has singularity as $\rho \to 0$ for any angle α.

The asymptotic formulae (2.24) and (2.25) show that estimate (2.18) is exact in the following sense. For $\pi < \alpha < 2\pi$ we have from (2.18) that any $\kappa < \pi/\alpha$ may be taken for characterizing the growth of $\mu(\rho)$ near a corner point. So $\mu(\rho)$ tends to infinity not slower than $\rho^{(\pi - \alpha)/\alpha}$, and the latter expression gives the exact behavior of $\mu(\rho)$ as (2.25) demonstrates. By comparing (2.18) with (2.24), we arrive at the same conclusion for $0 < \alpha < \pi$.

2.2. Conditions of Uniqueness for All Frequencies

In this section, conditions providing the uniqueness are obtained for the problem describing water waves in a layer W of variable depth that may contain a totally submerged body as well. For this purpose a rather general technique is applied. It is based on auxiliary integral identities involving the velocity potential, its derivatives, and also an arbitrary vector field and an arbitrary function.

A special choice of these entities gives geometric restrictions on bodies that represent the identity as a sum of nonnegative integrals equal to zero. This leads to a conclusion that the homogeneous water-wave problem has only a trivial solution for all values of ν when bodies satisfy these restrictions. The corresponding uniqueness criteria have simple geomertic interpretations.

We begin this section with a short subsection, Subsection 2.2.1, demonstrating that any solution to the homogeneous problem describes waves having finite kinetic and potential energy. In fact, many proofs of uniqueness for the water-wave problem rely on this property (in particular, those given in the present section). In the next subsection, Subsection 2.2.2, we illustrate the general method on a simple example by using a special version of the integral identity. The general auxiliary integral identities for the three- and two-dimensional cases are derived in Subsection 2.2.3. In Subsection 2.2.4, we present the most spectacular applications of the auxiliary identities, and further uniqueness results for various geometries involving totally submerged bodies as well as bottom topographies (a combination of both types of obstacles is also included) are given in Subsection 2.2.5.

2.2.1. On the Energy of Waves in the Homogeneous Problem

Here we show that *any solution to the homogeneous water-wave problem describes waves having finite kinetic and potential energy.*

Let W denote a three-dimensional water domain containing totally submerged as well as surface-piercing bodies, and bounded below by a bottom B, which is a curved surface coinciding with $\{y = -d\}$ at infinity. The case of deep water as well as the two-dimensional problem will be also considered. As usual, S denotes the union of all wetted surfaces of immersed bodies, and $S \cap B = \emptyset$. For the sake of simplicity we assume ∂W to be smooth enough.

Let a be a positive number, such that $W \setminus W_a$ has the constant depth d where $W_a = W \cap \{|x| < a\}$. Let $\chi(|x|)$ be an infinitely differentiable function that is equal to one for $|x| \geq a + 1$ and that vanishes for $a \geq |x| \geq 0$. If u is a solution to the homogeneous water-wave problem, then we have (see Subsection 1.3.2):

$$u(x, y) = A(\theta, y)|x|^{-1/2} e^{ik_0|x|} + r(x, y),$$

where

$$|r| + |\nabla r| = O\big(|x|^{-3/2}\big) \quad \text{as } |x| \to \infty.$$

It is clear that $A(\theta, y) = \beta(\theta) \cosh k_0(y + d)$, but this is not used in what follows.

Now let us write Green's formula for W_a:

$$0 = \int_{W_a} (\bar{u}\nabla^2 u - u\nabla^2 \bar{u})\,\mathrm{d}x\mathrm{d}y = \int_{\partial W_a} \left(u\frac{\partial \bar{u}}{\partial n} - \bar{u}\frac{\partial u}{\partial n} \right) \mathrm{d}S,$$

where $\mathbf{n}$ is directed into W_a. By the boundary conditions on the free surface and on the bottom this takes the form

$$\int_{\mathcal{C}_a} \left(\bar{u}\, u_{|x|} - u\bar{u}_{|x|}\right) \mathrm{d}S = \int_S \left(u\frac{\partial \bar{u}}{\partial n} - \bar{u}\frac{\partial u}{\partial n} \right) \mathrm{d}S = 2i \,\mathrm{Im} \int_S f\,\bar{u}\,\mathrm{d}S.$$

Here f is the right-hand term in the Neumann condition on S (the union of wetted immersed surfaces), and $\mathcal{C}_a = W \cap \{|x| = a\}$. From the asymptotic formula for u we see that the integral over $\mathcal{C}_a$ is equal to

$$2ik_0 \int_0^{2\pi} \int_{-d}^0 |A(\theta, y)|^2 \,\mathrm{d}\theta\mathrm{d}y + O(a^{-1}) \quad \text{as } a \to \infty.$$

Letting $a \to \infty$, we arrive at

$$k_0 \int_0^{2\pi} \int_{-d}^0 |A(\theta, y)|^2 \,\mathrm{d}\theta\mathrm{d}y = \mathrm{Im} \int_S f\,\bar{u}\,\mathrm{d}S.$$

Thus $A = 0$ when $f = 0$, and the asymptotics of u is given by the remainder term $r(x, y)$. This proves the assertion, because r is squarely integrable over the free surface, and its gradient is squarely integrable over W.

The same proof applies to the case of deep water, but a slightly different estimate for the remainder term,

$$|r| + |\nabla r| = O\big(e^{\nu y}(1 + |x|)^{-3/2} + (|x|^2 + y^2)^{-1}\big) \quad \text{as } |x|^2 + y^2 \to \infty,$$

arises from the theorem proved in Subsection 1.1.1.3.

Similar results for two-dimensional domains of finite and infinite depth follow from asymptotic formulae obtained in Subsection 1.2.1 for G. Omitting proofs, we give only the corresponding formulae where $0 < d \leq \infty$ and $k_0 = \nu$ for $d = \infty$:

$$u(x, y) = A_\pm(y)e^{ik_0|x|} + r_\pm(x, y) \quad \text{as } |x| \to \infty.$$

Here

$$k_0 \int_{-d}^0 [|A_+(y)|^2 + |A_-(y)|^2]\,\mathrm{d}y = \mathrm{Im} \int_S f\bar{u}\,\mathrm{d}S,$$

and

$$|r_\pm| + |\nabla r_\pm| = O(|x|^{-1}) \quad \text{as } |x| \to \infty, \text{ if } d < \infty;$$

$$|r_\pm| = O\big([x^2 + y^2]^{-1/2}\big), |\nabla r_\pm| = O([x^2 + y^2]^{-1}) \quad \text{as } x^2 + y^2 \to \infty,$$

when $d = \infty$.

2.2.2. Domain With Starlike Horizontal Cross Sections

In this subsection we derive a simple auxiliary integral identity providing uniqueness in the water-wave problem when a layer of variable depth satisfies a clear geometric condition. Our aim is to provide the reader with a simple example of the technique before considering the general auxiliary integral identity in Subsection 2.2.3. More general geometric conditions, containing the present one as a particular case and providing uniqueness for layers of variable depth, are developed in Subsection 2.2.5.

Let the free surface F coincide with the whole plane $\{y = 0\}$, that is,

$$F = \{x \in \mathbb{R}^m,\ y = 0\}, \quad m = 1, 2,$$

and let the bottom B be a continuously differentiable surface (line) for $m = 2$ ($m = 1$) at a finite distance from F. Moreover, we assume that B coincides with $\{y = -d\}$ at infinity. Thus a water domain having a bounded bottom obstruction is under consideration here.

We begin the proof of uniqueness with the following identity:

$$\begin{aligned}&\mathrm{Re}\left[\left(|x|\bar{u}_{|x|} + \bar{u}\right)\nabla^2 u\right]\\ &\quad = \mathrm{Re}\,\nabla\cdot\left[\left(|x|\bar{u}_{|x|} + \bar{u}\right)\nabla u\right] - |\nabla_x u|^2 - \nabla_x\cdot(\mathbf{x}|\nabla u|^2)/2,\end{aligned}$$

which is straightforward to verify; here $\mathbf{x} = (x_1, x_2, 0)$. We integrate the identity over $W_a = W \cap \{|x| < a\}$ where a is large enough (so that B is flat outside W_a). Since u satisfies the Laplace equation, we get, after using the divergence theorem,

$$\begin{aligned}&\mathrm{Re}\int_{F_a}\left(|x|\bar{u}_{|x|} + \bar{u}\right)u_y\,\mathrm{d}x - \int_{W_a}|\nabla_x u|^2\,\mathrm{d}x\mathrm{d}y\\ &\quad + \frac{1}{2}\int_{B_a}\mathbf{x}\cdot\mathbf{n}\,|\nabla u|^2\,\mathrm{d}S = \mathrm{Re}\int_{B_a}\left(|x|\bar{u}_{|x|} + \bar{u}\right)\partial_n u\,\mathrm{d}S\\ &\quad + \int_{C_a}\left[\frac{|x|}{2}|\nabla u|^2 - \mathrm{Re}\left(|x|\bar{u}_{|x|} + \bar{u}\right)u_{|x|}\right]\mathrm{d}S,\end{aligned} \tag{2.26}$$

where F_a, B_a are the portions of F, B respectively within $\{|x| < a\}$.

The free surface boundary condition allows us to transform the first term in the left-hand side:

$$\begin{aligned}\mathrm{Re}\int_{F_a}\left(|x|\bar{u}_{|x|} + \bar{u}\right)u_y\,\mathrm{d}x &= \nu\left[\mathrm{Re}\int_{F_a}\bar{u}_{|x|}u|x|^2\,\mathrm{d}|x|\,\mathrm{d}\theta + \int_{F_a}|u|^2\,\mathrm{d}x\right]\\ &= \frac{\nu a^2}{2}\int_{\partial F_a}|u|^2\,\mathrm{d}\theta.\end{aligned}$$

Assuming that u and its gradient are $O(|x|^{-3/2})$ as $|x| \to \infty$ (which is true for a solution to the homogeneous problem as is shown in Subsection 2.2.1), we see that the last integral, and the integrals over C_a in (2.26), tend to zero as $a \to \infty$. If we suppose that $\partial u/\partial n$ has a compact support on B, then the integrals over B have limits as well. Thus we get from (2.26) that *if u is a finite energy solution to the water-wave problem, then the following identity*

$$2\int_W |\nabla_x u|^2 \,\mathrm{d}x\mathrm{d}y - \int_B \mathbf{x}\cdot\mathbf{n}\,|\nabla u|^2 \,\mathrm{d}S = -2\,\mathrm{Re}\int_B \left(|x|\bar{u}_{|x|} + \bar{u}\right)\partial_n u \,\mathrm{d}S$$

holds. Hence, *the water-wave problem has at most one solution in W when*

$$\mathbf{x}\cdot\mathbf{n} \le 0 \quad \text{on } B. \tag{2.27}$$

The last condition simply means that for every negative constant C, a plane domain $W \cap \{y = C\}$ is starlike with respect to the point $(0, 0, C)$.

2.2.3. Auxiliary Integral Identities

In this subsection we derive the auxiliary integral identities for the three- and two-dimensional problems. Here W denotes a water domain containing totally submerged bodies and bounded below by a bottom B, which is a curved surface (line in two dimensions) coinciding with $\{y = -d\}$ at infinity. As in Section 2.1, S denotes the union of all wetted boundaries of immersed bodies, such that $S \cap B = \emptyset$. It is assumed (for the sake of simplicity) that ∂W is smooth enough.

2.2.3.1. Derivation of the Integral Identity in Three Dimensions

Let $\mathbf{V} = (V_1, V_2, V_3)$ be a vector field on $\bar{W}$ (V_3 is its projection on the y axis), whose components are real and uniformly Lipschitz functions on $\bar{W}$. We denote by H a real function on $\bar{W}$ having uniformly Lipschitz first derivatives. We recall that a function of one variable $f(t)$ satisfies the Lipschitz condition uniformly on $[a, b]$, if there exists a constant C, such that for arbitrary $t_1, t_2 \in [a, b]$ the inequality $|f(t_1) - f(t_2)| \le C|t_1 - t_2|$ holds. This definition can be easily generalized to functions of an arbitrary number of variables. It is well known that Lipschitz functions are differentiable almost everywhere.

The following identity,

$$\begin{aligned}2\,\mathrm{Re}\{(\mathbf{V}\cdot\nabla u + Hu)\nabla^2\bar{u}\} &= 2\,\mathrm{Re}\,\nabla\cdot\{(\mathbf{V}\cdot\nabla u + Hu)\nabla\bar{u}\}\\ &\quad + (Q\nabla\bar{u})\cdot\nabla u - \nabla\cdot[|\nabla u|^2\mathbf{V}\\ &\quad + |u|^2\nabla H] + |u|^2\nabla^2 H,\end{aligned} \tag{2.28}$$

can be verified directly. Here the matrix Q has the following elements:

$$Q_{ij} = (\nabla\cdot\mathbf{V} - 2H)\delta_{ij} - \left(\partial_{x_j} V_i + \partial_{x_i} V_j\right), \quad i, j = 1, 2, 3 \text{ and } x_3 = y,$$

and δ_{ij} is the Kronecker delta.

Let us integrate (2.28) over $W_a = W \cap \{|x| < a\}$ where a is so large that S is contained within the cylinder $\{|x| < a\}$. Then, assuming that u satisfies the Laplace equation, we get

$$\int_{W_a} [(Q\nabla\bar{u}) \cdot \nabla u + |u|^2 \nabla^2 H] \,\mathrm{d}x\mathrm{d}y + \int_{S \cup B_a} [|\nabla u|^2 \mathbf{V} \cdot \mathbf{n} + |u|^2 \partial_n H] \,\mathrm{d}S$$
$$- \int_{F_a} [|\nabla u|^2 V_3 + |u|^2 H_y] \,\mathrm{d}x + 2\,\mathrm{Re} \int_{F_a} (\mathbf{V} \cdot \nabla u + Hu)\,\bar{u}_y \,\mathrm{d}x$$
$$= \int_{C_a} \left[|\nabla u|^2 \mathbf{V} \cdot \frac{\mathbf{x}}{a} + |u|^2 H_{|x|} \right] \mathrm{d}S - 2\,\mathrm{Re} \int_{C_a} (\mathbf{V} \cdot \nabla u + Hu)\,\bar{u}_{|x|} \,\mathrm{d}S$$
$$+ 2\,\mathrm{Re} \int_{S \cup B_a} [\mathbf{V} \cdot \nabla u + Hu]\, \partial_n \bar{u} \,\mathrm{d}S, \tag{2.29}$$

where B_a, F_a are the portions of B, F contained respectively within $\{|x| < a\}$, and $C_a = W \cap \{|x| = a\}$; $\mathbf{x}$ denotes $(x_1, x_2, 0)$.

The free surface boundary condition allows us to write

$$2\,\mathrm{Re} \int_{F_a} (\mathbf{V} \cdot \nabla u + Hu)\,\bar{u}_y \,\mathrm{d}x - \int_{F_a} (|\nabla u|^2 V_3 + |u|^2 H_y) \,\mathrm{d}x$$
$$= \int_{F_a} [(\nu^2 V_3 + 2\nu H - H_y)\,|u|^2 - V_3\,|\nabla_x u|^2] \,\mathrm{d}x + 2\nu\,\mathrm{Re} \int_{F_a} \bar{u}\,\mathbf{V} \cdot \nabla_x u \,\mathrm{d}x.$$

Integrating by parts in the last integral, we obtain

$$-\nu \int_{F_a} |u|^2\, \nabla_x \cdot \mathbf{V} \,\mathrm{d}x + \nu \int_0^{2\pi} |u(a\cos\theta, a\sin\theta, 0)|^2\, \mathbf{V} \cdot \mathbf{x} \,\mathrm{d}\theta. \tag{2.30}$$

Assuming that u and its gradient are $O(|x|^{-3/2})$ as $|x| \to \infty$ (which is true for a solution to the homogeneous problem as is shown in Subsection 2.2.1), the last integral in (2.30) and the integrals over C_a in (2.29) tend to zero as $a \to \infty$. So we get from (2.29) that, *if u is a finite energy solution to the water-wave problem, then the following identity*

$$\int_W [(Q\nabla\bar{u}) \cdot \nabla u + |u|^2 \nabla^2 H] \,\mathrm{d}x\mathrm{d}y + \int_{S \cup B} [|\nabla u|^2 \mathbf{V} \cdot \mathbf{n} + |u|^2 \partial_n H] \,\mathrm{d}S$$
$$+ \int_F (\nu^2 V_3 + \nu[2H - \nabla_x \cdot \mathbf{V}] - H_y)|u|^2 \,\mathrm{d}x - \int_F V_3\,|\nabla_x u|^2 \,\mathrm{d}x$$
$$= 2\,\mathrm{Re} \int_{S \cup B} [\mathbf{V} \cdot \nabla u + Hu]\, \partial_n \bar{u} \,\mathrm{d}S \tag{2.31}$$

holds. It will be referred to as the *auxiliary identity.* In the literature, it is often referred to as Maz'ya's identity.

If W has infinite depth, then replacing the integral over $S \cup B$ by the integral over S in (2.31), one gets the auxiliary identity for this case when the integral over W should be understood as improper.

2.2.3.2. The Auxiliary Identity for an Axisymmetric Field and in Two Dimensions

Let $\mathbf{V}$ be axisymmetric with respect to the y axis; that is, $\mathbf{V} = (V_r, V_y, 0)$ in the cylindrical coordinates (r, y, θ), and the radial and vertical components (V_r and V_y, respectively, where V_y coincides with V_3 in the previous notations) are independent of the polar angle θ in the plane $\{y = 0\}$. Then the matrix Q has a peculiar structure. The following block

$$Q_* = \begin{bmatrix} (r^{-1}V_r - 2H) - (\partial_r V_r - \partial_y V_y) & -\partial_y V_r - \partial_r V_y \\ -\partial_y V_r - \partial_r V_y & (r^{-1}V_r - 2H) + (\partial_r V_r - \partial_y V_y) \end{bmatrix} \tag{2.32}$$

stands in the upper left corner, the last diagonal element is equal to

$$\partial_r V_r + \partial_y V_y - r^{-1} V_r - 2H,$$

and all other elements are equal to zero. To obtain this form of Q one has to use the cylindrical coordinates when integrating by parts in

$$\mathrm{Re} \int_W (\mathbf{V} \cdot \nabla u + Hu)\, \nabla^2 u \, \mathrm{d}x\mathrm{d}y.$$

The considerations in Subsection 2.2.3.1 remain valid in two dimensions. For this case we write down the auxiliary identity only for the homogeneous problem:

$$\int_W [(Q\nabla\bar{u}) \cdot \nabla u + |u|^2 \nabla^2 H]\, \mathrm{d}x\mathrm{d}y + \int_{S \cup B} [|\nabla u|^2 \mathbf{V} \cdot \mathbf{n} + |u|^2 \partial_n H]\, \mathrm{d}S$$
$$+ \int_F [\nu^2 V_2 + \nu(2H - \partial_x V_1) - H_y]|u|^2 \, \mathrm{d}x - \int_F V_2 |u_x|^2 \, \mathrm{d}x = 0. \tag{2.33}$$

Here the matrix is as follows:

$$Q = \begin{bmatrix} -\partial_x V_1 + \partial_y V_2 - 2H & -(\partial_y V_1 + \partial_x V_2) \\ -(\partial_y V_1 + \partial_x V_2) & \partial_x V_1 - \partial_y V_2 - 2H \end{bmatrix},$$

and $\mathbf{V} = (V_1, V_2)$ where V_2 is the projection on the y axis.

2.2.4. Sufficient Conditions for Uniqueness

In order to prove the uniqueness theorem for the water-wave problem using the auxiliary identity, we have to choose H and $\mathbf{V}$ so that the following statements are true.

1. All terms in the left-hand-side in (2.31) or (2.33) are nonnegative.
2. At least one of them is strictly positive for a nontrivial u.

Then assuming that u is a solution to the homogeneous problem, we arrive at a contradiction that implies the uniqueness.

2.2.4.1. The Three-Dimensional Problem

Let us show that *putting* $H = -1$, *and* $\mathbf{V} = (V_r, V_y, 0)$, *where*

$$V_r = r\frac{y^2 - r^2}{y^2 + r^2}, \quad V_y = \frac{-2yr^2}{y^2 + r^2}, \tag{2.34}$$

the auxiliary identity takes the form

$$\int_W \left\{ \left| u_r \frac{2yr}{y^2 + r^2} + u_y \frac{y^2 - r^2}{y^2 + r^2} \right|^2 + |\nabla u|^2 \frac{y^2}{y^2 + r^2} \right\} \mathrm{d}x\mathrm{d}y$$
$$+ \frac{1}{2}\int_{S \cup B} |\nabla u|^2 \frac{r}{y^2 + r^2} \{(y^2 - r^2)\partial_n r - 2yr\partial_n y\} \,\mathrm{d}S = 0, \tag{2.35}$$

if u *satisfies the homogeneous water-wave problem.*

Substituting $H = -1$ and $V_3(r, 0) = V_y(r, 0) = 0$ in (2.31) and taking into account the homogeneous Neumann condition on $S \cup B$, we get

$$\int_W [(Q\nabla\bar{u}) \cdot \nabla u] \,\mathrm{d}x\mathrm{d}y + \int_{S \cup B} |\nabla u|^2 \mathbf{V} \cdot \mathbf{n} \,\mathrm{d}S$$
$$- \nu \int_F [2 + \nabla_x \cdot \mathbf{V}] \, |u|^2 \,\mathrm{d}x = 0. \tag{2.36}$$

The last integral vanishes because

$$\nabla_x \cdot \mathbf{V}(x, 0) = -\nabla_x \cdot \mathbf{x} = -2,$$

where $\mathbf{x} = (x_1, x_2, 0)$.

Since $\mathbf{V}$ is axisymmetric, Q has the structure described in Subsection 2.2.3.2. For calculating the elements of Q_* given by (2.32) and Q_{33}, we introduce the complex variable $\rho = r + iy = (y^2 + r^2)^{1/2} e^{i\vartheta}$, and the function

$$w(\rho) = V_r + iV_y = -\frac{\rho + \bar{\rho}}{2} \frac{\rho^2}{|\rho|^2} = -re^{2i\vartheta}.$$

Then we find

$$Q_{11} = -\cos 2\vartheta - 2\,\mathrm{Re}\,\frac{\partial w}{\partial \bar{\rho}} + 2 = 2 - \cos 2\vartheta - \cos 4\vartheta$$

$$= 2(\sin^2\vartheta + \sin^2 2\vartheta) = 2\left[\frac{y^2}{y^2+r^2} + \left(\frac{2yr}{y^2+r^2}\right)^2\right],$$

$$Q_{22} = -\cos 2\vartheta + 2\,\mathrm{Re}\,\frac{\partial w}{\partial \bar{\rho}} + 2 = 2 - \cos 2\vartheta + \cos 4\vartheta$$

$$= 2(\sin^2\vartheta + \cos^2 2\vartheta) = 2\left[\frac{y^2}{y^2+r^2} + \left(\frac{y^2-r^2}{y^2+r^2}\right)^2\right],$$

$$Q_{12} = Q_{21} = -2\,\mathrm{Im}\,\frac{\partial w}{\partial \bar{\rho}} = -\sin 4\vartheta$$

$$= -2\sin 2\vartheta\cos 2\vartheta = 4\frac{yr}{y^2+r^2}\,\frac{y^2-r^2}{y^2+r^2},$$

$$Q_{33} = \cos 2\vartheta + 2\,\mathrm{Re}\,\frac{\partial w}{\partial \bar{\rho}} + 2 = 1 - \cos 2\vartheta = \frac{2y^2}{y^2+r^2}.$$

Thus the first integral in (2.36) is equal to

$$\int_W (Q_{11}|u_r|^2 + 2\,\mathrm{Re}\{Q_{12}u_r u_y\} + Q_{22}|u_y|^2 + Q_{33}|u_\theta|^2 r^{-2})\,dx dy$$

$$= 2\int_W \left\{\left|u_r\frac{2yr}{y^2+r^2} + u_y\frac{y^2-r^2}{y^2+r^2}\right|^2 + |\nabla u|^2\frac{y^2}{y^2+r^2}\right\} dx dy.$$

Inserting (2.34) in the integral over $S \cup B$ in (2.36), we complete the proof, thus obtaining from (2.35) the following uniqueness theorem.

The homogeneous water-wave problem has only a trivial solution when

$$(y^2 - r^2)\frac{\partial r}{\partial n} - 2yr\frac{\partial y}{\partial n} \geq 0 \quad \text{on } S \cup B. \tag{2.37}$$

Geometrically this condition means that the vector field (2.34) makes the angle not exceeding $\pi/2$ with the field of interior normals on $S \cup B$. It is easy to see that integral curves of (2.34) are semicircles

$$r^2 + (y + c^2)^2 = c^2, \quad \theta = \text{const}, \quad c > 0, \tag{2.38}$$

beginning at $(0, -2c)$, and ending at the origin. Hence, (2.37) is equivalent to the fact that *all transversal intersections of curves* (2.38) *with* $S \cup B$ *are points of entry into* W.

There is one more interpretation of (2.37). The inversion with respect the unit sphere centered at the origin maps the semicircles (2.38) into the rays

$$y = (2c)^{-1}, \quad \theta = \text{const},$$

emanating from the y axis. So (2.37) means that the inversion maps W into a domain whose cross sections by planes orthogonal to the y axis are starlike plane domains.

2.2.4.2. The Unique Solvability in Two Dimensions

Substituting $H = -1/2$ and the vector field

$$\mathbf{V}(x, y) = \left(x\frac{y^2 - x^2}{x^2 + y^2}, \frac{-2x^2 y}{x^2 + y^2} \right)$$

into (2.33), we arrive at

$$\frac{1}{2}\int_{S \cup B} |\nabla u|^2 \frac{x}{x^2 + y^2}\left\{ (y^2 - x^2)\frac{\partial x}{\partial n} - 2xy\frac{\partial y}{\partial n} \right\} \mathrm{d}S$$
$$+ \int_W |u_x\, 2xy + u_y\,(y^2 - x^2)|^2 \frac{\mathrm{d}x\mathrm{d}y}{(x^2 + y^2)^2} = 0.$$

It is assumed that u satisfies the homogeneous water-wave problem. This identity leads to the uniqueness theorem for the two-dimensional problem.

Let

$$(y^2 - x^2)\frac{\partial x}{\partial n} - 2xy\frac{\partial y}{\partial n} \geq 0 \quad \text{on } S \cup B. \tag{2.39}$$

Then the homogeneous two-dimensional water-wave problem in W has only a trivial solution.

The geometric interpretation of inequality (2.39) is the same as that of (2.37). Sometimes it is difficult to check the criterion for a particular body, but in some cases this can be done. An example of such geometry is given by Hulme [120]. Let S be an ellipse in deep water given as follows:

$$\frac{x^2}{(\lambda b)^2} + \frac{(y + h)^2}{b^2} = 1, \quad \lambda > 0,\, h > b.$$

Describing it parametrically by

$$x = b\lambda\cos\varphi, \quad y = -h + b\sin\varphi, \quad 0 \leq \varphi \leq 2\pi,$$

we get that the left-hand side in (2.39) is nonnegative simultaneously with

$$b\lambda\cos^2\varphi[(-h + b\sin\varphi)^2 - (b\lambda\cos\varphi)^2 - 2b\lambda(-h + b\sin\varphi)\lambda\sin\varphi]$$
$$= b\lambda\cos^2\varphi[(1 - \lambda^2)(h - b\sin\varphi)^2 + \lambda^2(h^2 - b^2)],$$

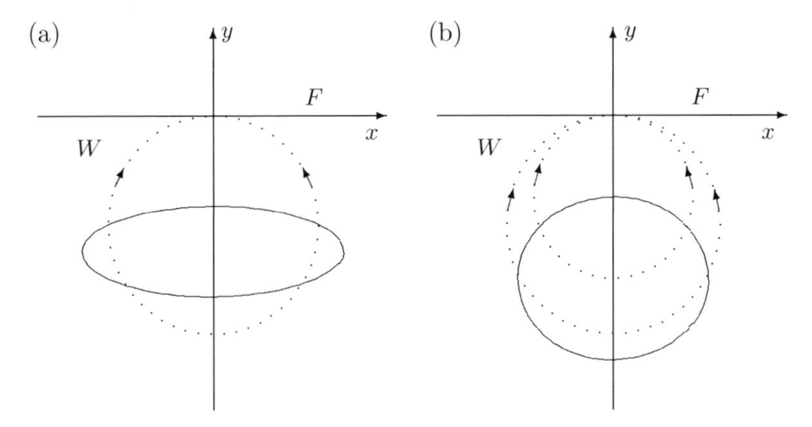

Figure 2.1.

which occurs when

$$h/b \geq \max\{1, 2\lambda^2 - 1\}. \tag{2.40}$$

This is obviously true if the major axis is vertical ($0 < \lambda \leq 1$), but if the major axis is horizontal ($\lambda > 1$), then the sum in square brackets has a minimum for $\varphi = 3\pi/2$. This minimum is equal to

$$(1 - \lambda^2)(h + b)^2 + \lambda^2(h^2 - b^2),$$

which is nonnegative when $h + b \geq 2b\lambda^2$, yielding (2.40).

An example of a "bad" ellipse having $\lambda = h/b = 3$ is shown in Fig. 2.1(a). A "good" ellipse having $\lambda = 6/5$ and $h/b = 2$ is plotted in Fig. 2.1(b). Figures 2.2(a) and 2.2(b) show other examples of cylinders violating (2.39).

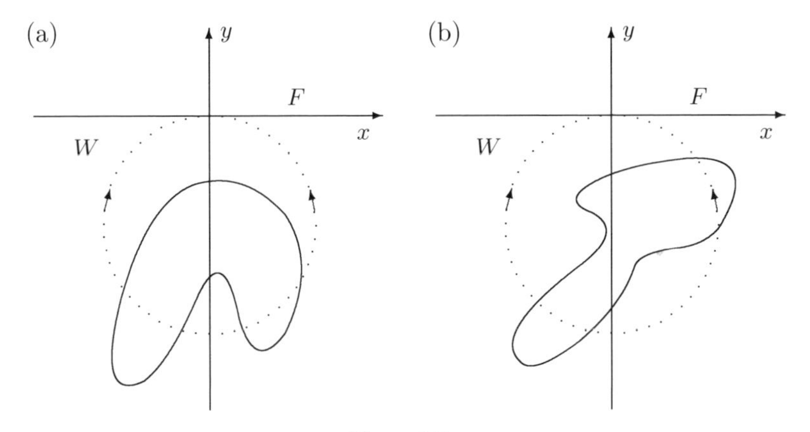

Figure 2.2.

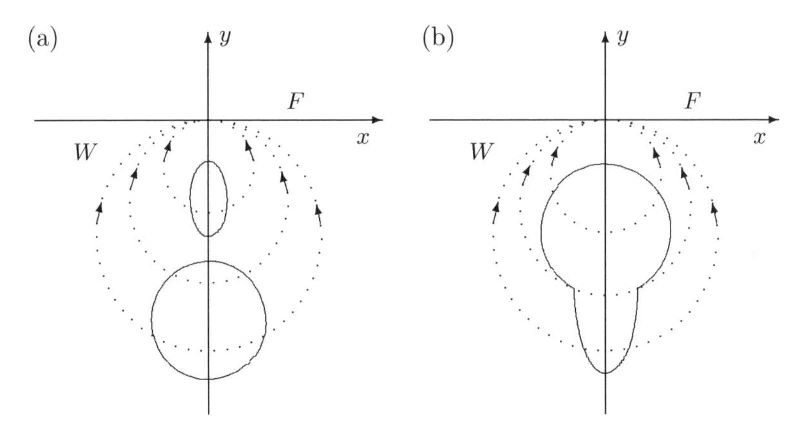

Figure 2.3.

Figures 2.3(a) and 2.3(b) illustrate some examples of cylinders satisfying (2.39). This condition holds for either curve in Fig. 2.3(a). The curve in Fig. 2.3(b) is composed of circular and elliptic arcs, either of which satisfy (2.39).

2.2.5. Further Uniqueness Results

In Subsection 2.2.4, some three- and two-dimensional conditions providing uniqueness were given – see inequalities (2.31) and (2.33) guaranteeing uniqueness. They were derived from the auxiliary integral identities by choosing the vector field $\mathbf{V}$ and the function H. Varying $\mathbf{V}$ and H, one might obtain other conditions of uniqueness for submerged bodies as well as for different types of bottom topography. Here we present more examples of this kind, generalizing vector fields considered above and considering new ones. We consider totally submerged bodies, first beginning with the simpler two-dimensional geometries in Subsection 2.2.5.1; in Subsection 2.2.5.2 we deal with an example in three dimensions obtained by rotation of a two-dimensional vector field about the y axis. In Subsections 2.2.5.3 and 2.2.5.4 we give examples of variable bottom topography providing uniqueness in two and three dimensions, respectively, and again, examples in three dimensions are obtained by rotation of two-dimensional vector fields.

2.2.5.1. Two-Dimensional Examples

As in Subsection 2.2.4, we choose H and $\mathbf{V}$ so that all terms on the left-hand side in (2.33) are nonnegative and at least one of them is strictly positive for a nontrivial u. The main difficulty is to verify that Q is a nonnegative matrix.

Its form implies that Q is nonnegative when $H \leq 0$ and

$$\det Q = 4H^2 - (\partial_x V_1 - \partial_y V_2)^2 - (\partial_y V_1 + \partial_x V_2)^2 \geq 0;$$

that is, Q has a nonnegative determinant.

Example 1

We put $N^2 = (y^2 - x^2 - a^2)^2 + 4x^2y^2$, and

$$\mathbf{V} = (x(y^2 - x^2 - a^2)N^{-1}, -2x^2yN^{-1}), \quad H = -1/2. \tag{2.41}$$

Here a is a nonnegative parameter (the case $a = 0$ was considered in Subsection 2.2.4.2). Then direct calculation gives

$$Q_{jj} = 1 - (-1)^j(y^2 + x^2 - a^2)[N^2 - 2(y^2 - x^2 - a^2)]N^{-3},$$
$$Q_{ij} = 4xy[(y^2 - a^2)^2 - x^4]N^{-3}, \quad i \neq j,$$

and

$$\det Q = 4x^2a^2N^{-2}.$$

Therefore Q is positive for $(x, y) \in \mathbb{R}^2_-$ and $a > 0$.

Substituting (2.41) in (2.33), we get

$$\int_W (Q\nabla\bar{u}) \cdot \nabla u \, dx dy + \int_{S \cup B} |\nabla u|^2 \, \mathbf{V} \cdot \mathbf{n} \, dS = 0, \tag{2.42}$$

where

$$\mathbf{V} \cdot \mathbf{n} = N^{-1}[(y^2 - |x|^2 - a^2)\partial_n |x|^2 - 2|x|^2 \partial_n y^2]. \tag{2.43}$$

Since Q is positive in (2.42), ∇u vanishes identically in W when (2.43) is nonnegative on $S \cup B$. Taking into account the result in Subsection 2.2.4.2, we arrive at the following theorem.

Let W be a domain such that (2.43) *is nonnegative on $S \cup B$ for a certain $a \geq 0$. Then the homogeneous water-wave problem has only a trivial solution in W.*

Geometrically, (2.43) is nonnegative when the angle between the vector field (2.41) and interior normals on $S \cup B$ does not exceed $\pi/2$. It is easy to see that the integral curves of (2.41) are semicircles

$$|x|^2 + \left[y + (a^2 + c^2)^{1/2}\right]^2 = c^2, \quad c > 0, \tag{2.44}$$

beginning at $(0, -c - (a^2 + c^2)^{1/2})$, and ending at $(0, c - (a^2 + c^2)^{1/2})$. So they are coordinate lines of a bipolar system having poles at $(0, \pm a)$. Thus, the nonnegativeness of (2.43) is equivalent to the following assertion.

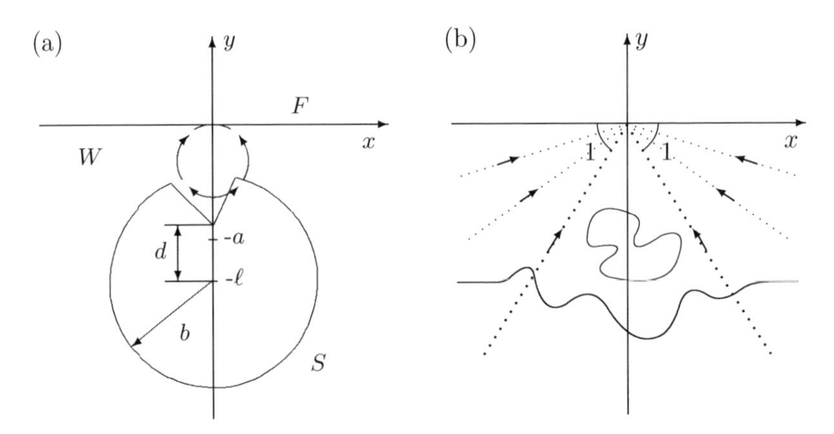

Figure 2.4.

All transversal intersections of the curves (2.44) *with* $S \cup B$ *are points of entry into* W.

Let us describe a configuration satisfying this assertion, and for the sake of simplicity let W have an infinite depth. The contour S shown in Fig. 2.4(a) bounds a disk of radius b centered at $(0, -\ell)$ with a deleted angle. The disk's boundary is given by (2.44), where $a = (\ell^2 - b^2)^{1/2}$ and $c = b$; it is a coordinate line of the bipolar system enclosing the pole at $(0, -a)$. We suppose that $b > d \geq \ell - (\ell^2 - b^2)^{1/2}$. Then (2.43) is nonnegative on S. In fact, it vanishes on the round part of S and is positive on the entering segments because circles (2.44) with radii less than b enter W across these segments. At the same time, there are circles with diameters smaller than $\ell - d$ tangent to the x axis at the origin that enter the domain inside S through the segments. Thus for this geometry *the vector field* (2.41) *with* $a = (\ell^2 - b^2)^{1/2} > 0$ *provides nonnegativeness of* (2.43) (and hence uniqueness), but when $a = 0$ the same vector field *fails to guarantee uniqueness.*

It is obvious that (2.43) is positive on a flat bottom placed under submerged bodies. Hence, *if* (2.43) *provides uniqueness for a deep water geometry, then the uniqueness holds for the same bodies immersed in water of constant depth.*

Example 2

Let ϑ be the polar angle that varies between $-\pi$ and 0 in $\mathbb{R}^2_-$; we put

$$H = -1/2, \quad \mathbf{V} = \alpha(\vartheta)(x, y), \tag{2.45}$$

where

$$\alpha(\vartheta) = \begin{cases} -(1+\vartheta) & \text{when } -1 \leq \vartheta \leq 0, \\ 0 & \text{when } -\pi+1 \leq \vartheta \leq -1, \\ \pi - 1 + \vartheta & \text{when } -\pi \leq \vartheta \leq -\pi+1. \end{cases}$$

We assume that there are no surface-piercing bodies and the curve S bounding totally submerged bodies lies within the angle

$$\left|\vartheta + \frac{\pi}{2}\right| \leq \frac{\pi}{2} - 1.$$

Then substituting (2.45) into (2.33) we get (2.42), and since $\mathbf{V}$ vanishes inside the above angle, the second integral involves only the portion of B outside this angle. If the inequality

$$\partial(x^2 + y^2)/\partial n \leq 0$$

holds on the described portion of B, then $\mathbf{V} \cdot \mathbf{n} \geq 0$ on $S \cup B$. Furthermore, one easily finds that

$$\det Q = 1 - [\alpha'(\vartheta)]^2 \geq 0,$$

which implies nonnegativeness of Q in $\mathbb{R}^2_-$. Thus the assumptions made imply that the homogeneous water-wave problem has only a trivial solution in W. Of course, for the flat bottom these assumptions are more restrictive than those obtained at the end of Subsection 3.2.2.1 (see Fig. 3.2). However, one can hardly expect that the method developed in Subsection 3.2.2.1 is applicable in the case of a bottom that is uneven outside a certain angle.

The vector field (2.45) is illustrated in Fig. 2.4(b). The field vanishes between two bold-dotted rays inclined at the angle $\pi/2 - 1$ to the vertical, and between these rays and the x axis the field is nonzero. The uniqueness theorem is true for any number of bodies having arbitrary shape and placed between the bold-dotted rays. Also, an arbitrary bottom profile is admissible there. Beyond the described angle, all vectors of (2.45) are directed to the origin and must enter the water domain for the uniqueness theorem to hold.

2.2.5.2. A Three-Dimensional Example

As we pointed out in Subsection 2.2.3.2, the matrix Q has a simple structure for an axisymmetric (with respect to the y axis) vector field $\mathbf{V}$. The main part of Q, denoted by Q_*, is given by (2.32), and all other elements vanish except for the diagonal one given by the formula next to (2.32). Moreover, since

$$\det Q_* = (2H - r^{-1}V_r)^2 - (\partial_r V_r - \partial_y V_y)^2 - (\partial_y V_r + \partial_r V_y)^2,$$

the matrix Q is nonnegative if $\det Q_* \geq 0$ and

$$r^{-1}V_r - 2H \geq 0, \quad \partial_r V_r + \partial_y V_y - r^{-1}V_r - 2H \geq 0,$$

where the last inequality is written for the lower-right-corner element.

Taking this into account, let us consider the vector field obtained by rotation of that in Example 1, Subsection 2.2.5.1. Then V_r and V_y are the first and the second components, respectively, in formulae (2.41) for the two-dimensional field with x replaced by r.

In order to avoid superfluous calculation we note that

$$\det Q_* - (2H - r^{-1}V_r)^2$$

coincides with $\det Q - 4H^2$, where Q is the matrix for the two-dimensional case with x replaced by r. Of course, these two expressions contain, generally speaking, different functions H. Thus $\det Q_*$ can be easily obtained from the known result for $\det Q$ in two dimensions.

In particular, for $H = -1$ (instead of $H = -1/2$ applied in two dimensions) and the vector field obtained by rotation of (2.41) about the y axis, we get

$$\det Q_* = \frac{[2 + (y^2 - r^2 - a^2)]^2}{N^4} - \frac{(y^2 + r^2 - a^2)^2}{N^2}.$$

Moreover,

$$-2H + \frac{V_r}{r} = 2 + \frac{y^2 - r^2 - a^2}{N} \geq 1 \quad \text{and} \quad \frac{(y^2 + r^2 - a^2)^2}{N^2} \leq 1.$$

Hence $\det Q_* \geq 0$. Taking into account that $Q_{33} = 2 - r^2/N \geq 0$, we see that Q is a nonnegative matrix.

On substitution of H and $\mathbf{V}$ into (2.31), we find that for a nontrivial u every integral on the right-hand side vanishes, except for the first two, which are nonnegative. This proves the following theorem.

If $S \subset \mathbb{R}^3_-$ and there exists $a \geq 0$ such that (2.43) *is nonnegative on $S \cup B$, then the water-wave problem has at most one solution in W for arbitrary* $\nu > 0$.

The geometric interpretation of this theorem is similar to that of the corresponding two-dimensional theorem. The integral curves of $\mathbf{V}$ are vertical semicircles lying on the sphere (2.44).

2.2.5.3. Strip of Variable Depth

We continue the search for geometric conditions providing uniqueness in the water-wave problem. The simple condition (2.27) valid for all positive ν was obtained for a layer of variable depth in Subsection 2.2.2. In fact, (2.27) follows from the identity that is a particular case of the general auxiliary identity; one simply has to substitute $\mathbf{V} = -(x, 0)$ and $H = -1$. Now we are going to enlarge the number of geometries providing uniqueness. However, in some situations the relevant range of frequencies (expressed in terms of ν) does not cover the whole positive half-axis.

Two examples of $\mathbf{V}$ and H are considered here for the two-dimensional case. They guarantee Q to be nonnegative in (2.33) and so lead to some geometric conditions on B. The first example generalizes the condition presented in Subsection 2.2.2.

Example 1

Let us put

$$H = \text{const}, \quad \mathbf{V} = (-x, -k(y + a)), \tag{2.46}$$

where a and k are parameters. We assume that $a \in [0, +\infty)$, and the ranges for k and H will be determined below. The matrix

$$Q = \begin{bmatrix} 1 - 2H - k & 0 \\ 0 & -1 - 2H + k \end{bmatrix}$$

is nonnegative if

$$1 - 2H \geq k \geq 1 + 2H, \tag{2.47}$$

and hence $H \leq 0$. Furthermore, the integrals over F in (2.33) are nonnegative if

$$ka \geq 0 \quad \text{and } \nu(1 + 2H) - ka\nu^2 \geq 0.$$

This implies that

$$1 + 2H \geq ka\nu \geq 0, \tag{2.48}$$

and so $H \geq -1/2$. Therefore $-1/2 \leq H \leq 0$, and comparing (2.47) with (2.48) we find that

$$0 \leq k \leq 2 \quad \text{and } a\nu \leq 1 \text{ when } k \neq 0.$$

Thus the following theorem is proved.

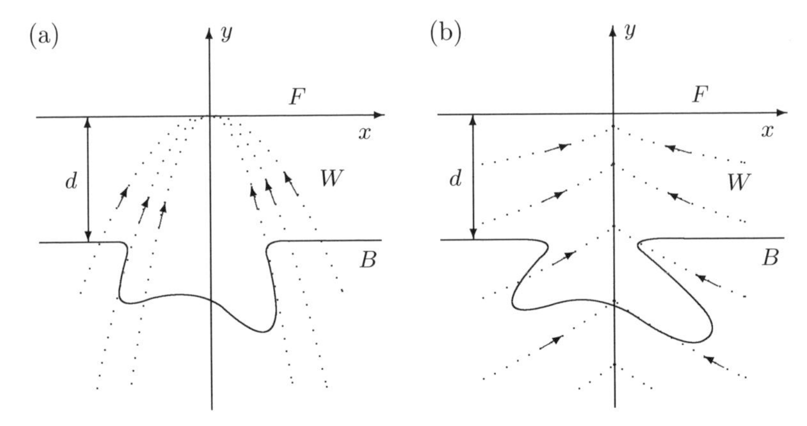

Figure 2.5.

Let k, H, a, and ν belong to $[0, 2]$, $[-1/2, 0]$, $[0, +\infty)$, *and* $(0, +\infty)$, *respectively. If* (2.47) *and* (2.48) *hold and*

$$\frac{\partial |x|^2}{\partial n} + k \frac{\partial (y+a)^2}{\partial n} \leq 0 \quad \text{on } B, \tag{2.49}$$

then the homogeneous water-wave problem has only a trivial solution in W.

As in Subsection 2.2.4, we see that geometrically (2.49) means that angles between the field of interior normals on B and the field (2.46) do not exceed $\pi/2$. The integral curves of the vector field (2.46) are lines

$$|y + a| = c|x|^k, \quad c \geq 0, \tag{2.50}$$

directed to $(0, -a)$ for $k \neq 0$ and to the y axis for $k = 0$.

The case when $k = 0$ and $H = -1/2$ is the two-dimensional analog of the theorem considered in Subsection 2.2.2 and providing that the solution is unique for all $\nu > 0$. The geometric meaning of the corresponding condition is that each horizontal line either does not intersect B or intersects B twice. Using (2.50), one easily interprets (2.49) for $k = 1$ – by (2.47) this implies that $H = 0$ – as the condition that W is a starlike domain with respect to the point $(0, -a)$, but by (2.48) the uniqueness of solution is proved only for $\nu \in (0, a^{-1}]$ in this case. Figure 2.5(a) shows a curve B satisfying (2.49) for $k = 2$, where dotted lines are integral curves of the corresponding vector field. In this case $H = -1/2$, $a = 0$, and a solution to the water-wave problem is unique for all $\nu > 0$. For $k = 3/2$ we can use $H = -1/4$ and arbitrary $a \geq 0$, such that $(0, -a) \in \bar{W}$. So the theorem implies that the uniqueness holds for $\nu \in (0, (3a)^{-1}]$.

Example 2

Let us put $N_0^2 = |x| + (x^2 + y^2)^{1/2}$ and

$$\mathbf{V} = \left(\frac{-x N_0}{\sqrt{2}(x^2 + y^2)^{1/4}}, \frac{-y|x|}{\sqrt{2}(x^2 + y^2)^{1/4} N_0} \right), \quad H = \frac{-1}{2}. \tag{2.51}$$

Substituting (2.51) into (2.33) we get (2.42). Moreover, Q defined by (2.51) is a nonnegative matrix for $(x, y) \in \mathbb{R}^2_-$, which follows from $\det Q \geq 0$ as was noted in Subsection 2.2.5.1. It is simpler to verify the last inequality by using polar coordinates (ρ, ϑ): $x = \rho \cos \vartheta$, $y = \rho \sin \vartheta$. Then we have

$$V_1 = -2^{-1/2} (1 + |\cos \vartheta|)^{1/2} \rho \cos \vartheta, \quad V_2 = \frac{-\rho \sin \vartheta |\cos \vartheta|}{\sqrt{2}(1 + |\cos \vartheta|)^{1/2}},$$

and

$$\begin{aligned}\det Q &= 1 - (\partial_\rho V_1 - \rho^{-1} \partial_\vartheta V_2)^2 - (\partial_\rho V_2 + \rho^{-1} \partial_\vartheta V_1)^2 \\ &= \frac{3}{4} \cos^2 \vartheta = \frac{3x^2}{4(x^2 + y^2)},\end{aligned}$$

and this is positive almost everywhere. Now analyzing $\mathbf{V} \cdot \mathbf{n}$, we arrive at the following theorem.

The homogeneous water-wave problem has only a trivial solution when

$$\left[|x| + (|x|^2 + y^2)^{1/2}\right] \frac{\partial |x|^2}{\partial n} + |x| \frac{\partial y^2}{\partial n} \leq 0 \quad \text{on } B. \tag{2.52}$$

Let us consider the geometric meaning of (2.52). We consider the family of integral lines of $\mathbf{V}$ given by (2.51):

$$2c|x| = y^2 - c^2, \quad c \geq 0,$$

that is, the parts of parabolas that are symmetric about the y axis and go from infinity to their ends on the y axis. Inequality (2.52) holds when these lines enter W at points of transversal intersection with B. An example of W satisfying this condition is shown in Fig. 2.5(b) – the dotted lines are integral curves of $\mathbf{V}$.

2.2.5.4. Layer of Variable Depth

Here we use vector fields obtained by rotating those given in Subsection 2.2.5.3 for the two-dimensional case.

Example 1

Rotating the two-dimensional field (2.46), one obtains the field leading to a diagonal matrix Q with elements

$$Q_{11} = Q_{33} = -k - 2H \quad \text{and } Q_{22} = -k - 2 - 2H.$$

This matrix is nonnegative when

$$-2H \geq k \geq 2 + 2H, \tag{2.53}$$

and so $H \leq -1/2$. Moreover, the integrals in (2.31) are nonnegative if $ka \geq 0$ and $\nu(2H + 2) - ka\nu^2 \geq 0$; that is,

$$2(H + 1) \geq ka\nu \geq 0. \tag{2.54}$$

From here $H \geq -1$, and so $H \in [-1, -1/2]$, which implies that $k \in [0, 2]$. Also, the integrals over S and B must be nonnegative and this leads to the following theorem generalizing the result obtained in Subsection 2.2.2.

Let the parameters k, H, a, and ν belong to $[0, 2]$, $[-1, -1/2]$, $[0, +\infty)$, *and* $(0, +\infty)$, *respectively. If inequalities* (2.53) *and* (2.54) *hold, then the homogeneous water-wave problem has only a trivial solution for the indicated values of ν when* (2.49) *holds on B.*

The geometric interpretation of this theorem is similar to that of the first theorem in Subsection 2.2.5.3. The corresponding integral curves lie in the vertical cross sections of the surface of rotation given by (2.50).

Example 2

For $H = -1$ one more field can be obtained by rotating the field (2.51) about the y axis. Then we get that

$$\det Q_* = \left\{ 2 - \frac{\left[r + (r^2 + y^2)^{1/2}\right]^{1/2}}{\sqrt{2}(r^2 + y^2)^{1/4}} \right\}^2 - \frac{r^2 + 4y^2}{4(r^2 + y^2)},$$

which is easily seen to be nonnegative. Since

$$-2H + \frac{V_r}{r} = 2 - \frac{\left[r + (r^2 + y^2)^{1/2}\right]^{1/2}}{\sqrt{2}(r^2 + y^2)^{1/4}} \geq 1,$$

and

$$Q_{33} = 2 - \frac{r\left[r + (r^2 + y^2)^{1/2}\right]^{1/2}}{2\sqrt{2}(r^2 + y^2)^{3/4}} \geq 0,$$

the matrix Q is nonnegative. Moreover, all integrals on the right-hand side of (2.31) except for the first two are equal to zero. Similarly to the two-dimensional case, (2.52) implies that *the homogeneous three-dimensional water-wave problem has only a trivial solution.*

Figure 2.5(b) in Subsection 2.2.5.3 shows a vertical cross section of a geometry satisfying this theorem.

2.3. Unique Solvability Theorems

We recall that Kochin's theorem (see Section 2.1) provides the unique solvability of the water-wave problem for all values of $\nu > 0$ with the possible exception of a finite number of them. Uniqueness theorems proved in Section 2.2 give geometric criteria for absence of these exceptional values. Now the reference to Fredholm's alternative for the integral equation (2.7) yields the unique solvability theorem for the nonhomogeneous water-wave problem in the case of a totally submerged body (see Subsection 2.3.1 for the theorem's formulation).

For the water-wave problem in a layer of variable depth, we have to consider problems about pulsating sources placed at a point $(\xi, \eta) \in W$ or $F \cup B$ before applying Fredholm's alternative. For this purpose, we give a comprehensive treatment in Subsection 2.3.2 to the problem having a nonzero right-hand-side term in the differential equation; that is, the Poisson equation is used in the water-wave problem. Above all, this allows us to embrace the problem of water waves generated by a time-periodic surface pressure $\text{Re}\,\{p(x)e^{-i\omega t}\}$, where ω is the radian frequency. This problem involves the following boundary condition,

$$u_y - \nu u = i\omega(\rho g)^{-1} p(x) \quad \text{on } F, \tag{2.55}$$

instead of the homogeneous free surface condition used previously. Here ρ and g are the water density and the acceleration caused by gravity, respectively.

The result of Subsection 2.3.2 allows us to investigate the problems about sources in a layer of variable depth and to formulate the main result on the unique solvability (see Subsection 2.3.3). Similar results are formulated for the two-dimensional case in Subsection 2.3.3.3.

2.3.1. Theorem for a Submerged Body

Combining the uniqueness theorem proved in Subsection 2.2.4.1 with the solvability theorem from Subsection 2.1.2.2, one immediately arrives at the following theorem on unique solvability.

Let the three- or two-dimensional water domain W have constant depth, and let for a certain $a \geq 0$ the inner product (2.43) *be nonnegative on S. Then* *the water-wave problem in* W *has the unique solution that can be found in the form of the single-layer potential* (2.1) *whose density* μ *satisfies the integral equation* (2.7).

We impose the condition on S only because it obviously holds on the horizontal bottom B (or there is no B at all when the depth is infinite). Furthermore, using sources constructed in Subsection 2.3.3, one can easily extend the theorem to the case of a locally curved bottom satisfying the same condition as S.

2.3.2. Problem for the Poisson Equation

Let us turn to waves that are due to either a surface pressure applied to the free surface

$$F = \{x \in \mathbb{R}^m, y = 0\}, \quad m = 1, 2,$$

or to a source placed in W, or on one of the components of ∂W, that is, on the free surface F or on the bottom B. The latter is assumed to be a continuously differentiable surface (line) for $m = 2$ ($m = 1$) at a finite distance from F, and coinciding with $\{y = -d\}$ at infinity. Thus a water domain having a bounded bottom obstruction is under consideration in the present subsection. For this purpose we first investigate the problem in which the Laplace equation is replaced by the Poisson equation. The fact of the unique solvability of this problem will be used in what follows for establishing the unique solvability of the water-wave problem with nonhomogeneous boundary conditions.

2.3.2.1. Statement of the Problem; Reduction to an Operator Equation

Here we assume that the domain $W \subset \mathbb{R}^3$ is obtained from the layer L of constant depth d by using a one-to-one C^2-mapping $\psi : \bar{L} \mapsto \bar{W}$, such that it coincides with the identity mapping for $|x| = r > a = \text{const}$ and leaves F fixed. By $H^s(Q)$ where Q is a certain domain we denote the Sobolev space of functions whose derivatives of order up to s are square integrable over Q. Let $H^0_c(W)$ $[H^0_c(L)]$ be the subspace of $H^0(W)$ $[H^0(L)]$ consisting of functions vanishing for $|x| > c$. Thus if $W_a = W \cap \{|x| < a\}$, then $H^0_a(W)$ consists of functions in $H^0(W_a)$ continued by zero to $W \setminus \bar{W}_a$.

Let $u \in H^2(W_c)$ for every $c > 0$ be a solution to the following problem:

$$\nabla^2 u = f \quad \text{in } W, \tag{2.56}$$

$$u_y - \nu u = 0 \quad \text{on } F, \tag{2.57}$$

$$\partial u / \partial n = 0 \quad \text{on } B, \tag{2.58}$$

where $f \in H_b^0(W)$ for some $b < \infty$. The aim of the present section is to prove the following theorem.

If B satisfies (2.27), *then for any $f \in H_b^0(W)$ problem* (2.56)–(2.58) *has one and only one solution satisfying the radiation condition.*

The uniqueness of the solution was proved in the previous section. Now we show the existence. Without loss of generality we assume that the mapping ψ has the following properties:

1. ψ is the identity mapping for $y \geq -\varepsilon$, where ε is a certain positive number.
2. ψ preserves normals; that is, the image of any normal to the plane $y = -d$ is a normal to B.

These properties follow from our assumption and the fact that every non-degenerate mapping maps the field of normals to $y = -d$ into a nontangential vector field on B. Furthermore, we can assume without loss of generality that $b = a$, and ψ is the identity mapping for $|x| > a - 1$; in fact, one can replace a and b by $1 + \max\{a, b\}$.

Changing variables (x, y) by $\psi^{-1}(x, y)$ in (2.56)–(2.58), we get the equivalent problem

$$(\nabla^2 + D)v = f \quad \text{in } L, \tag{2.59}$$

$$v_y - \nu v = 0 \quad \text{when } y = 0, \tag{2.60}$$

$$v_y = 0 \quad \text{when } y = -d, \tag{2.61}$$

$$v_{|x|} - ik_0 v = o\big(|x|^{-1/2}\big) \quad \text{as } |x| \to \infty, \tag{2.62}$$

where the coefficients of the differential operator $D(x, y, \nabla)$ vanish outside L_{a-1}. Here and in what follows for any $c > 0$ we denote by L_c the intersection $L \cap \{|x| < c\}$.

Let I_ψ be the absolute value of the Jacobian of ψ. It is obvious that the operator $\nabla^2 + D_1 = I_\psi(\nabla^2 + D)$ with the boundary conditions (2.60) and (2.61) is formally self-adjoint; that is, putting $\langle u, v\rangle = \int_L u\bar{v}\, dxdy$ we have

$$\langle(\nabla^2 + D_1)\varphi_1, \varphi_2\rangle = \langle\varphi_1, (\nabla^2 + D_1)\varphi_2\rangle,$$

where φ_i belongs to $H^2(L_c)$ for any $c > 0$, satisfies (2.60) and (2.61), and vanishes in a neighborhood of infinity. The coefficients of D_1 also vanish outside of L_{a-1}.

Replacing D by D_1 and f by $f_1 = I_\psi f$ in (2.59), we arrive at a problem that is equivalent to (2.59)–(2.62). In fact, the latter problem is obtained from the former one by multiplying both sides in (2.59) by a function I_ψ that is nonzero everywhere. So retaining previous notations (D and f instead of D_1 and f_1 respectively), we may assume that the operator $\nabla^2 + D$ in (2.59) with boundary conditions (2.60) and (2.61) is formally self-adjoint:

$$\langle(\nabla^2 + D)\varphi_1, \varphi_2\rangle = \langle\varphi_1, (\nabla^2 + D)\varphi_2\rangle. \tag{2.63}$$

Since (2.63) holds for ∇^2 (that is, when $D = 0$), this formula implies that $\langle D\varphi_1, \varphi_2\rangle = \langle\varphi_1, D\varphi_2\rangle$. Noting that the coefficients of D vanish outside of L_{a-1}, we get

$$\langle D\varphi_1, \varphi_2\rangle = \langle\varphi_1, D\varphi_2\rangle \tag{2.64}$$

for $\varphi_i \in H^2(L_{a-1/2})$, and satisfying (2.60) and (2.61) for $|x| < a - 1/2$.

By G we denote the operator mapping any $f \in H_a^0(L)$ into the solution of (2.59)–(2.62) with $D = 0$. It is clear that G is given explicitly as an integral operator having Green's function $G(x, y; \xi, \eta)$ as the kernel. We note that

$$G : H_a^0(L) \mapsto H^2(L_c) \tag{2.65}$$

is a bounded operator for every $c < \infty$. Thus, $DG : H_a^0(L) \mapsto H_a^0(L)$ is a bounded operator.

Let us seek a solution to (2.59)–(2.61) in the form $v = Gq$ where $q \in H_a^0(L)$ is an unknown function. Then $v \in H^2(L_c)$ *for arbitrary* $c > 0$, *and* v *is a solution to* (2.59) – (2.61) *when* q *satisfies*

$$q + DGq = f, \quad \text{where } f \in H_a^0(L). \tag{2.66}$$

The converse is also true, that is, *if* $v \in H^2(L_c)$ *for arbitrary* $c > 0$, *and* v *is a solution to* (2.59)–(2.61), *then it can be written in the form* $v = Gq$ *where* $q = -Dv + f$, *and* q *satisfies* (2.66). Thus (2.56)–(2.58) are reduced to the operator equation (2.66) in the space $H_a^0(L)$.

2.3.2.2. Investigation of the Operator Equation

To conclude the proof of theorem formulated in the previous subsection, we find it sufficient to show that (2.66) is solvable. First, we prove that the range of $I + DG$ (I denotes the identity operator) is a closed set. Let v be a solution to (2.59)–(2.62). A well-known local a priori estimate for solutions to elliptic boundary value problems (see, for example, Gilbarg and Trudinger's book [94], Chapter 9) gives

$$\|v\|_{H^2(L_{a-1/2})} \leq c\left[\|v\|_{H^0(L_a)} + \|f\|_{H_a^0(L)}\right], \tag{2.67}$$

where c is a certain constant. Then, since

$$v = G(-Dv + f) \tag{2.68}$$

and operator (2.65) is bounded, inequality (2.67) implies that

$$\|v\|_{H^2(L_a)} \leq c\left[\|v\|_{H^0(L_a)} + \|f\|_{H_a^0(L)}\right]. \tag{2.69}$$

A standard argument based on the uniqueness theorem for (2.59)–(2.62) allows us to drop the first term on the right-hand side of (2.69). Assuming the contrary, one can find a sequence $\{v_n\}$ of solutions to (2.59)–(2.62) with $f = f_n$, such that $\|v_n\|_{H^2(L_a)} = 1$, and $\|f_n\|_{H^0(L_a)} \to 0$ as $n \to \infty$. Since the embedding operator $H^2(L_a) \mapsto H^0(L_a)$ is compact, there is a subsequence $\{v_{n_k}\}$, such that

$$v_{n_k} \to v_0 \quad \text{in } H^0(L_a) \text{ as } k \to \infty.$$

Then (2.69) with $v = v_n$ and $f = f_n$ implies that $v_{n_k} \to v_0$ in $H^2(L_a)$, and hence, it follows from (2.68) that v_0 is given on L and is a solution to (2.59)–(2.62) with $f = 0$. This contradicts the uniqueness theorem, since $\|v_0\|_{H^2(L_a)} = 1$. Thus,

$$\|v\|_{H^2(L_a)} \leq c\|f\|_{H^0(L)}. \tag{2.70}$$

Now, let $\{v_n\}$ be solutions of (2.59)–(2.62) with $f = f_n \to f_0$ in $H_a^0(L)$ as $n \to \infty$. Then (2.70) implies that $v_n \to v_0$ in $H^2(L_a)$ as $n \to \infty$. So letting $n \to \infty$ in (2.68) with $v = v_n$ and $f = f_n$, one justifies (2.68) for $v = v_0$ and $f = f_0$. Therefore, v_0 can be extended to the whole L as a solution to (2.59)–(2.62) with $f = f_0$. Thus the set of functions $f \in H_a^0(L)$, for which (2.59)–(2.62) has a solution, is closed. As we pointed out above, the solvability of (2.59)–(2.62) is equivalent to that of (2.66). This leads to a conclusion that the range of $I + DG$ is a closed set.

Let us show that equation

$$w + (DG)^* w = 0, \quad w \in H_a^0(L), \tag{2.71}$$

has only a trivial solution that completes the proof of solvability of (2.66). We first verify that

$$(DG)^* \varphi = \overline{DG}\varphi \quad \text{for } |x| < a, \tag{2.72}$$

if

$$\varphi \in H_a^0(L) \cap H^2\left(L_{a-1/2}\right), \quad \text{and } \varphi \text{ satisfies (5.7), (5.8) for } |x| < a - 1/2. \tag{2.73}$$

Here and below we use the same notation for $\varphi \in H_a^0(L)$ and for a restriction of this function to L_c, $c \leq a$.

Let $f_i \in H_a^0(L)$ and $v_i = Gf_i, i = 1, 2$; that is, $v_i \in H^2(L_c)$ for any $c > 0$ and v_i is a solution to (2.59)–(2.62) with $D = 0$ and $f = f_i$. By Green's formula we have

$$\langle \nabla^2 v_1, \overline{v_2} \rangle = \langle v_1, \overline{\nabla^2 v_2} \rangle,$$

and so

$$\langle f_1, \overline{G f_2} \rangle = \langle G f_1, \overline{f_2} \rangle.$$

Substituting f for f_1 and $\bar{\varphi}$ for f_2, we get

$$\langle Gf, \varphi \rangle = \langle f, \bar{G}\varphi \rangle \quad \text{for } f, \varphi \in H^0_a(L). \tag{2.74}$$

Now, if conditions (2.73) hold, then it follows from (2.64) and (2.74) that

$$\langle DGf, \varphi \rangle = \langle Gf, D\varphi \rangle = \langle f, \bar{G} D\varphi \rangle,$$

and this proves (2.72) because D has real coefficients.

Let w be a solution of (2.71) and let ψ have the following properties: $\psi \in C^2(\bar{L})$, $\psi = 0$ for $|x| > a$, and ψ satisfies (2.60) and (2.61) for $|x| < a$. Since $G\nabla^2 \psi = \psi$, (2.71) implies that

$$0 = \langle w + (DG)^* w, \nabla^2 \psi \rangle = \langle w, \nabla^2 \psi + DG\nabla^2 \psi \rangle = \langle w, \nabla^2 \psi + D\psi \rangle.$$

Thus, w is a generalized solution (in the sense of the integral identity) to the following problem:

$$(\nabla^2 + D) w = 0 \quad \text{in } L_a,$$

and w satisfies (2.60) and (2.61) when $|x| < a$. It is well known (see, for example, the book by Kozlov, Maz'ya, and Rossmann [149], Section 3.3.2), that this yields $w \in H^2(L_{a-1/2})$ and (2.60) and (2.61) hold in the strong sense for $|x| < a - 1/2$. Hence, we have (2.72) for w, and then (2.71) leads to the relation

$$\bar{w} + GR\bar{w} = 0 \quad \text{in } L_a,$$

which allows us to extend $\bar{w}$ to L as a solution of the homogeneous problem (2.59)–(2.62). So $\bar{w} = 0$, and consequently (2.71) has only a trivial solution. The proof is complete.

2.3.3. Sources in a Layer of Variable Depth

The aim of the present subsection is to prove the following unique solvability theorem.

Let B satisfy (2.27) *and be sufficiently smooth. Then the water-wave problem has the unique solution for any localized disturbance applied to the free surface or bottom, that is, for any p in* (2.55) *and for any f in the Neumann condition on B, such that p and f have compact supports on F and on B, respectively.*

The uniqueness is established in Section 2.2. To prove the existence we demonstrate that there are unique Green's functions; that is, velocity potentials that are due to sources placed on F or on B. Then the solution to the water-wave problem is a single-layer potential having the density $i\omega(\rho g)^{-1}p$ or f and the corresponding Green's function as the kernel.

2.3.3.1. *Statement of Boundary Value Problems for Sources*

Let (ξ, η) be a point in $\bar{W}$. We recall that $x = (x_1, x_2)$, $\xi = (\xi_1, \xi_2)$, and $R = [|x - \xi|^2 + (y - \eta)^2]^{1/2}$ is the distance between (x, y) and (ξ, η). By $\mathcal{B}_\varepsilon(x, y)$ we denote a ball of radius ε centered at (x, y). We put

$$E^{(1)} = \frac{1}{R} \quad \text{when } (\xi, \eta) \in W, \quad E^{(2)} = \frac{-1}{R} + \nu \log(R - y) \quad \text{when } \eta = 0.$$

Let (α, β, γ) be Cartesian coordinates in $\mathbb{R}^3$ having the origin at $(\xi, \eta) \in B$, and the γ axis going into W along $\mathbf{n}_{(\xi,\eta)}$. We assume that in these coordinates the bottom B is given as follows:

$$\gamma = C_1\alpha^2 + C_2\beta^2 + O\big((\alpha^2 + \beta^2)^{3/2}\big)$$

in a neighborhood of (ξ, η). Then for $(\xi, \eta) \in B$ we put

$$E^{(3)} = \chi(r)\left[\frac{1}{R} + \frac{(C_1 - C_2)(R - \gamma)}{4(R + \gamma)} \cos 2\varphi + \frac{C_1 + C_2}{2} \log(R + \gamma)\right],$$

where φ is the polar angle in the (α, β) plane, and the function $\chi \in C^\infty(\overline{\mathbb{R}_+})$ is equal to one for $R < \delta$ and to zero for $R > 2\delta$. Here the constant $\delta > 0$ is chosen so small that $\mathcal{B}_{2\delta} \cap \{y = 0\} = \emptyset$, and the negative part of the γ axis does not intersect $\bar{W}$ within $\mathcal{B}_{2\delta}$. It easily follows from the last condition that $E^{(3)} \in C^2(\bar{W} \setminus \{(\xi, \eta)\})$.

We say that $G^{(j)}(x, y; \xi, \eta)$ $(j = 1, 2, 3)$ is called the velocity potential that is due to a source pulsating at a point (ξ, η) placed either within the water domain $(j = 1)$, in the free surface $(j = 2)$, or on the bottom $(j = 3)$, if $G^{(j)}$ satisfies the following conditions:

1. The point (ξ, η) belongs to W, F, or B, respectively.
2. $G^{(j)} - E^{(j)} \in H^2(W_c)$ for every $c < \infty$ $(j = 1, 2, 3)$.
3. For any $\varepsilon > 0$,

$$\nabla^2 G^{(j)} = 0 \quad \text{in } W \setminus \mathcal{B}_\varepsilon(\xi, \eta), \tag{2.75}$$

$$G^{(j)}_y - \nu G^{(j)} = 0 \quad \text{on } F \setminus \mathcal{B}_\varepsilon(\xi, \eta), \tag{2.76}$$

$$\partial G^{(j)} / \partial n = 0 \quad \text{on } B \setminus \mathcal{B}_\varepsilon(\xi, \eta), \tag{2.77}$$

$$G^{(j)}_{|x|} - ik_0 G^{(j)} = o\left(|x|^{-1/2}\right) \quad \text{as } |x| \to \infty. \tag{2.78}$$

Instead of (2.75)–(2.77) one can write formulae valid for the whole W, F, and B, respectively. For this purpose one has to put Dirac's measure in one of the relations and zero in the rest of them.

2.3.3.2. Existence and Uniqueness of the Source Potentials

Here we prove the following theorem.

If W satisfies (2.27), *then $G^{(j)}(x, y; \xi, \eta)$ exists and is unique* ($j = 1, 2, 3$).

We consider the case $j = 1$ first. Let χ be the function defined in the previous subsection, and $\delta > 0$ be so small that $\mathcal{B}_\delta(\xi, \eta) \subset W$. We seek $G^{(1)}$ in the form of $\chi(R)E^{(1)} - u$. Then u must belong to $H^2(W_c)$ for every $c < \infty$. Moreover, it should satisfy boundary conditions on the free surface and on the bottom, the radiation condition as well as

$$\nabla^2 u = f \quad \text{in } W \backslash \mathcal{B}_\varepsilon(\xi, \eta) \text{ for any } \varepsilon > 0. \tag{2.79}$$

Here $f = \nabla^2(\chi E^{(1)})$, and since $\nabla^2 E^{(1)} = 0$ in $W \backslash \mathcal{B}_\varepsilon(\xi, \eta)$, we get that

$$f = E^{(1)} \nabla^2 \chi(R) + 2\left[\nabla E^{(1)} \cdot \nabla \chi(R)\right].$$

This function belongs to $H_b^0(W)$ for a certain $b < \infty$. Moreover, $u \in H^2(W_c)$ for every $c < \infty$, and then $\nabla^2 u \in H^0(W_c)$. Therefore, (2.79) means that

$$\nabla^2 u = f \in H_b^0(W) \quad \text{in } W,$$

and this equality is understood as an equality in $H^0(W_c)$. So, u satisfies (2.56)–(2.58). According to Section 2.2, such u exists and is unique, and so the theorem is proved for $j = 1$.

Now let the source be placed in the free surface. We seek $G^{(2)}(x, y; \xi, \eta)$ in the form of $\chi(R)E^{(2)} - v$, where χ is the same function as above, and $\delta > 0$ is a constant, such that $B \cap \mathcal{B}_\varepsilon(\xi, \eta) = \emptyset$. Since $\nabla^2 E^{(2)} = 0$ in W, and

$$(\partial/\partial y - \nu)E^{(2)} = -\nu^2 \log R \quad \text{when } y = 0,\ R > \varepsilon,$$

the same argument as in the case $j = 1$ implies that $v \in H^2(W_c)$ for every $c > 0$. Also, we note that v satisfies the homogeneous Neumann condition on B, the radiation condition, the equation

$$\nabla^2 v = E^{(2)} \nabla^2 \chi(R) + 2\left[\nabla E^{(2)} \cdot \nabla \chi(R)\right] \in H_b^0(W) \quad \text{in } W,$$

and the boundary condition

$$v_y - \nu v = E^{(2)} \chi_y - \nu^2 \chi \log R \quad \text{on } F. \tag{2.80}$$

One can verify directly that the function

$$v_1 = y\left[E^{(2)} \chi_y - \nu^2 \chi \log(R - y)\right]$$

belongs to $H^2(W_c)$ for any $c < \infty$, has a compact support on F, and satisfies (2.80) there. Moreover, the homogeneous Neumann condition on B and the radiation condition hold. Hence, if we seek v in the form of $v_1 + u$, then we arrive at the auxiliary problem for u as in the case $j = 1$. The existence and uniqueness of u is thus established in the previous section. Therefore, the theorem is proved for $j = 2$.

The case of a bottom source can be treated in exactly the same way as the above two cases. The proof is complete.

2.3.3.3. The Two-Dimensional Case

We denote by W a water domain whose boundary consists of the x axis (the free surface F) and of a C^1 curve B (the bottom) at a finite distance from F. We also assume that B coincides with $\{y = -d\}$ at infinity.

As in Subsection 2.3.3.2, the existence theorem for the two-dimensional water-wave problem is an immediate corollary of the fact that the velocity potentials do exist for sources placed either in water, on the free surface, or on the bottom. The unique solvability of boundary value problems whose solutions are these potentials can be demonstrated in the same way as in three dimensions. As above, one has to use the uniqueness result from Section 2.2 and to apply the assumption that W is obtained from

$$L = \{-\infty < x < \infty, -d < y < 0\}$$

by a one-to-one C^2 mapping $\psi : \bar{L} \to \bar{W}$ that has the property that $\psi(x, 0) \in F$ for every x and is the identity mapping for $|x| > a$ where $a = \text{const} < \infty$.

We leave the details of considerations to the reader, noting only that the auxiliary functions $G^{(j)}_{(\xi,\eta)}$, $j = 1, 2, 3$ used in Subsection 2.3.3.2 should be replaced by the following ones:

$$E^{(j)} = -\log R, \quad j = 1, 3 \text{ for } (\xi, \eta) \in W \cup B,$$

$$E^{(2)} = -\log R - y \log R - (x - \xi) \arctan \frac{y}{x - \xi} - \frac{\pi}{2}|x - \xi| \quad \text{for } (\xi, \eta) \in F,$$

where $R = [(x - \xi)^2 + (y - \eta)^2]^{1/2}$.

We conclude the present section with the formulation of the theorem on unique solvability in two dimensions.

If the inequality $\mathbf{x} \cdot \mathbf{n} \leq 0$ *holds on* B *that is sufficiently smooth, then the two-dimensional water-wave problem has a unique solution for any disturbance having a compact support and applied to the free surface or bottom.*

2.4. Bibliographical Notes

2.1.1. References are given in the text.

2.1.2. Considerations in Subsection 2.1.2.1 are based on two observations. First, Kochin [142, 143] discovered that the integral equation (2.7) and hence the water-wave problem are solvable for sufficiently large ($m = 2, 3$) and sufficiently small ($m = 3$) values of ν. Second, Vainberg and Maz'ya noted that T_ν depends on ν analytically, and that combining the invertibility theorem with the solvability of the integral equation for some values of the parameter one obtains that it is solvable for all $\nu > 0$, except possibly for a set of isolated points.

In Subsection 2.1.2.2, we follow the standard method, which can be found in the classical books of Kellogg [136], Mihlin [246], and Vladimirov [348], where it is applied to the exterior Neumann problem for the Laplace equation in $\mathbb{R}^m$ ($m \geq 2$). Of more recent works, those by Colton and Kress [40] and Maz'ya [224] could be recommended.

The material in Subsection 2.1.2.3 was not published earlier.

2.1.3. In Subsections 2.1.3.1–2.1.3.4 the approach of Subsection 2.1.2.1 is combined with techniques developed by Carleman [36], who considered Dirichlet and Neumann problems for the Laplace equation. Unfortunately, this work is not widely known and here we present one of Carleman's results. His method is also worked out for the three-dimensional case when S is a surface with disjoint closed edges, such that at no edge-point do tangent semi-planes to S coincide. It is possible to consider more general irregular curves and surfaces. In two dimensions the corresponding integral equation method was developed by Radon [295] almost contemporaneously with Carleman. During the 1960s and 1970s it was generalized to three and more dimensions. These results are given in the books by Burago and Maz'ya [33] and by Král [151]. Later on, the method of integral equations was developed for domains with Lipschitz boundaries (see references in Chapter 2 of Kenig's book [137]).

The results presented in Subsection 2.1.3.5 are based on the general theory of elliptic boundary value problems in domains with a piecewise smooth boundary. This theory originates from the paper [145] by Kondratyev published in 1967. The present state of the art and further references one can find in the books by Kozlov et al. [149, 150] and Nazarov and Plamenevsky [260]. In particular, the asymptotic formulae (2.19), (2.20), (2.22), and (2.23) can be justified with the help of techniques developed there. The general method for finding coefficients in asymptotics was proposed by Maz'ya and Plamenevsky [225]. Asymptotic formulae for solutions of boundary integral equations were deduced from the corresponding corner asymptotics for solutions of boundary value problems by Zargaryan and Maz'ya [367].

2.2.1. The theorem on finiteness of kinetic and potential energy for waves described by a homogeneous problem was proved by Vainberg and Maz'ya [347] in the case of bottom obstruction. Maz'ya [222] proved the theorem for submerged bodies.

2.2.2. The uniqueness theorem is from Vainberg and Maz'ya [347].

2.2.3. The auxiliary identities (2.31) and (2.33) in three and two dimensions, respectively, were proposed by Maz'ya [222] (see also Maz'ya [223]).

2.2.4. Conditions (2.37) and (2.39) providing the unique solvability in three and two dimensions, respectively, were obtained by Maz'ya [222] (see also Maz'ya [223]). Examples at the end of Subsection 2.2.4.2 are borrowed from Hulme's paper [120]. Earlier the unique solvability for all ν was established by Ursell [322] for a submerged circular cylinder by using the method of multipole expansions. The same technique was applied by Livshits [210] for proving unique solvability in the axisymmetric problem for a submerged sphere.

2.2.5. Example 1 in Subsection 2.2.5.1, its three-dimensional counterpart in Subsection 2.2.5.2, and all examples in Subsections 2.2.5.3 and 2.2.5.4 were considered by Kuznetsov [155]. Example 2 in Subsection 2.2.5.1 is borrowed from Weck's paper [351].

A survey of uniqueness results given by Simon and Ursell [307] contains a reference to an unpublished manuscript, *The problem of uniqueness in the theory of small amplitude surface waves* by Kershaw (1983). Using the integral equation technique, he shows that the uniqueness theorem holds for a strictly convex body provided it is submerged deeply enough.

Earlier, Kreisel [152] discussed the two-dimensional water-wave problem in a layer of variable depth. He proved the unique solvability theorem for this problem under the assumption that the layer is obtained from a uniform strip by means of a conformal mapping that is close to the identity in a certain sense. Kreisel's work was extended by Fitz-Gerald and Grimshaw [88] who, in particular, treated the case of a layer having different asymptotic depths at $x = \pm\infty$.

Other results obtained for two-dimensional layers of variable depth involve restrictions either on the range of frequency or on the layer depth. Uniqueness is established by Fitz-Gerald [87] for a general bottom topography when the frequency is sufficiently high or low, and for a given frequency when the horizontal length scale of depth variations is large in a certain sense. In contrast, the result of McIver [231] (see Subsection 3.2.3.2 for details) provides uniqueness for arbitrary bottom obstructions and frequencies satisfying the inequality $\nu d_{\max} \leq 1$, where $d_{\max}$ stands for the maximum depth of the layer.

2.3.2 and 2.3.3. All results presented in these subsections are from Vainberg and Maz'ya [347].

Other results. Along with integral equations of the second kind, there are other techniques allowing us to solve the water-wave problem for submerged bodies. The use of multipoles should be mentioned first, and it is Ursell who pioneered in applying this method and obtained with its help many interesting results (see [322] and [337]). This method has also proved to be useful in channel problems (see a survey by Linton [202] and the work [364] by Wu). Another useful tool is the method of singular integral equations, which is particularly convenient for the solution of the water-wave problems involving plates, barriers, and other thin obstacles (see papers by Evans [66, 72], Chakrabarti [37], and Mandal and Banerjea [213]). A first-kind integral equation and a variational approach were used by Staziker, Porter, and Stirling [311] for the investigation of two-dimensional water-wave scattering by a local elevation of arbitrary shape in an otherwise horizontal bottom.

An approach to the water-wave problem as a weakly formulated problem was developed by Lenoir and Jami [188] and by Doppel and Schomburg [58], who considered the case of a totally submerged body. Another approach to the problem in the framework of functional analysis is outlined in the brief note [112] by Hazard.

Uniqueness theorems for the water-wave problem are used in a series of papers by Angell, Hsiao, and Kleinman [12, 14–16]. In these works, the authors investigate the problem of finding the shape of a smooth submerged body so that a certain quantity, which is a domain functional (for example, added mass), must be optimized.

In the case of waves in water of variable depth, a useful tool is the ray method reviewed by Shen [303].

3
Semisubmerged Bodies, I

In the present chapter, the first of two chapters dealing with surface-piercing bodies, we impose an essential restriction that no bounded part of the free surface is separated from infinity. For the three-dimentional problem, this means that the free surface is a connected two-dimensional region (possibly multiply connected). In two dimensions, the assumption requires that there is only one surface-piercing body. However, a finite number of totally submerged bodies might be present in both cases. Supplementing this general restriction by one condition of technical nature or another, a method was developed (essentially by John) for proving the uniqueness theorem for various geometries and all values of $\nu > 0$ (see Section 3.2). Provided the uniqueness is established, the machinery of integral equations developed in Section 3.1 leads to the unique solvability of the water-wave problem. Without the assumption about uniqueness, the integral equations method possibly does not guaranee the solvability for a certain sequence of values tending to infinity. Moreover, application of integral equations is rather tricky for semisubmerged bodies even when the uniqueness holds because of so-called *irregular frequencies*, which are also investigated in Section 3.1.

3.1. Integral Equations for Surface-Piercing Bodies

The essential point in application of the integral equation techniques to the case of a surface-piercing body is that the wetted boundary S is not a closed surface (contour) in three (two) dimensions, and it is bounded by a curve (a finite set of points) along the body's intersection with the free surface. This separating set will be referred to as the *water-line* and is denoted as ∂S. The presence of ∂S results in the fact that even for smooth surface-piercing bodies the operators arising in integral equations are not compact in the general case. This is related to the fact that the velocity field has singularities near the water-line that are similar to those arising near vertices of a piecewise smooth totally submerged body (see Subsection 2.1.3.5). Noncompactness of integral operators follows from the behavior of three- and two-dimensional Green's functions near the free surface (see Subsections 1.1.1.4, 1.1.2.2 and 1.2.1), and it manifests itself in the fact that the principal part of the integral operator can be interpreted as an integral over a closed surface (curve) formed by the union of S and its mirror reflection in $\{y = 0\}$. The only situation leading to equations with compact operators is that when the wetted body's boundary and its reflection in $\{y = 0\}$ form a smooth surface (curve). Nevertheless, the noncompactness of integral operators can be overcome in the same manner as in Subsection 2.1.3, that is, seeking the unknown source density in a class of functions unbounded in a vicinity of the water-line (see Subsection 3.1.1.1).

Another difficulty occurring for surface-piercing bodies is connected with the irregular frequencies that are the discrete values of the parameter ν such that the integral equation arising from a source distribution is not uniquely solvable despite the fact that the boundary value problem has at most one solution. The existence of these frequencies was recognized by John [126]. He was forced to adopt a more complicated representation for solution of the water-wave problem, leading to a uniquely solvable system of integro-algebraic equations for these anomalous values of ν. His method explicitly involves the eigenfunctions of the integral operator. The question of deriving integral equations less complicated than those of John and free of irregular frequencies has attracted much attention for decades, because such equations are important for numerical calculations.

The irregular frequencies are considered in Subsection 3.1.1.2, where we demonstrate that their nature is the same as in the exterior acoustical problems. In order to avoid the irregular frequencies, one has to modify the solution's representation; this can be done in many ways. We consider one of them in detail in Subsection 3.1.1.3, and we give a survey of other approaches in Subsection 3.1.2.

3.1.1. Integral Equation and Irregular Frequencies

In this subsection we use the single-layer potential introduced in Subsection 2.1.1 for a totally submerged body. Since many properties of the potential and the corresponding integral operators described in Section 2.1 still hold in the present situation, we do not derive them again but rewrite the formulae, where necessary. Thus it is reasonable to refresh the material of Section 2.1 before the present subsection is read.

Unless the contrary is stated, we treat two- and three-dimensional problems simultaneously, denoting the dimension of W by $m = 2, 3$. For the sake of simplicity we assume that water has a constant depth and there is only one partially immersed body. Let the wetted part S of its boundary belong to the class C^2, and let S be transversal to $\{y = 0\}$ at every point $(x, 0) \in \partial S$. In other words, this means that

$$0 \leq \beta_0 < \pi, \quad \text{where } \beta_0 = \max\{|\pi - 2\beta(x, 0)| : (x, 0) \in \partial S\}, \qquad (3.1)$$

where $\beta(x, 0)$ is the (dihedral) angle between the tangent to S at $(x, 0)$ and $\{y = 0\}$, and this angle is directed into W.

Since the velocity field may have singularities near the water-line, we recall that we consider only solutions having locally finite energy; that is, (2.14) must hold. The last condition guarantees that for any solution u and a certain positive constant δ the estimate holds:

$$|\nabla u(x, y)| = O(r^{\delta - 1}) \quad \text{as } r \to 0,$$

where r is the distance from $(x, y) \in W$ to ∂S. This estimate, following from the results obtained in the theory of elliptic boundary value problems for domains with a nonsmooth boundary (see, for example, the works of Kondratyev [145], Kozlov et al. [149], and Nazarov and Plamenevsky [260]), justifies integration by parts and application of Green's formula.

3.1.1.1. Application of the Invertibility Theorem

Assumption (3.1) allows us to define the space $C_\kappa(S)$ (cf. Subsection 2.1.3) consisting of continuous functions on $S \backslash \partial S$ having finite the following norm:

$$\|\mu\|_\kappa = \sup\{|y|^{1-\kappa} |\mu(x, y)| : (x, y) \in S \backslash \partial S\}, \quad 0 < \kappa < 1.$$

We recall that $C_\kappa(S)$ is continuously embedded into $L^p(S)$ for $p < (1 - \kappa)^{-1}$.

As in Subsection 2.1.1, we seek a solution to the water-wave problem in the form of the single-layer potential (2.1):

$$(V\mu)(x, y) = [(m - 1)\pi]^{-1} \int_S \mu(\xi, \eta)\, G(x, y; \xi, \eta)\, \mathrm{d}S,$$

but the unknown density μ belongs to $C_\kappa(S)$ now. Nevertheless, it is obvious that properties 1–4 listed in Subsection 2.1.1.1 still hold for $V\mu$. The remaining properties 5–8 require some additional considerations. In particular, the theorem formulated in Subsection 2.1.3.1 yields that $V\mu$ is continuous throughout

$$\bar{L} = \{x \in \mathbb{R}^{m-1}, -d \leq y \leq 0\},$$

where $d \leq \infty$, and d denotes the water depth.

Formula (1.16) established in Subsection 1.1.1.4 for $d = \infty$ and $d < \infty$, as well as the analogous formulae for $m = 2$ show that

$$(m-1)\pi(V\mu)(x, y) = \int_{S \cup S_0} \mu(\xi, \eta)\, E(x, y; \xi, \eta)\, \mathrm{d}S + \int_S \mu(\xi, \eta)\, G_0(x, y; \xi, \eta)\, \mathrm{d}S. \tag{3.2}$$

Here S_0 is the mirror reflection of S in $\{y = 0\}$, $\mu(x, y)$ is extended to S_0 as an even function with respect to y, and $E(x, y; \xi, \eta)$ is defined by (2.6). Since $|\nabla G_0| = O(R_0^{-1})$, the theorem formulated in Subsection 2.1.3.1 guarantees that the second term in (3.2) is a continuously differentiable function in $\bar{L}$–(1.45) and (1.49) give the analogous assertion for $m = 2$. Furthermore, a regular normal derivative of $V\mu$ exists on every compact subset of $S \setminus \partial S$ and is given by (2.3), implying (2.4) as well.

Assuming that $\mu \in C_\kappa(S)$, where $\kappa \in (0, 1)$ for $m = 2$ and $\kappa \in (1/4, 1)$ for $m = 3$, the theorem formulated in Subsection 2.1.3.1 can be applied to the first term in (3.2). Thus this term satisfies (2.14); that is, it has locally finite energy. Hence, the single-layer potential (2.1) with an arbitrary density μ belonging to the class described above meets all conditions of the water-wave problem except for the Neumann condition on S. The latter holds if and only if μ is a solution of the integral equation (2.7). For convenience of the reader, we recall that it has the following form:

$$-\mu(x, y) + (T\mu)(x, y) = f(x, y), \quad \text{for } (x, y) \in S \setminus \partial S,$$

where

$$(T\mu)(x, y) = \frac{1}{(m-1)\pi} \int_S \mu(\xi, \eta) \frac{\partial G}{\partial n_{(x,y)}}(x, y; \xi, \eta)\, \mathrm{d}S.$$

The theorem formulated in Subsection 2.1.3.1 also implies that T maps $\mu \in C_\kappa(S)$ to a continuous function on $S \setminus \partial S$, and the behavior of $T\mu$ near ∂S will be considered below.

From (3.2) it follows that T consists of two terms, and the main term is similar to the operator N introduced in Subsection 2.1.3.3 for a two-dimensional totally submerged body. So in the present situation, (3.1) means the same as the assumption that $\alpha_\pm$ is not equal to 0 or 2π in Subsections 2.1.3.3 and 2.1.3.4, and the same considerations show that the Fredholm alternative holds for (2.7) in $C_\kappa(S)$ when

$$\kappa < \pi/(\pi + \beta_0), \tag{3.3}$$

which is another form of (2.18).

In three dimensions, similar considerations are also applicable if a surface-piercing surface S satisfies (3.1). In essence, it was done by Carleman [36], (Chapter 1, Section 4), who combined the possibility of separating the integration along ∂S with the considerations used in Subsection 2.1.3, and who demonstrated that (3.3) implies $|T| < 1$ in $C_\kappa(S)$ (it should be mentioned that Carleman did not apply the term *essential norm* explicitly). Another complication of the three-dimensional case lies in the fact that (2.14) guaranteeing the possibility of applying the divergence theorem, and hence Green's formulae, imposes the additional assumption on κ that must be greater than $1/4$, but it is obvious that there exist values $\kappa > 1/4$ satisfying (3.3).

Since Fredholm's alternative holds for the integral equation in an appropriate space $C_\kappa(S)$ for $m = 2, 3$, we can investigate the solvability of this equation. First, we do not assume that a solution to the water-wave problem is unique. To be specific we suppose that $m = 3$ and $d = \infty$. Then we can apply the considerations from Subsection 2.1.2.1, but a reference to the invertibility theorem formulated in Subsection 2.1.3.2 should be made because the integral equation (2.7) is considered in $C_\kappa(S)$ instead of $L^2(S)$.

As at the end of Subsection 2.1.2.1, we consider (2.7) with T replaced by T_0 and μ, f extended to S_0 as even functions of y. The latter equation corresponds to the exterior Neumann problem in the domain external to $S \cup S_0$, and so it is uniquely solvable as was shown by Carleman [36], Chapter 1, Section 4. By (1.30) and (1.33) the norm of $T - T_0$ is small, when ν is sufficiently close to zero. In order to show this, we have to split $T - T_0$ into a sum of two integral operators, multiplying the integrand by

$$\alpha(\xi, \eta) + [1 - \alpha(\xi, \eta)],$$

where $\alpha(\xi, \eta)$ is equal to one for $|y| + |\eta| < \varepsilon$ and vanishes elsewhere. According to (1.33), the norm of the operator with the kernel multiplied by α can be made small by the appropriate choice of ε. Then after applying (1.30), we make the norm of the second operator also small for sufficiently small ν. Now the invertibility theorem leads to the following result.

For all $\nu > 0$ except possibly for a sequence of values tending to infinity, the integral equation (2.7) *is uniquely solvable in $C_\kappa(S)$, where κ satisfies* (3.3).

An immediate corollary is the solvability theorem for the water-wave problem.

For all $\nu > 0$ except possibly for a sequence tending to infinity, the water-wave problem with arbitrary $f \in C(S)$ has a solution of the form (2.1), *where μ satisfies* (2.7).

Now the question of existence of the exceptional values arises. There are two different reasons for their existence:

1. If the homogeneous water-wave problem has nontrivial solutions, then usually some condition should be imposed on f for solvability of the nonhomogeneous problem, and hence for solvability of the integral equation.
2. If the homogeneous integral equation has nontrivial solutions, then the nonhomogeneous integral equation is not solvable for all right-hand terms.

Thus the lack of uniqueness for a certain value of ν for both the water-wave problem and in the integral equation (2.7) means that this particular value of ν is exceptional. For surface-piercing bodies both types of non-uniqueness do actually exist. The non-uniqueness in the water-wave problem will be considered in Chapter 4. In the present section we proceed with the investigation of the second type of non-uniqueness, which is usually referred to as the problem of irregular frequencies and arises despite the fact that the uniqueness theorem holds for the boundary value problem. In order to separate the question of irregular frequencies form that of non-uniqueness in the boundary value problem, we assume in the rest of this subsection that the water-wave problem has at most one solution (see Sections 3.2 and 4.2 for conditions guaranteeing this property).

3.1.1.2. Irregular Frequencies

As in the previous subsection we treat two- and three-dimensional cases ($m = 2, 3$) simultaneously, and water can have finite or infinite depth. We recall that $L = \{x \in \mathbb{R}^{m-1},\ d < y < 0\}$, where $d \leq \infty$.

Let μ_0 be a nontrivial solution of the homogeneous integral equation (2.12). The corresponding single-layer potential $V\mu_0$ vanishes identically in $\bar{W}$, since we assumed that the water-wave problem has at most one solution, and the properties 1–4 hold for $V\mu_0$ defined by (2.1); see Subsection 2.1.1.1. Then $V\mu_0$ must be nontrivial in $L\backslash\bar{W}$ because otherwise (2.4) contradicts the assumption that μ_0 is a nontrivial function on S, and so property 5 implies the following assertion.

If ν is an irregular value, then the following boundary value problem,

$$\nabla^2 v = 0 \text{ in } L\backslash\bar{W}, \qquad v_y - \nu v = 0 \quad \text{on } \{y = 0\}\backslash\bar{F}, \qquad v = 0 \quad \text{on } S, \tag{3.4}$$

has a nontrivial solution.

The existence of a nontrivial solution to (3.4) means that ν belongs to the point spectrum of the so-called Dirichlet–Neumann operator D_N. It maps φ given on $\{y = 0\}\backslash\bar{F}$ into $D_N\varphi = v_y|_{y=0}$, where v should be found from the following Dirichlet problem:

$$\nabla^2 v = 0 \text{ in } L\backslash\bar{W}, \qquad v = \varphi \quad \text{on } \{y = 0\}\backslash\bar{F}, \qquad v = 0 \quad \text{on } S, \quad (3.5)$$

which is uniquely solvable. Furthermore, D_N is a positive self-adjoint operator in $L^2(\{y = 0\}\backslash\bar{F})$ – the proof is given, for example, in Aubin's book [10], Chapter 7, Section 1. An operator of the same type arises in the sloshing problem (see, for example, Fox and Kuttler [89], p. 671), but the boundary value problem for sloshing has the Neumann condition on S; that is, it is a mixed problem. Some authors (for example, Kopachevskiy, Krein, and Ngo Zuy Can [146], Chapter 3, Section 3, and Moiseev [249], Chapter 1, Section 1) use the inverse Neumann–Dirichlet operator in the sloshing problem. The advantage of the inverse operator follows from its *compactness*, and it is usually referred to as the operator of kinetic energy in the sloshing problem. In the present case $D_N{}^{-1}$ is also a compact operator, but no physical meaning can be attributed to it.

Now, let us show that *a given value* ν *is irregular if and only if* (3.4) *has a nontrivial solution for it.*

We have to demonstrate only the first assertion. Let v be a nontrivial solution of (3.4) for a certain ν. Then for $(x, y) \in L\backslash\bar{W}$ we have

$$v(x, y) = \frac{1}{2(m-1)\pi} \int_{\partial(L\backslash\bar{W})} \left[\frac{\partial v}{\partial n_{(\xi,\eta)}} G - v(\xi, \eta) \frac{\partial G}{\partial n_{(\xi,\eta)}} \right] \mathrm{d}S,$$

where $G(x, y; \xi, \eta)$ is Green's function for the layer L. The integral over "the free surface" $\{y = 0\}\backslash\bar{F}$ vanishes because of the boundary conditions imposed on v and G. The same is true for the second term in the integral over S. So we have the following for $(x, y) \in L\backslash\bar{W}$:

$$v(x, y) = \frac{1}{2(m-1)\pi} \int_S \frac{\partial v}{\partial n_{(\xi,\eta)}} G(x, y; \xi, \eta) \, \mathrm{d}S. \qquad (3.6)$$

Using (2.3) we find that $\partial v/\partial n_{(\xi,\eta)}$ is a solution to the homogeneous equation (2.12), which looks as follows:

$$-\frac{\partial v}{\partial n} + T\left(\frac{\partial v}{\partial n}\right) = 0.$$

For demonstrating that $\partial v/\partial n_{(\xi,\eta)}$ is nontrivial, we note that if the contrary is true then v vanishes identically in $L\backslash\bar{W}$ by (3.6), but $v = 0$ contradicts the assumption made. Thus (2.12) has a nontrivial solution, and so the value ν is irregular. The proof is complete.

Furthermore, the boundary value problem (3.4) is equivalent to the following operator equation:

$$D_N v = \nu v.$$

Since the Dirichlet–Neumann operator is positive, self-adjoint, and has a compact inverse operator, the set of irregular values of ν is a sequence

$$0 < \nu_1 \leq \nu_2 \leq \cdots \leq \nu_j \leq \cdots \quad \text{such that } \nu_j \to \infty.$$

The elements of this sequence can be found by using the variational procedure (see, for example, Birman and Solomyak [27], Chapter 10, Section 1). Since Green's formula for v satisfying (3.5) can be written as follows,

$$\int_{\{y=0\}\setminus\bar{F}} v D_N v \, \mathrm{d}x = \int_{L\setminus\bar{W}} |\nabla v|^2 \, \mathrm{d}x\mathrm{d}y,$$

the variational quotient for the principal eigenvalue takes the following form:

$$\nu_1 = \min_{H^1_S(L\setminus\bar{W})} \int_{L\setminus\bar{W}} |\nabla v|^2 \, \mathrm{d}x\mathrm{d}y \bigg/ \int_{\{y=0\}\setminus\bar{F}} |v|^2 \, \mathrm{d}x. \tag{3.7}$$

By $H^1_S(L\setminus\bar{W})$ we denote the subspace in $L^2(L\setminus\bar{W})$ consisting of functions that vanish on S and have all first derivatives in $L^2(L\setminus\bar{W})$. For ν_2 to be obtained, the same minimum should be taken over the subspace of $H^1_S(L\setminus\bar{W})$ orthogonal to the first eigenfunction v_1. Here orthogonality is understood with respect to the inner product in $L^2(\{y=0\}\setminus\bar{F})$. Further eigenvalues can be obtained in the same way.

Let us find a lower bound for ν_1. Since $v \in H^1_S(L\setminus\bar{W})$ vanishes on S, we have

$$v(x,0) = \int_{-b(x)}^{0} v_y(x,y)\, \mathrm{d}y$$

for almost every $x \in \{y=0\}\setminus\bar{F}$. Here $(x, -b(x))$ denotes the first common point of S and the vertical line through $(x,0)$. Then the Schwarz inequality gives

$$\int_{\{y=0\}\setminus\bar{F}} |v(x,0)|^2 \, \mathrm{d}x \leq h \int_{L\setminus\bar{W}} |v_y|^2 \, \mathrm{d}x\mathrm{d}y \leq h \int_{L\setminus\bar{W}} |\nabla v|^2 \, \mathrm{d}x\mathrm{d}y,$$

where $h = \sup_{x\in\{y=0\}\setminus\bar{F}}\{b(x)\}$. Comparing the last inequality with (3.7), we get that $\nu_1 \geq h^{-1}$. Hence *irregular frequencies* $\omega_j = (g\nu_j)^{1/2}$ *are not smaller than* $(g/h)^{1/2}$.

3.1.1.3. Uniquely Solvable System of Two Integral Equations

It was demonstrated in Subsection 3.1.1.2 that irregular values of ν actually exist for the integral equation (2.7) because the Dirichlet–Neumann operator has a discrete spectrum for the interior boundary value problem. From the acoustic diffraction theory, it is well known that irregular frequencies arise because of an improper representation of solution. However, other representations might lead to a uniquely solvable integral equation or a system of equations. Here we consider a simple approach that allows us to avoid irregular frequencies in the water-wave problem.

Let us seek a solution in the following form:

$$u(x, y) = (V\mu)(x, y) + (U\rho)(x, y), \tag{3.8}$$

where $V\mu$ is defined by (2.1) and

$$(U\rho)(x, y) = \frac{1}{2(m-1)\pi} \int_{L\setminus\bar{W}} \rho(\xi, \eta)\, G(x, y; \xi, \eta)\, \mathrm{d}\xi \mathrm{d}\eta$$

is the volume (or area, in two dimensions) Green's potential with a density ρ that is continuous in $\overline{L\setminus W}$. By the theorem formulated in Subsection 2.1.3.1, $U\rho$ is a continuously differentiable function throughout $\bar{L}$. Furthermore, it is a harmonic function in W and satisfies the Poisson equation

$$-\nabla^2(U\rho) = \rho \quad \text{in } L\setminus\bar{W} \tag{3.9}$$

in the sense of the distribution theory (see Sanchez-Palencia's book [302], Chapter 15, Section 4).

For an arbitrary density $\mu \in C_\kappa(S)$ (here $1/4 < \kappa < 1$ for $m = 3$) and $\rho \in C(\overline{L\setminus W})$, the function given by (3.8) meets all conditions of the water-wave problem except for the Neumann condition on S. The latter is satisfied if and only if

$$-\mu(x, y) + (T\mu)(x, y) + (Q\rho)(x, y) = f(x, y), \quad (x, y) \in S\setminus\partial S. \tag{3.10}$$

Here we take into account (2.3) and the differentiability of $U\rho$, and by Q we denote the intergal operator

$$(Q\rho)(x, y) = \frac{1}{2(m-1)\pi} \int_{L\setminus\bar{W}} \rho(\xi, \eta) \frac{\partial G}{\partial n_{(x,y)}}(x, y; \xi, \eta)\, \mathrm{d}\xi \mathrm{d}\eta.$$

However, equation (3.10) in $C_\kappa(S)$ does not allow us to determine two unknown functions uniquely, and so we complement it by requiring that

$$-\nabla^2 u = iu \quad \text{in } L\setminus\bar{W}. \tag{3.11}$$

In view of harmonicity of $V\mu$ in $L\backslash\bar{W}$ and (3.9), we obtain from (3.11) that

$$-\rho(x, y) + i(V\mu)(x, y) + i(U\rho)(x, y) = 0, \quad (x, y) \in L\backslash\bar{W}. \quad (3.12)$$

Equations (3.10) and (3.12) form a system for determining μ and ρ.

The theorem formulated in Subsection 2.1.3.1 guarantees that except for T all operators in this system are compact operators in appropriate pairs of spaces, namely (a) Q is is a compact operator from $C(\overline{L\backslash W})$ to $C(S)$ and hence to $C_\kappa(S)$; (b) iV is a compact operator from $C_\kappa(S)$ to $C(\overline{L\backslash W})$; and (c) iU is a compact operator from $C(\overline{L\backslash W})$ to $C(\overline{L\backslash W})$. Moreover, we have that $\|T\| < 1$ in $C_\kappa(S)$, when (3.3) holds. Hence the matrix operator of system (3.10) and (3.12) has the essential norm less than one in $C_\kappa(S) \times C(\overline{L\backslash W})$ when (3.3) holds. Thus the following theorem is proved.

If (3.3) *holds, then the Fredholm alternative is true for system* (3.10) *and* (3.12) *in* $C_\kappa(S) \times C(\overline{L\backslash W})$.

According to Fredholm's alternative system (3.10), (3.12) is uniquely solvable for arbitrary right-hand terms when the corresponding homogeneous system has only a trivial solution. In order to prove this fact, let μ_0 and ρ_0 denote a solution to the homogeneous system. Substituting them into (3.8), we get a solution u_0 of the homogeneous water-wave problem. Under the assumption that the water-wave problem has at most one solution, we obtain that u_0 vanishes identically in $\bar{W}$.

Furthermore, u_0 satisfies (3.11). Multiplying this equation by $\overline{u_0}$ and integrating over $L\backslash\bar{W}$, we have

$$-\int_{L\backslash\bar{W}} \overline{u_0}\, \nabla^2 u_0 \,\mathrm{d}x\mathrm{d}y = i \int_{L\backslash\bar{W}} |u_0|^2 \,\mathrm{d}x\mathrm{d}y.$$

Without loss of generality we can take μ_0 from $C_\kappa(S)$ with $\kappa \in (1/4, 1)$ if $m = 3$. Then we can integrate by parts in the last equation, obtaining

$$\int_{L\backslash\bar{W}} |\nabla u_0|^2 \,\mathrm{d}x\mathrm{d}y - \int_{\{y=0\}\backslash\bar{F}} \overline{u_0}\, \partial u_0/\partial y \,\mathrm{d}x = i \int_{L\backslash\bar{W}} |u_0|^2 \,\mathrm{d}x\mathrm{d}y. \quad (3.13)$$

Here the boundary condition on S is taken into account.

Since u_0 is a combination of potentials, the boundary condition holding for G on $\{y = 0\}$ implies

$$\partial u_0/\partial y = \nu u_0 \quad \text{on } \{y = 0\}\backslash\bar{F}.$$

Therefore (3.13) takes the form

$$\int_{L\backslash\bar{W}} |\nabla u_0|^2 \,\mathrm{d}x\mathrm{d}y - \nu \int_{\{y=0\}\backslash\bar{F}} |u_0|^2 \,\mathrm{d}x = i \int_{L\backslash\bar{W}} |u_0|^2 \,\mathrm{d}x\mathrm{d}y.$$

Since all the integrals here are real, we have a contradiction unless u_0 vanishes identically in $\overline{L\backslash W}$. Then ρ_0 also vanishes because $u_0 = i\rho_0$ by (3.8), (3.9), and (3.11). Now $u_0 = V\mu_0$ in $L\backslash\bar{W}$, and so (2.3) gives

$$\mu_0(x, y) + (T\mu_0)(x, y) = 0, \quad (x, y) \in S\backslash\partial S.$$

Comparing this with the homogeneous equation (3.10), where $\rho_0 = 0$, we find that $\mu_0 = 0$. Thus the following theorem is proved.

If the water-wave problem has no more than one solution for a surface-piercing body, then the solution can be obtained in the form of (3.8), *where* μ *and* ρ *satisfy the uniquely solvable system* (3.10) *and* (3.12).

3.1.2. Survey of Equations Without Irregular Frequencies

In 1991, Angell, Hsiao, and Kleinman [13] outlined some results (mainly obtained in the 1970s and 1980s) concerning various aspects of the water-wave problem. Among other topics, the authors gave a review of uniquely solvable (that is, having no irregular frequencies) integral equations. In the present subsection we restrict ourselves to a part of their material. However, in what follows we give more details than [13] contains.

3.1.2.1. Integral Equations Involving a Hypersingular Operator

First, we recall some formulae from Subsection 2.1.1.2 necessary in what follows. We begin with Green's representation of the velocity potential:

$$u(\xi, \eta) = \frac{1}{4\pi} \int_S \left[u(x, y) \frac{\partial G}{\partial n_{(x,y)}}(x, y; \xi, \eta) - G(x, y; \xi, \eta) \frac{\partial u}{\partial n_{(x,y)}} \right] \mathrm{d}S, \tag{3.14}$$

where $(\xi, \eta) \in W$. Letting $(\xi, \eta) \to S$, we arrive at the following integral equation:

$$-u(\xi, \eta) + \frac{1}{2\pi} \int_S u(x, y) \frac{\partial G}{\partial n_{(x,y)}}(x, y; \xi, \eta)\, \mathrm{d}S$$
$$= \frac{1}{2\pi} \int_S f(x, y) G(x, y; \xi, \eta)\, \mathrm{d}S, \quad (\xi, \eta) \in S. \tag{3.15}$$

Here the surface jump of a double-layer potential and the boundary condition on S are taken into account. One immediately notes that the integral operator on the left-hand side is $\bar{T}^*$, that is, the adjoint operator to $\bar{T}$ containing the complex conjugate Green's function $\bar{G}$ (compare with T used in Subsection 3.1.1.1). So according to Fredholm's theory, the homogeneous equation corresponding to (3.15) has nontrivial solutions for the same real values of ν as the homogeneous integral equation in Subsection 3.1.1.2. Thus (3.15) suffers the same drawback, resulting in the presence of irregular frequencies, as the integral equation of the source method.

In order to improve (3.15), we can apply the following procedure. Formally calculating the normal derivative of u in (3.14), we get

$$f(\xi, \eta) = \frac{1}{4\pi} \frac{\partial}{\partial n_{(\xi,\eta)}} \int_S u(x, y) \frac{\partial G}{\partial n_{(x,y)}} (x, y; \xi, \eta) - \frac{1}{2}[(T - I)f](\xi, \eta),$$

where I is the identity operator. Also, the boundary condition on S, (2.3), and the definition of T are used. Now a hypersingular operator $D_n u$ arising on the left-hand side, that is, the normal derivative of the double-layer potential – and for convenience, $D_n u$ is defined with the factor $(2\pi)^{-1}$ – should be defined somehow. However, formally we have

$$(D_n u)(\xi, \eta) = f(\xi, \eta) + (T f)(\xi, \eta), \quad (\xi, \eta) \in S. \tag{3.16}$$

Let λ be a certain complex number with $\operatorname{Im} \lambda \neq 0$. Multiplying (3.16) by λ and summing up the result with (3.15), we arrive at the required integral equation:

$$-u(\xi, \eta) + (\bar{T}^* u)(\xi, \eta) + \lambda(D_n u)(\xi, \eta) = \{[V + \lambda(T + I)]f\}(\xi, \eta), \tag{3.17}$$

where $(\xi, \eta) \in S$. This equation was derived by Kleinman [139], who also established the absence of irregular frequencies (that is, the uniqueness theorem) for it. Lee and Sclavounos [184], Lau and Hearn [181], and Liapis [196] developed numerical methods for solving (3.17). Their papers contain a substantial body of computational results for the added mass, damping coefficients, and exiting forces illustrating the effectiveness of numerical procedures.

Let us turn to the question of solvability for equation (3.17). It was considered by Wienert [361]. We begin with describing how D_n is understood in his approach. The idea is to regularize (3.17), reasoning similarly to the following formal regularization of (3.16). Applying Green's representation to $(\bar{T}^* u)(x, y)$ in $L \setminus \bar{W}$, one gets that

$$(V D_n u)(x, y) = (\bar{T}^{*2} - I)u(x, y) \quad (x, y) \in S.$$

Thus using V for regularization of (3.16) leads to the necessity of evaluating only the double-layer potential but not its normal derivative. However, this method gives rise to some difficulties in the case in which the water-line is an edge of $S \cup S_0$, and S_0 is the mirror reflection of S by $\{y = 0\}$. In order to overcome these difficulties, another operator V_+ was introduced in [361]. It involves Green's function with a suitable ν_+ replacing ν (we shall denote

Green's function in V_+ by G_+ instead of G). A combination of Green's representation for G_+ in W with the procedure used for derivation of the last equality gives

$$(V_+ D_n u)(x, y) = \{[(\bar{T}^*_+ + I)(\bar{T}^* - I) + (\nu_+ - \nu)V_+^{(i)} K]u\}(x, y),$$

where $(x, y) \in S$ and

$$(V_+^{(i)}\phi)(x, y) = \frac{1}{2\pi}\int_{\{y=0\}\setminus\bar{F}} \phi(\xi, \eta) G_+(x, y; \xi, \eta)\, \mathrm{d}\xi \mathrm{d}\eta \quad (x, y) \in S,$$

$$(K\phi)(\xi, \eta) = \frac{1}{2\pi}\int_S \phi(x, y)\frac{\partial G}{\partial n_{(x,y)}}(x, y; \xi, \eta)\, \mathrm{d}S \quad (\xi, \eta) \in \{y = 0\}\setminus\bar{F}.$$

On the basis of this formal regularizing procedure, an integral equation was rigorously derived in [361]. For this purpose one has to use (3.14) and a similar equality for the interior domain, which looks as follows:

$$(Ku)(\xi, \eta) = (Vf)(\xi, \eta), \quad (\xi, \eta) \in \{y = 0\}\setminus\bar{F}. \tag{3.18}$$

Let λ be a pure imaginary number with $\mathrm{Im}\,\lambda > 0$. Applying operators

$$[I + \lambda V_+(\bar{T}^*_+ + I)] \quad \text{and}\ \lambda(\nu_+ - \nu)V_+ V_+^{(i)}$$

to (3.15) and (3.18), respectively, and summing up the results, one obtains the following integral equation:

$$\begin{aligned}(\{-I + \bar{T}^* + \lambda V_+[(\bar{T}^*_+ + I)(\bar{T}^* - I) + (\nu_+ - \nu)V_+^{(i)} K]\}u)(x, y)\\ = f_0(x, y), \quad (x, y) \in S.\end{aligned} \tag{3.19}$$

Here

$$f_0(x, y) = (\{I + \lambda V_+\{(\bar{T}^*_+ + I) + (\nu_+ - \nu)\}V_+^{(i)}\}Vf)(x, y) \tag{3.20}$$

is a continuous function on $\bar{S}$.

Under assumption that the uniqueness theorem holds for the water-wave problem, Wienert [361] proved the following three assertions: *(i) the homogeneous equation corresponding to* (3.19) *has only a trivial solution;* (*ii*) *equation* (3.19) *is solvable, uniquely by* (*i*), *and its solution is a continuous function on S;* (*iii*) *the water-wave problem is uniquely solvable, and its solution is given by* (3.14) *where the trace of u on S must be found from the integral equation* (3.19) *having* (3.20) *as the right-hand side term.*

Let us compare (3.19) with the system of two integral equations considered in Subsection 3.1.1.3. The obvious advantage of (3.19) is that, on one hand, it is a boundary integral equation, whereas the system involves integral operators over $L\backslash\bar{W}$. On the other hand, (3.19) involves rather complicated compositions of boundary integral operators given on different parts of $\partial(L\backslash\bar{W})$. This makes (3.19) too difficult for numerical calculations, and so (3.17) was used for this purpose in [181] and [184].

To conclude the present subsection, we note that apart from (3.17), Kleinman [139] derived another boundary integral equation involving the hypersingular operator D_n. It is referred to as "the layer equation" by Angell et al. [13] because the associated representation of solution to the water-wave problem has the following form:

$$u(x, y) = \frac{1}{2\pi}\int_S \mu(\xi, \eta)\left[G(x, y; \xi, \eta) + \lambda \frac{\partial G}{\partial n_{(\xi,\eta)}}(x, y; \xi, \eta)\right] \mathrm{d}S, \quad \operatorname{Im}\lambda \neq 0.$$

The corresponding equation

$$-\mu + T\mu + \lambda D_n \mu = f$$

is simply a modified form of the integral equation given in Subsection 3.1.1.1. Kleinman proved the uniqueness theorem for the layer equation, but the existence of a solution for it is still uncertain.

3.1.2.2. Integral Equations Involving the Free Surface

In [11], Angell, Hsiao, and Kleinman applied the following simplified Green's function,

$$G_d(x, y; \xi, \eta) = R^{-1} + R_d^{-1},$$

to the case of water bounded below by the flat bottom $B = \{y = -d\}$. Here $R^2 = |x - \xi|^2 + (y - \eta)^2$ and $R_d^2 = |x - \xi|^2 + (y + 2d + \eta)^2$, and so G_d satisfies the homogeneous Neumann condition on B, but the free surface boundary condition does not hold on $\{y = 0\}$.

Using G_d instead of G in the derivation of Green's representation, see (3.15), leads to a new boundary integral equation for the velocity potential u:

$$\begin{aligned} -\alpha(x, y)u(x, y) &+ \int_S u(\xi, \eta)\frac{\partial G_d}{\partial n_{(\xi,\eta)}}(x, y; \xi, \eta)\, \mathrm{d}S \\ &+ \int_F u(\xi, \eta)\left[\frac{\partial G_d}{\partial n_{(\xi,\eta)}}(x, y; \xi, \eta) + \nu G_d(x, y; \xi, \eta)\right] \mathrm{d}\xi\, \mathrm{d}\eta \\ &= \int_S f(\xi, \eta) G_d(\xi, \eta)\, \mathrm{d}S, \quad (x, y) \in \bar{S} \cup F. \end{aligned} \tag{3.21}$$

Here

$$\alpha(x, y) = \lim_{\epsilon \to 0} \int_{W \cap \partial \mathcal{B}_\epsilon(x,y)} \frac{\partial G_d}{\partial n_{(\xi,\eta)}}(x, y; \xi, \eta)\, \mathrm{d}S_{(\xi,\eta)},$$

and $\partial \mathcal{B}_\epsilon(x, y)$ denotes the sphere of radius ϵ centered at (x, y).

Let us formulate the theorem established in [11], demonstrating that (3.21) has no irregular frequencies.

Let u have the following properties: (i) satisfies the homogeneous equation (3.21) *for all* $(x, y) \in \bar{S} \cup F$; *(ii) is continuous on* $\bar{S} \cup F$; *and (iii) has the behavior* (*cf. Subsections* 1.3.2 *and* 2.2.1)

$$u(x, y) = e^{ik_0|x|}|x|^{-1/2}[\beta(\theta)\cosh k_0(y + d) + O(|x|^{-1})] \quad \text{as } |x| \to \infty.$$

Then u vanishes identically.

The important feature of this theorem is the fact that it does not require the uniqueness of solution to the water-wave problem. Comparing (3.21) with the system of two integral equations considered in Subsection 3.1.1.3, we see that (3.21) has the advatage being a boundary integral equation. However, integration over the infinite free surface F sacrifices this advantage. Originally (3.21) was derived by Bai and Yeung [18] (see also Yeung [365]), who restricted themselves to the numerical treatment of this equation.

Liu [209] derived an intgral equation similar to (3.21) in two dimensions. However, his proof of uniqueness for the integral equation is quite different from that in [11] and requires the unique solvability of the water-wave problem.

3.1.2.3. Integral Equations Involving Modified Green's Functions

The idea of modifying Green's function for obtaining an integral equation without irregular frequencies was brought to the theory of time-harmonic water waves from acoustics, where it had demonstrated its fruitfulness (see, for example, papers by Jones [128], Kleinman and Roach [140], and Ursell [333]). A modified Green's function for the two-dimensional water-wave problem was introduced by Ursell [334] as follows:

$$G_M(x, y; \xi, \eta) = G(x, y; \xi, \eta) + \sum_{|l|=1}^{\infty} a_l\, \phi_l(x, y)\phi_l(\xi, \eta). \tag{3.22}$$

Here G is the usual Green's function considered in Subsection 1.2.1, and the standard multi-index notation is used in subscripts:

$$l = (n, j); \quad n = 0, 1, 2, \ldots; \quad j = 1, 2; \quad |l| = n + j.$$

The coefficients a_l are subject to resrictions described below, and ϕ_l are the so-called multipole potentials located at the coordinate origin assumed to belong to $\{y = 0\}\setminus\bar{F}$. If water has finite depth d, then these potentials in two dimensions have the following form:

$$\phi_{01}(x, y) = \int_{\ell_-} \frac{\cosh k(y+d) \cos kx}{k \sinh kd - \nu \cosh kd} \, \mathrm{d}k, \quad \phi_{02}(x, y) = -\frac{\partial \phi_{01}}{\partial x};$$

$$\phi_{n1}(x, y) = (-1)^n \left[\frac{\sin 2n\vartheta}{r^{2n}} + \frac{\nu}{2n-1} \frac{\cos(2n-1)\vartheta}{r^{2n-1}} \right] + \frac{1}{(2n-1)!}$$

$$\times \int_{\ell_-} \frac{e^{-kd}(\nu + k) k^{2m-2} (\nu \sinh ky + k \cosh ky) \cos kx}{k \sinh kd - \nu \cosh kd} \, \mathrm{d}k,$$

$$\phi_{n2}(x, y) = -\frac{1}{2n} \frac{\partial \phi_{n1}}{\partial x}, \quad n > 0.$$

Here ℓ_- denotes the contour defined in Subsection 1.1.1.1; (r, ϑ) are the usual polar coordinates in the (x, y) plane, and $\vartheta \in [-\pi, 0]$ in the lower half-plane. Letting $d \to \infty$, one easily obtains the set of the two-dimensional multipole potentials for water of infinite depth. The latter set of $\{\phi_l\}$ was established to be complete in $L^2(S)$ by Martin [217], who supposed S to satisfy certain rather stringent smoothness conditions. Another important property of $\{\phi_l\}$ demonsrated by Ursell [334] in two dimensions says that these potentials arise in certain expansions of Green's functions.

Two essential restrictions on a_l are as follows: (i) $\operatorname{Im} a_l > 0$; and (ii) $|a_l|$ must be sufficiently small for the series in (3.22) to converge when (x, y) and (ξ, η) belong to $\bar{W}$; for example, it is sufficient to require that

$$|a_l| \leq \frac{\text{const}}{n^2 M_l^2} \quad \text{for } n > 0,$$

where $M_l = \sup_{(x,y) \in \bar{W}} |\phi_l(x, y)|$.

In three dimensions, (3.22) remains valid, but another multi-index notation should be applied:

$$l = (n, m, j); \quad n, m = 0, 1, 2, \ldots; \quad j = 1, 2; \quad |l| = n + m + j.$$

In addition, $\phi_l(x, y)$ must be defined as follows:

$$\phi_l(x, y) = \psi_{nm}(x, y) \left[j \cos m\theta + (1 - j) \sin m\theta \right].$$

Here θ is the polar angle in the x plane, and

$$\psi_{0m}(|x|, y) = \int_{\ell_-} \frac{\cosh k(y+d) J_m(k|x|)}{k \sinh kd - \nu \cosh kd} \, \mathrm{d}k,$$

$$\psi_{nm}(|x|, y) = \frac{1}{(2n)!} \int_0^\infty k^{m+2n-1} (k+\nu) e^{ky} J_m(k|x|) \, \mathrm{d}k$$
$$- \frac{1}{(2n)!} \int_{\ell_-} \frac{k^{m+1} \cosh k(y+d) J_m(k|x|)}{k \sinh kd - \nu \cosh kd} \, \mathrm{d}k, \quad n > 0.$$

Green's function (3.22) can be used instead of G in the integral equation (3.15) as well as in the integral equation of the source method (see Subsection 3.1.1.1). The corresponding uniqueness theorems follow from Ursell's [334] proof. The existence of solution for (3.15) with G_M is ensured by means of the same argument as described in Subsection 3.1.2.2, whereas the existence theorem for the integral equation based on a single-layer potential with G_M as the kernel is still an open question. Martin [219] generalized Ursell's result to the two-dimensional problem of interaction between two surface-piercing cylinders. However, this is done under the assumption that the uniqueness theorem holds for the boundary value problem (see Section 4.2 for such theorems). Generally speaking, this is not true for all geometries, as examples in Section 4.1 show.

3.1.2.4. Null-Field Equations

Similarly to the technique based on modified Green's functions, the null-field equations were originally developed for the exterior acoustical problems (see, for example, Martin's paper [216]). Then this method was extended to the theory of time-harmonic water waves in a series of works by Martin [215, 217, 218] (see also Martin and Ursell [220]). This method does not strictly lead to an integral equation, but it involves an infinite set of moment equations. Two uniquely solvable systems of null-field equations were obtained. One of them has the following form:

$$\int_S u(\xi, \eta) \frac{\partial \phi_l}{\partial n}(\xi, \eta) \, \mathrm{d}S = \int_S f(\xi, \eta) \phi_l(\xi, \eta) \, \mathrm{d}S, \quad |l| = 1, 2, \ldots,$$

and the corresponding solution to the water-wave problem is given by Green's representation:

$$u(x, y) = \frac{1}{4\pi} \int_S u(\xi, \eta) \frac{\partial G}{\partial n_{(\xi,\eta)}}(x, y; \xi, \eta) \, \mathrm{d}S + \frac{1}{2}(V f)(x, y).$$

The derivation of these null-field equations uses Green's identity for $L \setminus \bar{W}$. Another approach to null-field equations combines the boundary integral equation with Green's representation evaluated at particular points in $L \setminus \bar{W}$. However, if only a finite number of such points is used, then the uniqueness theorem cannot be established for all values of ν.

3.2. John's Theorem on the Unique Solvability and Other Related Theorems

This section is mainly concerned with uniqueness theorems for the water-wave problem. There are two reasons for this. First, the integral-equations technique developed in the previous section demonstrates that the solvability is a consequence of the uniqueness theorem for the problem. Second, the examples of non-uniqueness constructed in Chapter 4 show that one has to impose geometrical restrictions (often combined with restrictions on frequencies) in order to guarantee the uniqueness. Above all, these assumptions depend on the method of proof.

A powerful tool for proving uniqueness theorems was suggested by John [126]. It is based on investigation of the simple wave of order zero presenting the principal part of the velocity potential in the homogeneous water-wave problem. As a result one obtains an inequality between the potential and kinetic energy, which leads to a contradiction unless the solution is trivial. We present John's original theorem in Subsection 3.2.1 for the weak solution that is understood in the sense of integral identity. In Subsection 3.2.2, we give some extensions of John's method that involve weaker geometrical assumptions for the two-dimensional and axisymmetric obstacles. Further criteria of uniqueness for geometries involving surface-piercing bodies are given in Subsection 3.2.3. In particular, McIver [231] proposed another way of using the fact that the simple wave of order zero vanishes [see (3.64) in Subsection 3.2.3.2] for establishing uniqueness in the case in which the depth is sufficiently small.

Since a surface-piercing body meets the free surface transversally, the boundary of water domain, generally speaking, is not smooth. Thus the reader has to take into account the description of singular behavior of the velocity field in a vicinity of the water-line (see the introductory remarks to Subsection 3.1.1).

3.2.1. Theorem of John

To be specific, we consider in detail the three-dimensional case in which the water domain contains at least one semi-immersed body and is bounded below by a flat bottom $B = \{x \in \mathbb{R}^2, y = -d\}$. Other situations in which the same technique provides the uniqueness are described briefly.

John's method relies essentially on the following geometrical assumptions: (i) *any straight segment parallel to the* y *axis, and connecting the free surface* F *with the bottom* B, *lies within the water domain* W *except for its ends; and* (ii) *the free surface* F *is a connected subset of the plane* $\{y = 0\}$. These assumptions allow us to investigate the simple wave of order zero for a solution of the homogeneous problem. For this purpose a uniqueness theorem holding for the Helmholtz equation in the plane domain F is applied, and it gives that the simple wave vanishes identically. The latter fact is then used for deriving an inequality between the kinetic and potential energy. This proves the uniqueness when Green's identity is taken into account. Note that John's assumptions impose no restriction on the number of bounded bodies submerged totally or partially, if they are located strictly below the surface-piercing bodies.

3.2.1.1. The Uniqueness Theorem for the Helmholtz Equation

Let $w(x)$ satisfy

$$\nabla_x^2 w + k^2 w = 0 \quad \text{in } F, \tag{3.23}$$

and F is assumed to be a connected set with a finite boundary. In any annulus $a_0 \leq |x| \leq a_1$ contained in F, the function w can be expanded into the Fourier series

$$w(x) = \sum_{m=-\infty}^{m=+\infty} Z_m(k|x|)e^{im\theta},$$

$$Z_m(k|x|) = (2\pi)^{-1} \int_{-\pi}^{+\pi} w(|x| \cos\theta, |x| \sin\theta) e^{-im\theta} \, d\theta.$$

From (3.23) we have $Z_m = \alpha_m J_m + \beta_m H_m^{(1)}$, where J_m $(H_m^{(1)})$ is the Bessel (Hankel) function of order m. The Schwarz inequality gives

$$a|Z_m(ka)|^2 \leq \int_{|x|=a} |w|^2 \, ds.$$

Let

$$\lim_{a\to\infty} \int_{|x|=a} |w|^2 \, ds = 0; \tag{3.24}$$

then $\alpha_m = \beta_m = 0$, and $Z_m(ka) = 0$ $(m = 0, \pm 1, \ldots)$, because otherwise we have a contradiction with the asymptotic behavior of J_m and $H_m^{(1)}$ at infinity. So (3.24) implies that $w = 0$ in any annulus contained in F. Since w is an analytic function, $w = 0$ in the whole connected set F. Thus the following lemma is proved.

If w is a solution to (3.23) *satisfying* (3.24), *then $w = 0$ in F.*

3.2.1.2. Investigation of the Simple Wave of Order Zero

First we define a generalized solution to the homogeneous water-wave problem. Such a solution u must have locally finite energy; that is,

$$\int_{W_a} |\nabla u|^2 \, dx dy + \int_{F_a} |u|^2 \, dx < \infty \quad \text{for any } a > 0, \tag{3.25}$$

and W_a and F_a are the portions of W and F, respectively, within the cylinder $\{|x| < a\}$. Furthermore, u must satisfy the radiation condition

$$\lim_{a \to \infty} \int_{C_a} \left| u_{|x|} - i k_0 u \right|^2 dS = 0, \tag{3.26}$$

where $C_a = W \cap \{|x| = a\}$, and k_0 is a unique positive root of $k \tanh kd = \nu$. At last, the following integral identity,

$$\int_W \nabla u \cdot \nabla \bar{v} \, dx dy = \nu \int_F u \bar{v} \, dx, \tag{3.27}$$

must hold for any $v \in H^1(W)$ with a compact support in $\bar{W}$; as usual, $H^1(W)$ denotes the Sobolev space.

Let us demonstrate that for u satisfying (3.25)–(3.27), the equality

$$\operatorname{Im} \int_{C_a} u_{|x|} \, \bar{u} \, dS = 0 \tag{3.28}$$

holds for sufficiently large values of a.

By $\zeta_a(|x|)$ we denote a function belonging to $C^\infty([0, \infty))$ and such that it is equal to one for $0 \leq |x| \leq a$ and to zero for $|x| \geq a + 1$. Let a be so large that S lies within the cylinder $\{|x| < a\}$. Putting $v(x, y) = u(x, y)\zeta_a(|x|)$, we get from (3.27)

$$\int_W |\nabla u|^2 \zeta_a \, dx dy + \int_{W \setminus W_a} \bar{u} \nabla u \cdot \nabla \zeta_a \, dx dy = \nu \int_F |u|^2 \zeta_a \, dx. \tag{3.29}$$

It is well known (see, for example, Chapter 8 in the book [94] by Gilbarg and Trudinger), that the generalized solution u satisfies the Laplace equation in W, and the following boundary conditions

$$u_y - \nu u = 0 \quad \text{on } F, \qquad u_y = 0 \quad \text{on } B$$

hold pointwise. Hence one can integrate by parts in the second integral in (3.29), obtaining

$$-\int_{W\setminus W_a} |\nabla u|^2 \zeta_a \,\mathrm{d}x\mathrm{d}y + \int_{F\setminus F_a} |u|^2 \zeta_a \,\mathrm{d}x - \int_{\mathcal{C}_a} u_{|x|}\,\bar{u}\,\mathrm{d}S.$$

Substituting this into (3.29), we get

$$\int_{W_a} |\nabla u|^2 \,\mathrm{d}x\mathrm{d}y - \nu \int_{F_a} |u|^2 \,\mathrm{d}x = \int_{\mathcal{C}_a} u_{|x|}\,\bar{u}\,\mathrm{d}S, \tag{3.30}$$

which immediately yields (3.28).

Assumption (i) allows us to consider

$$w(x) = \int_{-d}^{0} u(x, y) \cosh k_0(y + d)\,\mathrm{d}y \quad \text{for } x \in F,$$

where u is a generalized solution to the homogeneous water-wave problem. So $w(x)\cosh k_0(y + d)$ is the simple wave of order zero that corresponds to u (see Subsection 1.3.2 for the definition). The Laplace equation for u implies that

$$\nabla_x^2 w = -\int_{-d}^{0} u_{yy} \cosh k_0(y + d)\,\mathrm{d}y.$$

Integrating by parts twice on the right-hand side, we obtain

$$\nabla_x^2 w = u(x, 0) k_0 \sinh k_0 d - u_y(x, 0) \cosh k_0 d - k_0^2 w(x) = -k_0^2 w(x).$$

Here the definition of k_0 and the boundary condition for u on F are taken into account. So w is a solution to (3.23).

The lemma proven in Subsection 3.2.1.1 gives that $w = 0$ in F when conditions (i) and (ii) hold and (3.24) is true. In order to verify (3.24) we apply the Schwarz inequality:

$$\begin{aligned}|w|^2 &\leq \left(\int_{-d}^{0} |u|^2 \,\mathrm{d}y\right)\left[\int_{-d}^{0} \cosh^2 k_0(y + d)\,\mathrm{d}y\right] \\ &= (2k_0)^{-1}\left(k_0 d + \sinh k_0 d \cosh k_0 d\right) \int_{-d}^{0} |u|^2 \,\mathrm{d}y.\end{aligned}$$

Integrating this over $\{|x| = a\}$, we get

$$\int_{|x|=a} |w|^2 \,\mathrm{d}s \leq (2k_0)^{-1}(k_0 d + \sinh k_0 d \cosh k_0 d) \int_{\mathcal{C}_a} |u|^2 \,\mathrm{d}S.$$

The last integral tends to zero as $a \to \infty$ in view of (3.26), (3.28), and

$$2k_0 \mathrm{Im} \int_{C_a} u_{|x|}\, \bar{u}\, \mathrm{d}S = \int_{C_a} \left(\left|u_{|x|}\right|^2 + k_0^2 |u|^2 \right) \mathrm{d}S - \int_{C_a} \left|u_{|x|} - i k_0 u\right|^2 \mathrm{d}S.$$

Thus the following assertion is proved.

Let $u(x, y)$ be a generalized solution to the homogeneous water-wave problem. If geometrical conditions (i) and (ii) hold, then the simple wave of order zero corresponding to u vanishes identically in F.

3.2.1.3. John's Theorem on the Unique Solvability

Let us integrate by parts in the integral defining w. From the last assertion and the boundary condition on B we have

$$u(x, 0) \sinh(k_0 d) = \int_{-d}^{0} u_y(x, y) \sinh k_0 (y + d)\, \mathrm{d}y.$$

The Schwarz inequality gives

$$\begin{aligned} |u(x, 0) \sinh(k_0 d)|^2 &\leq \left(\int_{-d}^{0} |u_y|^2\, \mathrm{d}y \right) \left[\int_{-d}^{0} \sinh^2 k_0 (y + d)\, \mathrm{d}y \right] \\ &= \frac{1}{2} \left(\frac{\sinh 2k_0 d}{2k_0} - d \right) \int_{-d}^{0} |u_y(x, y)|^2\, \mathrm{d}y \\ &= \frac{1}{2} \left(\frac{\sinh^2 k_0 d}{\nu} - d \right) \int_{-d}^{0} |u_y(x, y)|^2\, \mathrm{d}y. \end{aligned}$$

The latter equality is a consequence of the definition of k_0.

Since $\nu^{-1} \sinh^2 k_0 d - d < \nu^{-1} \sinh^2 k_0 d$,

$$2\nu |u(x, 0)|^2 \leq \int_{-d}^{0} |u_y(x, y)|^2\, \mathrm{d}y, \quad x \in F$$

holds. Integrating this over F_a, we obtain

$$2\nu \int_{F_a} |u(x, 0)|^2\, \mathrm{d}x \leq \int_{W_a'} |u_y(x, y)|^2\, \mathrm{d}x\mathrm{d}y \leq \int_{W_a'} |\nabla u|^2\, \mathrm{d}x\mathrm{d}y.$$

Here W_a' is the part of W_a that lies strictly under F_a. It follows from the last inequality that

$$\nu \int_{F_a} |u|^2\, \mathrm{d}x \leq \int_{W_a} |\nabla u|^2\, \mathrm{d}x\mathrm{d}y - \nu \int_{F_a} |u|^2\, \mathrm{d}x = \int_{C_a} u_{|x|} \bar{u}\, \mathrm{d}S,$$

where the equality is simply (3.30). Considerations in Subsection 3.2.1.2

show that the integral over C_a tends to zero as $a \to \infty$, and so

$$\nu \int_F |u|^2 \, dx = 0.$$

Consequently,

$$\int_W |\nabla u|^2 \, dx dy = 0,$$

and the last two equalities yield that u vanishes identically in W. Thus the following theorem is proved.

Let u be a generalized solution to the homogeneous water-wave problem. If geometric conditions (*i*) *and* (*ii*) *hold, then u vanishes identically in W.*

It is obvious that this theorem remains true for domains having an uneven bottom satisfying (i) and (ii); that is, the uneven portions are located strictly under the surface-piercing bodies. Moreover, the result is valid for deep water, if in condition (i) one replaces vertical segments by rays emanating from F downward. Then the considerations in Subsections 3.2.1.2 and 3.2.1.3 hold for w defined as follows:

$$w(x) = \int_{-\infty}^{0} u(x, y) e^{\nu y} \, dy.$$

It can easily be seen that John's method is applicable to the two-dimensional problem as well. However, in this case only one surface-piercing body is admissible.

Combining the uniqueness theorem obtained here with the solvability result in Subsection 3.1.1.3, one arrives at the following theorem on unique solvability.

If geometrical conditions (i) and (ii) hold, then the water-wave problem has a unique solution that can be obtained in the form of (3.8) *with densities satisfying the system of integral equations* (3.10) *and* (3.12).

3.2.2. Extensions of John's Uniqueness Conditions

The aim of this subsection is to show that both assumptions (i) and (ii) made in Subsection 3.2.1 may be relaxed in the proof of the uniqueness, but it should be mentioned that results obtained below for obstacles satisfying (ii) and a weakened condition (i) are distinguished essentially from those in which (ii) does not hold. In the first case, it is possible to prove uniqueness for all positive values of ν, whereas if (ii) is violated, then in the proof of uniquenes some frequency intervals must be excluded. This may occur because of the existence of non-uniqueness examples considered in Section 4.1.

In Subsections 3.2.2.1–3.2.2.3, we treat the two-dimensional problem, and in Subsection 3.2.2.4 we are concerned with the axisymmetric problem.

It was pointed out at the end of the previous subsection that a combination of the uniqueness theorem with the integral equation techniques (see Sections 2.1 and 3.1 for the cases of submerged and surface-piercing obstacles, respectively) leads to the theorem guaranteeing the existence of a unique solution.

3.2.2.1. The Two-Dimensional Problem

Let W denote a water domain of infinite depth; that is, $\mathbb{R}^2_- \setminus \bar{W}$ is the union of a finite number of bounded, simply connected domains presenting immersed bodies. We recall that the essential condition (3.1) must hold on ∂W.

Let u be a solution to the homogeneous problem. Without loss of generality we assume u to be real (otherwise we can consider $\operatorname{Re} u$ and $\operatorname{Im} u$ separately). Since u has finite kinetic and potential energy, Green's formula can be applied to the whole water domain W, yielding

$$\int_W |\nabla u|^2 \, dx dy - \nu \int_F |u|^2 \, dx = 0. \tag{3.31}$$

If an inequality contradicting (3.31) can be proved for u, then the uniqueness theorem holds for the problem. To obtain such an inequality we impose some geometrical restrictions on W.

By $\ell_{\pm b}$ $(b \geq 0)$ we denote the ray $\{(x, y) : y = \pm b \mp x, \ \pm x > b\}$ emanating from the point $(\pm b, 0)$ at the angle $\pi/4$ to the vertical. It is clear that ℓ_b (ℓ_{-b}) goes to the right (to the left). We suppose that for any point $(\pm b, 0) \in F$ the whole ray $\ell_{\pm b}$ lies in W.

Let us consider $\psi(x, y) = \exp\{\nu(y + ix)\}$, which is obviously bounded and harmonic on $\mathbb{R}^2_-$ and satisfies the boundary condition

$$\psi_y - \nu\psi = 0 \quad \text{when } y = 0. \tag{3.32}$$

Furthermore, $\psi_x = i\psi_y$, and hence

$$\frac{\partial \psi}{\partial n} = i \frac{\partial \psi}{\partial s} \quad \text{on } \ell_b, b > 0, \tag{3.33}$$

where $\mathbf{n}$ and $\mathbf{s}$ are unit vectors defined as follows. The vector $\mathbf{s}$ is directed along ℓ_b from infinity to $(b, 0)$, and $(\mathbf{n}, \mathbf{s})$ form a right-hand pair of vectors.

If there are no bodies between ℓ_b and $x = +\infty$, then applying Green's theorem to u and ψ we get by (3.32) and harmonicity of ψ that

$$\int_{\ell_b} \left(u \frac{\partial \psi}{\partial n} - \psi \frac{\partial u}{\partial n} \right) ds = 0. \tag{3.34}$$

From (3.33) and (3.34) by virtue of integration by parts we obtain

$$\int_{\ell_b} \psi \frac{\partial u}{\partial n} \, \mathrm{d}s = i \int_{\ell_b} u \frac{\partial \psi}{\partial s} \, \mathrm{d}s = i u(b, 0) e^{i \nu b} - i \int_{\ell_b} \psi \frac{\partial u}{\partial s} \, \mathrm{d}s.$$

So we have

$$u(b, 0) = \int_{\ell_b} \left(\frac{\partial u}{\partial s} - i \frac{\partial u}{\partial n} \right) e^{\nu[y + i(x-b)]} \, \mathrm{d}s.$$

Since u is a real function,

$$|u(b, 0)| \leq \int_{\ell_b} |\nabla u| e^{\nu y} \, \mathrm{d}s.$$

By the definition of ℓ_b the last integral is equal to

$$\sqrt{2} \int_{\ell_b} |\nabla u| e^{\nu y} \, \mathrm{d}y.$$

Then the Schwarz inequality yields

$$\nu |u(b, 0)|^2 \leq 2\nu \left(\int_{-\infty}^{0} e^{2\nu y} \, \mathrm{d}y \right) \left(\int_{\ell_b} |\nabla u|^2 \, \mathrm{d}y \right) = \int_{\ell_b} |\nabla u|^2 \, \mathrm{d}y.$$

In a similar way the same inequality can be derived for $-b$. After integration we arrive at

$$\nu \int_F |u|^2 \, \mathrm{d}x \leq \int_{W_c} |\nabla u|^2 \, \mathrm{d}x \mathrm{d}y, \tag{3.35}$$

where $W_c \subset W$ and is covered either with rays of the family $\{\ell_b : (b, 0) \in F\}$ or with rays of the family $\{\ell_{-b} : (-b, 0) \in F\}$. Inequality (3.35) does not contradict (3.31) only if $u = 0$ in $W \setminus W_c$. Since u is an analytic function, it vanishes identically in W when it is zero in $W \setminus W_c$. Thus the following theorem is proved.

Let no finite part of the free surface F be isolated from infinity. If any ray $\ell_{\pm b}$ belongs to W for $(\pm b, 0) \in F$, then the homogeneous water-wave problem has only a trivial solution.

The proof given allows for generalization in two directions. First, one can replace straight rays by an arbitrary family of nonintersecting lines $x = x(y, \pm b)$ parameterized by the end point $(\pm b, 0)$, and such that

$$\nu \int_{\ell_{\pm b}} e^{2\nu y} \left| \frac{\mathrm{d}s}{\mathrm{d}y} \right|^2 \mathrm{d}y \leq \inf_y \left\{ \frac{\partial x}{\partial |b|} \right\} \quad \text{for all } (\pm b, 0) \in F. \tag{3.36}$$

Here $\ell_{\pm b}$ denotes a curve belonging to the family with end point $(\pm b, 0)$. It is easy to see that (3.35) remains valid under the last condition, but now W_c denotes the subset of W covered with curves $\ell_{\pm b}$.

The second generalization is concerned with water of finite depth. Let W be the same as above, but bounded below by a bottom B that coincides with $\{x \in \mathbb{R}, y = -d\}$ everywhere except for a finite part. By Green's formula a solution u to the homogeneous water-wave problem in W satisfies (3.31). Now inequality (3.35) contradicting (3.31) for a nontrivial u can be obtained with the help of a family of segments inclined at an angle β to the vertical:

$$\ell_{\pm b}(\beta) = \{(x, y) : y = (\pm b \mp x) \cot \beta, x > \pm b, -d < y < 0\}.$$

Here $b \geq 0$ and $0 \leq \beta < \pi/2$. If $\beta \neq 0$, then the segment ℓ_b (ℓ_{-b}) goes to the right from $(b, 0)$ [goes to the left from $(-b, 0)$]. If $\beta = 0$, then $\ell_{\pm}$ goes downward from $(\pm b, 0)$. The following theorem is similar to that proved for deep water.

Let no finite part of the free surface F be isolated from infinity, and let any segment belonging to the families

$$\{\ell_b(\beta) : (b, 0) \in F\}, \quad \text{and } \{\ell_{-b}(\beta) : (-b, 0) \in F\} \tag{3.37}$$

lie within W and have the second end on B. If $\beta = \arctan p$ where p is a positive root of

$$\frac{2}{1 + p^2} = \frac{[p \sinh(2k_0 d)]^2 - 2[1 - \cos(2k_0 dp)]}{p[p \sinh(2k_0 d) + \sin(2k_0 dp)] \sinh(2k_0 d)}, \tag{3.38}$$

then the homogeneous water-wave problem has only a trivial solution.

For the proof it is sufficient to demonstrate that (3.38) implies (3.35), where W_c is the subset of W covered with segments of families (3.37), but unlike the proof given for deep water, two auxiliary harmonic functions

$$\psi(x, y) = \cosh[k_0(y + d)] \exp[-ik_0(x - b)],$$
$$\tilde{\psi}(x, y) = -i \sinh[k_0(y + d)] \exp[-ik_0(x - b)]$$

are needed now. One can verify directly that

$$\psi_x = -ik_0\psi = \tilde{\psi}_y, \quad \psi_y = -ik_0\tilde{\psi} = -\tilde{\psi}_x;$$
$$\psi_y = 0, \quad \tilde{\psi} = 0 \quad \text{when } y = -d;$$
$$\psi_y - \nu\psi = 0 \quad \text{when } y = 0;$$
$$\partial\psi/\partial n = \partial\tilde{\psi}/\partial s \quad \text{on } \ell_b;$$

here **n** and **s** are defined in the same way as above, and the last equation is crucial for the proof. Details are left to the reader or can be found in the paper [307] by Simon and Ursell.

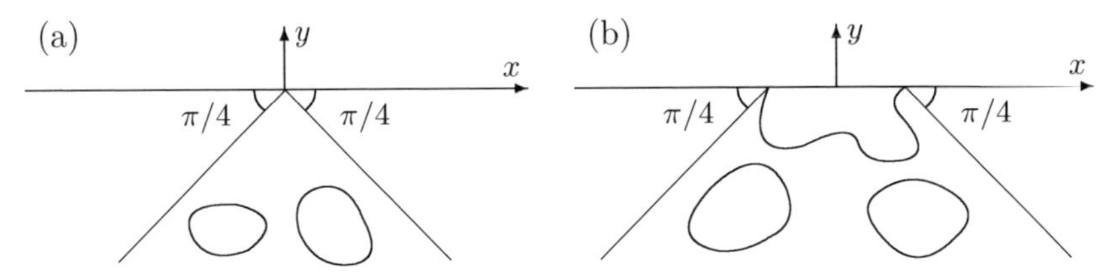

Figure 3.1.

A numerical solution of (3.38) shows that the inequality (3.35) (and hence, the uniqueness theorem for the water-wave problem) holds for all $k_0 d$ provided $0^\circ \leq \beta \leq 44\,1/3^\circ$. As for deep water, one can use curves instead of straight segments. Of course, these curves must satisfy a condition similar to (3.36).

3.2.2.2. Some Examples

Here we illustrate the criteria obtained in Subsection 3.2.2.1. A couple of geometries for which uniqueness is proved for deep water are shown in Fig. 3.1. Any finite number of totally submerged bodies is allowed in the right angle shown in Fig. 3.1(a), or between the rays enclosing a surface-piercing body as shown in Fig. 3.1(b). Figure 3.2 demonstrates similar geometries for the case of constant depth when the angle to the vertical guaranteeing uniqueness for all frequencies is smaller than that for deep water. Moreover, for constant depth the uniqueness theorem holds when there are no submerged or surface-piercing bodies, but the bottom is uneven within the angle shown in Fig. 3.3(a). Further examples of geometries guaranteeing uniqueness can be readily obtained by combining the previous ones. For example, the obstacle in Fig. 3.3(b) is a combination of those in Fig. 3.2(a) and (b), and in Fig. 3.3(a). The crucial point for all these examples is the absence of bounded isolated portions of the free surface.

If the bottom is flat, then for totally submerged bodies the uniqueness condition given in Subsection 3.2.2.1 is less restrictive than that obtained in Subsection 2.2.5.1, Example 2. In that example, the radian measure of the

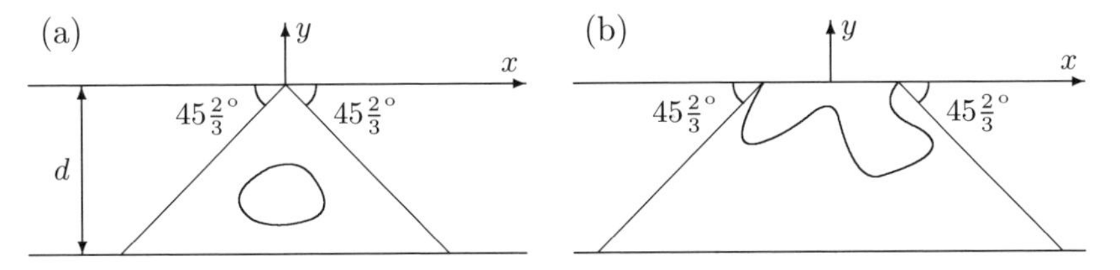

Figure 3.2.

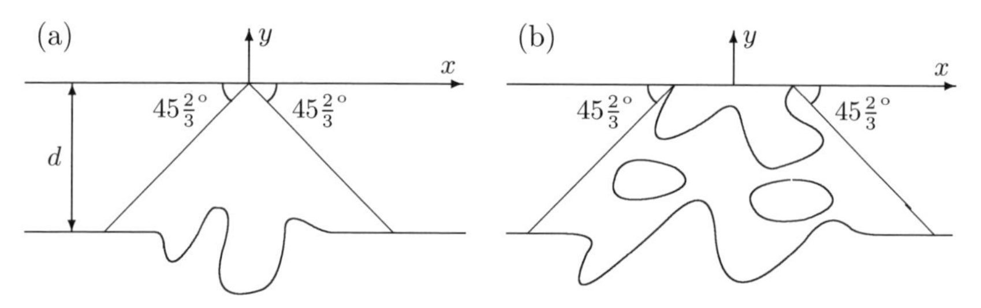

Figure 3.3.

angle between the free surface and each line bounding the immersed bodies is equal to one, and so there is more room for bodies in Fig. 3.3. However, the condition from Example 2 in Subsection 2.2.5.1 has an advantage when the bottom is uneven beyond the angle shown in Fig. 3.3.

If the geometry of obstacles is described analytically, then the condition of uniqueness can be expressed as an inequality. Let us consider the ellipse

$$\frac{x^2}{(\lambda b)^2} + \frac{(y+h)^2}{b^2} = 1, \quad \lambda > 0, h > b$$

submerged in deep water. The theorem proved in Subsection 3.2.2.1 provides uniqueness when

$$h/b \geq (1+\lambda^2)^{1/2}. \tag{3.39}$$

This inequality imposes a restriction on the depth of submergence for any λ, whereas the result in Subsection 2.2.4.2 guarantees uniqueness at any depth of submergence when $0 < \lambda \leq 1$, that is, the major axis is vertical. In contrast, for $\lambda > \sqrt{5}/2$ the inequality (3.39) is an improvement of the estimate

$$h/b \geq \max\{1, 2\lambda^2 - 1\}$$

obtained in Subsection 2.2.4.2.

3.2.2.3. A Pair of Symmetric Bodies Submerged in Deep Water

It was shown in Subsection 3.2.2.1 – see Fig. 3.1(a) – that the uniqueness theorem holds for a pair of arbitrarily shaped submerged bodies confined within a right angle having the vertex at the origin and symmetric about the y axis. However, the technique developed in Subsection 3.2.2.1 can be combined with the original method of John in a way allowing us to prove the uniqueness for *a pair of submerged bodies symmetric about the y axis but having parts outside of the mentioned angle. However, certain restrictions on ν should be imposed in this case.*

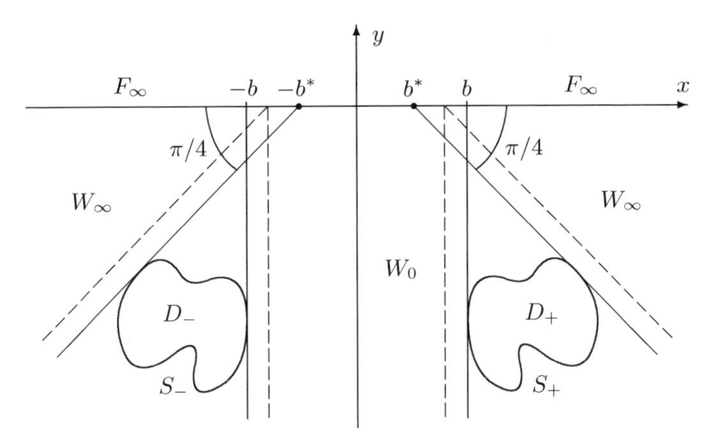

Figure 3.4.

We denote by S_+ a piecewise smooth contour submerged in deep water so that any vertical line on the left of $x = b$ $(b > 0)$ does not intersect S_+, and vertical line $x = c$ intersects this contour for any c, belonging to a certain finite interval with the left end point b. Furthermore, we assume that S_+ lies below $y = -x + b_0$, where $0 < b_0 \leq b$ (the case $b_0 = 0$ was considered in Subsection 3.2.2.1). We put $b^* = \min b_0$, where the minimum is taken over all possible values of b_0 such that $y = -x + b_0$ is above S_+. In order to generalize the theorem proved in Subsection 3.2.2.1, we assume that $b^* > 0$. Let S_- be the reflection of S_+ in the y axis, and $W_0 = \{|x| < b_0, -\infty < y < 0\}$. By W_∞ we denote the union of two parts of W: one lying to the right of $y = -x + b_0$, and the other one to the left of $y = x - b_0$. The geometry described is shown in Fig. 3.4.

Let u denote a solution to the homogeneous water-wave problem. Using the symmetry of the water domain W, we can decompose u into the sum of a symmetric part $u^{(+)}$ and an antisymmetric part $u^{(-)}$ defined as follows:

$$u^{(\pm)}(x, y) = \pm u^{(\pm)}(-x, y).$$

It is obvious that

$$u_x^{(+)}(0, y) = 0, \qquad u^{(-)}(0, y) = 0, \tag{3.40}$$

and that Green's formula (3.31) is true for $u^{(\pm)}$. The considerations in Subsection 3.2.2.1 [see (3.35)] imply that

$$\nu \int_{F_\infty} \left|u^{(\pm)}\right|^2 \mathrm{d}x < \int_{W_\infty} \left|\nabla u^{(\pm)}\right|^2 \mathrm{d}x\mathrm{d}y, \tag{3.41}$$

where $F_\infty = \{|x| > b_0, y = 0\}$.

For the sake of brevity, it is convenient to consider $u^{(+)}$ and $u^{(-)}$ simultaneously. Thus in what follows each formula containing $\pm$ must be considered as two formulae for two different problems.

As in Subsection 3.2.1.2 we define

$$w^{(\pm)}(x) = \int_{-\infty}^{0} u^{(\pm)}(x, y) e^{\nu y}\, dy$$

on $\{-b < x < b\}$. It is shown in Subsection 3.2.1.2 that

$$w_{xx}^{(\pm)} + \nu^2 w^{(\pm)} = 0, \quad \text{for } -b < x < b.$$

Then (3.40) implies that

$$w^{(\pm)}(x) = C_{\pm} \cos\left(\nu x - \frac{\pi}{4} \pm \frac{\pi}{4}\right),$$

where $C_{\pm}$ are real constants. Integration by parts in

$$C_{\pm} \cos\left(\nu x - \frac{\pi}{4} \pm \frac{\pi}{4}\right) = \int_{-\infty}^{0} u^{(\pm)}(x, y) e^{\nu y}\, dy \tag{3.42}$$

leads to

$$u^{(\pm)}(x, 0) = \nu C_{\pm} \cos\left(\nu x - \frac{\pi}{4} \pm \frac{\pi}{4}\right) + \int_{-\infty}^{0} u_y^{(\pm)}(x, y) e^{\nu y}\, dy,$$

from which

$$\left|u^{(\pm)}(x, 0)\right|^2 \le 2\left\{\nu^2 C_{\pm}^2 \cos^2\left(\nu x - \frac{\pi}{4} \pm \frac{\pi}{4}\right) + \left[\int_{-\infty}^{0} u_y^{(\pm)}(x, y) e^{\nu y}\, dy\right]^2\right\}.$$

Applying the Schwarz inequality to the last integral, we get

$$\nu\left|u^{(\pm)}(x, 0)\right|^2 \le 2\nu^3 C_{\pm}^2 \cos^2\left(\nu x - \frac{\pi}{4} \pm \frac{\pi}{4}\right) + \int_{-\infty}^{0} \left|u_y^{(\pm)}(x, y)\right|^2 dy. \tag{3.43}$$

In contrast, we have from (3.42)

$$-\nu C_{\pm} \sin\left(\nu x - \frac{\pi}{4} \pm \frac{\pi}{4}\right) = \int_{-\infty}^{0} u_x^{(\pm)}(x, y) e^{\nu y}\, dy,$$

which implies that

$$2\nu^3 C_{\pm}^2 \sin^2\left(\nu x - \frac{\pi}{4} \pm \frac{\pi}{4}\right) \le \int_{-\infty}^{0} \left|u_x^{(\pm)}(x, y)\right|^2 dy. \tag{3.44}$$

Let us assume that

$$\pi(m + 1/4 \pm 1/4) \leq \nu b_0 \leq \pi(m + 3/4 \pm 1/4), \quad m = 0, 1, \ldots, \tag{3.45}$$

which is equivalent to

$$\int_0^{b_0} \left[\cos^2 \left(\nu x - \frac{\pi}{4} \pm \frac{\pi}{4} \right) - \sin^2 \left(\nu x - \frac{\pi}{4} \pm \frac{\pi}{4} \right) \right] \mathrm{d}x \leq 0.$$

Now, let us integrate (3.43) and (3.44) over $F_0 = \{-b_0 < x < b_0, \, y = 0\}$ and apply (3.45). Then we arrive at

$$\nu \int_{F_0} \left| u^{(\pm)}(x, 0) \right|^2 \mathrm{d}x \leq \int_{W_0} \left| \nabla u^{(\pm)} \right|^2 \mathrm{d}x \mathrm{d}y.$$

Adding this inequality to (3.41) produces

$$\nu \int_F \left| u^{(\pm)}(x, 0) \right|^2 \mathrm{d}x \leq \int_{W_0 \cup W_\infty} \left| \nabla u^{(\pm)} \right|^2 \mathrm{d}x \mathrm{d}y.$$

This contradicts (3.31) unless $u^{(\pm)} \equiv 0$ in W. Thus *a symmetric (antisymmetric) solution is unique when* ν *satisfies* (3.45), *where the plus (minus) sign is taken and* b_0 *belongs to* $[b^*, b]$.

Dividing (3.45) by b_0, we see that putting $b_0 = b$ we get the best lower bound for ν, and it is equal to $\pi b^{-1}(m + 1/4 \pm 1/4)$. Similarly, putting $b_0 = b^*$, we get the best upper bound for ν, and it is equal to $\pi {b^*}^{-1}(m + 3/4 \pm 1/4)$. Thus the following result is obtained.

The uniqueness is guaranteed for a symmetric (antisymmetric) solution when the inequality

$$\pi(m + 1/4 \pm 1/4) \leq \nu b \leq \pi(m + 3/4 \pm 1/4) b/b^* \tag{3.46}$$

holds with plus (minus), where $m = 0, 1, \ldots.$

If $b/b^* > 1$, then the latter assertion implies that non-uniqueness might occur only for ν belonging to a finite number of intervals. From (3.46) we see that if m_+ is the smallest nonnegative integer such that

$$(m_+ + 1) b/b^* \geq m_+ + 3/2, \tag{3.47}$$

then the right-hand end of the uniqueness interval (3.46) with $m = m_+$ for symmetric solutions belongs to the similar interval with $m = m_+ + 1$, and so we obtained the following corollary.

If m_+ *is the smallest nonnegative integer satisfying* (3.47), *then the non-uniqueness of symmetric modes might occur only for* ν *belonging to the first* $m_+ + 1$ *intervals defined by* (3.46) *with the minus sign. In particular,* *when* $b \geq 3b^*/2$, *a symmetric solution is unique for all* $\nu > 0$ *with a possible exception for the interval* $(0, \ \pi(2b)^{-1})$.

Similarly, we have the following assertion:

If m_- *is the smallest nonnegative integer satisfying*

$$(m_- + 1/2)b/b^* \geq m_- + 1,$$

then the non-uniqueness of antisymmetric modes might occur only for ν *belonging to the first* m_- *intervals defined by* (3.46) *with the plus sign. In particular, when* $b \geq 2b^*$, *an antisymmetric solution is unique for all* $\nu > 0$.

3.2.2.4. *The Axisymmetric Problem in Deep Water*

The method developed in Subsection 3.2.2.1 relies on a relation of the Cauchy–Riemann type – see (3.33). Since similar relations exist in the axisymmetric case, one may expect the same idea to work in the axisymmetric problem. Thus we assume W to be an axisymmetric water domain having no cusps and zero-angled edges on ∂W. Moreover, let the free surface $F = \partial W \cap \{y = 0\}$ be a connected plane region.

Let the velocity potential of the form $u(|x|, y)$ be a solution to the homogeneous problem. Without loss of generality we assume u to be real. Then we arrive at (3.31) in the same way as in Subsection 3.2.2.1. Our aim is to derive an inequality contradicting (3.31), and this implies that the uniqueness theorem holds for the problem.

Consider the conical surface S_b generated by revolving the line

$$\ell_b(\beta) = \{(x, y) : y = (b - x_1)\cot\beta, x_1 > b, x_2 = 0, y < 0\} \qquad (3.48)$$

around the y axis, so that S_b has a water-line intersection $|x| = b$. Provided any bodies are inside S_b we have

$$0 = \int_{S_b} \left(u\frac{\partial \phi}{\partial n} - \phi\frac{\partial u}{\partial n} \right) \mathrm{d}S = 2\pi \int_{\ell_b} \left(u\frac{\partial \phi}{\partial n} - \phi\frac{\partial u}{\partial n} \right) |x|\, \mathrm{d}s. \qquad (3.49)$$

Here $\phi(|x|, y)$ is an axisymmetric harmonic function in the water domain that satisfies the free surface boundary condition, and it is bounded as $|x|^2 + y^2 \to \infty$. Let ψ be related to ϕ through the following equations:

$$\frac{\partial \phi}{\partial |x|} = -\frac{\partial \psi}{\partial y}, \quad \frac{\partial \phi}{\partial y} = |x|^{-1}\frac{\partial(|x|\psi)}{\partial |x|}, \qquad (3.50)$$

which are similar to relations between the velocity potential and stream function in the axisymmetric case. These equations lead to

$$|x|\frac{\partial \phi}{\partial n} = -\frac{\partial(|x|\psi)}{\partial s} \quad \text{on } \ell_b,$$

where **n** and **s** are defined in Subsection 3.2.2.1. Then we get from (3.49) that

$$\int_{\ell_b} \phi \frac{\partial u}{\partial n} |x| \, ds = \int_{\ell_b} u \frac{\partial \phi}{\partial n} |x| \, ds = -\int_{\ell_b} u \frac{\partial(|x|\psi)}{\partial s} \, ds$$
$$= -bu(b,0)\psi(b,0) + \int_{\ell_b} \psi \frac{\partial u}{\partial s} |x| \, ds.$$

As in Subsection 3.2.2.1, an integration by parts is possible after substitution of ψ instead of ϕ. The result is the expression for $u(b, 0)$ in the form of the integral along ℓ_b:

$$bu(b,0)\psi(b,0) = \int_{\ell_b} \left(\phi \frac{\partial u}{\partial n} - \psi \frac{\partial u}{\partial s} \right) |x| \, ds. \tag{3.51}$$

Now we have that $F = \{|x| > b_{\min}, y = 0\}$ for some $b_{\min}$ that will be zero if all bodies are totally submerged, but that will be nonzero if there is a body intersecting the free surface. We want to bound

$$\nu \int_F |u|^2 \, dx = 2\pi \nu \int_{b_{\min}}^{\infty} |u(b,0)|^2 b \, db$$

in terms of $\int_{W_c} |\nabla u|^2 \, dx dy$, where $W_c \subset W$ and W_c is covered by conical surfaces from the family $\{S_b\}$. Then according to (3.51) we must bound

$$\nu b |u(b,0)|^2 = \nu \frac{\left| \int_{\ell_b} (\phi \partial_n u - \psi \partial_s u) \, |x| \, ds \right|^2}{b |\psi(b,0)|^2} \tag{3.52}$$

in terms of $\int_{\ell_b} |\nabla u|^2 |x| \, dy$. By ∂_n and ∂_s we denote the corresponding directional derivatives.

It is appropriate to put

$$\phi(|x|, y) = e^{\nu y} H_0^{(1)}(\nu |x|) \quad \text{and} \quad \psi(|x|, y) = e^{\nu y} H_1^{(1)}(\nu |x|),$$

where $H_0^{(1)}$ and $H_1^{(1)}$ are the Hankel functions. Formulae 9.1.30 in Abramowitz and Stegun [1] guarantee that ϕ and ψ satisfy (3.50). Now taking into account (3.48), we obtain

$$\frac{1}{\sin^2 \beta} \left| \int_{\ell_b} \left[\frac{H_0^{(1)}(\nu|x|)}{H_1^{(1)}(\nu|x|)} \frac{\partial u}{\partial n} + \frac{\partial u}{\partial s} \right] e^{\nu y} \frac{|x|^{1/2} H_1^{(1)}(\nu|x|)}{b^{1/2} H_1^{(1)}(\nu b)} |x|^{1/2} \, dy \right|^2$$

for the left-hand side in (3.52).

Let us prove that *$k|H_1^{(1)}(k)|^2$ is a monotonically decreasing function.* This result is formulated in 8.478 in Gradshteyn and Ryzhik [96], but the proof is not widely known.

Since $m(k) = k|H_1^{(1)}(k)|^2 = k\{[J_1(k)]^2 + [Y_1(k)]^2\}$ (by Y_j we denote the Bessel function of the second kind as in [1], Chapter 9), formula 6.664.4 in [96] yields that

$$m(k) = \frac{8k}{\pi^2}\int_0^\infty K_0(2k\sinh\mu)\cosh 2\mu\,\mathrm{d}\mu,$$

where K_0 is the modified Bessel function. So we have

$$m'(k) = I_1 + I_2, \tag{3.53}$$

where

$$\begin{aligned} I_1 &= \frac{8}{\pi^2}\int_0^\infty K_0(2k\sinh\mu)\cosh 2\mu\,\mathrm{d}\mu,\\ I_2 &= \frac{8k}{\pi^2}\int_0^\infty K_0'(2k\sinh\mu)\,2\sinh\mu\cosh 2\mu\,\mathrm{d}\mu,\\ &= \frac{8}{\pi^2}\int_0^\infty \tanh\mu\cosh 2\mu\,\mathrm{d}K_0(2k\sinh\mu). \end{aligned}$$

Integration by parts in the last integral gives

$$I_2 = \frac{-8}{\pi^2}\int_0^\infty K_0(2k\sinh\mu)\,(\mathrm{sech}^2\mu\cosh 2\mu + 2\tanh\mu\sinh 2\mu)\,\mathrm{d}\mu.$$

Substituting I_1 and I_2 into (3.53), and taking into account that $1 - \mathrm{sech}^2\mu = \tanh^2\mu$, we get that

$$m'(k) = \frac{8}{\pi^2}\int_0^\infty K_0(2k\sinh\mu)\cosh 2\mu\tanh\mu(\tanh\mu - 2\tanh 2\mu)\,\mathrm{d}\mu.$$

One easily finds that $s\tanh s\mu$ is a monotonically increasing function of s for $s > 0$ and fixed μ, and so $2\tanh 2\mu > \tanh\mu$. This and the last formula for $m'(k)$ prove that $m'(k) < 0$.

This lemma and the Schwarz inequality imply that

$$\nu b|u(b,0)|^2 \leq \frac{1}{2\sin^2\beta}\int_{\ell_b}\left|\frac{H_0^{(1)}(\nu|x|)}{H_1^{(1)}(\nu|x|)}\frac{\partial u}{\partial n} + \frac{\partial u}{\partial s}\right|^2 |x|\,\mathrm{d}y. \tag{3.54}$$

This means that we have to estimate

$$\left|\frac{H_0^{(1)}(\nu|x|)}{H_1^{(1)}(\nu|x|)}\frac{\partial u}{\partial n} + \frac{\partial u}{\partial s}\right|^2 \left|\frac{\partial u}{\partial s} + i\frac{\partial u}{\partial n}\right|^{-2}, \tag{3.55}$$

which is equivalent to finding

$$\sup_{X\in\mathbb{R}} \frac{|X+p+iq|^2}{|X+i|^2} = 1 + \sup_{X\in\mathbb{R}} \frac{2pX+c}{X^2+1}, \tag{3.56}$$

where

$$\frac{H_0^{(1)}(\nu|x|)}{H_1^{(1)}(\nu|x|)} = p+iq \quad \text{and } c = p^2+q^2-1. \tag{3.57}$$

It is obvious that $p = [J_0J_1 + Y_0Y_1]/|H_1^{(1)}|^2$, and

$$q = [J_1Y_0 - J_0Y_1]/\left|H_1^{(1)}\right|^2 = 2/\left\{\pi\nu|x|\left|H_1^{(1)}(\nu|x|)\right|^2\right\}. \tag{3.58}$$

The last equality is a consequence of 9.1.16 in [1] for the Wronskian of Bessel functions, and this equality implies that $0 < q < 1$.

The maximum in (3.56) occurs at X_m, such that

$$2pX_m + c = (c^2+4p^2)^{1/2}, \quad X_m^2 + 1 = 2 - \frac{c}{p}X_m = (c^2+4p^2)^{1/2}\frac{X_m}{p},$$

and so

$$\sup_{X\in\mathbb{R}} \frac{2pX+c}{X^2+1} = \frac{c+(c^2+4p^2)^{1/2}}{2}.$$

Substituting c from (3.57) we find that the maximum in (3.56) is equal to

$$\begin{aligned} &1 + 1/2\left\{p^2+q^2-1+[(p^2+q^2-1)^2+4p^2]^{1/2}\right\} \\ &\quad = 1/2\left\{p^2+q^2+1+[(p^2+q^2+1)^2-4q^2]^{1/2}\right\}. \end{aligned}$$

Substituting p^2+q^2 from (3.57) and q from (3.58) into the latter expression, we get a bound for (3.55). This, together with (3.54), gives the following inequality:

$$\nu b|u(b,0)|^2 \le \frac{M}{2\sin^2\beta}\int_{\ell_b} |\nabla u|^2|x|\,dy,$$

where

$$M = \frac{1}{2}\sup_{X\in\mathbb{R}} \left|H_1^{(1)}(X)\right|^{-2}\left(\left|H_0^{(1)}(X)\right|^2 + \left|H_1^{(1)}(X)\right|^2 \right.$$
$$\left. + \left\{\left[\left|H_0^{(1)}(X)\right|^2 + \left|H_1^{(1)}(X)\right|^2\right]^2 - \frac{16}{(\pi X)^2}\right\}^{1/2}\right). \tag{3.59}$$

From the last inequality we get that

$$\nu \int_F |u|^2 \,\mathrm{d}x \leq \frac{M}{2\sin^2\beta} \int_{W_c} |\nabla u|^2 \,\mathrm{d}x\mathrm{d}y. \tag{3.60}$$

If $M/(2\sin^2\beta) \leq 1$ (the equality here is admissible because $W_c \neq W$), then this contradicts (3.31) unless u vanishes identically in W. Thus the following uniqueness theorem is proved.

Let W be an axisymmetric water domain, such that the free surface F is a connected plane region. Let any cone S_b obtained by rotation of the line (3.48) *about the y axis belong to W when $\{|x| = b, y = 0\} \in F$, and let $M/(2\sin^2\beta) \leq 1$, where M is defined by* (3.59). *Then the homogeneous axisymmetric water-wave problem has only a trivial solution.*

It is obvious that the inequality $M < 2\sin^2\beta$ is true for $\beta = \pi/2$, because in this case the theorem of John proven in Subsection 3.2.1.3 guarantees the uniqueness. The constant M can be evaluated numerically from (3.59), which gives $M \approx 1.2$, and this maximum occurs at $X \approx 0.8$. Hence there exists β_0 such that $0.6 \approx \sin^2\beta_0$, that is $\beta_0 \approx 51.7°$, and if $\beta > \beta_0$, then the uniqueness theorem holds.

3.2.3. *Further Uniqueness Results*

In this subsection we give a number of geometrical conditions providing the uniqueness in the water-wave problem. In some cases, they are combined with restrictions on the range of frequencies. In Subsection 3.2.3.1, we demonstrate how the auxiliary integral identity extended to geometries including surface-piercing bodies leads to examples for which the uniqueness holds. In Subsection 3.2.3.2, we outline uniqueness results for two-dimensional domains having sufficiently small depth nondimensionalized by ν.

As in Section 2.2 the water domain W is bounded below by the bottom B, which is a curved smooth surface (line) coinciding with $\{y = -d\}$ at infinity, but here we assume that there are surface-piercing bodies in water. As usual, S denotes the union of all wetted smooth surfaces (contours) of immersed bodies, and ∂S denotes the water-line, that is, the union of closed smooth curves (a finite set of points) belonging to $\{y = 0\}$ and separating S from the free surface F.

As was pointed out in the introductory remarks to Subsection 3.1.1, condition (3.1) guarantees that for any solution u to the water-wave problem the following estimate holds:

$$|\nabla u(x, y)| = O(r^{\delta - 1}),$$

where r is the distance from $(x, y) \in W$ to ∂S, and δ is a certain positive constant. This condition justifies intergation by parts when the auxiliary integral identity must be derived.

3.2.3.1. *The Extended Auxiliary Integral Identity and Its Applications*

For water domains described above, the derivation of the auxiliary integral identities given in Subsection 2.2.3 remains true. The only new point is that an integral over ∂S (a certain difference) arises when integrating by parts in the integral over F_a in three dimensions (in two dimensions). We restrict ourselves to the final identities, which are as follows:

$$\int_W [(Q\nabla\bar{u})\cdot\nabla u + |u|^2\nabla^2 H]\,\mathrm{d}x\mathrm{d}y + \int_{S\cup B}\left[|\nabla u|^2\mathbf{V}\cdot\mathbf{n} + |u|^2\frac{\partial H}{\partial n}\right]\mathrm{d}S$$
$$+ \int_F [\nu^2 V_3 + \nu(2H - \nabla_x\cdot\mathbf{V}) - H_y]|u|^2\,\mathrm{d}x - \int_F V_3|\nabla_x u|^2\,\mathrm{d}x$$
$$- \nu\int_{\partial S}|u|^2\mathbf{V}\cdot\mathbf{n}_0\,\mathrm{d}s = 0, \tag{3.61}$$

and

$$\int_W [(Q\nabla\bar{u})\cdot\nabla u + |u|^2\nabla^2 H]\,\mathrm{d}x\mathrm{d}y + \int_{S\cup B}\left[|\nabla u|^2\mathbf{V}\cdot\mathbf{n} + |u|^2\frac{\partial H}{\partial n}\right]\mathrm{d}S$$
$$+ \int_F \left[\nu^2 V_2 + \nu\left(2H - \frac{\partial V_1}{\partial x}\right) - H_y\right]|u|^2\,\mathrm{d}x - \int_F V_2|u_x|^2\,\mathrm{d}x$$
$$- \nu\sum_{m=1}^{M}[V_1|u|^2]_{a_m}^{b_m} = 0, \tag{3.62}$$

in three and two dimensions, respectively. They will be referred to as *the extended auxiliary identities* for solutions to the corresponding homogeneous water-wave problems. In (3.61), we have $\mathbf{V} = (V_1, V_2, V_3)$, where V_3 is the projection on the y axis. By $\mathbf{n}_0$ we denote the unit normal to ∂S lying in $\{y = 0\}$ and directed in F. In (3.62), we have $\mathbf{V} = (V_1, V_2)$, where V_2 is the projection on the y axis. By $[a_m, b_m]$ we denote the intersection of mth surface-piercing body with the x axis, that is, $\partial S = \{a_1, b_1, \ldots, a_M, b_M\}$, where $a_1 < b_1 < a_2 < b_2 < \cdots < a_M < b_M$.

We remind the reader that if $\mathbf{V}$ is either axisymmetric about the y axis or two dimensional, then the matrix Q has the simplified structure described in Subsection 2.2.3. Let us proceed with a few applications of (3.61) and (3.62). We begin with the latter one, which is simpler.

Example 1

Let there be at most one surface-piercing body, that is, $M = 1$ in (3.62). Then, without loss of generality, we assume that the free surface is symmetric about the origin. Putting $-a_1 = b_1 = b$, we substitute

$$H = \text{const}, \qquad \mathbf{V} = (-x, -k(y+a)), \tag{3.63}$$

where a and k are parameters (these H and $\mathbf{V}$ were used in Example 1, Subsection 2.2.5.3), into (3.62) and note that the last term on the left-hand side is strictly positive unless $u(\pm b, 0) = 0$. Since the positiveness of other terms on the left-hand side was established in Subsection 2.2.5.3, we obtain the result similar to that in Example 1, Subsection 2.2.5.3.

Let k, H, a, and ν belong to $[0, 2]$, $[-1/2, 0]$, $[0, +\infty)$, *and* $(0, +\infty)$, *respectively. If* (2.47) *and* (2.48) *hold and*

$$\frac{\partial |x|^2}{\partial n} + k\frac{\partial (y+a)^2}{\partial n} \leq 0 \quad \text{on } B \cup S,$$

then the homogeneous water-wave problem has only a trivial solution in W.

As in Subsection 2.2.5.3 the geometrical meaning of the last inequality is that the angles between the field of interior normals on $B \cup S$ and the field (2.46) do not exceed $\pi/2$. The integral curves of the vector field (2.46) are given by (2.50).

As in Subsection 2.2.5.3, if $k = 1$ [by (2.47) this implies that $H = 0$], then the assumptions of the theorem mean that W is a starlike domain with respect to the point $(0, -a)$, and by (2.48) the uniqueness of solution holds for $\nu \in (0, a^{-1}]$. For $k = 3/2$, we can use $H = -1/4$ and arbitrary $a \geq 0$ satisfying the condition $(0, -a) \in \bar{W}$. Then the theorem implies the uniqueness for $\nu \in (0, (3a)^{-1}]$. Figure 3.5(a) shows a geometry for which we have the uniqueness in this case.

Example 2

Modifying the vector field (2.45), we find it possible to give another proof of John's uniqueness theorem in two dimensions. Let $F = \{|x| > b\}$, where $b > 0$, and let a surface-piercing body be contained within the vertical semistrip $\{-b \leq x \leq b, y \leq 0\}$. We put

$$H = -1/2, \qquad \mathbf{V} = (m_b(x), 0),$$

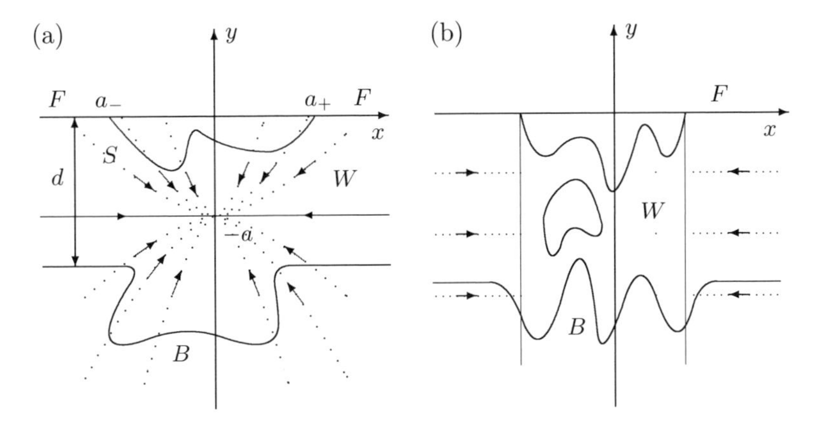

Figure 3.5.

where

$$m_b(x) = \begin{cases} -x + b & \text{when } x \geq b, \\ 0 & \text{when } -b \leq x \leq b, \\ -x - b & \text{when } x \leq -b. \end{cases}$$

Then Q is the identity matrix for $-b < x < b$, and

$$Q = \begin{bmatrix} 2 & 0 \\ 0 & 0 \end{bmatrix}$$

outside the semistrip $\{-b \leq x \leq b, \, y \leq 0\}$. For these H and $\mathbf{V}$ the extended auxiliary identity takes the form of (2.42). Moreover, $\mathbf{V} \cdot \mathbf{n} = 0$ for arbitrary S and B within the semistrip, and so *the uniqueness holds if there are no bodies under the free surface* ($S \cap \{|x| > b\} = \emptyset$) *and*

$$\mathbf{V} \cdot \mathbf{n} = m_b(x) \, \partial x / \partial n \geq 0 \quad \text{on } B \text{ for } |x| > b.$$

It is obvious that the last inequality is equivalent to

$$x \, \cos(n, x) \leq 0 \quad \text{on } B \text{ for } |x| > b.$$

Geometrically this means that any horizontal ray emanating from the y axis to $\pm\infty$ goes out of W at any point of transversal intersection with $B \cap \{|x| > b\}$. Hence in the presence of a surface-piercing body satisfying John's condition the bottom must not necessarily be flat beneath the free surface for validity of the uniqueness theorem. This is illustrated in Fig. 3.5(b).

Example 3

Let us consider the field obtained by rotating the two-dimensional field (3.63). Then we arrive at the following theorem generalizing the result obtained in Subsection 2.2.5.4.

Let the parameters k, H, a, and ν belong to the intervals $[0, 2], [-1, -1/2]$, $[0, +\infty)$, *and* $(0, +\infty)$, *respectively; and let the inequalities* (2.53) *and* (2.54) *hold. Then the homogeneous water-wave problem has only a trivial solution for the indicated values of ν, if* (2.49) *holds on $B \cup S$, and when $\mathbf{x} \cdot \mathbf{n}_0 \geq 0$ on ∂S.*

The geometrical interpretation of this theorem is similar to that formulated in Example 1. The corresponding integral curves lie in the vertical cross sections of the surface of rotation given by (2.50).

3.2.3.2. Two-Dimensional Domains of Sufficiently Small Depth

Let W be a water domain having constant depth d at infinity and containing a finite number of surface-piercing bodies, each satisfying John's condition. Also, let the origin be chosen so that there are no bodies and depth variations in the region $\{x \geq 0\}$. As in Subsection 3.2.2.1, a solution u to the homogeneous water-wave problem can be taken to be real without loss of generality.

The real auxiliary harmonic function

$$\psi(x, y) = \cosh k_0(y + d) \sin k_0(x - b),$$

where k_0 denotes a unique positive root of $\nu = k_0 \tanh k_0 d$ and $b > 0$ is a parameter, is the imaginary part of the auxiliary function applied at the end of Subsection 3.2.2.1. Green's theorem applied to u and ψ in $W \cap \{x > b\}$ gives

$$\int_{-d}^{0} u(b, y) \cosh k_0(y + d)\, dy = 0. \tag{3.64}$$

Since $\cosh k_0(y + d) > 0$ and $u(b, y)$ is a continuous function of y,

$$u(b, y_0) = 0 \quad \text{for some } y_0(b) \in (-d, 0).$$

For every $b > 0$ at least one such value may be found, and so there is no less than one nodal line described by

$$u(x, y) = 0 \quad \text{in } W \cap \{x > 0\}$$

because of the harmonicity of u. An argument presented in Subsection 4.1.1.2 (it is used there for deriving properties of stream lines, but it is applicable to nodal lines of the velocity potential as well) shows that nodal lines cannot terminate in W, but must either end on ∂W or go to infinity.

In a similar manner to the expansion of the finite depth Green's function (see Subsection 1.1.2.2), we have for $x > 0$

$$u(x, y) = \sum_{n=1}^{\infty} a_n \cos k_n(y + d)e^{-k_n x}, \tag{3.65}$$

where k_n is the nth positive root of $k_n \tan k_n d = \nu$. Also, the asymptotics of u given in Subsection 2.2.1 should be taken into account. The first nonzero term dominates in the expansion (3.65) for sufficiently large x, and so

$$u(x, y) = a_m \cos k_m(y + d)e^{-k_m x} + O(e^{-k_{m+1} x}) \quad \text{as } x \to +\infty,$$

for some $m \geq 1$. Therefore, there is a nodal line of u that asymptotes to the straight line $y = -h$, where h is the smallest positive root of $\cos k_m \times (d - h) = 0$. Such a root exists because $k_m d \geq k_1 d > \pi/2$.

Let us suppose that the nodal line, which asymptotes to $y = -h$ as $x \to +\infty$, either goes to infinity as $x \to -\infty$ or ends on the bottom. Then there is a subdomain of W, W_0 say, bounded by this nodal line and the bottom. Since the homogeneous Neumann condition is imposed on the bottom, after applying Green's theorem to u in W_0 we get

$$\int_{W_0} |\nabla u|^2 \, \mathrm{d}x\mathrm{d}y = \int_{\partial W_0} u \frac{\partial u}{\partial n} \, \mathrm{d}S = 0. \tag{3.66}$$

Then u is a constant in W_0, and since a part of ∂W_0 is the nodal line, $u = 0$ in W_0. By analytical continuation we find that $u = 0$ throughout W.

There remains a possibility for the nodal line under consideration to end on the free surface or on a surface-piercing body. Now by W_0 we denote the region contained between the nodal line and the union of the free surface and bodies' contours. Since John's condition holds for surface-piercing bodies, we can apply the same considerations as at the end of Subsection 3.1.1.2 (they were used for obtaining the upper bound for irregular frequencies there). We have for every x such that $(x, 0) \in \partial W_0$:

$$u(x, 0) = \int_{-h(x)}^{0} u_y(x, y) \, \mathrm{d}y,$$

where $h(x)$ is the smallest value of $|y|$ such that the point $(x, -h(x))$ belongs to the nodal line. Then the Schwarz inequality gives

$$|u(x, 0)|^2 \leq h(x) \int_{-h(x)}^{0} |u_y(x, y)|^2 \, \mathrm{d}y < d_{\max} \int_{-h(x)}^{0} |u_y(x, y)|^2 \, \mathrm{d}y,$$

where $d_{\max}$ is the maximum depth of W. Substutution of the boundary conditions on the free surface, on bodies' contours, and on the nodal line into the first equation (3.66) gives

$$\int_{W_0} |\nabla u|^2 \,\mathrm{d}x\mathrm{d}y = \nu \int_{\partial W_0 \cap \{y=0\}} |u(x,0)|^2 \,\mathrm{d}x.$$

If $\nu d_{\max} \leq 1$, then combining the last equality with the previous inequality leads to a contradiction unless u is a constant in W_0. Now by using the argument applied above and using the fact that ∂W_0 contains the nodal line combined with the analytical continuation, we prove the following theorem.

Let $\nu d_{\max} \leq 1$, where $d_{\max}$ is the maximum depth of the water domain W, and let a finite number of surface-piercing bodies, each satisfying John's condition, be immersed into water. Then the homogeneous water-wave problem has only a trivial solution in W.

This result also extends to the cases in which there is a single surface-piercing body of arbitrary shape or a single totally submerged body.

3.3. Bibliographical Notes

3.1.1. In Subsection 3.1.1.1 we apply the same method as in Subsection 2.1.2.1. The notion of irregular frequency was introduced into the theory of water waves by John [126]. Results in Subsection 3.1.1.2 concerning these frequencies are obtained with the help of standard theory of elliptic boundary value problems (see, for example, the book [94] by Gilbarg and Trudinger). The lower estimate for ν_1 was obtained in [126] in a slightly less general form. In Subsection 3.1.1.3, we follow Kuznetsov's work [156]. For the acoustical boundary value problems, Werner [357] developed the same techniques in the 1960s (see also Chapter 15, Section 5 in the book [302] by Sanchez-Palencia).

3.1.2. References to this section are given in the text. Further references can be found in the works of Angell et al. [13] and Martin [218, 219].

3.2.1. The theorem on the unique solvability presented in this subsection was obtained by John in 1950 [126].

3.2.2. There are several extensions of John's result using nonvertical lines in the two-dimensional problem. The theorems in Subsection 3.2.2.1 are borrowed from Simon and Ursell [307]. Their paper contains a more general form of the uniqueness theorem that involves a family of curves instead of straight lines. In addition, in [307] one can find more examples of geometries providing uniqueness. The two-dimensional uniqueness theorem in Subsection 3.2.2.3 is obtained by Kuznetsov [306]. Simon [306] proposed an extension of John's theorem to the axisymmetric problem described in Subsection 3.2.2.4.

3.2.3. The extended auxiliary identity and examples in Subsection 3.2.3.1 are given by Kuznetsov [155]. The result on uniqueness in the two-dimensional problem for water of finite depth presented in Subsection 3.2.3.2 is from McIver [231].

Other works. Among other methods used for the investigation of the water-wave problem, the method of multipole expansions is one of the oldest (see surveys [332] and [337] by Ursell). Further development of this method in combination with conformal mapping was given in a series of papers: see Athanassoulis [4, 5], and Athanassoulis and Politis [9].

Here we list several other techniques used for the study of water waves. The limiting absorption principle well known in the mathematical acoustics is also applicable to the water-wave problem as was shown by Lenoir and Martin [189] and Doppel and Hochmuth [56]. Friis, Grue, and Palm [91] applied the Fourier transform to the wave-diffraction problem. Geometrical theory was applied to surface waves by Ludwig [212].

There are several works in which the water-wave problem is treated by using various forms of weak solution. We restrict ourselves to the following list: Doppel [55], Doppel and Hsiao [57], Doppel and Schomburg [59, 60], Hsiao, Kleinman, and Roach [117], and Lenoir and Tounsi [190].

Two limiting cases of the water-wave problem, namely, short waves and low-frequency waves, were considered by a number of authors. For example, Davis [44], Leppington [192, 193], Rhodes-Robinson [297], and others treated short waves. Results concerning low-frequency waves were surveyed by McIver [236] and by Athanassoulis, Kakilis, and Politis [6].

4

Semisubmerged Bodies, II

It was demonstrated in Section 3.1 that in the presence of a surface-piercing obstacle the water-wave problem is solvable for an arbitrary right-hand-side term in the Neumann condition on the obstacle's surface. However, there is an uncertainty about the set of frequencies providing the solvability. According to the proof given in Subsection 3.1.1, a sequence $\nu_n \to \infty$ $(n = 1, 2, \ldots)$ possibly exists such that for these exceptional values the solvability could be violated for some data given on the obstacle's surface. In particular, this must occur for values ν_n that are point eigenvalues of the water-wave problem embedded in the continuous spectrum (the latter is known to be the whole positive half-axis as is shown in the Examples section of the Introduction). If a value of the spectral parameter ν belongs to the point spectrum, then the homogeneous problem possesses a nontrivial solution with finite energy, or in other words, there is no uniqueness of solution for the nonhomogeneous problem.

In this chapter (see Section 4.1), we give examples of such non-uniqueness for the two-dimensional and axisymmetric problems, and so the exceptional values of ν do exist at least for some obstacle geometries. Moreover, for every $\nu > 0$ a certain family of obstacles exhibiting the non-uniqueness property can be obtained. An essential point in all these examples is the presence of an isolated portion of the free surface inside the obstacle where the eigenmode waves are trapped. However, there is an evidence obtained numerically that the presence of an isolated portion of the free surface is not crucial for constructing non-uniqueness examples.

The examples of trapped modes provoke many new questions, such as (1) whether trapped waves exist for an arbitrary obstacle isolating a portion of the free surface or whether geometric restrictions on the obstacle should be imposed to provide non-uniqueness, and (2) whether a trapped mode can exist at more than one value ν for a given obstacle. The first of these questions will be answered in Section 4.2 by establishing that the uniqueness holds for all $\nu > 0$ for a particular class of surface-piercing obstacles isolating a portion of the free surface. The second question is still unanswered, and the list of open questions can easily be continued.

Since examples of trapped modes constructed in Section 4.1 involve surface-piercing structures, isolating a bounded component of the free surface from infinity, the situation with uniqueness for such structures is more complicated than for geometries considered in Chapters 2 and 3. Moreover, the method developed in Chapter 2 for proving uniqueness theorems fails in the case in which there is an isolated part of the free surface. However, a modification of John's method, developed in Section 4.2 and similar to that in Subsection 3.2.2.3, allows us to show that there are complementary intervals of uniqueness for symmetric, and antisymmetric solutions when the two-dimensional geometry has a vertical axis of symmetry. Similar results are also obtained for each azimuthal mode when an axisymmetric toroidal surface-piercing body is immersed in water. In the absence of symmetry, other techniques are applicable, but they provide the uniqueness for pairs of surface-piercing cylinders and toroidal bodies only in a frequency interval bounded above by a certain number depending on the geometry.

4.1. Trapped Waves

A significant moment for obtaining non-uniqueness examples is the use of the so-called inverse method. Presumably, it was Troesch [319] who pioneered in applying this method in the water-wave theory for the so-called sloshing problem. The inverse method is a powerful tool, replacing attempts to find a solution to a given problem by determining a physically acceptable water region for a given solution. In the problem under consideration this method works as follows. A potential with special properties is constructed by placing sources in the free surface of water. The geometric locus of sources is chosen depending on ν in order to cancel waves at infinity. An investigation of the streamlines of the flow (they do exist for the two-dimensional and axisymmetric problems in question) reveals that enclosing the source points is a family of streamlines connecting the free surface on either side of the sources. Since streamlines may be interpreted as boundaries of rigid bodies, any two such streamlines represent a surface-piercing structure for which the homogeneous problem has a nontrivial solution.

4.1.1. Trapped Waves in Two Dimensions

In Subsections 4.1.1.1–4.1.1.4, we are concerned with waves trapped in deep water. The corresponding examples are investigated in detail. Some numerical examples for the case of finite depth are considered in Subsection 4.1.1.4 as well.

4.1.1.1. Construction of Trapping Geometries

We begin with the case of water unbounded below. Let us consider the following velocity potential:

$$u_n^{(\pm)}(x, y) = 1/2\,[G(z, a_\pm) \pm G(z, -a_\pm)]\,, \tag{4.1}$$

where

$$\nu a_\pm = \pi\,(n - 1/4 \mp 1/4)\,, \quad n = 1, 2, \ldots, \tag{4.2}$$

and $G(z, \pm a_\pm)$ is the two-dimensional Green's function (see Subsection 1.2.1) describing waves in $\mathbb{R}^2_-$ that are due to a source of frequency $(\nu g)^{1/2}$ placed at $(\pm a_\pm, 0)$. Using $u_n^{(+)}[u_n^{(-)}]$, we shall construct obstacles trapping symmetric (antisymmetric) modes. Equation (1.44) reduces (4.1) to

$$u_n^{(\pm)}(x, y) = \int_{\ell_-} \frac{\cos k(x - a_\pm) \pm \cos k(x + a_\pm)}{k - \nu}\, e^{ky} \mathrm{d}k.$$

Note that (4.2) implies that the numerator vanishes at $k = \nu$. Thus,

$$u_n^{(\pm)}(x, y) = \int_0^\infty \frac{\cos k(x - a_\pm) \pm \cos k(x + a_\pm)}{k - \nu}\, e^{ky} \mathrm{d}k, \tag{4.3}$$

and $u_n^{(\pm)}$ is a real-valued harmonic function in $\mathbb{R}^2_-$. Moreover, the asymptotics of G obtained in Subsection 1.2.1 gives that $u_n^{(\pm)}$ has finite energy outside some neighborhood of the sources.

The potential $u_n^{(\pm)}$ represents a nontrivial solution to the homogeneous boundary value problem for a certain structure of two (or more) surface-piercing bodies if the corresponding stream function $v_n^{(\pm)}$, a conjugate to $u_n^{(\pm)}$ harmonic function in $\mathbb{R}^2_-$, has a certain property expressed in terms of its streamlines, that is, in terms of level lines $v_n^{(\pm)}(x, y) = \text{const}$. Since $\nabla u_n^{(\pm)}$ is tangent to streamlines, they are smooth with the possible exception of points at which $\nabla u_n^{(\pm)} = 0$.

The required property can be formulated as follows. At least one of streamlines of $v_n^{(\pm)}$ connects parts of the free surface on either side of the source point at $(a_+, 0)$, and another streamline similarly surrounds the other source point at $(-a_\pm, 0)$ (some other streamlines connecting different points of the x axis might exist). If such a curve S exists, consisting of two (or more) arcs, then it may be interpreted as the rigid boundary where the homogeneous Neumann condition (the no-flow condition) holds for the velocity potential $u_n^{(\pm)}$.

Thus, we consider

$$v_n^{(\pm)}(x, y) = \int_0^\infty \frac{\sin k(x - a_\pm) \pm \sin k(x + a_\pm)}{\nu - k} e^{ky} \mathrm{d}k, \tag{4.4}$$

which is the stream function corresponding to $u_n^{(\pm)}$ [we omit an arbitrary constant term in $v_n^{(\pm)}$]. It is clear that $v_n^{(+)}(x, y)$ [$v_n^{(-)}(x, y)$] *is an odd (even) function in* x, *and*

$$v_n^{(+)}(0, y) = 0 \quad \text{when } -\infty < y \leq 0. \tag{4.5}$$

Hence, it is sufficient to investigate the behavior of $v_n^{(\pm)}(x, y)$ only in the quadrant $\{x \geq 0, y \leq 0\}$. In the subsequent analysis it is convenient to denote the portions of the free surface between the origin and $(a_\pm, 0)$, and to the right of $(a_\pm, 0)$, as $F_0^{(\pm)}$ and $F_\infty^{(\pm)}$, respectively. The properties of $v_n^{(\pm)}$ providing the existence of non-uniqueness example can be formulated as follows.

For $x \in F_\infty^{(\pm)}$ *the range of* $v_n^{(\pm)}(x, 0)$ *is the interval* $(0, \mathrm{Si}(2\pi n - \pi/2 \mp \pi/2))$, *and this interval is contained in the range of* $v_n^{(\pm)}(x, 0)$ *for* $x \in F_0^{(\pm)}$. *To every value* $V_\pm \in (0, \mathrm{Si}\,(2\pi n - \pi/2 \mp \pi/2))$ *there corresponds only one stream line* $S\,(V_\pm)$, *having one end point on* $F_\infty^{(\pm)}$, *the other end point on* $F_0^{(\pm)}$, *and all internal points in* $\{x > 0, y < 0\}$, *on which* $v_n^{(\pm)}(x, y) = V_\pm$.

Here Si and Ci are the sine and cosine integral, respectively, defined by

$$\mathrm{Si}(X) = \int_0^X \frac{\sin k}{k} \mathrm{d}k, \qquad \mathrm{Ci}(X) = -\int_X^\infty \frac{\cos k}{k} \mathrm{d}k.$$

See Chapter 5 in Abramowitz and Stegun [1], where properties of these functions are described. For instance, $\mathrm{Si}(X) > 0$ for $X > 0$, $\mathrm{Ci}(X)$ has logarithmic singularity at $X = 0$, and so on.

The proof of the above assertion and further properties of streamlines are given in Subsection 4.1.1.2, and examples of streamlines $S\,(V_\pm)$ are calculated numerically in Subsection 4.1.1.4. The immediate consequence of the assertion is the following corollary.

For every $n = 1, 2, \ldots$ *the potential* $u_n^{(\pm)}$ *describes waves trapped by structure with the boundary*

$$S_\pm = S\big(V_\pm^{(1)}\big) \cup S^-\big(V_\pm^{(2)}\big), \tag{4.6}$$

where $S^-(V_\pm^{(2)})$ *is obtained by reflection of* $S(V_\pm^{(2)})$ *in the* y *axis, and* $V_\pm^{(1)}$, $V_\pm^{(2)}$ *are arbitrary values in* $[0, \mathrm{Si}(2\pi n - \pi/2 \mp \pi/2)]$. *Symmetric (antisymmetric) modes are given by* $u_n^{(+)}$ $(u_n^{(-)})$.

The structure of (4.6) *has a bulbous profile facing infinity in the horizontal direction; that is, vertical straight lines exist, intersecting* $F_\infty^{(\pm)}$ *and* $S(V_\pm^{(i)})$ *twice* ($i = 1, 2$).

In Subsection 4.2.1 we obtain intervals of uniqueness for values of a nondimensional spectral parameter. These intervals are complementary for symmetric and antisymmetric solutions to the water-wave problem in the presence of two surface-piercing bodies. Moreover, each interval where the symmetric solution is unique contains a subinterval such that antisymmetric trapped modes do exist. They are trapped by structures of the form of (4.6). The same is also shown for some intervals, where an antisymmetric solution is unique.

It is possible that a structure containing more than two surface-piercing bodies traps waves of the frequency $(\nu g)^{1/2}$. Such structures can be obtained with the help of all potentials (4.1) except for $u_1^{(+)}$, and the existence of such structures depends on the behavior of the so-called *nodal streamlines*. In Subsection 4.1.1.2 we prove the existence of nodal lines and consider their simplest properties. A further investigation of nodal lines is made in Subsection 4.1.1.3 and is illustrated numerically in Subsection 4.1.1.4.

4.1.1.2. Proof of Stream Line Properties

Applying formula 3.722.5 from Gradshteyn and Ryzhik [96] to

$$v_n^{(\pm)}(x, 0) = \int_0^\infty \frac{\sin k(x - a_\pm) \pm \sin k(x + a_\pm)}{\nu - k} \, \mathrm{d}k,$$

we express this function for $x \geq 0$ in terms of sine and cosine integrals as follows:

$$\begin{aligned} v_n^{(\pm)}(x, 0) = {} & \sin \nu(x - a_\pm) \,\mathrm{Ci}\,(\nu|x - a_\pm|) - \cos \nu(x - a_\pm) \,\mathrm{Si}\,[\nu(x - a_\pm)] \\ & \pm \{\sin \nu(x + a_\pm) \,\mathrm{Ci}\,[\nu(x + a_\pm)] - \cos \nu(x + a_\pm) \,\mathrm{Si}\,[\nu(x + a_\pm)]\} \\ & - \pi/2[\mathrm{sign}(x - a_\pm) \cos \nu(x - a_\pm) \pm \cos \nu(x + a_\pm)]. \end{aligned} \tag{4.7}$$

The sum of first two terms here is a continuous function because the logarithmic singularity in $\mathrm{Ci}\,(\nu|x - a_\pm|)$ at $x = a_\pm$ is suppressed by the first-order zero of $\sin \nu(x - a_\pm)$ at $x = a_\pm$. Hence, $v_n^{(\pm)}(x, 0)$ has a jump equal to $-\pi$ at $x = a_\pm$. By (4.2) we have

$$v_n^{(+)}(x, 0) \to \mathrm{Si}(2\pi n - \pi) + \pi/2 \mp \pi/2 \quad \text{as } x \to a_+ \pm 0, \tag{4.8}$$

$$v_n^{(-)}(x, 0) \to \mathrm{Si}(2\pi n) + \pi/2 \mp \pi/2 \quad \text{as } x \to a_- \pm 0. \tag{4.9}$$

Here $\mathrm{Si}(\pi) \approx 1.852$, $\mathrm{Si}(2\pi) \approx 1.418$, and $\mathrm{Si}(\pi k) \to \pi/2$ as $k \to \infty$ (see Chapter 5 in Abramowitz and Stegun [1]).

Combining (4.7) with 3.354.1 in [96], we obtain

$$v_n^{(\pm)}(x,0) = \int_0^\infty \frac{e^{-|x-a_\pm|k\nu}\operatorname{sign}(x-a_\pm) \pm e^{-(x+a_\pm)k\nu}}{1+k^2}\,\mathrm{d}k$$
$$+(-1)^{n-1}2\pi H(a_\pm - x)\sin\left(\nu x - \frac{\pi}{4} \pm \frac{\pi}{4}\right). \quad (4.10)$$

Here (4.2) is taken into account, and H denotes the Heaviside function that is equal to zero when $x < 0$ and to one when $x > 0$.

By (4.10) $v_n^{(\pm)}(x,0) \to 0$ as $x \to \infty$. This and (4.8) and (4.9) imply that $(0, \mathrm{Si}\,(2\pi n - \pi/2 \mp \pi/2))$ is contained in the range of $v_n^{(\pm)}(x,0)$ for $x \in F_\infty^{(\pm)}$. Furthermore, differentiating (4.10), we get the following for $x > a_\pm$:

$$\frac{\partial v_n^{(\pm)}}{\partial x}(x,0) = -\nu \int_0^\infty \left[e^{-(x-a_\pm)k\nu} \pm e^{-(x+a_\pm)k\nu}\right] \frac{k\,\mathrm{d}k}{1+k^2} < 0, \quad (4.11)$$

which means that $v_n^{(\pm)}(x,0)$ decreases strictly monotonically on this half-axis. Hence, the intervals defined above coincide with the range of $v_n^{(\pm)}$ for $(x,0) \in F_\infty^{(\pm)}$. Also, by (4.5) and (4.8) the interval $(0, \mathrm{Si}(2\pi n - \pi))$ is contained in the range of $v_n^{(+)}(x,0)$ for $(x,0) \in F_0^{(+)}$.

The analogous fact is true for $v_n^{(-)}(x,0)$, because for every $n = 1, 2, \ldots$, this function has at least one zero in $(0, a_-)$. In fact, (4.10) with $0 \le x < a_-$ gives

$$v_n^{(-)}(x,0) = -\int_0^\infty \frac{e^{-(a_- - x)k\nu} + e^{-(a_- + x)k\nu}}{1+k^2}\,\mathrm{d}k + (-1)^n 2\pi \cos \nu x.$$

We see that the last integral is a positive convex function of x, which increases strictly monotonically and is strictly smaller than π. However, the behavior of $\cos \nu x$ for $x \in (0, a_-)$, where a_- is defined by (4.2), implies that $v_n^{(-)}(x,0)$ has n zeros on $F_0^{(-)}$. Similarly, the number of zeros of $v_n^{(+)}(x,0)$ on $F_0^{(+)}$ is equal to $n-1$. This can be summarized as follows.

The function $v_n^{(\pm)}(x,0)$ is strictly monotonic for $|x| > a_\pm$ (positive for $x > a_\pm$) and tends to zero as $|x| \to \infty$. Between $-a_\pm$ and $+a_\pm$, it has $2n - 1/2 \mp 1/2$ zeros.

We are now able to show the following assertion.

Every value $V_\pm$ in the range of $v_n^{(\pm)}(x,0)$ for $x \in [0, a_\pm) \cup (a_\pm, \infty)$, with an exception for values corresponding to local extrema of this function, is not an isolated point in the range of $v_n^{(\pm)}(x,y)$ when $(x,y) \in \overline{\mathbb{R}^2_-}$.

Let $x_\pm^{(0)} \in [0, a_\pm) \cup (a_\pm, \infty)$ be a point such that $v_n^{(\pm)}(x_\pm^{(0)}, 0) = V_\pm$. Since $x_\pm^{(0)}$ is not a point of local extremum, there exists $\delta_\pm > 0$ such that $v_n^{(\pm)}(x,0)$ is strictly monotonic for $x \in (x_\pm^{(0)} - \delta_\pm, x_\pm^{(0)} + \delta_\pm)$. Let $\ell^{(\pm)}$ be an arbitrary arc with the end points $(x_\pm^{(0)} \pm \delta_\pm, 0)$ and internal points in $\mathbb{R}^2_-$. Considering the variation of $v_n^{(\pm)}(x, y)$ along $\ell^{(\pm)}$, we see that the function being continuous must take all values between $v_n^{(\pm)}(x_\pm^{(0)} - \delta_\pm, 0)$ and $v_n^{(\pm)}(x_\pm^{(0)} + \delta_\pm, 0)$, and in particular the value $V_\pm$. As $\delta_\pm$ and the length of the arc may be made arbitrarily small, this means that there are points in $\mathbb{R}^2_-$ arbitrarily close to $(x_\pm^{(0)}, 0)$, at which $v_n^{(\pm)} = V_\pm$.

Thus, *a streamline on which $v_n^{(\pm)} = V_\pm$ emanates from $(x_\pm^{(0)}, 0)$ into water.* In particular, zero is not an isolated point; hence, *a nodal line of $v_n^{(\pm)}$ emanates from every point on the x axis where $v_n^{(\pm)}(x, 0)$ vanishes.*

Let us show that *the streamline cannot end in $\mathbb{R}^2_-$*. Assume that it terminates at $z_0 \in \mathbb{R}^2_-$ and $\delta > 0$ be such a radius that any circle $|z - z_0| < \delta$ lies beneath the x axis ($z = x + iy$, $z_0 = x_0 + iy_0$). The mean value theorem for harmonic functions implies that either $v_n^{(\pm)}(x, y) = v_n^{(\pm)}(x_0, y_0)$ everywhere on the circle $|z - z_0| = \delta$, or there are values both greater than $v_n^{(\pm)}(x_0, y_0)$ and less than $v_n^{(\pm)}(x_0, y_0)$. Thus, by continuity, $v_n^{(\pm)}$ takes this value at least twice on the circle. In either case, only one point can be on the terminating line. The presence of the second point contradicts the supposition that the line ends in $\mathbb{R}^2_-$.

It was shown that $v_n^{(\pm)}(x, 0) \to 0$ as $|x| \to \infty$. It follows from (4.4) that $v_n^{(\pm)}(x, y) \to 0$ as $|z| \to \infty$ (cf. the investigation of asymptotic behavior of G in Subsection 1.2.1). Thus we have the following assertion:

Nodal lines are the only streamlines that can go to infinity. All other streamlines emanating from the x axis must reenter it at some other point.

The streamline cannot reenter the x axis at the point where it left, because otherwise it would be closed, and by the maximum principle the stream function would be constant inside. Therefore, it would be constant in $\mathbb{R}^2_-$, which is impossible in view of the behavior of $v_n^{(\pm)}(x, 0)$.

Let us derive another representation for $v_n^{(\pm)}(x, y)$. From (4.4) we have

$$\frac{\partial v_n^{(\pm)}}{\partial y} - \nu v_n^{(\pm)} = \int_0^\infty [\sin k(a_\pm - x) \mp \sin k(a_\pm + x)]\, e^{ky} dk$$
$$= \frac{a_\pm - x}{y^2 + (a_\pm - x)^2} \mp \frac{a_\pm + x}{y^2 + (a_\pm + x)^2}.$$

The solution of this differential equation is

$$v_n^{(\pm)}(x, y) = e^{\nu y} \left\{ v_n^{(\pm)}(x, 0) - \int_y^0 \left[\frac{a_\pm - x}{k^2 + (a_\pm - x)^2} \mp \frac{a_\pm + x}{k^2 + (a_\pm + x)^2} \right] e^{-k\nu} dk \right\}. \quad (4.12)$$

By changing variables, we can write the last integral in the following form:

$$-\operatorname{sign}(x-a_\pm)\int_{y/|x-a_\pm|}^{0}\frac{e^{-|x-a_\pm|k\nu}}{1+k^2}\,dk \mp \int_{y/(a_\pm+x)}^{0}\frac{e^{-(a_\pm+x)k\nu}}{1+k^2}\,dk.$$

Then use of (4.10) gives

$$\begin{aligned} v_n^{(\pm)}(x,y) &= (-1)^{n-1}2\pi e^{\nu y}H(a_\pm - x)\sin\left(\nu x - \frac{\pi}{4} \pm \frac{\pi}{4}\right) \\ &\quad + e^{\nu y}\left[\operatorname{sign}(x-a_\pm)\int_{y/|x-a_\pm|}^{\infty}\frac{e^{-|x-a_\pm|k\nu}}{1+k^2}\,dk \right. \\ &\quad \left. \pm \int_{y/(x+a_\pm)}^{\infty}\frac{e^{-(x+a_\pm)k\nu}}{1+k^2}\,dk\right]. \end{aligned} \tag{4.13}$$

This formula is convenient for analyzing $v_n^{(-)}(0, y)$, which has different behavior depending on the parity of n. We have

$$v_n^{(-)}(0,y) = (-1)^n 2\pi e^{\nu y} - 2e^{\nu y}\int_{y/a_-}^{\infty}\frac{e^{-k\nu a_-}}{1+k^2}\,dk. \tag{4.14}$$

The first term in (4.14) decays exponentially as $y \to -\infty$. To analyze the behavior of the second term we integrate by parts:

$$-2e^{\nu y}\int_{y/a_-}^{\infty}\frac{e^{-k\nu a_-}}{1+k^2}\,dk = -\frac{2a_-}{\nu(y^2+a_-^2)} + \frac{4e^{\nu y}}{\nu a_-}\int_{y/a_-}^{\infty}\frac{ke^{-k\nu a_-}}{(1+k^2)^2}\,dk.$$

The absolute value of the second term is less than

$$\frac{4a_-^2 e^{\nu y}}{\nu|y|^3}\int_{y/a_-}^{\infty} e^{-k\nu a_-}\,dk = \frac{4a_-}{\nu^2|y|^3}.$$

Thus, as $y \to -\infty$,

$$-2e^{\nu y}\int_{y/a_-}^{\infty}\frac{e^{-k\nu a_-}}{1+k^2}\,dk = -\frac{2a_-}{\nu(y^2+a_-^2)} + O(|y|^{-3}).$$

This result in conjunction with (4.14) shows that $v_n^{(-)}(0, y)$ *is negative for sufficiently large negative values of* y. Moreover, by (4.14),

$$v_{2m-1}^{(-)}(0,y) < 0 \quad \text{for } -\infty < y \le 0,\ m = 1, 2, \ldots.$$

Since there are $2m-1$ zeros on the half-axis $x > 0$, the latter inequality implies that *at least one nodal line of* $v_{2m-1}^{(-)}$ *goes to infinity in the quadrant* $\{x > 0, y < 0\}$.

By (4.7) and (4.2)

$$v_n^{(-)}(0,0) = (-1)^n\left[\pi + 2\operatorname{Si}(\pi n)\right],$$

so that $v_{2m}^{(-)}(0, 0) > 0$, and $v_{2m}^{(-)}(0, y)$ changes sign on the negative y axis. Furthermore, (4.14) implies that $e^{-\nu y} v_n^{(-)}(0, y)$ is a strictly monotonic function of y. Consequently $v_{2m}^{(-)}(0, y)$ vanishes only once, and the only nodal line of $v_{2m}^{(-)}$, emanating from the positive x axis, intersects the y axis. Since nodal lines emanate from $2m$ points, *at least one of them must go to infinity.* On account of (4.5) we formulate the following proposition.

For each stream function $v_n^{(\pm)}$ there exists at least one nodal line emanating from $\overline{F_0^{(\pm)}}$, which goes to infinity.

Now, let us show that a streamline $v_n^{(\pm)}(x, y) = V_\pm > 0$, having one end point on $F_\infty^{(\pm)}$, has the other end point on $F_0^{(\pm)}$. It was shown that it cannot go to infinity or reenter the same point. Since $v_n^{(\pm)}(x, 0)$ is strictly monotonic for $x > a_\pm$, the streamline cannot reenter another point of $F_\infty^{(\pm)}$. The presence of a nodal line, emanating from $\overline{F_0^{(\pm)}}$ and going to infinity, prevents the streamline under consideration from reentering the negative x axis. The remaining two possibilities are to end on $F_0^{(\pm)}$ or at $x = a_\pm$. Let us demonstrate that the latter is impossible, which proves our assertion.

The proof of this fact is based on the representation of $v_n^{(\pm)}$ in terms of the exponential integral, defined as (see 5.1.1 in Abramowitz and Stegun [1])

$$E_1(z) = \int_z^\infty \frac{e^{-k}}{k}\, dk, \quad |\arg z| < \pi,$$

and there is the following relationship between this function and the sine and cosine integrals:

$$E_1(-iX) = -i\,[\mathrm{Si}(X) - \pi/2] - \mathrm{Ci}(X), \quad X > 0, \tag{4.15}$$

which follows from 5.2.21 and 5.2.23 in [1].

By deforming the contour of integration into the positive or negative imaginary half-axis, depending on the sign of X, we may show the following for $X \neq 0$, $y < 0$:

$$\int_0^\infty \frac{e^{k(y-iX)}}{\nu - k}\, dk = e^{\nu(y-iX)}[\pi i \,\mathrm{sign}(X) - E_1(\nu\{y - iX\})].$$

This, (4.4), and (4.2) yield the following for $x \geq 0$:

$$\begin{aligned} v_n^{(\pm)}(x, y) = \mathrm{Im}\big\{ & e^{\nu[y-i(x-a_\pm)]}[\pi i\{1 - \mathrm{sign}(x - a_\pm)\} \\ & + E_1(\nu[y - i(x - a_\pm)]) - E_1(\nu[y - i(x + a_\pm)])]\big\}, \end{aligned} \tag{4.16}$$

and $v_n^{(\pm)}(x, y)$ is continuous on $\{x = a_\pm, y < 0\}$. Substituting the expansion (see 5.1.11 in [1]),

$$E_1(z) = -\gamma - \log z + O(z) \quad \text{as } |z| \to \infty, |\arg z| < \pi,$$

where γ is Euler's constant, into (4.16) gives

$$v_n^{(\pm)}(x, y) = \pi[1 - \operatorname{sign}(x - a_\pm)] - \arg(\nu\{y - i(x - a_\pm)\}) + \operatorname{Si}\left(2\pi n - \frac{\pi}{2} \mp \frac{\pi}{2}\right) - \frac{\pi}{2} + o(1) \quad \text{as } (x - a_\pm)^2 + y^2 \to 0,\ y \leq 0,$$

where (4.2) and (4.15) have been used. Thus, the values of $v_n^{(\pm)}$ on the streamlines that enter the source point $(a_\pm, 0)$ from $\mathbb{R}^2_-$ are in the closed interval

$$\left[\operatorname{Si}\left(2\pi n - \frac{\pi}{2} \mp \frac{\pi}{2}\right), \operatorname{Si}\left(2\pi n - \frac{\pi}{2} \mp \frac{\pi}{2}\right) + \pi\right],$$

which does not contain $V_\pm$. Hence, the streamline $v_n^{(\pm)}(x, y) = V_\pm$ cannot enter the point $(a_\pm, 0)$.

Finally, we shall prove that the structure (4.6) is bulbous on the right-hand side (by symmetry it is also true for the left-hand side). It is sufficient to demonstrate that $y'_\pm(x) < 0$ for $x > a_\pm$ and $y = 0$, where $y_\pm(x)$ is defined by $v_n^{(\pm)}(x, y) = V_\pm^{(1)}$. Since for such points

$$y'_\pm(x) = -\frac{\partial v_n^{(\pm)}}{\partial x}(x, 0) \Big/ \frac{\partial v_n^{(\pm)}}{\partial y}(x, 0),$$

and the numerator is negative by (4.11), it is sufficient to verify that the denominator is negative also. From (4.13) we get the following for $x > a_\pm$:

$$\begin{aligned}\frac{\partial v_n^{(\pm)}}{\partial y}(x, 0) &= \nu \int_0^\infty \frac{e^{-(x-a_\pm)k\nu} \pm e^{-(x+a_\pm)k\nu}}{1 + k^2}\, dk - \left(\frac{1}{x - a_\pm} \pm \frac{1}{x + a_\pm}\right) \\ &= \nu \int_0^\infty \left\{\frac{e^{-(x-a_\pm)k\nu} \pm e^{-(x+a_\pm)k\nu}}{1 + k^2} - \left[e^{-(x-a_\pm)k\nu} \pm e^{-(x+a_\pm)k\nu}\right]\right\} dk,\end{aligned}$$

which is obviously negative. The proof is complete.

4.1.1.3. *Properties of Nodal Lines*

It was shown in Subsection 4.1.1.2 that the nodal lines defined by $v_n^{(\pm)}(x, y) = 0$ are the only streamlines that can go to infinity. Moreover, for every stream function $v_n^{(\pm)}$ there exists at least one nodal line going to infinity in the quadrant $\{x \geq 0, y \leq 0\}$ – for example, the negative y axis for $v_n^{(+)}$. We remind the reader that $v_n^{(\pm)}(x, 0)$ vanishes n times for $x \geq 0$, and the function $v_{2m}^{(-)}$ has at least one nodal line having both ends on the x axis and symmetric about the y axis. Here we prove further properties of nodal lines. It is convenient to consider $v_n^{(+)}$ and $v_n^{(-)}$ separately, and we investigate the latter function first.

If $v_n^{(-)}(x, 0) < 0$ on a certain interval, then from (4.12), $v_n^{(-)}(x, y)$ is negative in the whole vertical semistrip below this interval. From (4.13) and (4.2) there are $[(n+1)/2]$ such semistrips separating nodal lines in the quadrant $\{x > 0, y < 0\}$ ($[p]$ denotes the integer part of p). Furthermore, the nodal line intersecting the negative y axis emanates from the first positive zero of $v_{2m}^{(-)}(x, 0)$.

The two integrals in (4.13) can be evaluated asymptotically (as $y \to -\infty$) on the same way as the second term on the right-hand side in (4.14). Therefore $v_n^{(-)}(x, y) < 0$ for every x and sufficiently large negative values of y.

Hence, *all nodal lines of $v_n^{(-)}$, except for the extreme left and right ones, reenter the x axis*. Since $v_n^{(-)}(x, 0)$ has $2n$ zeros, *the number of nodal lines with both ends on the x axis is equal to $n - 1$ for $v_n^{(-)}$*.

Let us turn to the asymptotic behavior of nodal lines that go to infinity. From 5.1.51 in Abramowitz and Stegun [1],

$$E_1(z) \sim \frac{e^{-z}}{z}[1 + O(z^{-1})] \quad \text{as } |z| \to \infty, |\arg z| < \frac{3\pi}{2}.$$

Substituting this into the equation of nodal line $v_n^{(-)} = 0$, where $v_n^{(-)}$ is expressed by (4.16), gives

$$\operatorname{Im}\{[y - i(x - a_-)]^{-1} - [y - i(x + a_-)]^{-1}\} \sim 0,$$

or, after simple manipulation,

$$\frac{x^2}{a_-^2} - \frac{y^2}{a_-^2} \sim 1 \quad \text{as } |z| \to \infty.$$

Thus, *two nodal lines of $v_n^{(-)}$ go to infinity along equilateral hyperbola*.

We begin the investigation of nodal lines of $v_n^{(+)}$ ($n = 2, 3, \ldots$) by noting that their behavior at infinity is different from that of the nodal lines of $v_n^{(-)}$. So, from (4.13),

$$v_n^{(+)}(x, y) > 0 \quad \text{when } x > a_+.$$

Hence, *all nodal lines of $v_n^{(+)}$ are confined within the semistrip*

$$\{-a_+ \le x \le +a_+, y \le 0\}$$

and reenter the x axis except for the nodal line coincident with the negative y axis.

The latter assertion can be proved as follows. We write

$$\chi_n(x, y) = \frac{v_n^{(+)}(x, 0)}{x} - 2\int_y^0 \frac{[a_+^2 - x^2 - k^2]\, e^{-k\nu}\, dk}{[k^2 + (a_+ - x)^2][k^2 + (a_+ + x)^2]}. \tag{4.17}$$

From (4.12) this function has the same nodal lines as $v_n^{(+)}$ with the exception for the negative y axis. This follows from (4.10), which implies that

$$\chi_n(0, 0) = \lim_{x \to +0} \frac{v_n^{(+)}(x, 0)}{x} = (-1)^{n-1} 2\pi \lim_{x \to +0} \frac{\sin \nu x}{x} = (-1)^{n-1} 2\pi \nu,$$

and hence, $\chi_n(0, y)$ does not vanish identically.

An asymptotic evaluation of the integral in (4.17), which can be done in a manner similar to that described in Subsection 4.1.1.2, shows that as $y \to -\infty$ the function $\chi_n(x, y)$ tends to $+\infty$ uniformly for $x \in [0, a_+]$. Thus, the all nodal lines reenter the x axis for $v_n^{(+)}$.

This property complicates the behavior of nodal lines for $v_n^{(+)}$ (see numerical examples in Subsection 4.1.1.4). Nevertheless, some other properties can be proved. For every fixed $x \in [0, a_+)$,

$$\frac{\partial \chi_n}{\partial y}(x, y) = \frac{2(a_+^2 - x^2 - y^2)\, e^{-\nu y}}{[y^2 + (a_+ - x)^2][y^2 + (a_+ + x)^2]}$$

vanishes only once on the negative y axis. Hence, $\chi_n(x, y)$ has only one extremum (minimum) there. Consequently, if $x \neq 0$ and $v_n^{(+)}(x, 0) \leq 0$, then $\chi_n(x, y)$ vanishes only once for $0 > y > -\infty$. Then, *there is only one nodal line intersecting every vertical semistrip having $v_n^{(+)}$ negative on $y = 0$.* If $v_n^{(+)}(x, 0) > 0$, then $\chi_n(x, y)$ can have two zeros or no zeros at all, or the minimum of $\chi_n(x, y)$ is equal to zero, in which case the zero is unique. In any case, *no more than two nodal lines intersect any vertical line $x = x_0 > 0$.*

Furthermore,

$$\chi_n(0, y) = (-1)^{n-1}\pi\nu - 2\int_y^0 \frac{[a_+^2 - k^2]\, e^{-k\nu}\, dk}{[k^2 + a_+^2]^2},$$

which implies that $\chi_{2m}(0, y) < 0$ for small negative values of y. Hence, $\chi_{2m}(0, y)$ vanishes only once. On the contrary, $\chi_{2m-1}(0, y) > 0$ near the origin and this function has two zeros on the negative y axis. Thus, *one (a couple of) nodal line(s) of $v_{2m}^{(+)}$ ($v_{2m-1}^{(+)}$) intersects the y axis.*

By (4.17), $\chi_n(x, y) < 0$, when $v_n^{(+)}(x, 0) < 0$ and $y^2 < a_+^2 - x^2$. Since the last inequality holds near the x axis, *every nodal line enters the x axis from a vertical semistrip bounded from above by an interval where $v_n^{(+)}(x, 0) > 0$.*

In order to prove that every such semistrip is intersected by two nodal lines, one has to verify that the minimum of $v_n^{(+)}(x, y)$ as a function of y is negative if $v_n^{(+)}(x, 0) > 0$. However, numerical examples in Subsection 4.1.1.4 show that the absolute value of this minimum is very small, which makes the proof very difficult.

4.1.1.4. Numerical Examples

To give an idea of the streamline behavior, we present some figures calculated numerically. In our examples the spatial variables are nondimensionalized by the wave number ν. Hence, the actual position of the trapping bodies changes as ν varies.

Typical patterns corresponding to symmetric modes are illustrated in Figs. 4.1 and 4.2, where the behavior of $v_3^{(+)}$ and of $v_4^{(+)}$ is shown respectively. The behavior of $v_n^{(+)}$ becomes more and more complicated as n grows up.

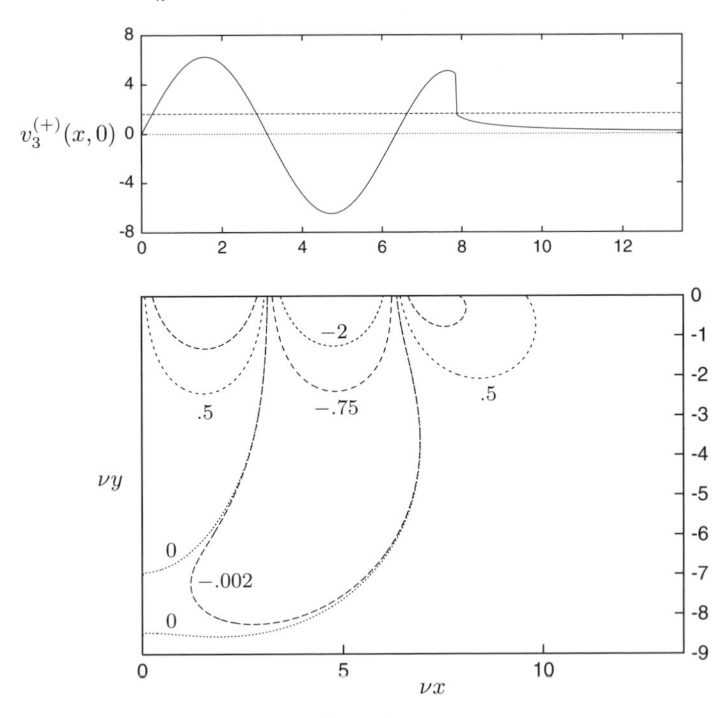

Figure 4.1.

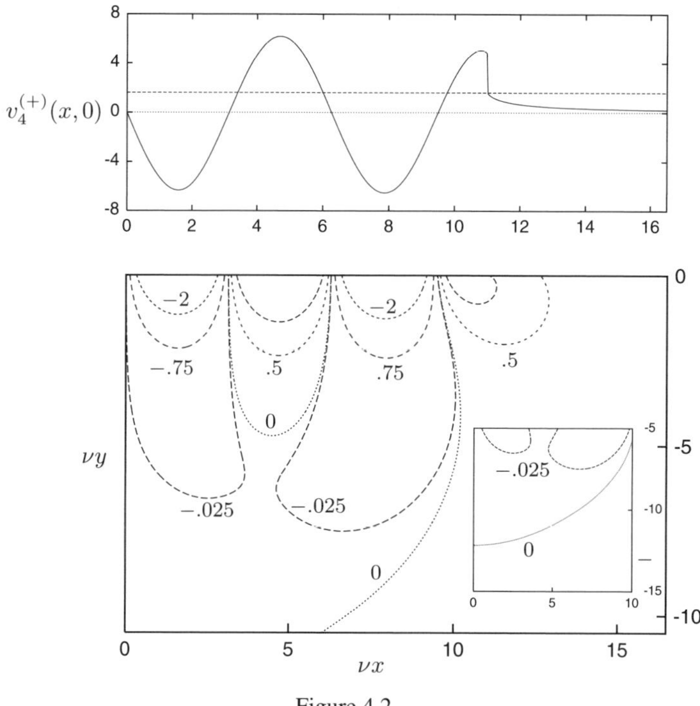

Figure 4.2.

To make the figures clearer we plot the curve of $v_n^{(+)}(x, 0)$ for $x \geq 0$ as well as the pattern of streamlines in the quadrant $\{x \geq 0, y \leq 0\}$. The critical level $\mathrm{Si}(2\pi n - \pi/2 \mp \pi/2)$ for $v_n^{(+)}(x, 0)$ is shown by a heavy dashed line. The extreme right branch of the corresponding streamline terminates in the source point as well as all streamlines between this one and the x axis. The inset on Fig. 4.2 shows how the nodal line intersects the y axis. Any one of the streamlines surrounding those terminating in the source point may be interpreted as a contour of a rigid body giving half of a two-body trapping structure. Figure 4.3, where $v_2^{(-)}$ is plotted, demonstrates a typical pattern of an antisymmetric mode. All above remarks should be repeated in this case with simple corrections.

We see from Figs. 4.1–4.3 that there are additional streamlines that might be interpreted as surface-piercing rigid contours. For example, we may choose a minimum of two and a maximum of eight bodies in the trapping structure generated by $v_4^{(+)}$ shown in Fig. 4.2. For $v_2(-)$ the number of bodies in the trapping structure that can be proved rigorously may vary between two and five. The corresponding lines are either the nodal lines of $v_n^{(\pm)}$ (labeled with

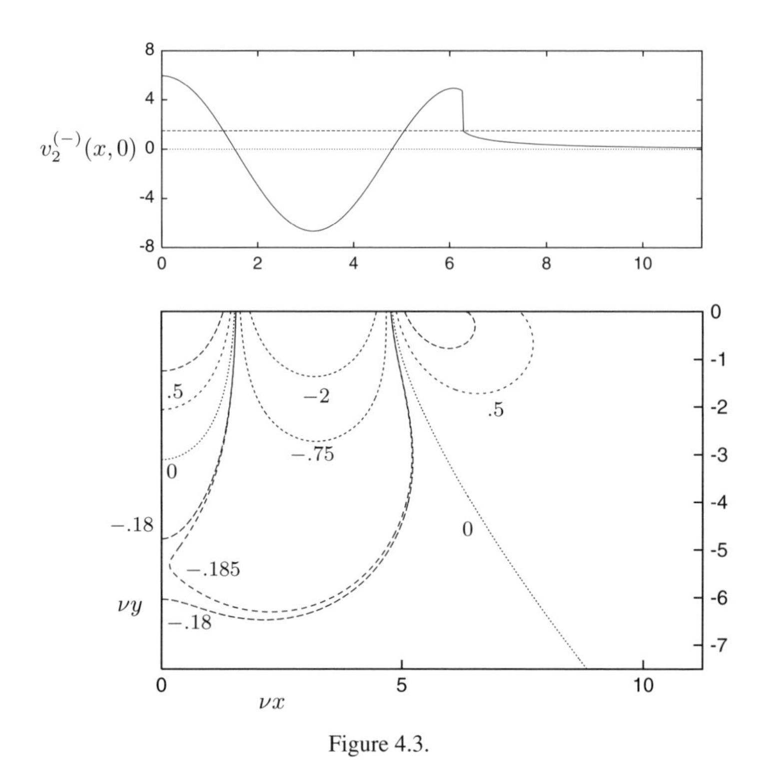

Figure 4.3.

a zero and denoted in the figures by a dotted line) or are placed between the nodal lines of $v_n^{(\pm)}$. Moreover, the extreme right nodal line of $v_n^{(-)}$ goes to infinity like a hyperbola, whereas for $v_n(+)$ $(n \geq 3)$ there is a nodal line surrounding all other curvilinear nodal lines. Thus, we can consider waves trapped between the shore and a floating body by interpreting the hyperbolic nodal line as the bottom in the antisymmetric case, or the y axis as the cliffs, for symmetric potentials. In the latter case it is possible to combine the y axis with a nodal line intersecting the y axis. This leads to different types of cliffs, including overhanging ones.

For further examples of trapping geometries, we consider water of finite depth. In this case we restrict ourselves to consideration of the simplest symmetric and antisymmetric potentials that allow us to construct non-uniqueness examples. In a similar fashion to (4.1) we write

$$u_\pm(x, y) = \frac{1}{2}[G(x, y; a_\pm, 0) \pm G(x, y; -a_\pm, 0)]$$
$$= \int_{\ell_-} \frac{\cos k(x - a_\pm) \pm \cos k(x + a_\pm)}{k \sinh kd - \nu \cosh kd} \cosh k(y + d)\, \mathrm{d}k, \quad (4.18)$$

where

$$k_0 a_\pm = (3 \mp 1)\pi/4 \tag{4.19}$$

and k_0 is the only positive root of the denominator in (4.18). The integral representation for $u_\pm$ is a consequence of (1.42), where $q(k)$ is given by (1.31) and the symmetry of $G(z, \zeta)$ should be taken into account. It is clear that (4.18) tends to the corresponding infinite depth potential – see (4.3) – as $d \to \infty$.

As a result of the source placement specified in (4.19), the integral over indentation of ℓ_- cancels, and we can write

$$u_\pm(x, y) = \int_0^\infty \frac{\cos k(x - a_\pm) \pm \cos k(x + a_\pm)}{k \sinh kd - \nu \cosh kd} \cosh k(y + d)\, dk, \tag{4.20}$$

where in the integrand the effect of the singularity in the denominator is suppressed by the occurrence of a zero in the numerator. Thus, $u_\pm$ is a real harmonic function in the strip $\{-\infty < x < +\infty, -d < y < 0\}$. By (1.42) and (1.35) $u_\pm$ decays exponentially as $|x| \to \infty$. Following the scheme outlined in Subsection 4.1.1.1, we consider the streamlines of the flow corresponding to (4.20), that is, the level lines of a conjugate to $u_\pm$ harmonic function $v_\pm$, in order to verify that at least one of them connects the free surface on either side of the source point $(a_\pm, 0)$, and another streamline in the same way surrounds the other source point $(-a_\pm, 0)$. The curve S consisting of these two (at least) arcs may be interpreted as a contour of a rigid body.

Thus, we introduce the stream function

$$v_\pm(x, y) = \int_0^\infty \frac{\sin k(x - a_\pm) \pm \sin k(x + a_\pm)}{\nu \cosh kd - k \sinh kd} \sinh k(y + d)\, dk.$$

However, it can be investigated only numerically because methods applied in Subsection 4.1.1.2 fail for this integral. Figures 4.4 and 4.5 give examples of the streamlines of the flow in the case of the first symmetric and antisymmetric modes, respectively. As in the infinite depth case the axes are nondimensionalized by ν; the water depth d is taken to be equal to π/ν. As in the previous figures the variation of $v_\pm(x, 0)$ is plotted as well as the pattern of streamlines. Again, a critical level (denoted by a heavily dashed line) exists and all streamlines contained within this level go into the source point, whereas all exterior streamlines are possible rigid contours giving structures of non-uniqueness.

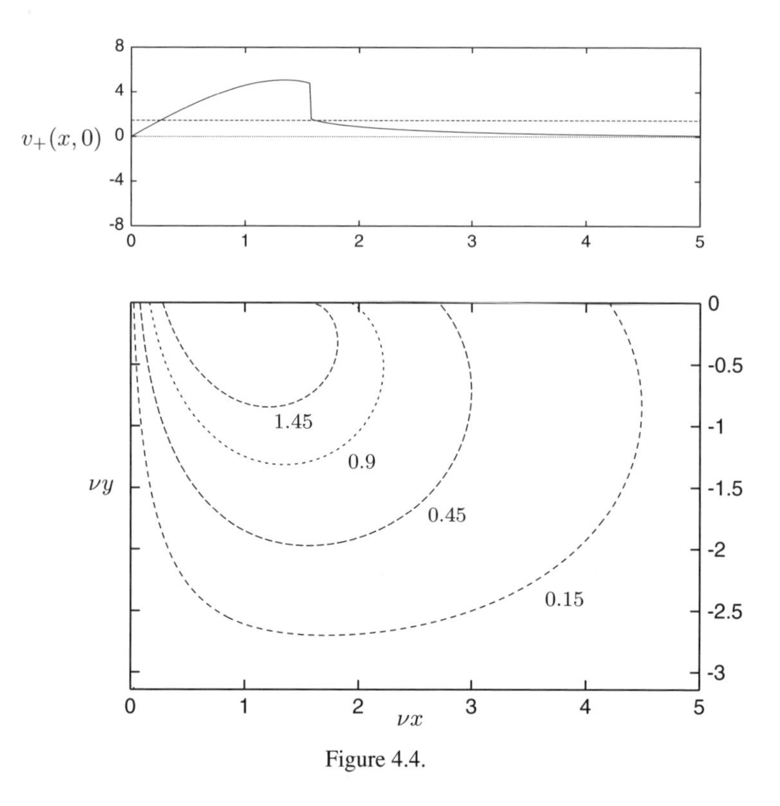

Figure 4.4.

4.1.2. Waves Trapped by Axisymmetric Structures

As in the two-dimensional case, we seek axisymmetric trapping structures by virtue of the inverse procedure, and in Subsection 4.1.2.1 we give a velocity potential providing trapping toroids and prove the existence of such toroids for the axisymmetric problem. Examples of axisymmetric toroids trapping modes of higher order are constructed numerically in Subsection 4.1.2.2.

4.1.2.1. Trapping Velocity Potentials

Solutions to the water-wave problem in deep water are sought in the form

$$u_n(r, y, \theta) = 2\pi c G^{(n)}(r, y; c, 0) \cos n\theta,$$

where the horizontal polar coordinates (r, θ) are defined by

$$x_1 = r \cos\theta, \qquad x_2 = r \sin\theta,$$

and $G^{(n)}$ stands for the ring Green's function of order n expressed by (1.48) or, equivalently, by (1.51) (see Subsection 1.2.2). From the latter formula we

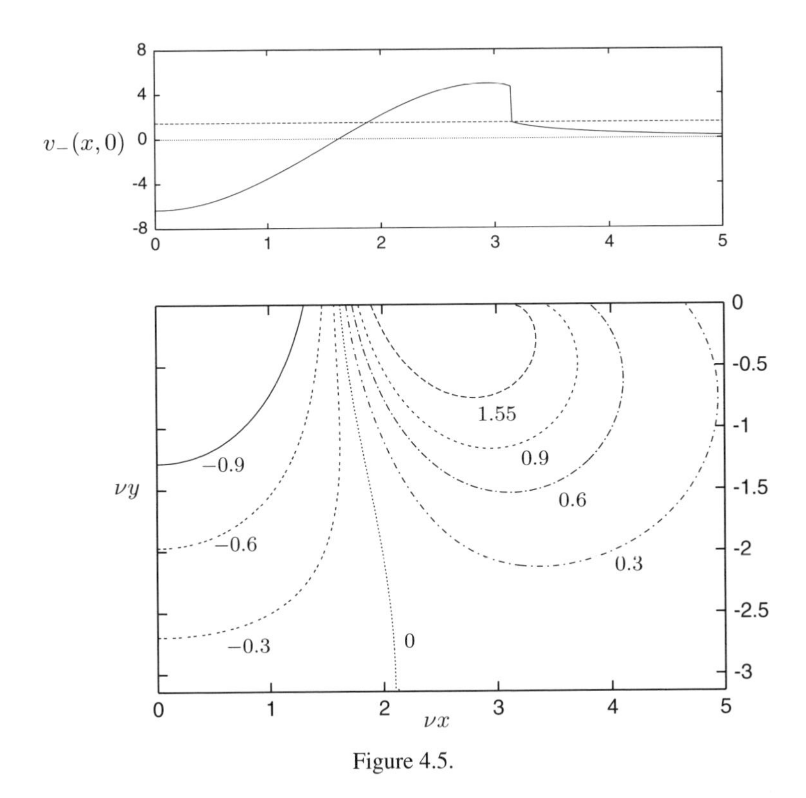

Figure 4.5.

have that u_n depends on (r, y) through

$$4\pi^2 i\nu c e^{\nu y} J_n(\nu r_<) H_n^{(1)}(\nu r_>) \\ + 8c \int_0^\infty (k \cos ky + \nu \sin ky) I_n(kr_<) K_n(kr_>) \frac{k\,\mathrm{d}k}{k^2 + \nu^2}, \qquad (4.21)$$

where $r_> = \max\{r, c\}$, $r_< = \min\{r, c\}$, and J_n, I_n, K_n, and $H_n^{(1)}$ denote standard Bessel, modified Bessel, and Hankel functions of order n. In general, at large radial distances, u_n behaves like outgoing waves satisfying the radiation condition as a result of the Hankel function in the first term, but the integral term decays like r^{-3} as $r \to \infty$. Radiating waves are annulled by taking c to satisfy $J_n(\nu c) = 0$; that is, νc is chosen to be a zero $j_{n,m}$ of the Bessel function J_n (see Section 9.5 in Abramowitz and Stegun [1] for properties of these zeros). Any surface in the water domain that is always parallel to the local velocity may be interpreted as the surface of a structure.

The purely axisymmetric case $n = 0$ can be considered in detail by using the Stokes stream function; that is, for a given nonradiating ring source a family of corresponding toroidal structures can be constructed explicitly so that they exclude the source from the water domain, thus establishing the existence of trapped mode solutions. In order to find the Stokes stream function $v_0(r, y)$ corresponding to $u_0(r, y)$ given by (4.21), one has to solve

$$\partial_r u_0 = r^{-1}\partial_y v_0, \qquad \partial_y u_0 = -r^{-1}\partial_r v_0,$$

and the result is as follows:

$$v_0(r, y) = -4\pi^2 icre^{\nu y} J_1(\nu r) H_0^{(1)}(\nu c) + 8cr \int_0^\infty (k \sin ky - \nu \cos ky) I_1(kr) K_0(kc) \frac{k\, dk}{k^2 + \nu^2}, \quad 0 \le r < c, \tag{4.22}$$

$$v_0(r, y) = -4\pi^2 icre^{\nu y} J_0(\nu c) H_1^{(1)}(\nu r) - 8cr \int_0^\infty (k \sin ky - \nu \cos ky) I_0(kc) K_1(kr) \frac{k\, dk}{k^2 + \nu^2}, \quad r > c. \tag{4.23}$$

For $r = c$ and $y < 0$ both expressions are valid, and constants of integration have been chosen so that $v_0(r, y)$ decays as $r \to \infty$ and is continuous across the half-line $\{r = c,\ y < 0\}$. The latter property is a consequence of formulae for Wronskians of Bessel functions (see 9.1.15 and 9.1.16 in [1]), and the former one follows in the same way as for (4.21).

Assuming that $\nu c = j_{0,m}$ $(m = 1, 2, \ldots)$, we see that the first terms in (4.21) and (4.22) do vanish, and so there are no outgoing waves. As in Subsection 4.1.1, it remains for us to investigate the level surfaces of $v_0(r, y)$ given by (4.22) and to show that there exists a surface such that it intersects the plane $\{y = 0\}$ along two circles on either side of the source ring $\{r = c,\ y = 0\}$. Then this level surface may be interpreted as the surface of a trapping structure because the homogeneous Neumann condition holds there for the velocity potential $u_0(r, y)$. Since level surfaces do not depend on θ, we will speak about streamlines as in the two-dimensional case.

First we note that (4.22) implies that $v_0(0, y) = 0$ for all $y \le 0$, and so any streamline corresponding to a positive level of v_0 cannot intersect the y axis.

Turning to the behavior of $v_0(r, 0)$, we get from (4.22)

$$v_0(r, 0) = 8cr \int_0^\infty I_0(kc) K_1(kr) \frac{k\, dk}{k^2 + \nu^2} \quad \text{for } r > c. \tag{4.24}$$

It is known that $K_1(kr)$ is a strictly monotonically decreasing function of r for a fixed $k > 0$ (see, for example, Section 9.6 in Abramowitz and Stegun [1]). Since the integrand in (4.24) is positive, $v_0(r, 0) > 0$ for $r \in (c, +\infty)$

decreases strictly monotonically and tends to zero as $r \to \infty$. Also, $v_0(r, 0)$ has a finite limit

$$v_0(c+0, 0) = 8c^2 \int_0^\infty I_0(kc) K_1(kc) \frac{k\,dk}{k^2+\nu^2} \quad \text{as } r \to c+0. \tag{4.25}$$

From (4.22), we have the following on the interval $0 \leq r \leq c$:

$$v_0(r, 0) = 4\pi^2 cr J_1(\nu r) Y_0^{(1)}(\nu c) - 8cr \int_0^\infty I_1(kr) K_0(kc) \frac{k\,dk}{k^2+\nu^2}. \tag{4.26}$$

Here the integral is a positive and strictly monotonically increasing function of r because $I_1(kr)$ has this property as a function of r for a fixed k and the integrand is positive. Moreover, the finite limit

$$v_0(c-0, 0) = 4\pi^2 c^2 J_1(\nu c) Y_0^{(1)}(\nu c) - 8c^2 \int_0^\infty I_1(kc) K_0(kc) \frac{k\,dk}{k^2+\nu^2}$$

does exist and is greater than $v_0(c+0, 0)$ because a jump in the value of $v_0(r, 0)$ across $r = c$ is negative in view of

$$\begin{aligned} &v_0(c+0, 0) - v_0(c-0, 0) \\ &\quad = 8c^2 \int_0^\infty [I_0(kc)K_1(kc) + I_1(kc)K_0(kc)] \frac{k\,dk}{k^2+\nu^2} - 4\pi^2 c^2 J_1(\nu c) Y_0^{(1)}(\nu c) \\ &\quad = -4\pi c. \end{aligned} \tag{4.27}$$

Again, the formulae for Wronskians of Bessel functions are applied when the last equality is derived. In the same way as in Subsection 4.1.1.2, these facts imply the following assertion.

Every value in the range of $v_0(r, 0)$, with an exception for values corresponding to local extrema of this function, is not an isolated point in the range of $v_0(r, y)$ when $x > 0$ and $y < 0$, and so a streamline emanates from $(r, 0)$, if $v_0(r, 0)$ is not a local extremum.

In addition, a modification of considerations in Subsection 4.1.1.2 allows us to demonstrate the following.

1. *All streamlines cannot terminate in $\{x > 0, y < 0\}$.*
2. *Every streamline corresponding to a nonzero level (that is, not a nodal line) and emanating from $(r_0, 0)$, where $r_0 > 0$ and $r_0 \neq c$, must reenter the half-line $\{r > 0\ y = 0\}$ at a certain point different from $(r_0, 0)$.*
3. *The local behavior of the stream function v_0 as $(r-c)^2 + y^2 \to 0$ is given by*

$$v_0(r, y) = v_0(c+0, 0) - 4c \arctan \frac{y}{r-c} + o(1),$$

and so the level values of the streamlines entering $(c, 0)$ *from* $\{r > 0, y < 0\}$ *are in the range* $(v_0(c+0,0), v_0(c+0,0)+4\pi c)$.

These assertions, combined with the fact that $v_0(r, 0)$ strictly monotonically decreases from $v_0(c+0,0) > 0$ to zero, lead to the following main result.

Every streamline emanating from $\{r > 0, y = 0\}$, *where* $r > c$, *reenters* $\{r > 0, y = 0\}$ *at a certain point with* $r < c$, *and so generates an axisymmetric toroidal trapping body.*

4.1.2.2. *Numerical Calculation of Trapping Structures for* $n > 0$

Since no stream function is available for $n \geq 1$, the evidence for the existence of trapped modes can be given only purely numerically. On the surface of any structure it is required that there is no flow in the normal direction. For axisymmetric structures, surfaces independent of θ are sought in the form $r = r(y)$ and the condition of no flow in the direction of the local normal $\mathbf{n}$ may be written as

$$\nabla u_n \cdot \mathbf{n} = (\partial_r u_n, r^{-1}\partial_\theta u_n, \partial_y u_n) \cdot (1, 0, -\mathrm{d}r/\mathrm{d}y) = 0$$

or

$$\mathrm{d}r/\mathrm{d}y = \partial_r u_n / \partial_y u_n. \tag{4.28}$$

This differential equation may be solved numerically by using standard procedures, and the derivatives of u_n are easily calculated numerically from (4.21). It is sometimes convenient to solve in terms of polar coordinates (ρ, χ) centered on $(r, y) = (r_0, y_0)$ and defined by

$$r = r_0 + \rho \sin\chi, \qquad y = y_0 + \rho\cos\chi.$$

In this case, the differential equation for the stream surfaces follows from the chain rule and is

$$\frac{\mathrm{d}\rho}{\mathrm{d}\chi} = \rho \frac{\partial_y u_n \cos\chi + \partial_r u_n \sin\chi}{-\partial_y u_n \sin\chi + \partial_r u_n \cos\chi}. \tag{4.29}$$

The accuracy of the numerical solutions was tested by making computations for axisymmetric modes ($n = 0$) and comparing with the results calculated directly from the stream function given in Subsection 4.1.2.1. For a stream surface traced from the free surface in $r > c$ to the free surface in $r < c$, better than four-figure agreement is obtained in all of the test calculations for the location of the ring where the stream surface returns to the free surface.

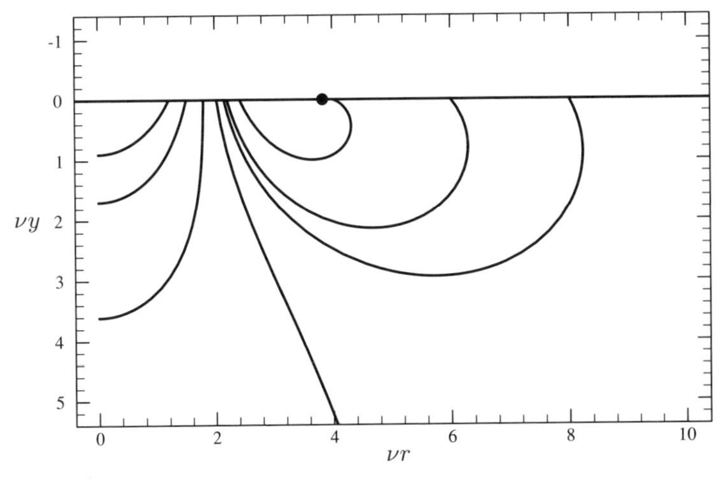

Figure 4.6.

A typical calculation is given in Fig. 4.6 for the case of azimuthal mode $n = 1$; the radius of the ring source is chosen as its smallest possible value $\nu c = j_{1,1}$ of the first zero of J_1. The three rightmost surfaces were plotted by solving (4.29) with the origin of the axial polar coordinates at $(\nu r_0, \nu y_0) = (\nu c, 0)$ and starting from $\nu r = \{4, 6, 8\}$ on the free surface. The three leftmost surfaces were plotted by solving (4.29) with $(\nu r_0, \nu y_0) = (0, 0)$ and starting from $\nu r = \{1.2, 1.5, 1.8\}$ on the free surface. The remaining stream surface was plotted by solving (4.28), starting from $(\nu r, \nu y) = (5, 10)$, and is close to the surface that divides surfaces surrounding the source point from those surrounding the origin.

One feature of the solution illustrated in Fig. 4.6 is that the stream surfaces leaving the free surface in $r > c$ do so with positive gradient $\mathrm{d}r/\mathrm{d}y$, giving a characteristic bulbous appearance. This may be demonstrated as follows. From (4.21) and (4.28), for a single ring source positioned so that $J_n(\nu c) = 0$ is satisfied, we have

$$\left.\frac{\mathrm{d}r}{\mathrm{d}y}\right|_{y=0} = \frac{\partial_r u_n|_{y=0}}{\partial_r u_n|_{y=0}} = \frac{\int_0^\infty I_n(kc)K_n'(kr)[k^3/(k^2+\nu^2)]\,\mathrm{d}k}{-\int_0^\infty I_n(kc)K_n(kr)[\nu k^2/(k^2+\nu^2)]\,\mathrm{d}k}$$

for $r > c$. By elementary properties of the modified Bessel functions (see 9.6.1 and 9.6.26 in Abramowitz and Stegun [1]), we have that

$$I_n(\mu) > 0, \qquad K_n(\mu) > 0, \qquad K_n'(\mu) < 0 \quad \text{for } \mu > 0,$$

and so the integrand in the numerator is always negative and the integrand in the denominator is always positive within the range of integration. Hence

$$\left.\frac{\mathrm{d}r}{\mathrm{d}y}\right|_{y=0} > 0 \quad \text{for } r > c.$$

This means that John's condition required for the uniqueness proof in Subsection 4.2.3 is not satisfied. Despite this, the trapped modes constructed are entirely consistent with the uniqueness results in that subsection.

4.2. Uniqueness Theorems

In Subsection 4.2.1 of the present section, we show that surface-piercing structures exist such that the uniqueness theorem holds for all values of $\nu > 0$ despite the fact that they isolate a bounded portion of the free surface from infinity. In three dimensions, these structures are vertical shells of arbitrary horizontal cross section, and the corresponding two-dimensional geometries are presented by two arbitrary or three symmetric vertical barriers.

In Subsections 4.2.2 and 4.2.3, we turn to two-dimensional geometries with the mirror symmetry and axisymmetric toroidal bodies. The reason for this stems from the fact that the intervals of nondimensional point eigenvalues (ν is nondimensionalized by the size of the bounded part of the free surface), corresponding to symmetric and antisymmetric (different azimuthal) trapped modes in the two-dimensional (axisymmetric) problem, do not overlap. Thus one might hardly expect that the uniqueness of symmetric and antisymmetric (different azimuthal) solutions holds simultaneously without some geometric restrictions. This conjecture is confirmed in Subsection 4.2.2, where we investigate the uniqueness conditions for a pair of surface-piercing infinite cylinders symmetric about a vertical plane. Under assumption that there are no parts of wetted contours beneath the isolated portion of the free surface, we apply John's method and prove the uniqueness of symmetric and antisymmetric solutions for values of a nondimensional spectral parameter belonging to complementary intervals. Moreover, for common ends of these intervals both solutions are unique.

Similar results are obtained in Subsection 4.2.3 for each azimuthal mode when an axisymmetric toroidal surface-piercing body is immersed in water. Again, John's condition is essential for the isolated part of the free surface. It requires that no submerged part of the body is contained within the cylinder extending downward from the inner circle of intersection between the surface of the body and the free surface.

The rest of the present section is concerned with uniqueness theorems of another type. In Subsection 4.2.4 (4.2.5) we demonstrate that for more restrictive geometries the uniqueness holds for pairs of surface-piercing cylinders (for toroidal bodies) in a frequency interval bounded above by a certain number depending on the geometry. At the same time, these intervals can overlap with the point spectrum of structures that are not subjected to these geometric restrictions. The method applied in Subsections 4.2.4 and 4.2.5 is based on estimating the potential energy of waves on the isolated component of the free surface through the combination of the kinetic energy and the potential energy on other components of the free surface. Different two-dimensional geometries are investigated in Subsection 4.2.4, including a single body in the presence of a coastline, and two or three bodies in finite-depth water. In each case a limited class of depth variations is allowed. An important feature of proofs in Subsections 4.2.4 and 4.2.5 that distinguishes the corresponding geometries from those in Subsections 4.2.2 and 4.2.3 is that no symmetry must be imposed on surface-piercing bodies and structures.

4.2.1. Uniqueness for Vertical Shells and Barriers

In Subsection 4.2.1.1, we give a detailed proof of uniqueness for a vertical shell and briefly further discuss three-dimensional geometries for which the uniqueness holds. The uniqueness results for vertical barriers are outlined in Subsection 4.2.1.2.

4.2.1.1. Vertical Shells

Let water occupy a layer

$$L = \{x \in \mathbb{R}^2, -d < y < 0\}$$

outside a shell S that (i) is a vertical cylindrical surface, that is, has its generators parallel to the y axis; (ii) has a boundary ∂S consisting of two horizontal edges belonging to planes $\{y = -a\}$ and $\{y = -b\}$, where $0 \leq a < b \leq d$, and equalities $a = 0$, $b = d$ cannot hold simultaneously; (iii) is assumed (for simplicity) to be smooth; that is, the projection of S onto the x plane is a simple closed C^2 curve ℓ, dividing $\mathbb{R}^2$ into a simply connected bounded domain F_0 and an infinite domain F_∞.

Thus $W = L \setminus S$ is the water domain; the free surface F coincides with $\{y = 0\}$ when $a > 0$, and $F = F_0 \cup F_\infty$ when $a = 0$; the bottom B coincides with $\{y = -d\}$ when $b < d$, and $B = \{x \in F_0 \cup F_\infty, y = -d\}$ when $b = d$.

Because of the presence of edges, a condition near ∂S should be imposed, and so we assume u to belong to the Sobolev space $H^1(W_a)$ for any finite $a > 0$, where $W_a = W \cap \{|x| < a\}$. It is well known (see, for example, Maz'ya and Rossmann [227]) that if $u \in H^1(W_a)$ for a large enough (so that $S \subset \overline{W_{a-1}}$), then u is continuous throughout $\bar{W}$, and near the immersed edge (where $a \neq 0, b \neq d$) we have

$$|\nabla u(x, y)| = O\big(\rho^{-1/2}\big) \quad \text{as } \rho \to 0. \tag{4.30}$$

We denote by ρ the distance of a point $(x, y) \in W$ from ∂S. When $a = 0$ or $b = d$, one can replace (4.30) by

$$|\nabla u(x, y)| = O(1) \quad \text{as } \rho \to 0. \tag{4.31}$$

The question of uniqueness reduces to the demonstration that u vanishes when it satisfies the homogeneous water-wave problem. It was shown (see Subsection 2.2.1) that for such u the total energy is finite; that is,

$$\int_W |\nabla u|^2 \,\mathrm{d}x\mathrm{d}y + \nu \int_F |u|^2 \,\mathrm{d}x < \infty.$$

Then, the equipartition of the kinetic and potential energy

$$\int_W |\nabla u|^2 \,\mathrm{d}x\mathrm{d}y - \nu \int_F |u|^2 \,\mathrm{d}x = 0 \tag{4.32}$$

is a consequence of Green's formula.

The main aim of the present subsection is to prove the following result.

Under assumptions (i)–(iii) the homogeneous water-wave problem has only a trivial solution.

As in Subsection 3.2.1, we consider the simple wave component of order zero,

$$w(x) = \int_{-d}^{0} u(x, y) \cosh k_0(y + d) \,\mathrm{d}y, \tag{4.33}$$

defined for $x \in F_0 \cup F_\infty$ under assumptions (i)–(iii). It was demonstrated in Subsection 3.2.1.2 that integrating by parts twice on the right-hand side of

$$\nabla_x^2 w = -\int_{-d}^{0} u_{yy}(x, y) \cosh k_0(y + d) \,\mathrm{d}y,$$

we get

$$\nabla_x^2 w + k_0^2 w = 0. \tag{4.34}$$

In what follows, two other assertions from Subsections 3.2.1.2 and 3.2.1.3

will be used: (i) from the radiation condition and (4.34) one obtains

$$w = 0 \quad \text{in } F_\infty; \tag{4.35}$$

(ii) for the water region $W_\infty = \{x \in F_\infty, -d < y < 0\}$, it follows from (4.35) that

$$\nu \int_{F_\infty} |u|^2 \,\mathrm{d}x \leq \frac{1}{2} \int_{W_\infty} |u_y|^2 \,\mathrm{d}x\mathrm{d}y. \tag{4.36}$$

Attention is now turned to obtaining a similar inequality between the potential and kinetic energy for the water region

$$W_0 = \{x \in F_0, -d < y < 0\} \quad \text{bounded above by} F_0.$$

First, let us prove that $\partial w/\partial n_0$ is continuous across ℓ (n_0 is the unit normal to ℓ). We note that by (4.30) and (4.31) the integrals defining $\nabla_x w$ converge absolutely and uniformly on either side of ℓ (in F_0 and F_∞), and the uniformity across ℓ also takes place. Moreover, $\partial u/\partial n$ is continuous across the vertical cylindrical surface having ℓ as the director (of course, only the part within W is considered, and edges of S should be excluded). On int S, the homogeneous Neumann condition yields that $\partial u/\partial n$ is continuous. Outside S, the continuity is a consequence of smoothness of solutions to the Laplace equation. Then $\partial w/\partial n_0$ is continuous across ℓ.

The last fact and (4.35) imply that

$$\partial w/\partial n_0 = 0 \quad \text{on } \ell, \tag{4.37}$$

where w is considered as a function in F_0. This and (4.34) show that the homogeneous water-wave problem may have a nontrivial solution only if $\nu = k_0 \tanh k_0 d$, where k_0^2 is an eigenvalue of (4.34) and (4.37) in F_0. Then for a solution of this eigenvalue problem Green's formula gives

$$\int_{F_0} |\nabla_x w|^2 \,\mathrm{d}x = k_0^2 \int_{F_0} |w|^2 \,\mathrm{d}x, \tag{4.38}$$

which is the crucial point for deriving an inequality similar to (4.36).

Integrating by parts in (4.33), we have the following for $x \in F_0$:

$$u(x,0) \sinh k_0 d = k_0 w(x) + \int_{-d}^{0} u_y(x,y) \sinh k_0(y+d) \,\mathrm{d}y.$$

Squaring this and using Cauchy's inequality with ϵ (its value is to be chosen later for convenience) and the Schwarz inequality, one gets

$$|u(x,0)\sinh k_0 d|^2 \leq (1+\epsilon)k_0^2|w(x)|^2$$
$$+(1+\epsilon^{-1})\left[\int_{-d}^{0}|u_y(x,y)|^2\,\mathrm{d}y\right]\left[\int_{-d}^{0}\sinh^2 k_0(y+d)\,\mathrm{d}y\right].$$

Let us calculate the last integral, integrate over F_0, and take into account (4.38), thus obtaining

$$\nu\int_{F_0}|u|^2\,\mathrm{dx} \leq \frac{\nu(1+\epsilon)}{\sinh^2 k_0 d}\int_{F_0}|\nabla_x w|^2\,\mathrm{d}x + \frac{(1+\epsilon^{-1})}{2}$$
$$\times\left(1-\frac{\nu d}{\sinh^2 k_0 d}\right)\int_{W_0}|u_y|^2\,\mathrm{d}x\mathrm{d}y. \tag{4.39}$$

On the other hand (4.33) gives

$$\nabla_x w(x) = \int_{-d}^{0}\nabla_x u(x,y)\cosh k_0(y+d)\,\mathrm{d}y.$$

Using the Schwarz inequality, we have

$$|\nabla_x w|^2 \leq \left[\int_{-d}^{0}|\nabla_x u(x,y)|^2\,\mathrm{d}y\right]\left[\int_{-d}^{0}\cosh^2 k_0(y+d)\,\mathrm{d}y\right].$$

After calculation of the last integral and integration over F_0, this inequality takes the form

$$\int_{F_0}|\nabla_x w|^2\,\mathrm{d}x \leq \frac{\sinh^2 k_0 d + \nu d}{2\nu}\int_{W_0}|\nabla_x u(x,y)|^2\,\mathrm{d}x\mathrm{d}y.$$

This and (4.39) combine to give

$$\nu\int_{F_0}|u|^2\,\mathrm{d}x \leq \frac{1+\epsilon}{2}\left(1+\frac{\nu d}{\sinh^2 k_0 d}\right)\int_{W_0}|\nabla_x u|^2\,\mathrm{d}x\mathrm{d}y$$
$$+\frac{1+\epsilon^{-1}}{2}\left(1-\frac{\nu d}{\sinh^2 k_0 d}\right)\int_{W_0}|u_y|^2\,\mathrm{d}x\mathrm{d}y.$$

The choice

$$\epsilon = \frac{\sinh^2 k_0 d - \nu d}{\sinh^2 k_0 d + \nu d},$$

which is positive by the definition of k_0, simplifies the last inequality to

$$\nu\int_{F_0}|u|^2\,\mathrm{d}x \leq \int_{W_0}|\nabla u|^2\,\mathrm{d}x\mathrm{d}y. \tag{4.40}$$

Finally, the combination of (4.40) with (4.36) produces

$$\nu \int_F |u|^2 \, \mathrm{d}x \leq \int_{W_0} |\nabla u|^2 \, \mathrm{d}x\mathrm{d}y + \frac{1}{2} \int_{W_\infty} |u_y|^2 \, \mathrm{d}x\mathrm{d}y.$$

Comparing this with (4.32), one immediately finds that $\nabla u = 0$ in W_∞, and as u is analytic ∇u vanishes throughout W. Then (4.32) shows that $u = 0$ on F, which substituted into the free surface boundary condition gives that $u_y = 0$ on F. Now, application of the uniqueness theorem for the Cauchy problem for the Laplace equation proves the theorem.

Let us discuss what novelty the theorem proven for any position of the shell – surface piercing, totally immersed, and bottom-standing – comprises when compared with other theorems. For a surface-piercing shell, the theorem shows that there are obstacles, separating a bounded portion of the free surface from infinity and such that the uniqueness theorem is valid for all frequencies in the three-dimensional problem. For two other cases mentioned above, the theorem also extends the uniqueness results given in Chapter 2, where uniqueness criteria are given for cases of totally submerged bodies and curved bottom.

Let us consider other geometries for which the method applied provides the uniqueness theorem. First we note that in the case of infinite depth, one simply has to use

$$w(x) = \int_{-\infty}^{0} u(x, y) \, e^{\nu y} \mathrm{d}y$$

instead of (4.33). Now we subject to further analysis the uniqueness conditions (i)–(iii). Condition (i) is crucial for defining w throughout $\mathbb{R}^2 \setminus \ell$, but it can be weakened in two directions as follows. Consider a finite number of rigid vertical cylinders extending throughout the depth, and having such smooth horizontal cross sections that their projections on the x plane are contained within ℓ. Then F_0 becomes a smooth multiply connected domain, but (4.38) still holds, and hence our considerations remain valid. Also, a finite number of shells is admissible if projections of their contours on the x plane are disjoint, and each of them lies outside the others.

Condition (ii) is not necessary, and any edge bounding the shell inside W might be an arbitrary smooth curve because (4.30) remains true in this case (see Maz'ya and Rossmann [227]). Even a shell extending throughout the depth is allowed, but such a shell must have a hole so that W is a connected fluid domain. Also, condition (iii) can be replaced by a requirement that ℓ is a piecewise smooth curve without cusps, but this involves more technical details.

4.2.1.2. Vertical Barriers

Here $L = \{-\infty < x < \infty; -d < y < 0\}$ and S is the boundary curve in the (x, y) plane of any vertical barriers within water; the water domain W is therefore that part of L outside $\bar{S}$. First, we consider two rigid vertical barriers, so that their wetted surfaces occupy the lines $S_\pm = \{x = \pm b; -a_\pm < y < 0\}$, $0 < a_\pm < d$. The free surface of the fluid is $F = \{|x| \neq b; y = 0\} = F_0 \cup F_\infty$, where the portion of the free surface that extends to infinity is $F_\infty = \{|x| > b; y = 0\}$, and the portion between the barriers is $F_0 = \{|x| < b; y = 0\}$.

For the particular case of two barriers, the homogeneous Neumann boundary condition on $S_\pm$ becomes

$$u_x(+b \pm 0, y) = 0 \quad \text{on } S_+, \qquad u_x(-b \pm 0, y) = 0 \quad \text{on } S_-. \tag{4.41}$$

As in Subsection 4.2.1.1, the boundary of the water domain is not smooth, and we have to take into account the condition that the kinetic and potential energy is locally finite. Then it follows from the general theory of elliptic boundary value problems in piecewise smooth domains (see [227]) that u is continuous throughout $\bar{W}$ and at the submerged barrier tips we have

$$|\nabla u(x, y)| = O\left(\rho^{-1/2}\right) \quad \text{as } \rho \to 0,$$

where ρ is the distance of a point $(x, y) \in W$ from that tip. Also, in each corner between a barrier and the free surface

$$|\nabla u(x, y)| = O(1) \quad \text{as } \rho \to 0,$$

where ρ is the distance of a point $(x, y) \in W$ from that corner.

As in Subsection 4.2.1.1, it is convenient to consider

$$w(x) = \int_{-d}^{0} u(x, y) \cosh k_0(y + d)\, dy,$$

which is given throughout F and satisfies $w_{xx} + k_0^2 w = 0$ there. The general solution of this equation is

$$w(x) = C_1 \cos k_0 x + C_2 \sin k_0 x,$$

and hence from the radiation condition we get (see Subsection 3.2.1)

$$w(x) = 0 \quad \text{for } x \in F_\infty.$$

Now the continuity of w_x across $x = \pm b$ follows as above from (4.41), the local asymptotics near tips and corner points, and harmonicity of u. Hence the boundary condition $w_x = 0$ for $|x| = b$ accompanies the equation for w

on F_0. Then the same considerations as in Subsection 4.2.1.1 lead to the following theorem.

For two surface-piercing barriers the homogeneous water-wave problem has only a trivial solution.

In fact, this result holds for more geometries than two surface-piercing barriers. The proof does not depend in any way on the positions of the barriers on the lines $x = \pm b$. Thus the water-wave problem has only a trivial solution for any configuration of barriers lying on these vertical lines. The only requirement is that at least one of the lines has one or more gaps as the analytic continuation argument breaks down without a gap in at least one of the lines. Moreover, this result holds in the case of infinite depth, for which the only known explicit solution for two surface-piercing structures was given by Levine and Rodemich [195], who considered the problem of scattering by a pair of identical barriers.

Let us turn to the problem of three symmetrically arranged rigid surface-piercing barriers. Their wetted surfaces occupy the lines

$$S_\pm = \{x = \pm b; -a < y < 0\}, \qquad S_0 = \{x = 0; -a_0 < y < 0\},$$

with $0 < a, a_0 < d$. The free surface of the fluid is

$$F = \{x \neq 0, \pm b; y = 0\} = F_\infty \cup F_{-0} \cup F_{+0},$$

where $F_\infty = \{|x| > b; y = 0\}$ is the portion of the free surface that extends to infinity, and $F_{\pm 0} = \{0 < \pm x < \pm b; y = 0\}$ are the portions between the barriers. The water domain is $W = L \setminus \{\bar{S}_- \cup \bar{S}_0 \cup \bar{S}_+\}$. The boundary conditions on the rigid barriers are

$$u_x(+b \pm 0, y) = 0 \quad \text{on } S_+, \qquad u_x(\pm 0, y) = 0 \quad \text{on } S_0,$$
$$u_x(-b \pm 0, y) = 0 \quad \text{on } S_-.$$

Again, conditions prescribing the asymptotic behavior near tips and corner points must hold.

Since the geometry is symmetric about the y axis, it is possible to split u into a sum of symmetric and antisymmetric potentials. It is obvious that the symmetric solution $u^{(+)}$ defined by $u^{(+)}(-x, y) = u^{(+)}(x, y)$ satisfies

$$u_x^{(+)}(0, y) = 0 \quad \text{on } -d < y < 0.$$

Hence the symmetric case is just a special case of the two-barrier problem in which one barrier, corresponding to the line of symmetry for the three-barrier problem, extends throughout the depth. Thus the following result is established immediately.

For a symmetric arrangement of three barriers, the homogeneous water-wave problem has no symmetric solution other than a trivial one.

4.2.2. Two Cylinders Symmetric About a Vertical Plane

Here we consider the two-dimensional problem for finite (as shown in Fig. 4.7) as well as infinite water depths. Let two surface-piercing bodies occupy the domains D_+ and D_- and be symmetric about the y axis (see Fig. 4.7). We assume that $D_\pm$ is contained between straight lines through the end points of the wetted boundaries $S_\pm$, a vertical line through the end point $(\pm b, 0)$ nearest to the y axis, and a line making an angle β with the free surface through the other point. This angle β is equal to $\pi/4$, if the depth is infinite. For the finite depth case $\beta = \pi/2 - \arctan p$, where p is a positive root of equation (3.38). For the sake of simplicity one can take $\beta = 45°2/3$ in the latter case, since this angle is larger than all angles arising from (3.38). In the next subsection we consider the case of finite depth in detail in the form that allows the subsequent tending of the depth to infinity.

As in Subsection 4.2.1, we have to note that, generally speaking, the boundary of the water domain is not smooth, since a surface-piercing body meets the free surface transversally. Thus the reader has to take into account the description of singular behavior of the velocity field in a vicinity of such an intersection (see the introductory remarks to Subsection 3.1.1).

4.2.2.1. Water of Finite Depth

As usual, W denotes the domain

$$\{-\infty < x < +\infty, -d < y < 0\} \setminus (\bar{D}_+ \cup \bar{D}_-), \quad 0 < d \leq +\infty,$$

occupied by water. That part of the free surface (water domain) contained between the vertical lines through $(\pm b, 0)$ is labeled F_0 (W_0). The remainder of the free surface exterior to the bodies is denoted by F_∞, and the part of

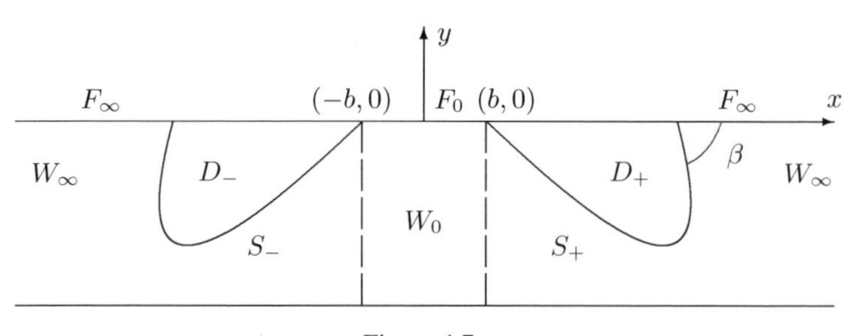

Figure 4.7.

W contained between F_∞, the bottom (if the depth is finite), and the lines making an angle β with the horizontal through the end points of $S_\pm$ is labeled W_∞.

As is shown in Subsection 2.2.1, a solution u to the homogeneous water-wave problem satisfies the following condition

$$\int_W |\nabla u|^2 \,\mathrm{d}x\mathrm{d}y + \nu \int_{F_0 \cup F_\infty} |u|^2 \,\mathrm{d}x < \infty,$$

which means that the kinetic and potential energy are finite. Moreover, without loss of generality u may be considered to be real.

Using the symmetry of W we can decompose u into the sum of a symmetric part $u^{(+)}$ and an antisymmetric part $u^{(-)}$ so that

$$u^{(\pm)}(x, y) = \pm u^{(\pm)}(-x, y).$$

It clearly follows that

$$u_x^{(+)}(0, y) = 0, \quad u^{(-)}(0, y) = 0. \tag{4.42}$$

Symmetric and antisymmetric solutions, $u^{(+)}$ and $u^{(-)}$, respectively, may be considered simultaneously; thus in what follows, each formula containing $\pm$ must be considered as two formulae for two different problems.

As is shown in Subsection 3.2.2.1, the condition that $D_+ \cup D_-$ lies between the lines at an angle β to the horizontal implies the inequality

$$\nu \int_{F_\infty} \left|u^{(\pm)}\right|^2 \mathrm{d}x \leq \int_{W_\infty} \left|\nabla u^{(\pm)}\right|^2 \mathrm{d}x\mathrm{d}y, \tag{4.43}$$

since $u^{(+)}$ and $u^{(-)}$ are solutions to the homogeneous water-wave problem. Applying Green's theorem to $u^{(\pm)}$ we get

$$\int_W \left|\nabla u^{(\pm)}\right|^2 \mathrm{d}x\mathrm{d}y = \nu \int_{F_0 \cup F_\infty} \left|u^{(\pm)}\right|^2 \mathrm{d}x. \tag{4.44}$$

As in Subsection 4.2.1, we consider a simple wave component of order zero for $u^{(\pm)}$, but only on the inner part of the free surface F_0. More specifically, we first assume that $d < \infty$, in which case the coefficient of the simple zero-order component is as follows:

$$w^{(\pm)}(x) = \int_{-d}^{0} u^{(\pm)}(x, y) \cosh k_0(y + d)\,\mathrm{d}y.$$

It was shown in Subsection 3.2.1 that this function satisfies $w_{xx}^{(\pm)} + k_0^2 w^{(\pm)} = 0$. Then (4.42) yields

$$w^{(\pm)}(x) = C_\pm \cos\left(k_0 x - \frac{\pi}{4} \pm \frac{\pi}{4}\right) \quad \text{for } -b < x < b,$$

where $C_\pm$ are arbitrary constants. Integration by parts in

$$C_\pm \cos\left(k_0 x - \frac{\pi}{4} \pm \frac{\pi}{4}\right) = \int_{-d}^{0} u(x, y) \cosh k_0(y + d)\, \mathrm{d}y \qquad (4.45)$$

leads to

$$u^{(\pm)}(x, 0) \sinh k_0 d = k_0 C_\pm \cos\left(k_0 x - \frac{\pi}{4} \pm \frac{\pi}{4}\right) + \int_{-d}^{0} u_y^{(\pm)}(x, y) \sinh k_0(d + y)\, \mathrm{d}y.$$

Applying Cauchy's inequality with ϵ to the right-hand side and the Schwarz inequality to the last integral, we obtain

$$\left|u^{(\pm)}(x, 0)\right|^2 \sinh^2 k_0 d \le (1 + \epsilon) k_0^2 C_\pm^2 \cos^2\left(k_0 x - \frac{\pi}{4} \pm \frac{\pi}{4}\right) + (1 + \epsilon^{-1}) \frac{\sinh^2 k_0 d - \nu d}{2\nu} \int_{-d}^{0} \left|u_y^{(\pm)}(x, y)\right|^2 \mathrm{d}y, \qquad (4.46)$$

where the following identity has been used:

$$\int_{-d}^{0} \sinh^2 k_0(y + d)\, \mathrm{d}y = \frac{\sinh^2 k_0 d}{2\nu} - \frac{d}{2}.$$

Alternatively, from (4.45), we have

$$-k_0 C_\pm \sin\left(k_0 x - \frac{\pi}{4} \pm \frac{\pi}{4}\right) = \int_{-d}^{0} u_x(x, y) \cosh k_0(y + d)\, \mathrm{d}y,$$

which, after application of the Schwarz inequality, implies that

$$k_0^2 C_\pm^2 \sin^2\left(k_0 x - \frac{\pi}{4} \pm \frac{\pi}{4}\right) \le \frac{\sinh^2 k_0 d + \nu d}{2\nu} \int_{-d}^{0} \left|u_x^{(\pm)}(x, y)\right|^2 \mathrm{d}y. \qquad (4.47)$$

Let us assume that the inequality

$$\mp \sin 2k_0 b \ge 0 \qquad (4.48)$$

is true. This is equivalent to the condition

$$\int_0^b \cos^2\left(k_0 x - \frac{\pi}{4} \pm \frac{\pi}{4}\right) \mathrm{d}x \le \int_0^b \sin^2\left(k_0 x - \frac{\pi}{4} \pm \frac{\pi}{4}\right) \mathrm{d}x.$$

Integrating (4.46) and (4.47) over F_0 and using the preceding inequality, we obtain

$$\nu \int_{F_0} \left|u^{(\pm)}(x,0)\right|^2 \mathrm{d}x \le (1+\epsilon)\frac{\sinh^2 k_0 d + \nu d}{2\sinh^2 k_0 d} \int_{W_0} \left|u_x^{(\pm)}\right|^2 \mathrm{d}x\mathrm{d}y$$
$$+ (1+\epsilon^{-1})\frac{\sinh^2 k_0 d - \nu d}{2\sinh^2 k_0 d} \int_{W_0} \left|u_x^{(\pm)}\right|^2 \mathrm{d}x\mathrm{d}y.$$

The choice

$$\epsilon = \frac{\sinh^2 k_0 d - \nu d}{\sinh^2 k_0 d + \nu d},$$

which is positive by the definition of k_0, simplifies the last inequality to

$$\nu \int_{F_0} \left|u^{(\pm)}(x,0)\right|^2 \mathrm{d}x \le \int_{W_0} \left|\nabla u^{(\pm)}\right|^2 \mathrm{d}x\mathrm{d}y,$$

which when added to (4.43) produces

$$\nu \int_{F_0\cup F_\infty} \left|u^{(\pm)}\right|^2 \mathrm{d}x \le \int_{W_0\cup W_\infty} \left|\nabla u^{(\pm)}\right|^2 \mathrm{d}x\mathrm{d}y.$$

This contradicts (4.44) unless $u^{(\pm)} \equiv 0$ in W. Thus, the following theorem is proved.

Let $u^{(\pm)}$ be a solution to the homogeneous water-wave problem satisfying (4.42). *If* (4.48) *holds with the upper (lower) sign, then $u^{(+)}$ $[u^{(-)}]$ vanishes identically in W, having finite depth and satisfying the conditions listed at the beginning of this section.*

Note that the assumption that the bottom is flat under $S_+ \cup S_-$ and under inclined lines bounding W_∞ is superfluous. These two straight segments can be replaced by arbitrary curves, which are symmetric about the y axis and have no parts within $W_0 \cup W_\infty$. Further discussion of (4.48) is given in the next subsection for the case of deep water.

4.2.2.2. Deep Water

Letting $d \to \infty$, we note that the left-hand side in (4.48) vanishes, and $k_0 \to \nu$. Thus, (4.48) takes the form $\mp \sin 2\nu b \ge 0$, which is equivalent to

$$\pi\,(m + 1/4 \pm 1/4) \le \nu b \le \pi\,(m + 3/4 \pm 1/4)\,, \quad m = 0, 1, \ldots. \tag{4.49}$$

Hence we arrive at the following theorem.

Let W have infinite depth, and the conditions listed above hold. If $u^{(\pm)}$ satisfies the homogeneous water-wave problem, (4.42) *is true, and νb satisfies* (4.49), *then $u^{(\pm)}$ is trivial.*

A direct proof of this theorem is similar to the proof given above, where

$$w(x) = \int_{-\infty}^{0} u(x, y) e^{\nu y} \, dy$$

should be used.

It immediately follows from the last theorem that *both* $u^{(+)}$ *and* $u^{(-)}$ *vanish identically in* W, *when*

$$\nu b = m\pi/2, \quad m = 1, 2, \ldots . \tag{4.50}$$

We note that values of νb given by (4.50) for which the homogeneous problem has only a trivial solution are exactly the eigenvalues of the two-dimensional sloshing problem in an infinitely deep basin between vertical walls $x = \pm b$.

Now, let us demonstrate that the last theorem cannot be essentially improved in view of the following assertion (see Subsection 4.1.1).

Every interval of uniqueness for symmetric solution contains a subinterval such that nontrivial antisymmetric solutions to the homogeneous problem do exist for certain geometries and values of νb *in this subinterval.*

By (4.49) $u^{(+)}$ is unique for

$$\pi(m + 1/2) \leq \nu b \leq \pi(m + 1), \quad m = 0, 1, \ldots . \tag{4.51}$$

According to the second theorem in Subsection 4.1.1, $u_{m+1}^{(-)}$ given by (4.1) and (4.2) provides the trapped mode potential for a structure, whose boundary is given by (4.6) where

$$V_{-}^{(1)} = V_{-}^{(2)} \in (0, \mathrm{Si}(2\pi m + 2\pi)).$$

Thus, we have a certain family of symmetric trapping structures. Denoting by $(b, 0)$ the right innermost end point of such structure, we see that it lies between the greatest zero, $x_{\max}$ of $v_{m+1}^{(-)}(x, 0)$, and the source point $(a_{-}, 0)$, where $\nu a_{-} = \pi\,(m + 1)$. By (4.10) $\nu x_{\max} > \pi(m + 1/2)$. Hence, the values of νb, producing non-uniqueness examples in the antisymmetric case, belong to a subinterval of (4.51).

The same result is valid for structures trapping the first three symmetric modes, and it can be verified in the same way. There is numerical evidence (see Figs. 4.1 and 4.2) that a similar result holds for higher symmetric modes.

4.2.2.3. On the Significance of John's Condition for Uniqueness

Here we are concerned with the following consequence of the uniqueness theorem proven in Subsection 4.2.2.2.

The interval $(\pi/2, \pi)$ *is free of point eigenvalues of the symmetric water-wave problem provided two geometric conditions hold:* (*i*) *John's condition* *on the inner part of the free surface and (ii) the condition on exterior parts of the free surface, which is similar to that of John with rays inclined at $\pi/4$ to the vertical instead of vertical ones.*

The aim of the present subsection is to investigate the proposition contrary to this assertion, which says the following.

If there is a nondimensionalized (multiplied by b) point eigenvalue of the symmetric water-wave problem in $(\pi/2, \pi)$, then either (i) or (ii) is violated.

In fact, we will be concerned with the violation of (i). The idea is to construct such a pair of surface-piercing bodies, trapping a symmetric mode with $\nu b \in (\pi/2, \pi)$, that this pair does not satisfy (i). For this purpose two dipoles are placed on the x axis, and, they are positioned so that the waves radiating from each cancel at infinity. Here we restrict ourselves to the interval $(\pi/2, \pi)$ for the sake of simplicity only; the same approach works for other intervals (4.49), but we leave details to the reader.

As in Subsection 4.1.1, two functions are needed for formulation of the main result. We define the first of them as follows:

$$u(x, y) = \frac{1}{2\nu}\left[G_x\left(x, y; -\frac{\pi}{\nu}, 0\right) - G_x\left(x, y; \frac{\pi}{\nu}, 0\right)\right], \tag{4.52}$$

where G is the two-dimensional Green's function defined in Subsection 1.2.1. The choice of the dipole points cancels the integrals along indentations in the definition of G in (4.52), and one immediately obtains that

$$\begin{aligned} u(x, y) &= \frac{1}{\nu}\int_0^\infty \frac{k}{k-\nu} e^{ky}\left[\sin k\left(x - \frac{\pi}{\nu}\right) - \sin k\left(x + \frac{\pi}{\nu}\right)\right] \mathrm{d}k \\ &= \int_0^\infty \frac{k}{k-1} e^{k\nu y}\left[\sin k(\nu x - \pi) - \sin k(\nu x + \pi)\right] \mathrm{d}k, \end{aligned}$$

where the integrands are bounded because the singularities in the denominators coincide with zeros of the numerators. Thus, $u(x, y)$ is a real harmonic function in the lower half-plane $\mathbb{R}^2_-$ even with respect to x, and the free surface boundary condition holds for it on $\{x \neq \pm\pi/\nu, y = 0\}$. Further, one immediately obtains that

$$\begin{aligned} u(x, y) = &\frac{1}{\nu}\left[\frac{x + \pi/\nu}{(x + \pi/\nu)^2 + y^2} - \frac{x - \pi/\nu}{(x - \pi/\nu)^2 + y^2}\right] \\ &+ \int_0^\infty \frac{\sin k(\nu x - \pi) - \sin k(\nu x + \pi)}{k-1} e^{k\nu y}\, \mathrm{d}k. \end{aligned}$$

The properties of last integral are similar to those derived for integrals in Subsection 4.1.1.2. In particular, the integral is bounded as $z \to \pm\pi/\nu$ and decays as $|z| \to \infty$. So u has finite kinetic and potential energies in every water domain W that does not contain a neighborhood of the dipole points $(\pm\pi/\nu, 0)$.

The second required function is defined as follows:

$$v(x, y) = \int_0^\infty \frac{k}{k-1} e^{k\nu y} \left[\cos k(\nu x - \pi) - \cos k(\nu x + \pi)\right] \mathrm{d}k$$
$$= \frac{1}{\nu}\left[\frac{y}{(x + \pi/\nu)^2 + y^2} - \frac{y}{(x - \pi/\nu)^2 + y^2}\right]$$
$$+ \int_0^\infty \frac{\cos k(\nu x - \pi) - \cos k(\nu x + \pi)}{k-1} e^{k\nu y}\, \mathrm{d}k; \qquad (4.53)$$

that is, v is a harmonic function conjugate to u. Equivalently, v is the stream function corresponding to the velocity potential u, and having an arbitrary constant term equal to zero.

The function v allows us to construct a family of water domains W, so that u satisfies the homogeneous water-wave problem in W, and this domain does not satisfy (i). In fact, any streamline (a level line of v) may be used as S_+ (see Fig. 4.7 for the notation), if it has the following two properties. It connects with the positive x axis on either side of the dipole point $(\pi/\nu, 0)$. The angle directed into W between the streamline and the positive x axis is acute on the left of $(\pi/\nu, 0)$. On Fig. 4.8(b) a number of streamlines having these properties are plotted (the bold point marks the dipole position);

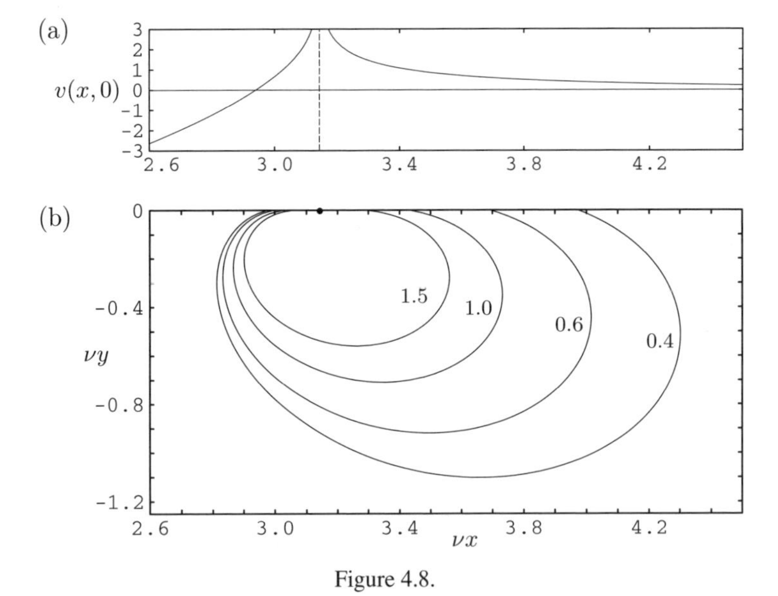

Figure 4.8.

on Fig. 4.8(a) the corresponding part of the graph of $v(x, 0)$ is shown for convenience. Since $v(x, y)$ is an odd function with respect to x, the reflection of S_+ in the y axis is also a streamline, which we take as S_-. Now, let us formulate the theorem concerning the existence of streamlines with these properties.

For every level $V > 0$ there exists only one streamline,

$$S_+(V) = \{(x, y) : v(x, y) = V\}, \tag{4.54}$$

with all internal points in $\{x > 0, y < 0\}$ and the end points $(x_V^{(\pm)}, 0)$ such that $x_V^{(\pm)} > 0$, $\pm(x_V^{(\pm)} - \pi/\nu) > 0$, and $x_V^{(-)}\nu > 2\pi/3$. For every streamline $S_+(V)$ John's condition does not hold on the left of $x_V^{(-)}$.

We note that $x_V^{(-)} = b$ for the water domain W having $S_+(V)$ and its reflection in the y axis as the wetted rigid contours. Thus we have

$$2\pi/3 < \nu b = \nu x_V^{(-)} < \pi$$

for W defined by this streamline. Since u given by (4.52) delivers a symmetric eigenfunction in this domain W, condition (i) is violated in the proposition under consideration. Numerical calculations demonstrate that some of constructed examples also violate (ii), whereas the others satisfy this condition (see Fig. 4.8). The question about the existence of an example, contradicting the uniqueness theorem and such that (i) holds and (ii) is violated, is still an open question.

We begin the proof of properties for stream lines by noting that these lines must have end points either on the free surface or at infinity (like those in Subsection 4.1.1.2). It follows from the asymptotics of G (see Subsection 1.2.1) that the nodal lines, that is, the loci, where $v = 0$ are the only streamlines going to infinity. These lines divide the water domain into subdomains, each containing a family of contours with the same properties. Thus we have to consider the behavior of the function $v(x, 0)$ and of the nodal lines of $v(x, y)$ in the quadrant $\{x > 0, y < 0\}$.

The theorem's proof uses a number of simple lemmas. We formulate them as needed, and their proofs are given at the end of the present subsection.

1. *The function $v(x, 0)$ has only one positive zero ξ_0. It belongs to $(2\pi/3\nu, \pi/\nu)$, and $\pm v(x, 0) > 0$ for $\pm(x - \xi_0) > 0$, $x \neq \pi/\nu$.*

This lemma is illustrated in Fig. 4.9, where the graph of $v(x, 0)$ is shown by a solid line, and the right arrow marks the point $2\pi/3$ in nondimensional coordinates.

2. *The function $v_x(x, 0)$ has only one positive zero ξ_1. It belongs to $(x^*, 2\pi/3\nu)$, where $\nu x^* = \arcsin(4e)^{-1}$. Moreover, $\pm v_x(x, 0) > 0$ for $\pm(x - \xi_1) > 0$, $x < \pi/\nu$, and $v_x(x, 0) < 0$ for $x > \pi/\nu$.*

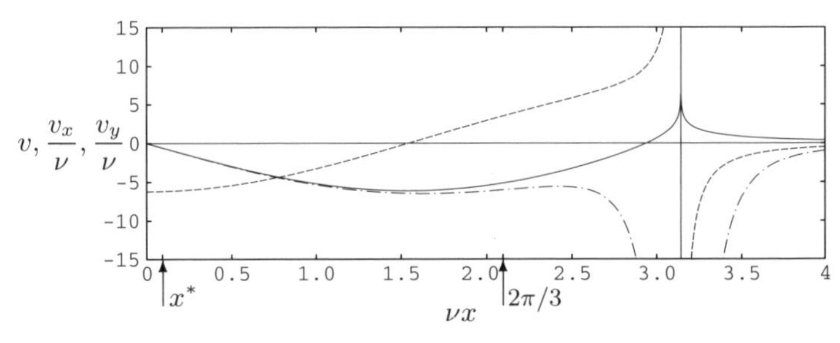

Figure 4.9.

This lemma is also illustrated in Fig. 4.9, where the graph of $\nu^{-1}v_x(x, 0)$ is shown by a dashed line, and the left arrow marks the point νx^* in non-dimensional coordinates.

From Lemma 2, it follows that $v(x, 0)$ increases and decreases strictly monotonically in intervals $(\xi_0, \pi/\nu)$ and $(\pi/\nu, +\infty)$, respectively. From Lemma 1 $v(x, 0) > 0$ only in these intervals of the positive x axis, and hence a streamline emanating from the x axis at a point of the first interval reenters this axis on the right of the dipole point and encloses this point inside. These streamlines form the family (4.54), which proves the first assertion of the theorem.

To complete theorem's proof we have to show that for the streamlines (4.54) John's condition does not hold on the left-hand side. The latter is true if

$$\frac{v_x(x, 0)}{v_y(x, 0)} < 0 \quad \text{when } x \in \left(\xi_0, \frac{\pi}{\nu}\right).$$

This is a consequence of Lemma 2 and the following lemma.

3. *The function $v_y(x, 0)$ is strictly negative for $x \in (0, \pi/\nu)$.*

See Fig. 4.9, where the graph of $\nu^{-1}v_y(x, 0)$ is shown by alternating dashes and points.

In addition to the results obtained above, it is of worth to consider the behavior of nodal lines $v(x, y) = 0$ at infinity. We remind the reader that $v(x, y)$ is an odd function with respect to x, and hence the negative y axis is a nodal line. The following lemma guarantees the absence of saddle points on the negative y axis.

4. *The function $v_x(0, y)$ is strictly negative for $y \leq 0$.*

This lemma does not allow the nodal line emanating from $(\xi_0, 0)$ to intersect the negative y axis. Thus this line is confined within the quadrant $\{x > 0, y < 0\}$, and its behavior at infinity can be derived from another representation for v:

$$v(x, y) = \frac{\nu y}{(\nu x + \pi)^2 + (\nu y)^2} - \frac{\nu y}{(\nu x - \pi)^2 + (\nu y)^2}$$
$$+ 2\pi \operatorname{Re} \int_0^\infty \frac{e^{ik} \, dk}{(k + \nu z)^2 - \pi^2}. \tag{4.55}$$

The latter is a consequence of (4.53) and the following lemma.

5. *For $y, \eta \leq 0$ we have*

$$\int_0^\infty \frac{\cos k(x - \xi)}{k - \nu} e^{k(y+\eta)} \, dk + \pi e^{\nu(y+\eta)} \sin \nu(x - \xi) = \operatorname{Re} \int_0^\infty \frac{e^{ik} \, dk}{k + \nu(z - \bar{\zeta})}.$$

Integrating by parts in (4.55) we get

$$v(x, y) = 4\pi \operatorname{Im} \int_0^\infty \frac{e^{ik}(k + \nu z) \, dk}{[(k + \nu z)^2 - \pi^2]^2}. \tag{4.56}$$

Another integration by parts leads to the asymptotic formula:

$$v(x, y) \sim \frac{4\pi x(x^2 - 3y^2)}{\nu^3(x^2 + y^2)^3} \quad \text{as } |z| \to \infty.$$

It means that the nodal line approaches the straight line $y = -3^{-1/2}x + c$ as $x \to +\infty$. We determine the coefficient c by using the two-term asymptotic formula for v:

$$v = 4\pi x \frac{(x^2 + y^2)(x^2 - 3y^2) - 12y(x^2 - y^2)}{(x^2 + y^2)^4} + O(|z|^{-5}),$$

which requires two integrations by parts in (4.56). Finally, the asymptote of the nodal line can be written as $y = -3^{-1/2}x - 1$ (see Fig. 4.10).

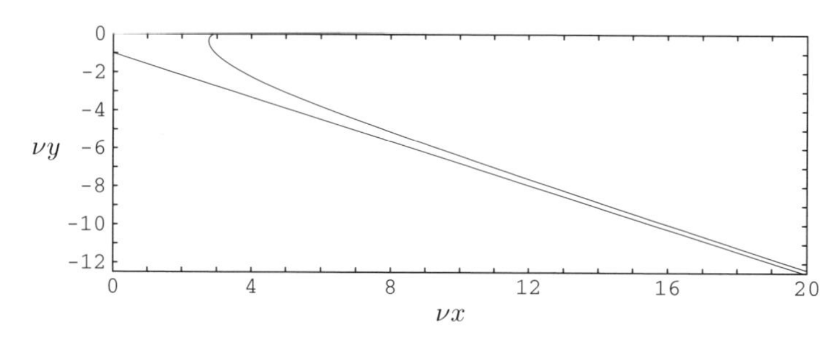

Figure 4.10.

Proof of Lemmas

Proof of Lemma 1. Let us demonstrate that $v(x, 0) < 0$ for $x \in (0, 2\pi/3\nu]$. From 3.354.2 and 3.722.7 in Gradshteyn and Ryzhik [96], we get

$$\int_0^{+\infty} \frac{\cos ak}{k-1}\, dk = \int_0^{+\infty} \frac{ke^{-ak}}{1+k^2}\, dk - \pi \sin a, \quad a > 0. \tag{4.57}$$

Then for $x \in (0, \pi/\nu)$ we have

$$v(x, 0) = I(x) - 2\pi \sin x\nu, \qquad I(x) = \int_0^{+\infty} \frac{k\left[e^{(x\nu-\pi)k} - e^{-(x\nu+\pi)k}\right]}{1+k^2}\, dk. \tag{4.58}$$

It is obvious that $I'(x) > 0$, and hence $I(x)$ is a nonnegative increasing function. Thus we get for $x \in (0, 2\pi/3\nu]$

$$I(x) \leq I(2\pi/3\nu) \leq e^{-1} \int_0^{+\infty} \frac{e^{(1-\pi/3)k}}{1+k^2}\, dk < e^{-1} \int_0^{+\infty} \frac{dk}{1+k^2} = \frac{\pi}{2e},$$

because $e^{k-1} \geq k$ for $k \geq 0$. Then $v(x, 0) < 0$ for $x \in [x^*, 2\pi/3\nu]$, where $\nu x^* = \arcsin(4e)^{-1}$, so that $2\pi \sin x\nu \geq \pi/2e$ in this interval.

Now, let us prove that $v(x, 0) < 0$ for $x \in (0, x^*]$. Assuming the contrary and taking into account that $v(0, 0) = 0$, we note that then there exists $\xi \in (0, x^*)$ such that $v_x(\xi, 0) = 0$. This is impossible because

$$v_x(x, 0) < 2\pi\nu\{[\pi^2 - (x^*\nu)^2]^{-1} - \cos x^*\nu\} < 0 \quad \text{for } x \in (0, x^*],$$

and this inequality follows from

$$v_x(x, 0) = 2\pi\nu\{[\pi^2 - (x\nu)^2]^{-1} - \cos x\nu\} - \nu \int_0^{+\infty} \frac{e^{(x\nu-\pi)k} + e^{-(x\nu+\pi)k}}{1+k^2}\, dk, \tag{4.59}$$

which is valid for $x \in (0, \pi/\nu)$.

We have that $v(x, 0) < 0$ for $x \in (0, 2\pi/3\nu]$ and $v(x, 0) \to +\infty$ as $x \to \pi/\nu$. Hence, $v(x, 0)$ vanishes at some point $\xi_0 \in (2\pi/3\nu, \pi/\nu)$.

To complete the proof we have to show that there is only one zero of $v(x, 0)$. Using (3.27) we get that

$$v(x, 0) = \int_0^{+\infty} \frac{k\left[e^{(\pi-x\nu)k} - e^{-(\pi+x\nu)k}\right]}{1+k^2}\, dk > 0 \quad \text{for } x > \pi/\nu.$$

Moreover, when $x\nu \in [2\pi/3, \pi)$, the inequalities

$$-\cos x\nu \geq \frac{1}{2}, \quad \int_0^{+\infty} \frac{e^{(x\nu-\pi)k} + e^{-(x\nu+\pi)k}}{1+k^2}\, dk \leq \pi,$$

and (4.59) imply that

$$v_x(x,0) \geq 2\pi\nu[\pi^2 - (x\nu)^2]^{-1} > 0.$$

This guarantees that there exists only one zero.

Proof of Lemma 2. It follows from the proof of Lemma 1 that $v_x(x,0) > 0$ for $x \in (2\pi/3\nu, \pi/\nu)$, and $v_x(x,0) < 0$ for $x \in (0, x^*) \cup (\pi/\nu, +\infty)$. Thus there exists at least one zero ξ_1 of $v_x(x,0)$ in the interval $(x^*, 2\pi/3\nu)$. To complete the proof we find it sufficient to show that $v_{xx} \neq 0$ in this interval. Differentiating (4.59), and comparing the result with $v(x,0)$, we get that

$$v_{xx}(x,0) = -\nu^2 v(x,0) + \frac{4\pi x\nu^3}{[\pi^2 - (x\nu)^2]^2}.$$

Since $v(x,0) < 0$ in $(x^*, 2\pi/3\nu)$,

$$v_{xx} > \frac{4\pi x\nu^3}{[\pi^2 - (x\nu)^2]^2} > 0$$

in this interval, which proves the assertion.

Proof of Lemma 3. Let us differentiate (4.53) with respect to y, and then put $y = 0$. Applying (4.57) to the result, we obtain that for $x \in (0, \pi/\nu)$,

$$v_y(x,0) = -\frac{4\pi\nu^2 x}{[\pi^2 - (x\nu)^2]^2} - 2\pi\nu\sin\nu x + \nu\int_0^\infty \frac{k\left[e^{(x\nu-\pi)k} - e^{-(x\nu+\pi)k}\right]}{1+k^2}\,dk. \tag{4.60}$$

From here and (4.58) it follows that

$$v_y(x,0) = \nu v(x,0) - \frac{4\pi\nu^2 x}{[\pi^2 - (x\nu)^2]^2}.$$

Then Lemma 1 implies that $v_y(x,0) < 0$ for $x \in (0, \xi_0]$.

Further, the integral in (4.60) can be estimated as follows:

$$\int_0^\infty \frac{k\left[e^{(x\nu-\pi)k} - e^{-(x\nu+\pi)k}\right]}{1+k^2}\,dk < \int_0^\infty \frac{ke^{(x\nu-\pi)k}}{1+k^2}\,dk$$
$$\leq \frac{1}{2}\int_0^\infty e^{(x\nu-\pi)k}\,dk = \frac{1}{2(\pi - x\nu)},$$

and we get that

$$v_y(x,0) < -\frac{4\pi\nu^2 x}{[\pi^2 - (x\nu)^2]^2} + \frac{\nu}{2(\pi - x\nu)} = \frac{\nu}{2[\pi^2 - (x\nu)^2]^2}P(x\nu), \tag{4.61}$$

where $P(t) = \pi^3 - t^3 + \pi^2 t - \pi t^2 - 8\pi t$. It is easy to see that

$$P'(t) = -8\pi + \pi^2 - 2\pi t - 3t^2 < 0 \quad \text{for } t > 0,$$

and $P(2\pi/3) = -16/3\pi^2 + 25/27\pi^3 < 0$. Then $P(x\nu) < 0$ for $x > 2\pi/3\nu > \xi_0$. This and (4.61) guarantee that $v_y(x, 0) < 0$ for $x \in (\xi_0, \pi/\nu)$, which completes the proof because it was shown above that $v_y(x, 0) < 0$ for $x \in (0, \xi_0]$.

Proof of Lemma 4. It follows from (4.53) that

$$v_{xy}(0, y) - \nu v_x(0, y) = 4\pi \nu^2 \frac{3(y\nu)^2 - \pi^2}{[\pi^2 + (y\nu)^2]^2}.$$

Considering $v_x(0, y)$ as unknown in this differential equation, we get

$$v_x(0, y) = e^{\nu y} \left\{ v_x(0, 0) + 4\pi \nu \int_{y\nu}^{0} \frac{\pi^2 - 3t^2}{(t^2 + \pi^2)^3} e^{-t} \, dt \right\}. \tag{4.62}$$

From (4.59) it follows that

$$v_x(0, 0) = \frac{2\nu}{\pi} - 2\pi\nu - 2\nu \int_0^\infty \frac{e^{-\pi k}}{1 + k^2} \, dk < -2\pi\nu. \tag{4.63}$$

The second term in braces in (4.62) can be estimated as follows:

$$4\pi\nu \int_{y\nu}^{0} \frac{\pi^2 - 3t^2}{(t^2 + \pi^2)^3} e^{-t} \, dt \leq 4\pi\nu \int_{-\pi/\sqrt{3}}^{0} \frac{\pi^2 - 3t^2}{(t^2 + \pi^2)^3} e^{-t} \, dt$$
$$\leq 4\pi\nu e^{\pi/\sqrt{3}} \int_{-\pi/\sqrt{3}}^{0} \frac{\pi^2 - 3t^2}{(t^2 + \pi^2)^3} \, dt = \nu e^{\pi/\sqrt{3}} \frac{3\sqrt{3}}{4\pi^2} < 2\pi\nu,$$

because the integrand is positive only for $t \in (-\pi/\sqrt{3}, 0)$. Combining the last estimate with (4.62) and (4.63) completes the proof.

Proof of Lemma 5. It obviously follows from the residue theorem that

$$\int_0^\infty \frac{\cos k(x - \xi)}{k - \nu} e^{k(y+\eta)} \, dk + \pi e^{\nu(y+\eta)} \sin \nu(x - \xi)$$
$$= \mathrm{Re} \int_{\ell_+} \frac{\exp\{-ik(z - \bar{\zeta})\}}{k - \nu} \, dk,$$

where ℓ_+ is the contour going along the positive half-axis and indented above at ν. By deforming ℓ_+ into the ray going to the origin through $k = -(z - \bar{\zeta})^{-1}$, we arrive at the required integral after changing the integration variable.

4.2.3. Axisymmetric Toroidal Bodies

A structure, axisymmetric about the y axis and with submerged volume D and wetted surface S, floats in the free surface of water (both infinite and finite depth will be considered). As usual, the water domain is denoted by W, and the free surface by F. The structure is toroidal in shape so that the free surface is in two distinct parts; the outer free surface is denoted by F_∞, and the inner free surface of radius b is denoted by F_0. The cross section of geometry is sketched in Fig. 4.7 (see Subsection 4.2.2), where D_+ and D_- are left and right cross section of D, respectively, and the horizontal bottom should be omitted for deep water. We will use notation shown in Fig. 4.7, taking into account that all plane domains in the free surface and volume domains in the lower half-space are axisymmetric. Hence, it is convenient to use horizontal polar coordinates (r, θ) defined by

$$x_1 = r\cos\theta, \qquad x_2 = r\sin\theta.$$

The value of a (dihedral) angle β will be specified below.

The uniqueness results for deep water are proved in Subsection 4.2.3.1. In Subsection 4.2.3.2, we compare them with examples of trapped modes constructed in Subsection 4.1.2. The case of water having finite depth is treated in Subsection 4.2.3.3.

4.2.3.1. Deep Water

First we assume that $\beta = \pi/2$. Consider modes of the form

$$u_n(r, \theta, y) = u^{(n)}(r, y)\cos n\theta, \quad n = 0, 1, \ldots, \tag{4.64}$$

satisfying the homogeneous water-wave problem. Separation of variables allows modes with $\sin n\theta$ variation, but these are simply rotations about the axis of symmetry of those in (4.64). Define

$$w_n(r, \theta) = \int_{-\infty}^{0} u_n(r, \theta, y)e^{\nu y}\,dy, \quad n = 0, 1, \ldots.$$

Then by the Laplace equation and the free surface boundary condition we get (cf. Subsection 3.2.1)

$$\left(\nabla_x^2 + \nu^2\right)w_n = 0 \quad \text{in } F. \tag{4.65}$$

As is shown in Subsection 3.2.1.2, $w_n = 0$ in F_∞, and so (see Subsection 3.2.1.3)

$$\nu\int_{F_\infty} |u_n|^2\,dx \le \frac{1}{2}\int_{W_\infty}\left|\frac{\partial u_n}{\partial y}\right|^2 dx dy \le \frac{1}{2}\int_{W_\infty} |\nabla u_n|^2\,dx dy. \tag{4.66}$$

Now, for modes of the form specified by equation (4.64),

$$w_n(r, \theta) = w^{(n)}(r) \cos n\theta, \quad n = 0, 1, \ldots,$$

and so from (4.65)

$$\frac{\mathrm{d}^2 w^{(n)}}{\mathrm{d}r^2} + \frac{1}{r}\frac{\mathrm{d}w^{(n)}}{\mathrm{d}r} + \left(\nu^2 - \frac{n^2}{r^2}\right) w^{(n)} = 0, \quad n = 0, 1, \ldots,$$

for $0 < r < b$. Thus, for solutions u_n that are nonsingular on $r = 0$,

$$w^{(n)}(r) = C_n J_n(\nu r), \quad n = 0, 1, \ldots,$$

where C_n is a constant, or, for $0 \leq r < b$,

$$C_n J_n(\nu r) = \int_{-\infty}^{0} u^{(n)}(r, y) e^{\nu y}\, \mathrm{d}y, \quad n = 0, 1, \ldots. \tag{4.67}$$

Here J_n denotes the Bessel function of order n.

The aim now is to obtain a further bound involving the kinetic energy of the fluid motion within W_0. This is proportional to

$$\begin{aligned} &\int_{W_0} |\nabla u_n|^2 \,\mathrm{d}x\mathrm{d}y \\ &= \int_{W_0} \left\{ \left|u_r^{(n)}\right|^2 \cos^2 n\theta + \frac{n^2}{r^2} \left|u^{(n)}\right|^2 \sin^2 n\theta + \left|u_y^{(n)}\right|^2 \cos^2 n\theta \right\} r\,\mathrm{d}r\,\mathrm{d}\theta\mathrm{d}y \\ &= (1 + \delta_{n0})\pi \int_0^b r\,\mathrm{d}r \int_{-\infty}^{0} \left\{ \left|u_r^{(n)}\right|^2 + \frac{n^2}{r^2}\left|u^{(n)}\right|^2 + \left|u_y^{(n)}\right|^2 \right\} \mathrm{d}y, \end{aligned} \tag{4.68}$$

where δ_{nm} is the Kronecker delta. For later use it is also noted that

$$\begin{aligned} &(1 + \delta_{n0})\pi \int_0^b r\,\mathrm{d}r \int_{-\infty}^{0} \left\{ \left|u_r^{(n)}\right|^2 + \frac{n^2}{r^2}\left|u^{(n)}\right|^2 + \left|u_y^{(n)}\right|^2 \right\} \mathrm{d}y \\ &\quad = \int_{W_0} \left\{ \left|u_r^{(n)}\right|^2 + \frac{n^2}{r^2}\left|u^{(n)}\right|^2 + \left|u_y^{(n)}\right|^2 \right\} \cos^2 n\theta\, r\,\mathrm{d}r\,\mathrm{d}\theta\mathrm{d}y, \end{aligned} \tag{4.69}$$

that is $\cos^2 n\theta$ and $\sin^2 n\theta$ yield the same result in the θ integration.

Bounds will be obtained separately for each of the three terms on the right-hand side of (4.68). Squaring both sides of (4.67) and applying the Schwarz inequality yields

$$[C_n J_n(\nu r)]^2 \leq \int_{-\infty}^{0} \left|u^{(n)}\right|^2 \mathrm{d}y \int_{-\infty}^{0} e^{2\nu y}\, \mathrm{d}y = \frac{1}{2\nu} \int_{-\infty}^{0} \left|u^{(n)}\right|^2 \mathrm{d}y. \tag{4.70}$$

Integration by parts in (4.67) gives

$$u^{(n)}(r,0) = \nu C_n J_n(\nu r) + \int_{-\infty}^{0} u_y^{(n)}(r,y) e^{\nu y}\,\mathrm{d}y,$$

and hence

$$\left|u^{(n)}(r,0)\right|^2 \le 2\left\{\nu^2\,[C_n J_n(\nu r)]^2 + \left[\int_{-\infty}^{0} u_y^{(n)}(r,y)e^{\nu y}\,\mathrm{d}y\right]^2\right\}.$$

Application of the Schwarz inequality to the last integral yields

$$\nu\left|u^{(n)}(r,0)\right|^2 \le 2\nu^3\,[C_n J_n(\nu r)]^2 + \int_{-\infty}^{0}\left|u_y^{(n)}\right|^2\mathrm{d}y. \tag{4.71}$$

Finally, differentiation of equation (4.67) with respect to r gives

$$\nu C_n J_n'(\nu r) = \int_{-\infty}^{0} u_r^{(n)}(r,y)e^{\nu y}\,\mathrm{d}y,$$

and after squaring and another application of the Schwarz inequality, this leads to

$$2\nu^3[C_n J_n'(\nu r)]^2 \le \int_{-\infty}^{0}\left|u_r^{(n)}\right|^2\mathrm{d}y. \tag{4.72}$$

The inequalities (4.70)–(4.72) may be combined to give

$$\begin{aligned}\int_{-\infty}^{0}\left\{\left|u_r^{(n)}\right|^2 + \frac{n^2}{r^2}\left|u^{(n)}\right|^2 + \left|u_y^{(n)}\right|^2\right\}\mathrm{d}y &\ge 2\nu^3[C_n J_n'(\nu r)]^2 \\ &\quad + \frac{n^2}{r^2}2\nu\,[C_n J_n(\nu r)]^2 + \nu\left|u^{(n)}(r,0)\right|^2 - 2\nu^3\,[C_n J_n(\nu r)]^2 \\ &= \nu\left|u^{(n)}(r,0)\right|^2 - 2\nu^3 C_n^2\left\{\left[1 - \frac{n^2}{(\nu r)^2}\right][J_n(\nu r)]^2 - [J_n'(\nu r)]^2\right\},\end{aligned}$$

and so, if it is now assumed that

$$\int_0^b r\left\{\left[1 - \frac{n^2}{(\nu r)^2}\right][J_n(\nu r)]^2 - [J_n'(\nu r)]^2\right\}\mathrm{d}r \le 0, \tag{4.73}$$

it follows that

$$\nu\int_0^b\left|u^{(n)}(r,0)\right|^2 r\,\mathrm{d}r \le \int_0^b r\,\mathrm{d}r\int_{-\infty}^{0}\left\{\left|u_r^{(n)}\right|^2 + \frac{n^2}{r^2}\left|u^{(n)}\right|^2 + \left|u_y^{(n)}\right|^2\right\}\mathrm{d}y.$$

Multiplying the last inequality by $\cos^2 n\theta$ and integrating over all θ yields the new inequality [see equations (4.68) and (4.69)]:

$$\nu\int_{F_0}|u_n|^2\,\mathrm{d}x \le \int_{W_0}|\nabla u_n|^2\,\mathrm{d}x\mathrm{d}y.$$

When this is combined with (4.66), it yields

$$\nu \int_F |u_n|^2 \, \mathrm{d}x \le \left(\int_{W_0} + \frac{1}{2} \int_{W_\infty} \right) |\nabla u_n|^2 \, \mathrm{d}x \mathrm{d}y. \tag{4.74}$$

An application of Green's theorem gives [cf. (4.44)]

$$\nu \int_F |u_n|^2 \, \mathrm{d}x = \int_W |\nabla u_n|^2 \, \mathrm{d}x \mathrm{d}y,$$

which contradicts (4.74) unless u_n vanishes identically.

Inequality (4.73) may be simplified as follows. Integration by parts gives

$$\begin{aligned}\int_0^{\nu b} \mu [J_n'(\mu)]^2 \, \mathrm{d}\mu &= \nu b J_n(\nu b) J_n'(\nu b) - \int_0^{\nu b} J_n(\mu)[J_n'(\mu) + \mu J_n''(\mu)] \, \mathrm{d}\mu \\ &= \nu b J_n(\nu b) J_n'(\nu b) - \int_0^{\nu b} \left(\frac{n^2}{\mu} - \mu \right) [J_n(\mu)]^2 \, \mathrm{d}\mu,\end{aligned}$$

where the differential equation for J_n has been used to simplify the integral. Using this last result in (4.73) yields

$$J_n(\nu b) J_n'(\nu b) \ge 0,$$

which is satisfied provided

$$j_{n,m} \le \nu b \le j'_{n,m+1}, \tag{4.75}$$

where $j_{n,m}$ denotes the mth zero of J_n and $j'_{n,m}$ denotes the mth zero of J_n' (see Section 9.5 in Abramowitz and Stegun [1]). If $n = 0$, then $m = 1, 2, \ldots$. If $n \ge 1$, then $m = 0, 1, \ldots,$ with $j_{n,0} = 0$.

The above calculation has led to the following theorem.

Consider the axisymmetric fluid domain W illustrated in Fig. 9.1 *with $\beta = \pi/2$; that is, the torus D is strictly bounded by two vertical cylinders that intersect D at the free surface F; the inner cylinder has radius b. For a given azimuthal mode number n, let* (4.75) *be satisfied for some value of the nondimensional frequency parameter νb. Then, for this value of νb, the water-wave problem has only trivial solutions in the form of* (4.64).

The zeros of the Bessel functions and their derivatives are tabulated in [1], Table 9.5. Further, the asymptotic behaviors of the zeros (formulae 9.5.12 and 9.5.13 in [1]) are

$$\begin{aligned} j_{n,m} = &\left(m + \frac{n}{2} - \frac{1}{4} \right) \pi \\ &- \frac{4n^2 - 1}{8\,(m + n/2 - 1/4)\,\pi} + O\left(\frac{1}{m^3} \right) \quad \text{as } m \to \infty, \end{aligned}$$

and

$$j'_{n,m+1} = \left(m + \frac{n}{2} + \frac{1}{4}\right)\pi$$
$$- \frac{4n^2 + 3}{8\,(m + n/2 + 1/4)\,\pi} + O\left(\frac{1}{m^3}\right) \quad \text{as } m \to \infty.$$

Hence, for a given azimuthal mode number n, the intervals in νb for which uniqueness has been established are asymptotically of length $\pi/2$. Conversely, the intervals in which trapped modes may be sought are also asymptotically of length $\pi/2$.

For nonaxisymmetric modes, $n \geq 1$, there is an interval of uniqueness for $\nu b \in [0, j'_{n,1}]$. The length of this interval increases asymptotically (see 9.5.16 in [1]) according to

$$j'_{n,1} = n\left[1 + O\left(n^{-2/3}\right)\right], \quad \text{as } n \to \infty.$$

For the axisymmetric mode ($n = 0$), the above theorem may be improved because from (3.60) with M defined by (3.59), we have the following for $\beta \in [\beta_0, \pi/2]$:

$$\nu \int_{F_\infty} |u_0|^2 \, dx \leq \int_{W_\infty} |\nabla u_0|^2 \, dx dy,$$

and as is noted in Subsection 3.2.2.4 one can take the angle 52° as β_0. Using the last inequality instead of (4.66) in (4.74) proves the theorem under weaker assumption than $\beta = \pi/2$. Thus, the uniqueness of the axisymmetric mode is established for a wider class of geometries than for other modes. Of course, this property is guaranteed only when νb satisfies (4.75) with $n = 0$.

4.2.3.2. Comparison of the Uniqueness Results with Examples of Trapped Modes

The latter are constructed in Subsection 4.1.2, and it is shown there that John's condition is not satisfied for these examples on the side directed to infinity. Despite this the trapped modes constructed are entirely consistent with the uniqueness assertions proved.

For the axisymmetric case, $n = 0$, this consistency may be readily demonstrated by using the Stokes stream function v_0; see equations (4.22) and (4.22) in Subsection 4.1.2. Stream surfaces that can be interpreted as the surface of a structure leave the free surface in $r > c$ and reenter the free surface again in $r < c$. Therefore, the inner free-surface radius (b in Fig. 4.7) of any structure must certainly be less than c. Further, in $r > c$ the stream function decreases monotonically from a positive value $v_0(c + 0, 0)$ [equation (4.25) in

Subsection 4.1.2] at the ring source to zero as $r \to \infty$. A stream surface emanating from the free surface in $r > c$ will therefore have a positive value of the stream function. This stream surface returns to the free surface in $r < c$ at another ring with the same positive value of the stream function. The smallest possible inner free surface radius corresponds to a zero value of the stream function, and a lower bound may be obtained as follows. On the free surface and inside the ring source, the stream function is [equation (4.26) in Subsection 4.1.2]

$$v_0(r,0) = 4\pi^2\nu^2 cr J_1(\nu r)Y_0(\nu c) - 8\nu^2 cr \int_0^\infty I_1(\mu r)K_0(\mu c)\frac{\mu\,\mathrm{d}\mu}{\mu^2+\nu^2},$$

where c satisfies $J_0(\nu c) = 0$. Now,

$$\int_0^\infty I_1(\mu r)K_0(\mu c)\frac{\mu\,\mathrm{d}\mu}{\mu^2+\nu^2} > 0 \quad \text{for } r > 0,$$

because $I_1(\mu r)$ and $K_0(\mu c)$ are both positive definite for $\mu r, \mu c > 0$, and $v^-(c) > 0$ from the results of Subsection 4.1.2 [equations (4.25) and (4.27)]. Also, the zeros of J_1 and Y_0 interlace with those of J_0 (see 9.5.2 in Abramowitz and Stegun [1]), and $J_1(\nu c)$ and $Y_0(\nu c)$ have the same sign so that $J_1(\nu c)Y_0(\nu c) > 0$. Hence, as r is reduced from the source radius c, $v_0(r,0)$ will certainly become negative when $J_1(\nu r)$ passes through a zero and so the inner radius νb must be greater than the first zero of J_1, or equivalently the first zero of J_0', below νc. Hence, for trapped modes to exist it is necessary that the inner radius b of the structure satisfies

$$j'_{0,m} < \nu b < j_{0,m}, \quad m = 1, 2, \ldots. \tag{4.76}$$

Thus, the following result has been proved.

Axisymmetric trapped modes of the form (4.21) *with* $n = 0$ *can be found only in the intervals* (4.76). *Moreover, for every* $m \geq 2$ *the interval* (4.76) *contains a subinterval adjacent to* $j'_{0,m}$ *which is free of such modes.*

Clearly, the intervals defined by (4.76) are complementary to those in (4.75), in which uniqueness has been established for geometries satisfying John's condition on the inner part of the free surface.

A more detailed picture is given in Fig. 4.11, which shows numerical calculations of the values of νb corresponding to the intervals of existence for trapped modes constructed by using a single ring source (dashed lines). Also shown are the intervals for uniqueness given in (4.75) (solid lines). The integer n is the azimuthal wave number. Axisymmetric modes ($n = 0$) are considered first. The left-hand point of each trapped-mode interval corresponds to a zero

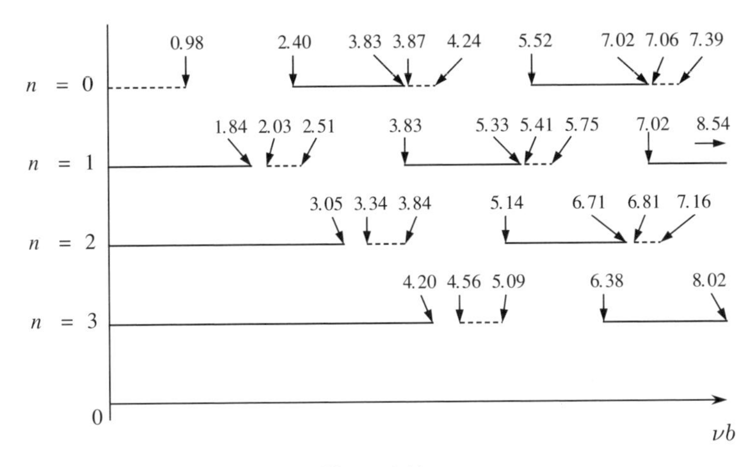

Figure 4.11.

in the free-surface stream function (see above and Subsection 4.1.2), and this is easily found numerically. The corresponding right-hand point of the interval corresponds to the stream surface that emerges from the ring source in the plane $y = 0$, and with the stream function having the value $v_0(c + 0, 0)$; see equation (4.25) in Subsection 4.1.2. The point on the free surface in $r < c$ with this value of the stream function is also easily located numerically.

For nonaxisymmetric modes ($n \geq 1$) the end points of the intervals are located by stream-surface tracing using equation (4.28) or (4.29). The left-hand end of the interval corresponds to the dividing surface that separates stream surfaces that surround the source point from those that do not (see Fig. 4.6). The dividing ring radius on the free surface is estimated by tracking a number of stream surfaces up to the free surface from great depth. This gives only an estimate of the appropriate ring radius, and the corresponding numbers in Fig. 4.11 may have errors of up to 0.02. As before, the corresponding right-hand point of the interval corresponds to the stream surface that emerges from the ring source in the plane $y = 0$, and this stream surface can be followed accurately. In Fig. 4.11, it may also be noted that, for $n = 1, 2, 3$, the length of the first interval containing trapped modes is approximately of length 0.5.

It is apparent from Fig. 4.11 that there are intervals in νb, for which there may be uniqueness for all modes (disregarding for the moment the requirement that John's condition must be satisfied). For example, for $n = 1$ no trapped modes have been found for $\nu b \in (2.51, 3.05)$, whereas uniqueness of the solution has been established in this interval for all other modes. However, it should be pointed out that in the equivalent two-dimensional problem of two-surface piercing bodies, Linton and Kuznetsov [207] have found evidence of modes trapped by bodies violating John's condition within the region for which uniqueness is predicted by the theory that requires John's condition.

4.2.3.3. Water of Finite Depth

Let a water domain W be bounded below by a horizontal bottom $\{y = -d\}$, where $d > \max\{-y : (x, y) \in S\}$, and let $\beta = \pi/2$. Then, the uniqueness theorem similar to that obtained for deep water is true. However, the nondimensional spectral parameter in (4.75) should be $k_0 b$, where k_0 is the unique positive root of $k_0 \tanh k_0 d = \nu$. To derive the corresponding inequality for modes of the form (4.64), one has to consider

$$w_n(r, \theta) = \int_{-d}^{0} u_n(r, \theta, y) \cosh k_0 y dy, \quad n = 0, 1, \ldots,$$

which gives $(\nabla_x^2 + k_0^2) w_n = 0$ instead of (4.65). Combining considerations applied in Subsection 4.2.2.1 with those in Subsection 4.2.1 (we leave the calculations to the reader), one arrives at (4.75), where νb is replaced by $k_0 b$, which proves the result.

4.2.4. Water Domains Without Mirror Symmetry

In the beginning of this chapter it was pointed out that in the absence of mirror symmetry of the water domain, uniqueness results for surface-piercing cylinders are much less general than for symmetric domains. This is a consequence of technical difficulties arising in this situation, which must be overcome in a rather tricky way. As in the previous subsection we begin with the case of deep water and provide (in Subsection 4.2.4.1) an upper bound for the interval of ν within which the uniqueness theorem holds. We discuss this bound in Subsection 4.2.4.2 and generalize it in Subsection 4.2.4.3, where different geometries, having finite depth and a limited class of depth variations, are considered. In all cases an upper bound of the uniqueness interval depends on the geometry of the water domain.

4.2.4.1. Upper Bound for the Uniqueness Interval: The Case of Deep Water

Let the two bodies occupy domains D_+ and D_-, and let them be contained inside semicircles of radius r_+ and r_-, respectively, each centered on the x axis. Moreover, we assume that each circumscribing semicircle begins at the inner water-line point of the corresponding body; thus the distance h between the semicircles is also the distance between the bodies. Label the outer

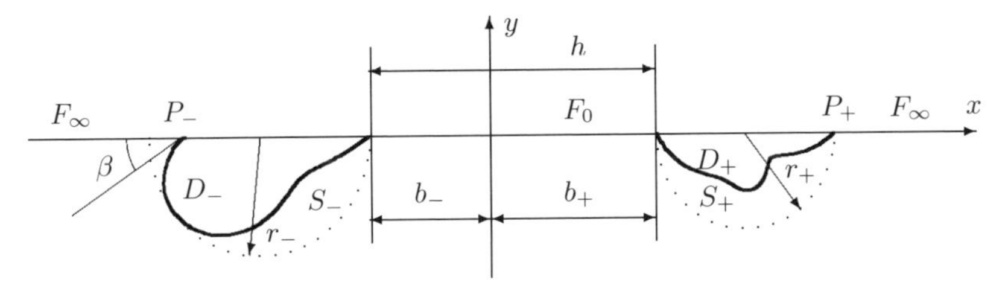

Figure 4.12.

water-line points by $P_\pm$. Let F_0 be the part of the free surface between the bodies, and that outside be F_∞; also let F_∞^e and W^e be the parts of F_∞ and W, respectively, that are external to to the semicircles bounding $D_\pm$ (see Fig. 4.12).

We can choose the origin in the x axis between the bodies so that these semicircles will coincide with coordinate τ lines of the bipolar system (σ, τ):

$$x = \frac{a \sinh \tau}{\cosh \tau - \cos \sigma}, \qquad y = \frac{a \sin \sigma}{\cosh \tau - \cos \sigma}. \tag{4.77}$$

For this purpose we have to find (see, for example, Section 10.1 in Morse and Feshbach [251]) positive constants $b_\pm$, $d_\pm$, and a such that

$$a = r_- \sinh d_- = r_+ \sinh d_+, \quad b_+ + b_- = h,$$
$$b_- + r_- = r_- \cosh d_-, \quad b_+ + r_+ = r_+ \cosh d_+.$$

One can easily verify that this system of equations has a unique solution such that

$$\cosh d_\pm - 1 = \frac{h^2 + 2hr_\pm}{2r_\pm(r_+ + r_- + h)}. \tag{4.78}$$

Without loss of generality we can assume that $r_- \geq r_+$, and so $d_- \leq d_+$. Our aim is to prove the following theorem.

Let $D_- \cup D_+$ be enclosed between two rays emanating from P_- and P_+ at an angle $\pi/2 - \beta$ to the vertical. Let also

$$D_\pm \subset \{[x \mp (b_\pm + r_\pm)]^2 + y^2 r_\pm^2 \; y < 0\}.$$

If the inequality

$$2 - \operatorname{cosec}^2 \beta > \frac{h^2 + 2hr_-}{2r_+(r_+ + r_- + h)} \tag{4.79}$$

holds, then the homogeneous water-wave problem has only a trivial solution *for all values $\nu > 0$, which do not exceed*

$$\frac{2(2 - \operatorname{cosec}^2)(r_+ + r_- + h)r_+ - (h^2 + 2hr_-)}{2\pi r_+ [h(h + 2r_-)(h + 2r_+)(h + r_+ + r_-)]^{1/2}} . \tag{4.80}$$

As usual we begin the proof by noting that

$$\int_W |\nabla u|^2 \, \mathrm{d}x\mathrm{d}y = \nu \int_{F_0 \cup F_\infty} |u|^2 \, \mathrm{d}x \tag{4.81}$$

for a solution of the homogeneous water-wave problem.

The mapping $x + iy \mapsto \sigma + i\tau$ [see (4.77)] conformally maps the lower half-plane into the strip $\{-\pi < \sigma < 0, -\infty < \tau < +\infty\}$. By the hypothesis of the theorem the rectangle $\Pi = \{-\pi < \sigma < 0, -d_- < \tau < +d_+\}$ is the image of W^e, which is contained in the image of W. Furthermore, the image of F_0 is the side $\{\sigma = \pi, -d_- < \tau < +d_+\}$, and $\{\sigma = 0, -d_- < \tau < +d_+\}$ is the image of F^e_∞, a subset of the image of F_∞.

Let us denote by $v(\sigma, \tau)$ the function obtained from $u(\sigma, \tau)$ by the conformal transformation (4.77). Then for $\tau \in (-d_-, d_+)$ we have

$$v(-\pi, \tau) = v(0, \tau) - \int_{-\pi}^{0} v_\sigma(\sigma, \tau) \, \mathrm{d}\sigma,$$

and so

$$|v(-\pi, \tau)|^2 \leq 2 \left\{ |v(0, \tau)|^2 + \left| \int_{-\pi}^{0} v_\sigma(\sigma, \tau) \, \mathrm{d}\sigma \right|^2 \right\}.$$

Applying the Schwarz inequality and integrating we find that

$$\int_{-d_-}^{+d_+} |v(-\pi, \tau)|^2 \, \mathrm{d}\tau \leq 2 \left\{ \int_{-d_-}^{+d_+} |v(0, \tau)|^2 \, \mathrm{d}\tau + \pi \int_\Pi |v_\sigma(\sigma, \tau)|^2 \, \mathrm{d}\sigma \mathrm{d}\tau \right\}. \tag{4.82}$$

Also

$$\nu \int_{F_0} |u|^2 \, \mathrm{d}x = \nu a \int_{-d_-}^{+d_+} \frac{|v(-\pi, \tau)|^2}{\cosh \tau + 1} \, \mathrm{d}\tau \leq \frac{\nu a}{2} \int_{-d_-}^{+d_+} |v(-\pi, \tau)|^2 \, \mathrm{d}\tau.$$

By (4.82) and the invariance of the Dirichlet integral with respect to a conformal mapping, we therefore obtain

$$\nu \int_{F_0} |u|^2 \, \mathrm{d}x \leq \nu a \left\{ \int_{-d_-}^{+d_+} |v(0, \tau)|^2 \, \mathrm{d}\tau + \pi \int_{W^e} |\nabla u|^2 \, \mathrm{d}x\mathrm{d}y \right\}. \tag{4.83}$$

As $d_+ \geq d_-$, we have

$$\nu a \int_{-d_-}^{+d_+} |v(0,\tau)|^2 \, d\tau \leq \nu a(\cosh d_+ - 1) \int_{-d_-}^{+d_+} \frac{|v(0,\tau)|^2}{\cosh \tau - 1} \, d\tau$$
$$= \nu(\cosh d_+ - 1) \int_{F_+^e} |u|^2 \, dx.$$

As u satisfies the homogeneous problem, we have $|u| = O(|x|^{-1})$ as $|x| \to \infty$ (see Subsection 2.2.1). Hence, $|v(0,\tau)| = O(|\tau|)$ as $\tau \to \infty$, and so the middle integration is proper.

Noting that straight vertical lines down from F_∞^e do not intersect S, we can apply John's inequality (see Subsection 3.2.1.3):

$$\nu \int_{F_\infty^e} |u|^2 \, dx \leq \frac{1}{2} \int_{W_\infty^e} |\nabla u|^2 \, dx dy,$$

where W_∞^e is the part of W^e covered by the above-mentioned vertical lines. Now (4.83) takes the form

$$\nu \int_{F_0} |u|^2 \, dx \leq \left[\frac{1}{2}(\cosh d_+ - 1) + \nu a \pi\right] \int_{W^e} |\nabla u|^2 \, dx dy. \tag{4.84}$$

We can use the estimate

$$\nu \int_{F_\infty} |u|^2 \, dx \leq \frac{1}{2} \operatorname{cosec}^2 \beta \int_{W^c} |\nabla u|^2 \, dx dy, \tag{4.85}$$

where W^c is the subset of W covered with two families of F_∞ (by the hypothesis made, such rays do not intersect S). To prove (4.85) we have to apply the method developed in Subsection 3.2.2.1, replacing the angle $\pi/4$ by β.

Substituting (4.84) and (4.85) into (4.81), we can write

$$\int_W |\nabla u|^2 \, dx dy \leq \frac{1}{2}[\operatorname{cosec}^2 \beta + \cosh d_+ - 1 + 2\nu a \pi] \int_W |\nabla u|^2 \, dx dy. \tag{4.86}$$

For $M = 2\pi \nu a + \cosh d_+ - 1$, there will be a contradiction in (4.86) if $M < 2 - \operatorname{cosec}^2 \beta$, unless $\nabla u \equiv 0$, and so $u \equiv 0$, since $u \to 0$ as $|x| \to \infty$. As W^c and W_∞^e are never the whole of W, such a conclusion will also be true if $M = 2 - \operatorname{cosec}^2 \beta$. This proves uniqueness provided ν does not exceed the value of (4.80), where the results of (4.78) and

$$a = [b_+(b_+ + 2r_+)]^{1/2}, \quad \text{with } b_+ = r_+(\cosh d_+ - 1),$$

have been used.

4.2.4.2. *Discussion of the Upper Bound (4.80)*

The main idea of the proof in Subsection 4.2.4.1 is to estimate the potential energy of the inner part F_0 of the free surface through the potential energy of F_∞^e, and the kinetic energy of W^e [see the inequalities (4.82)–(4.84)]. For this purpose the transformation (4.77) was used; other transformations may lead to other uniqueness theorems.

The theorem proved in Subsection 4.2.4.1 shows that the lower bound for eigenvalues for ν will depend on $r_\pm$, as well as h and β. Examples of non-uniqueness constructed in Subsection 4.1.1 confirm this assertion. Geometries in these examples violate the hypothesis made in Subsection 4.2.4.1, and the corresponding eigenvalues occur below the bound (4.80).

Another point to be emphasized is that for given two surface-piercing bodies $D_\pm$, it is not immediately clear how $r_\pm$ should be chosen so as to achieve the best (that is, the largest) bound for ν. Clearly the geometry imposes minimum values on $r_\pm$, and the constraint (4.79) must be satisfied, but in fact, it is always possible to increase the radii of the circumscribing semicircles. So, we now consider the optimization of the bound (4.80) over the allowable values of $r_\pm$ with $r_- \geq r_+$.

When $r_+ = r_- = r$, *the fraction* (4.80) *simplifies to*

$$\frac{2cr - h}{2\pi r(h^2 + 2hr)^{1/2}}, \quad \text{where } c = 2 - \operatorname{cosec}^2\beta. \tag{4.87}$$

It is straightforward to show that (4.87) *has a maximum at*

$$r = \bar{r}(c) = \frac{h\,[3 + (9 + 4c)]^{1/2}}{4c}.$$

It is also possible to show that the bound (4.80), when it is viewed as a function of r_+ for fixed r_- in the range $r_+ < r_-$, is maximized in one of two ways, and this depends on whether $r_- > hr^*(c)$ or $r_- \leq hr^*(c)$ where $r^*(c)$ is the positive root of the following cubic equation:

$$4cr^3 + (2c - 14)r^2 - 8r - 1 = 0.$$

For $r_- \leq hr^*$ the optimum value of (4.80) occurs for $r_+ = r_-$, and so (4.87) applies. However, for $r_- > hr^*$ the optimum value of r_0 of r_+ is determined by a cubic equation whose coefficients involve c and r_- in a complicated way. Calculations show that r_0 varies little over a large range of values of $r_- > hr^*$.

These results are illustrated in Fig. 4.13 for $c = 1$. Corresponding diagrams for other values of $c < 1$ look remarkably similar provided that cr_+ and cr_- are used as axes. The technique for locating the optimum values of $r_\pm$ is as

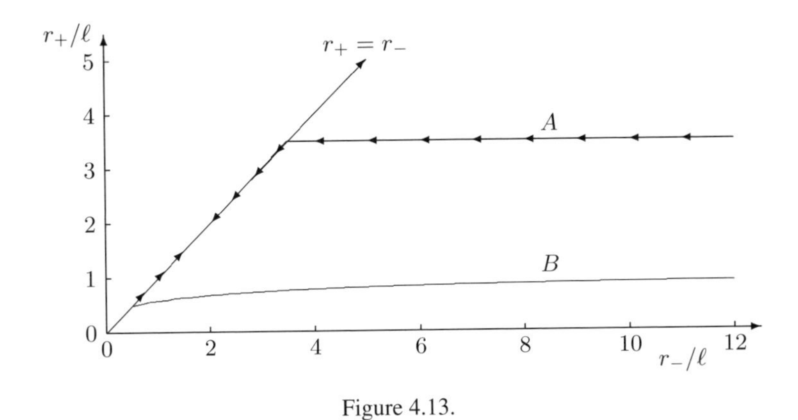

Figure 4.13.

follows. Given any first guess of $r_\pm$, adjust r_+ closer to the arrowed line, and then move in the direction of the arrows once on this line. The optimum is found when this procedure reaches the minimum value of r_+, r_-, or both imposed by the geometry.

Let us illustrate the improvement that is achieved by use of the optimum value r_0 of r_+ in (4.80), compared with the simple bound (4.87) for which $r_+ = r_- = r$. For this purpose the case $r_- = 12h$ has been studied for $c = 1$ and $c = 2/3$, that is, $\beta = \pi/2$ and $\beta = \pi/3$. For $c = 1$ with $r_0 = 3.50068h$, the bound is improved by over 18%; and for $c = 2/3$ for which $r_0 = 5.20445h$, the improvement is nearly 8.5%.

It is also possible to let r_- tend to infinity in (4.80); this is useful *for the problem of a single body in the presence of a (overhanging) cliff. This results in the bound*

$$\frac{cr_+ - h}{2\pi r_+(h^2 + 2hr_+)^{1/2}},$$

provided that $h < cr_+$. More details on the case of a single two-dimensional body in the presence of a straight coastline parallel to the body are given in Subsection 4.2.4.3.

Another limit that can be considered is to let h tend to zero; in this limit both the bounds (4.80) and (4.87) become infinite, and so uniqueness holds at all frequencies provided (4.79) is satisfied, which gives $\operatorname{cosec}^2\beta \leq 2$. This therefore recovers a special case of the uniqueness theorem proved in Subsection 3.2.1.

As the proof in Subsection 4.2.4.1 depends on the relative position and size of the semicircles circumscribing the pair of surface-piercing bodies, the bound (4.80) is the same for any pairs with the same water-line intersections also contained within the semicircles. Hence, the problem of two horizontal strips in the free surface has the same bound as the problem of two half-immersed circles of diameters equal to the strip lengths. The extended auxiliary integral identity obtained in Subsection 3.2.3 allows us to improve the method developed here, but this material is treated in detail below.

4.2.4.3. Further Results

Here we present several uniqueness theorems obtained with the help of the technique developed in Subsection 4.2.4.1 and its modifications.

Example 1: A Single Body in Water of Finite Depth Near a Coastline

The water depth is assumed to be equal to d everywhere except for a neighborhood of the origin (the shoreline), where the bottom B meets the x axis (see Fig. 4.14). The surface-piercing body D_+ with wetted surface S is contained in a semicircle of radius r_+ centered on the x axis. Moreover, we assume that the body is in contact with its bounding semicircle at the water-line position nearest the origin, that is, at $x = h$. The other water-line intersection of the body be denoted by P_+.

Let there exist a number $r_- > r_+ + h$ such that the semicircle of radius r_- centered on the x axis and with one end at the origin belongs to $\bar{W}$. The part of the free surface between the body and the shoreline is denoted F_0, and that part to the right of P_+ is F_∞. In addition, F_∞^e and W^e are the parts of F_∞ and W, respectively, that are between the semicircles of radius $r_\pm$. At last, we assume that for some $\beta \leq \pi/2$, a straight line down from P_+ at an angle β to F_∞ does not intersect D_+, and it intersects B outside the region of depth variations. We note that it is advantageous to use as large a value of β as possible.

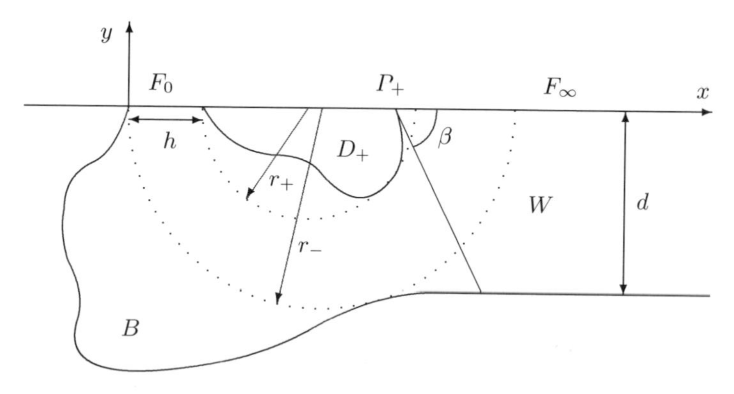

Figure 4.14.

Let the above-listed geometric conditions hold, and let

$$M = \left[1 - \frac{\sin(2k_0 d \cot\beta)}{\cot\beta \sinh 2k_0 d}\right]\left[1 + \frac{hr_-}{r_+(r_- - r_+ - h/2)}\right],$$

where for $\beta = \pi/2$ the expression in the first set of brackets should be understood as a limit as $\beta \to \pi/2$.

If the inequality $M < 2\sin^2\beta$ is valid, then the homogeneous water-wave problem has only a trivial solution for all $\nu > 0$, which do not exceed

$$\frac{(2 - M \operatorname{cosec}^2\beta)}{2\pi r_-}\left[\frac{(r_- - h/2)(r_- - r_+ - h/2)}{h(2r_+ + h)}\right]^{1/2}. \tag{4.88}$$

Let us discuss this upper bound and possible improvement resulting from another transformation.

1. It is easily verified that M is a monotonically decreasing function of r_-, and that (4.88) consequently increases with r_-. Hence the best upper bound is produced by choosing the largest possible value of r_-; that is, by expanding that semicircle until it touches the bottom B.

Let r_1 be the radius of the smallest semicircle, centered on the x axis, that circumscribes D_+. As a function of r_+, M has a minimum at

$$r_+ = (2r_- - h)/4 = r_2,$$

say, and thus the upper bound (4.88) is a decreasing function of r_+ for $r_+ \geq r_2$. Hence, if $r_1 \geq r_2$, then the optimum value of (4.88) is produced by choosing $r_+ = r_1$, whereas if $r_1 < r_2$, the optimum choice will lie in $[r_1, r_2)$.

It is interesting to consider two limiting cases of the problem in Fig. 4.14, first as the depth becomes large, and then as the body D_+ approaches the coastline. If we let d tend to infinity, then r_- can tend to infinity simultaneously, and from (4.88) we get the upper bound

$$\frac{2r_+ \sin^2\beta - (r_+ + h)}{2\pi r_+ (h^2 + 2hr_+)^{1/2}} \sin^2\beta$$

for the values of ν that are not point eigenvalues. One can easily see that M tends to

$$1 - \frac{\sin(2k_0 d \cot\beta)}{\cot\beta \sinh 2k_0 d} \quad \text{as } h \to 0.$$

In this case the upper bound (4.88) for the values ν not in the point spectrum tends to infinity, and in the limit as $h \to 0$ we have the special case of the uniqueness theorem in Subsection 3.2.1; the inequality for M takes the form

$$\frac{\sin(2k_0 d \cot\beta)}{\cot\beta \sinh 2k_0 d} > \cos 2\beta$$

in this situation.

2. It is worth noting that other coordinate transformations can be used in place of a bipolar system of the type of (4.77) in order to try to improve the bound (4.88). Such transformations do not have to be conformal mappings and we consider the transformation $(x, y) \mapsto (\rho, \theta)$ given by

$$x = A + \rho(B + \cos\theta), \quad y = \rho \sin\theta, \tag{4.89}$$

where A and B are constants at our disposal. The lower half-plane becomes $\{-\pi < \theta < 0, 0 < \rho < \infty\}$, and $\rho = \text{const}$ gives a circle of radius ρ centered at $(A + B\rho, 0)$. Referring back to Fig. 4.14, we see that we have a suitable mapping for the water domain with constants A and B that satisfy

$$A + Br_+ = r_+ + h, \qquad A + Br_- = r_-.$$

Thus we have

$$A = \frac{hr_-}{r_- - r_+}, \qquad B = \frac{r_- - r_+ - h}{r_- - r_+}.$$

It is easily shown that, for $r_- > r_+ + h/2$, the inequalities

$$h < A < h + 2r_+, \qquad -1 < B < 1$$

hold with B and $r_+ + h - A$ taking the same sign as $r_- - h - r_+$. Hence, we get the following for Jacobian in the lower half-plane:

$$\frac{\partial(x, y)}{\partial(\rho, \theta)} = \rho(1 + B\cos\theta) \neq 0.$$

Note that this Jacobian vanishes when $\rho = 0$, that is, at $(A, 0)$. If the body D_+ does not have the same water-line intersection as the smaller semicircle, it is possible that this point can be in F_∞. However, the transformation (4.89) is not needed inside the smaller semicircle, so the method will still work even if the transformation can be singular in the closure of the water domain.

We have $\partial x/\partial\rho = B + \cos\theta$, and $\theta = 0$ $(\theta = \pi)$ corresponds to F_∞ (F_0). Then

$$\nu \int_{F_0} |u|^2 \, \mathrm{d}x = (1 - B)\nu \int_{r_+}^{r_-} |v(\rho, -\pi)|^2 \, \mathrm{d}\rho,$$

$$\nu \int_{F_\infty^e} |u|^2 \, \mathrm{d}x = (1 + B)\nu \int_{r_+}^{r_-} |v(\rho, 0)|^2 \, \mathrm{d}\rho,$$

where $v(\rho, \theta) = u(x, y)$.

We follow the approach used in Subsection 4.2.4.1 for deriving

$$\nu \int_{F_0} |u|^2 \, \mathrm{d}x \le 2\nu \frac{1-B}{1+B} \int_{F^e_\infty} |u|^2 \, \mathrm{d}x + 2\pi \nu (1-B) \int_{r_+}^{r_-} \int_{-\pi}^{0} |v_\theta|^2 \, \mathrm{d}\theta \mathrm{d}\rho.$$

Since $|v_\theta|^2 \le [(x_\theta)^2 + (y_\theta)^2] |\nabla u|^2 = \rho^2 |\nabla u|^2$, we have

$$\int_{r_+}^{r_-} \int_{-\pi}^{0} |v_\theta|^2 \, \mathrm{d}\theta \mathrm{d}\rho \le \int_{r_+}^{r_-} \int_{-\pi}^{0} |\nabla u|^2 \frac{\partial(x, y)}{\partial(\rho, \theta)} \frac{\rho \, \mathrm{d}\rho \mathrm{d}\theta}{1 + B \cos\theta}$$

$$\le \frac{r_-}{1 \mp B} \int_{W^e} |\nabla u|^2 \, \mathrm{d}x\mathrm{d}y.$$

Here the upper (lower) sign corresponds to $r_- > r_+ + h$ $(r_- < r_+ + h)$ and so

$$\int_W |\nabla u|^2 \, \mathrm{d}x\mathrm{d}y \le \nu \left(1 + 2\frac{1-B}{1+B}\right) \int_{F^e_\infty} |u|^2 \, \mathrm{d}x + 2\pi \nu r_- \int_{W^e} |\nabla u|^2 \, \mathrm{d}x\mathrm{d}y.$$

Hence we arrive at the following assertion.

The uniqueness holds if

$$\frac{M'}{2 \sin^2 \beta} + 2\pi \nu r_- \le 1, \tag{4.90}$$

where

$$M' = \left[1 - \frac{\sin(2k_0 d \cot\beta)}{\cot\beta \sinh 2k_0 d}\right] \left(1 + 2\frac{1-B}{1+B}\right).$$

It is easily seen that M' is a monotonically increasing function of r_+ for $r_+ < r_- - h/2$, and that the upper bound for ν from (4.90) consequently decreases with r_+. Then, the best upper bound is achieved by using the smallest possible value of r_+, namely r_1.

As a function of r_-, M' is monotonically decreasing. Hence, the upper bound for ν is increasing for $r_+ + h/2 < r_- \le r_+ + h$. It can be shown that the optimum choice for r_-, ignoring constraints of the geometry, depends on the quantity

$$D = 1 - \frac{\sin(2k_0 d \cot\beta)}{\cot\beta \sinh 2k_0 d} - \frac{2h \sin^2 \beta}{4r_+ + 7h}.$$

If $D < 0$, then choose $r_- = r_+ + h$, whereas if $D > 0$, then the best upper bound occurs for some $r_- > r_+ + h$.

The transformation (4.89) does not fail for $r_- = r_+ + h$ (unlike bipolar coordinates). In this case $B = 0$ and (r, θ) are plane polar coordinates with center at $(r_-, 0)$, allowing us *to establish the uniqueness for*

$$0 < \nu \leq \frac{2\sin^2\beta - M'}{4\pi(r_+ + h)\sin^2\beta},$$

provided $M' < 2\sin^2\beta$, *where*

$$M' = 3\left[1 - \frac{\sin(2k_0 d \cot\beta)}{\cot\beta \sinh 2k_0 d}\right].$$

This is clearly an improvement over the previous corresponding result.

Example 2: Two Bodies in Water of Finite Depth

We consider the finite-depth water layer extending to both $x = \pm\infty$. For simplicity, we first restrict our attention to the case of uniform depth. Moreover, we consider only the case in which two bodies are symmetric about a vertical line (without loss of generality, it is the y axis). The right half of this geometry arises when one replaces the curvilinear part of B in Fig. 4.14 by two straight segments: the vertical one, B_v, having length d and directed along the negative y axis; and the horizontal one, constituting with the rest of B (it is horizontal) a half-line $B_h = \{x > 0, y = -d\}$.

Under this condition we can split the solution u of the homogeneous boundary value problem in the symmetric water domain into two parts: $u^{(+)}$ is symmetric, and $u^{(-)}$ is antisymmetric with respect to the y axis (cf. Subsection 4.2.2). These latter functions can be investigated in the half W of the whole water domain (W is shown in Fig. 4.14, where amendments described above should be made). The function $u^{(+)}$ must satisfy the homogeneous water-wave problem in W. The function $u^{(-)}$ must satisfy all the same conditions except for the bottom boundary condition. The latter is to be replaced by two conditions:

$$\partial u^{(-)}/\partial n = 0 \quad \text{on } B_h \cup S, \qquad u^{(-)} = 0 \quad \text{on } B_v. \tag{4.91}$$

For the function $u^{(+)}$, the theorem formulated for Example 1 is obviously valid. The same is true for the function $u^{(-)}$ because conditions (4.91) allow us to prove the equipartition of energy similarly to the Neumann boundary condition on B. Hence we obtain the following proposition.

Let two surface-piercing bodies be floating in water of constant depth d. Let these bodies be symmetric about a vertical line distance h from each of them. If the right-hand body satisfies the conditions of the first assertion for Example 1 with $r_- = d$, then the homogeneous water-wave problem has only *a trivial solution for all* $\nu > 0$ *that do not exceed*

$$\frac{(2 - M \operatorname{cosec}^2 \beta)}{2\pi d} \left[\frac{(d - h/2)(d - r_+ - h/2)}{h(2r_+ + h)} \right]^{1/2},$$

provided

$$M = \left[1 - \frac{\sin(2k_0 d \cot \beta)}{\cot \beta \sinh 2k_0 d} \right] \left[1 + \frac{hd}{r_+(d - r_+ - h/2)} \right]$$

is less than $2 \sin^2 \beta$.

This proposition essentially complements the result proven for $u^{(+)}$ in Subsection 4.2.2, because here the uniqueness is guaranteed in an interval beginning at zero, unlike the first interval provided by the inequality (4.48) with a minus sign. As far as $u^{(-)}$ is concerned, it is difficult to check whether the last proposition adds anything new to the first interval of uniqueness defined by (4.48) with a plus sign.

Example 3: Three Bodies in Water of Finite Depth

The bodies $D_\pm$ are contained in semicircles of radii $r_\pm$, centered on the x axis. The body D_0 lies between the arcs O_+Q_+ and O_-Q_-, which each have radius d, and intersect the x axis at right angles. As in Subsection 4.2.4.1, each body $D_\pm$ must be in contact with its bounding semicircle at one water-line intersection, which in this case is the one nearest the body D_0. The other water-line intersections are denoted by $P_\pm$.

The parts of the mean free surface between $D_\pm$ and D_0 are labeled $F_0^{(\pm)}$, and they are of length $h_\pm$, respectively. The remaining parts of the mean free surface between $P_\pm$ and $x = \pm\infty$ are denoted by $F_\infty^{(\pm)}$. Straight lines $P_\pm R_\pm$ at an angle $\beta_\pm$ to $F_\infty^{(\pm)}$ must not intersect $D_\pm$. Finally, any depth variations that exist must be between R_- and R_+, and they must be below the semicircles of radius d. The corresponding geometry is shown in Fig. 4.15 for the case of a flat bottom, and the following proposition holds for it.

Let the above-listed geometric conditions hold along with the inequalities $d > r_\pm + h_\pm/2$ *and* $(1 + A_\pm)C_\pm < 1$, *where*

$$A_\pm = \frac{dh_\pm}{r_\pm(d - r_\pm - h_\pm/2)}, \qquad C_\pm = \frac{1}{2 \sin^2 \beta_\pm} \left[1 - \frac{\sin(2k_0 d \cot \beta_\pm)}{\cot \beta_\pm \sinh 2k_0 d} \right].$$

Then the homogeneous water-wave problem has only a trivial solution for all positive values of ν, *which do not exceed*

$$\frac{1 - \max_\pm \{(1 + A_\pm)C_\pm\}}{\pi d B},$$

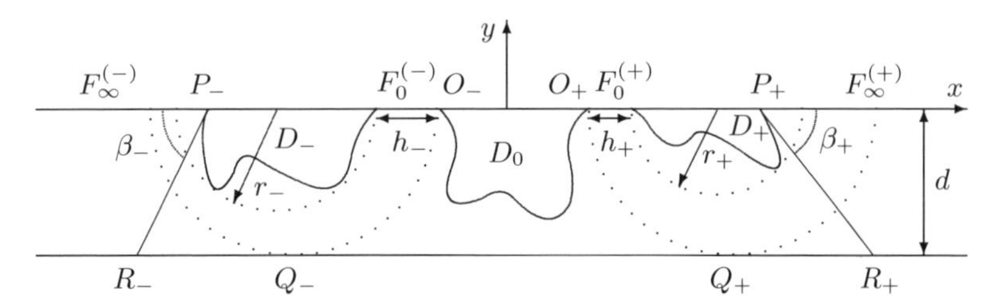

Figure 4.15.

where

$$B = \max_{\pm} \{B_\pm\} = \max_{\pm} \left[\frac{2h_\pm r_\pm + h_\pm^2}{(d - h_\pm/2)(d - r_\pm - h_\pm/2)} \right]^{1/2}.$$

Under the hypothesis of the proposition, the proof in Subsection 4.2.4.1 can be adapted to the present geometry by repeating the estimates twice. In the case that involves $D_\pm$, the semicircles of radii d and $r_\pm$ must replace r_- and r_+, respectively. Also, h is replaced by $h_\pm$.

Letting $d \to \infty$ in the obtained proposition, one gets an assertion for three surface-piercing bodies in deep water.

Let the bodies $D_\pm$ satisfy the assumptions made at the beginning of the present subsection, and let the body D_0 lie between the vertical lines through the points $O_\pm$. Let the water domain W have infinite depth. If the inequality

$$\max_{\pm} \left\{ \frac{r_\pm + h_\pm}{r_\pm \sin^2 \beta_\pm} \right\} < 2$$

holds, then the homogeneous water-wave problem has only a trivial solution for all $\nu > 0$, which do not exceed

$$\frac{1 - \max_{\pm}\{(r_\pm + h_\pm)/(2r_\pm \sin^2 \beta_\pm)\}}{\pi \max_{\pm}\left\{(2h_\pm r_\pm + h_\pm^2)^{1/2}\right\}}.$$

In conclusion, we note that since the two problems involving D_+ and D_- are dealt with separately, there is no reason why the (uniform) depths for large $|x|$ should not take different values $d_\pm$ at $\pm\infty$, provided always that the region of depth variation is between R_+ and R_-, and below the larger semicircle in each problem. Of course, this also means that k_0 takes different values $k_0^\pm$ in the two problems.

4.2.5. Toroidal Bodies Without Axial Symmetry

In this subsection we consider the three-dimensional water-wave problem for a surface-piercing toroidal body of arbitrary shape, that is, without assumption about axial symmetry. We use the technique developed in the previous section and based on coordinate transformations. We shall use two different transformations, because one of them is only appropriate to the infinite-depth case, but usually gives a better bound than the other. The second transformation can also be used in the finite-depth case. In Subsection 4.2.5.1, we start with obtaining an upper bound for the uniqueness interval in the case of the deep-water problem and discuss this bound in the end. Another bound for the case of deep water is given in Subsection 4.2.5.2; also, we compare it with the first bound and outline a modification needed for water of finite depth.

It is convenient to use horizontal polar coordinates (r, θ) defined by

$$x_1 = r\cos\theta, \quad x_2 = r\sin\theta.$$

We denote the surface-piercing toroidal body (its wetted surface) by D (S). Let D be contained in a half of a circular torus:

$$(r - r_+ - h)^2 + y^2 = r_+^2, \quad y < 0$$

for some positive constants r_+ and h. Thus, the half-torus has ring diameter $2r_+$, and it encloses a "pool" of diameter $2h$. As usual, the water domain is denoted by W, and F_∞ (F_0) denotes the external (internal) free surface.

We assume that the internal water-line intersection of the body is the same as that of the half-torus, namely the circle $r = h$. We also suppose that D satisfies John's condition; that is, no vertical line down from the free surface intersects the body (nor a region of depth variation in the finite-depth case). Notice, however, that it is *not* necessary to assume D to be axisymmetric.

4.2.5.1. Deep Water

The aim of this subsection is to prove the following uniqueness theorem.

Let the above geometric conditions hold, and let $h < r_+\sqrt{2}$. Then the homogeneous water-wave problem has only a trivial solution for all $\nu > 0$ less than or equal to

$$\frac{2r_+^2 - h^2}{2\pi r_+[h(2r_+ + h)^3]^{1/2}}. \tag{4.92}$$

In order to prove this theorem we use the toroidal coordinates (σ, θ, τ):

$$(x_1, x_2) = \frac{a\sinh\tau}{\cosh\tau - \cos\sigma}(\cos\theta, \sin\theta), \quad y = \frac{a\sin\sigma}{\cosh\tau - \cos\sigma} \tag{4.93}$$

(see, for example, Section 10.3 in Morse and Feshbach [251]). The mapping $(x_1, x_2, y) \mapsto (\sigma, \theta, \tau)$ maps the lower half-space $\{y < 0\}$ into the following semi-infinite parallelepiped

$$\{-\pi < \sigma < 0, 0 \leq \theta < 2\pi, 0 \leq \tau < \infty\},$$

and the surface $\tau =$ const is the image of torus

$$(r - a \coth \tau)^2 + y^2 = (a \operatorname{cosech} \tau)^2.$$

Thus, in order to map the half-torus, which contains D, into one of these coordinate surfaces, we have to find positive constants a and d_0 such that

$$r_+ = a \operatorname{cosech} d_0, \qquad h + r_+ = a \coth d_0$$

which gives $\cosh d_0 = 1 + h/r_+$, and $a = (h^2 + 2hr_+)^{1/2}$.

In what follows, it is convenient to write

$$A = \sinh \tau \sin \sigma, \qquad B = \cosh \tau \cos \sigma - 1, \qquad C = \cosh \tau - \cos \sigma.$$

Then, $A^2 + B^2 = C^2$, and taking into account that $x = (x_1, x_2)$, we have

$$x_\sigma = -\frac{aA}{C^2}(\cos\theta, \sin\theta), \quad x_\theta = \frac{a \sinh \tau}{C}(-\sin\theta, \cos\theta),$$

$$x_\tau = -\frac{aB}{C^2}(\cos\theta, \sin\theta), \quad y_\sigma = \frac{aB}{C^2}, \quad y_\theta = 0, \quad y_\tau = -\frac{aA}{C^2}.$$

It is well known that (σ, θ, τ) are orthogonal coordinates, and the Lamé constants h_σ, h_θ, and h_τ are given by

$$h_\sigma = \left(|x_\sigma|^2 + y_\sigma^2\right)^{1/2} = \frac{a}{C^2}(A^2 + B^2)^{1/2} = \frac{a}{C},$$

$$h_\theta = \left(|x_\theta|^2 + y_\theta^2\right)^{1/2} = \frac{a \sinh \tau}{C},$$

$$h_\tau = \left(|x_\tau|^2 + y_\tau^2\right)^{1/2} = \frac{a}{C^2}(B^2 + A^2)^{1/2} = \frac{a}{C}.$$

Hence,

$$\frac{\partial(x, y, z)}{\partial(\sigma, \theta, \tau)} = h_\sigma h_\theta h_\tau = \frac{a^3 \sinh \tau}{C^3}.$$

Also, $\{\sigma = 0\}$ is the image of $\{x^2 + y^2 > a^2, y = 0\}$, which contains F_∞, and $\{\sigma = -\pi\}$ is the image of $\{x^2 + y^2 < a^2, y = 0\}$, which contains F_0. Thus, we have the following on F:

$$\mathrm{d}x = \frac{a^2 \sinh \tau}{C^2}\, \mathrm{d}\theta \mathrm{d}\tau = \frac{a^2 \sinh \tau}{(\cosh \tau \mp 1)^2}\, \mathrm{d}\theta \mathrm{d}\tau,$$

where plus (minus) must be used on F_0 (F_∞). In fact, F_0 corresponds to

$$\{\sigma = -\pi,\ 0 \le \tau < d_0\},$$

and similarly

$$\{\sigma = 0,\ 0 \le \tau < d_0\}$$

corresponds to F_∞^e, which is the part of F_∞ external to the half-torus. In addition, the part of the water domain external to the half-torus will be denoted by W^e, and that corresponds to $\{-\pi < \sigma < 0, 0 \le \tau < d_0\}$. Thus,

$$\begin{aligned}\nu \int_{F_0} |u|^2\, \mathrm{d}x &= \nu a^2 \int_0^{d_0} \int_0^{2\pi} |u(-\pi, \theta, \tau)|^2 \frac{\sinh \tau}{(\cosh \tau + 1)^2}\, \mathrm{d}\theta \mathrm{d}\tau \\ &\le \frac{\nu a^2}{4} \int_0^{d_0} \int_0^{2\pi} |u(-\pi, \theta, \tau)|^2 \sinh \tau\, \mathrm{d}\theta \mathrm{d}\tau. \end{aligned} \tag{4.94}$$

As in Subsection 4.2.4.1, we have

$$|u(-\pi, \theta, \tau)|^2 \le 2 \left\{ |u(0, \theta, \tau)|^2 + \pi \int_{-\pi}^0 |u_\sigma|^2\, \mathrm{d}\sigma \right\}. \tag{4.95}$$

Hence,

$$\nu \int_{F_0} |u|^2\, \mathrm{d}x \le \frac{\nu a^2}{2} \int_0^{d_0} \int_0^{2\pi} |u(0, \theta, \tau)|^2 \sinh \tau\, \mathrm{d}\theta \mathrm{d}\tau \tag{4.96}$$

$$+ \frac{\pi \nu a^2}{2} \int_0^{d_0} \int_0^{2\pi} \int_{-\pi}^0 |u_\sigma|^2 \sinh \tau\, \mathrm{d}\sigma\ \mathrm{d}\theta \mathrm{d}\tau. \tag{4.97}$$

We can form an upper bound of (4.96) by

$$\begin{aligned} &\frac{\nu a^2}{2} (\cosh d_0 - 1)^2 \int_0^{d_0} \int_0^{2\pi} |u(0, \theta, \tau)|^2 \frac{\sinh \tau}{(\cosh \tau - 1)^2}\, \mathrm{d}\theta \mathrm{d}\tau \\ &= \frac{\nu}{2} (\cosh d_0 - 1)^2 \int_{F_\infty^e} |u|^2\, \mathrm{d}x. \end{aligned}$$

To bound (4.97) we observe that

$$|u_\sigma|^2 \le h_\sigma^2 |\nabla u|^2 = \frac{a^2}{C^2} |\nabla u|^2.$$

So, a suitable bound is as follows:

$$\begin{aligned} &\frac{\pi \nu a^4}{2} \int_0^{d_0} \int_0^{2\pi} \int_{-\pi}^0 |\nabla u|^2 \frac{\sinh \tau}{C^2}\, \mathrm{d}\sigma \mathrm{d}\theta \mathrm{d}\tau \\ &= \frac{\pi \nu a}{2} \int_0^{d_0} \int_0^{2\pi} \int_{-\pi}^0 |\nabla u|^2 C \frac{\partial(x, y, z)}{\partial(\sigma, \theta, \tau)}\, \mathrm{d}\sigma \mathrm{d}\theta \mathrm{d}\tau = \frac{\pi \nu a}{2} \int_{W^e} |\nabla u|^2 C\, \mathrm{d}x \mathrm{d}y \\ &\le \frac{\pi \nu a}{2} (\cosh d_0 + 1) \int_{W^e} |\nabla u|^2\, \mathrm{d}x \mathrm{d}y. \end{aligned} \tag{4.98}$$

As usual, Green's theorem shows that there is equipartition of energy in the homogeneous water-wave problem:

$$\int_W |\nabla u|^2 \,\mathrm{d}x\mathrm{d}y = \nu \int_F |u|^2 \,\mathrm{d}x. \tag{4.99}$$

Then (4.96)–(4.99) together give

$$\int_W |\nabla u|^2 \,\mathrm{d}x\mathrm{d}y \leq \nu \int_{F_\infty} |u|^2 \,\mathrm{d}x + \frac{\nu}{2}(\cosh d_0 - 1)^2 \int_{F_\infty^e} |u|^2 \,\mathrm{d}x$$
$$+ \frac{\pi \nu a}{2}(\cosh d_0 + 1) \int_{W^e} |\nabla u|^2 \,\mathrm{d}x\mathrm{d}y. \tag{4.100}$$

As the toroidal body satisfies John's condition, the estimate derived in Subsection 3.2.1 applies:

$$\nu \int_{F_\infty} |u|^2 \,\mathrm{d}x \leq \frac{1}{2} \int_{W^c} |\nabla u|^2 \,\mathrm{d}x\mathrm{d}y, \tag{4.101}$$

where W^c is the part of W covered by vertical lines down from F_∞. Hence, (4.100) becomes

$$\int_W |\nabla u|^2 \,\mathrm{d}x\mathrm{d}y < \left[\frac{1}{2} + \frac{1}{4}(\cosh d_0 - 1)^2 + \frac{\pi \nu a}{2}(\cosh d_0 + 1)\right]$$
$$\times \int_W |\nabla u|^2 \,\mathrm{d}x\mathrm{d}y,$$

since $F_\infty^e \subseteq F_\infty$, $W^c \subseteq W$, and $W^e \subseteq W$. The last inequality will clearly be a contradiction if

$$1/2(\cosh d_0 - 1)^2 + \pi \nu a(\cosh d_0 + 1) \leq 1, \tag{4.102}$$

unless $\nabla u \equiv 0$, and so $u \equiv 0$, because $u \to 0$ as $r \to \infty$. Note that equality is included in (4.102), since W^c is never the whole of W. Rearranging (4.102), we obtain the upper bound (4.92), which completes the proof.

Let us look at the bound (4.92) in the limit as $h \to 0$. The limit body would have no internal free surface. Then the water-wave problem should be unique for all frequencies by the result in Subsection 3.2.1. This is confirmed by the behavior of (4.92), which tends to infinity like

$$[8\pi r_+ h]^{-1/2} \quad \text{as } h \to 0.$$

The same behavior has the upper bound as $r_+ \to \infty$ for fixed h. Moreover, as (4.92) has zero for $h = r_+\sqrt{2}$, it is clear the bound achieves a maximum for some finite r_+. This maximum occurs for $r_+ = r_m = (1 + \sqrt{6}/2)h$. Hence, if the geometry allows it, the choice $r_+ = r_m$ should be made; otherwise the minimum possible r_+ should be chosen.

Note that the coordinate transformation (4.93) is singular at $\tau = \infty$, which is the polar circle $\{r = a,\ y = 0\}$, and part of this may be in the external free surface F_∞, if the body D does not have the same water-line area as the half-torus. However, the transformation is not needed inside the half-torus, so the method will still work even if τ can be infinite in the closure of the water domain.

There are two weaknesses of the simple bound (4.92). One is that it is nonpositive for a toroidal body with $h \geq r_+\sqrt{2}$, so that nothing can be said about uniqueness if the pool diameter is large enough. The second lies in the fact that the bound (4.95), and hence what follows, involves integrating with respect to σ for fixed θ and τ, and hence along a semicircle in a given half-plane of constant θ. In order to construct the bound on the potential energy for free-surface points close to the origin, the semicircle used must be very large, and must therefore involve large depth. Consequently, this construction cannot be used in any finite depth of water, no matter how large the depth is.

In order to get around these two difficulties, we use another construction of the bound on the potential energy.

4.2.5.2. Another Bound

Here we consider the coordinates (r, θ, α) given by

$$(x_1, x_2) = [A + \rho(B + \cos\alpha)]\,(\cos\theta, \sin\theta), \quad y = \rho\sin\alpha. \tag{4.103}$$

Here A and B are constants at our disposal. These coordinates can be derived by rotating around the vertical axis (and relabeling) the second set of coordinates [see (4.89)] used in Subsection 4.2.4 to construct a bound in the two-dimensional finite-depth case.

The surface $\rho = \text{const} > 0$ is the torus $(r - A - B\rho)^2 + y^2 = \rho^2$. Thus, the half-torus that contains D is the surface $\{\rho = r_+,\ -\pi < \alpha < 0\}$, provided $A + Br_+ = r_+ + h$. We also consider a torus of radius $r_- > r_+ + h/2$ that touches the vertical axis at the origin provided $A + Br_- = r_-$. We easily find that

$$A = \frac{hr_-}{r_- - r_+}, \qquad B = 1 - \frac{h}{r_- - r_+}, \tag{4.104}$$

and hence $h < A < h + 2r_+$ and $-1 < B < 1$, with B and $(r_+ + h - A)$ taking the same sign as $r_- - h - r_+$.

Under the transformation $(x_1, x_2, y) \mapsto (\rho, \theta, \alpha)$, the surface F_0 maps to

$$\{\alpha = -\pi,\ r_+ < \rho \leq r_-\},$$

and

$$\{\alpha = 0,\ r_+ < r < r_-\}$$

is the image of F_∞^e, which is, in this subsection, the part of F_∞ *between* the two tori. As usual, a bound for the potential energy at each point of F_0 except the origin can be constructed by integration along a semicircle in water corresponding to constant values of ρ and θ. The origin, which is a singularity of (4.103) and (4.93), can be deleted from F_0 without changing the potential energy integral.

We have

$$\frac{\partial(x_1, x_2, y)}{\partial(\rho, \theta, \alpha)} = \rho(1 + B\cos\alpha)[A + \rho(B + \cos\alpha)].$$

The region of interest here is $\{r_+ < \rho < r_-, -\pi < \alpha < 0\}$. As $|B| < 1$, this Jacobian does not vanish between the two tori; also

$$\mathrm{d}x = (1 \pm B)[A + \rho(B \pm 1)]\,\mathrm{d}\rho\mathrm{d}\theta \quad \text{on } F,$$

where plus (minus) should be taken on F_∞ (F_0). Now following the scheme given in Subsection 4.2.5.1 [see (4.94)–(4.97)], we obtain from (4.99)

$$\int_W |\nabla u|^2\,\mathrm{d}x\mathrm{d}y \le \left[\frac{1}{2} + \frac{h^2}{(2r_+ + h)(2r_- - 2r_+ - h)} + 2\pi\nu C r_-\right] \times \int_W |\nabla u|^2\,\mathrm{d}x\mathrm{d}y. \tag{4.105}$$

Therefore, we have the following theorem.

Let the geometric conditions described in Subsection 4.2.5.1 *hold. In addition, let there exist a number r_- such that*

$$2r_- > 2r_+ + h + \frac{2h^2}{2r_+ + h}, \tag{4.106}$$

and the half-torus $(r - r_-)^2 + y^2 = r_-^2$, $y \le 0$ belongs to the closure of the water region. Then the homogeneous water-wave problem has only a trivial solution for all $\nu > 0$, which do not exceed

$$\frac{1}{4\pi C r_-}\left[1 - \frac{2h^2}{(2r_+ + h)(2r_- - 2r_+ - h)}\right], \tag{4.107}$$

where $C = 1$ when $r_- \ge r_+ + h$ and

$$C = (1 - B)/(1 + B) = h/(2r_- - 2r_+ - h) \quad \text{when } r_- < r_+ + h.$$

Here B is given in (4.104).

This theorem is stated for deep water, in which case the geometry imposes no upper limit on the value of r_-. The lower limit on r_- given by (4.106) ensures that the bound (4.107) is positive, and (a fortiori), that the torus $\rho = r_-$ contains the torus $\rho = r_+$.

Let us compare the upper bounds (4.92) and (4.107), which both apply for deep water. The latter bound contains the parameter r_-, which can be chosen, subject to (4.106), and this is useful. Choosing $r_- = r_+ + h$, so that the tori have concentric cross section, we get that $B = 0$. Hence, uniqueness is proved for

$$0 < \nu \leq \frac{2r_+ - h}{4\pi(2r_+ + h)(r_+ + h)},$$

provided $r_+ > h/2$.

Even this simple bound can be used in cases in which (4.92) cannot, namely for a greater range of r_+/h. However, the bound (4.107) is not subject to any restriction on r_+ and gives a useful bound even for sufficiently small values of r_+. Furthermore, (4.107) can be optimized with respect to r_-, giving

$$0 < \nu \leq \nu_m = \frac{\left[(m^2 + 2h^2)^{1/2} - h\sqrt{2}\right]^2}{2\pi m^3}, \tag{4.108}$$

where $m = 2r_+ + h$. This optimum is achieved by using

$$\frac{2r_-}{m} = \left[1 - \left(\frac{2h^2}{m^2 + 2h^2}\right)^{1/2}\right]^{-1}.$$

This optimized upper bound (4.108) is particularly useful when (4.92) and (4.107) do not apply, that is, for $r_+ < h/2$. In fact, as $r_+ \to 0$ we have $m \to h$, so $\nu_m = (5 - 2\sqrt{6})/(2\pi h)$ is achieved by using $2r_- = (3 + \sqrt{6})h$.

The bound (4.108) can also be optimized as a function of r_+. A single maximum occurs for $m^2 = 6h^2$; that is, $r_+ = (\sqrt{6} - 1)h/2$ and this gives $\nu_m \leq (6\pi h\sqrt{6})^{-1}$.

Let us consider modifications needed for the finite-depth case. The bound (4.101) is in this case replaced by

$$\nu \int_{F_\infty} |u|^2 \,\mathrm{d}x \leq \frac{M_1}{2} \int_{W^c} |\nabla u|^2 \,\mathrm{d}x\mathrm{d}y,$$

where

$$M_1 = 1 - \frac{2k_0 d}{\sinh 2k_0 d},$$

and k_0 is the unique positive root of $k_0 \tanh k_0 d = \nu$. This revised bound changes (4.105), and so uniqueness will follow for all $\nu > 0$, which do not exceed

$$\frac{1}{4\pi C r_-}\left[2 - M_1\left\{1 + \frac{2h^2}{(2r_+ + h)(2r_- - 2r_+ - h)}\right\}\right].$$

This will be nonnegative provided r_- can be found such that

$$2r_- > 2r_+ + h + \frac{2M_1 h^2}{(2 - M_1)(2r_+ + h)}.$$

As the positions of the bottom B and the body D impose restrictions on r_-, it may occur that no value of r_- can be found to use in this proof of uniqueness. In particular, if $d \leq r_+ + h/2$, then this uniqueness proof cannot be used.

4.3. Bibliographical Notes

4.1. Presumably, the first application of the inverse method in the linear theory of water waves was made by Troesch [319]. He had found some conical and hyperboloid containers along with the corresponding frequencies and modes in the sloshing problem. Roseau [299], Chapter 8, developed the inverse method for the problem of water oscillations in an infinite channel of variable depth.

Using the inverse procedure, McIver [230] constructed the first example of trapped waves in the two-dimensional water-wave problem. Her example is given by our function $u_1^{(+)}$. Other examples considered in Subsection 4.1.1 have been considered by Kuznetsov, and Porter has supplied the numerical examples (these results are unpublished). McIver and McIver [240] generalized the construction from [230] to the axisymmetric problem (see Subsection 4.1.2). Structures trapping higher azimuthal modes were obtained numerically by Kuznetsov and McIver [169].

Other results on trapped modes. Recently, McIver [232] produced convincing numerical evidence that a pair of totally submerged two-dimensional bodies, which are symmetric about a vertical axis and support a trapped mode, does exist. Another example of a totally submerged trapping structure was obtained by McIver and Porter [234], who produced numerical evidence that trapping frequencies exist for submerged tori having elliptical meridional cross section. Numerical evidence of modes trapped by a pair of surface-piercing bodies in the presence of a submerged body is obtained by Evans and Porter [83].

Newman [266] numerically performed a hydrodynamic analysis of the McIver toroid constructed in [240] and discovered that added mass, damping, and elevation of the free surface in the moon pool inside the toroid demonstrate singular behavior in the resonant regime. McIver and Newman [241] modified the approach developed in [240] and [169] in order to construct nonaxisymmetric trapping structures in the three-dimensional water-wave problem.

4.2. The uniqueness theorems presented in Subsection 4.2.1 were published in papers by Kuznetsov and Maz'ya [168] (the case of shell) and Kuznetsov, McIver, and Linton [170] (the case of barriers). In connection with the latter paper, it should be mentioned that numerous papers treating various aspects of problems involving vertical barriers are surveyed by Evans [72], Evans and Porter [81], Chakrabarti [37], and Mandal and Banerjea [213].

The uniqueness theorem for water of infinite depth presented in Subsection 4.2.2.2 was published by Linton and Kuznetsov [207], and Kuznetsov generalized it for water of finite depth (this unpublished result is given in Subsection 4.2.2.1). The proposition in Subsection 4.2.2.3 that is contrary to a particular case of the uniqueness theorem for deep water was obtained by Motygin and Kuznetsov [256].

The uniqueness result in Subsection 4.2.3.1, valid for all azimuthal modes in deep water, is borrowed from the work [169] by Kuznetsov and McIver. The generalization of the uniqueness property in the axisymmetric problem to a wider class of geometries presented at the end of Subsection 4.2.3.1 is a straightforward consequence of the energy inequality published in a note by Kuznetsov and Simon [176]. The finite-depth uniqueness theorem in Subsection 4.2.2.3 has not been published before. The results in Subsection 4.2.4 were published in the papers [174, 175] by Kuznetsov and Simon, but the simpler upper bound (4.87) had originally appeared in the note [153] by Kuznetsov. In Subsection 4.2.5, the results of the preprint [306] by Simon and Kuznetsov are presented.

5

Horizontally Periodic Trapped Waves

The most well-known trapped mode in the theory of water waves is the simple exponential solution derived by Stokes [313]. It describes a wave over a uniformly sloping beach that can travel unchanged in the direction of the shoreline, and that decays exponentially to zero in the seaward direction. The Stokes edge wave is characterized by being confined to or trapped by the boundary, in this case the beach, despite the fact that the fluid region is unbounded. Little interest was shown in this solution for over a 100 years. Thus Lamb [179], p. 447, states "it does not appear that the type of motion here referred to is very important." Only in the last forty years or so have further examples of such trapped modes been discovered, and it was recognized that "there is now considerable evidence which indicates that edge waves are common in occurrence and practical use" (see Le Blond and Mysak's book [183], p. 227). Thus the aim of the present chapter is to review many papers representing a considerable progress in the investigation of edge waves and other types of trapped modes that have occurred during the past twenty-five years. The material here is to a great extent borrowed from a survey paper by Evans and Kuznetsov [75], but we also cover new results published since 1995, when that paper was written.

The Stokes solution is an eigenfunction corresponding to a point eigenvalue below the continuous spectrum of the Laplacian operator satisfying certain boundary conditions in an unbounded domain. General results on the nature of the spectrum in unbounded domains are few and involve deep analysis, and it is for this reason that the consideration of trapped modes has attracted the attention of many mathematicians.

Like the Stokes edge wave, all trapped modes we are going to describe in the present chapter are periodic (or can be continued periodically) in one of two horizontal directions. It is interesting to note that for some geometries the results of this chapter are the same as those obtained in the previous chapters for the two-dimensional problem. Besides, there are geometries for which some new effects arise.

We conclude our this introductory remarks by noting that the basis of our consideration is the same classical linearized theory of water waves as in the other four chapters of Part 1. Meanwhile, for sufficiently long waves it is possible to adopt the linearized shallow-water equations as a starting point, and this simplification enables a number of explicit solutions for trapped modes or edge or shelf waves to be constructed for particular bottom geometries. Such long waves are of considerable importance in oceanography, and many observations of these waves exist. We shall not be concerned with such solutions here because an excellent review of the subject was already given by Mysak [257].

The plan of the chapter is as follows. In Section 5.1, we describe notations (they are distinguished from those in the previous chapters) and remind the reader of the governing equations. After that we give a brief description of two distinct types of trapped modes: the first embracing generalizations of the Stokes edge waves to arbitrary bottom shapes and submerged horizontal cylinders and including modes for pairs of surface-piercing cylinders, and the second introducing the relatively new possibility of trapped modes near vertical cylinders in channels.

In Section 5.2, the latest results on edge waves are described, including a summary of recent existence and uniqueness proofs and a description of methods for estimating bounds on the allowable frequencies of edge waves.

Section 5.3 is concerned with trapped modes above submerged horizontal cylinders or protrusions on the bottom. A description of a recent existence proof and a method for determining lower bounds on the eigenfrequencies are presented. The latter method also provides the uniqueness theorem for a certain class of bottom obstructions. Similar questions of uniqueness and trapped modes for surface-piercing cylinders are considered in Section 5.4.

In Section 5.5, some recent papers dealing with trapped modes in wave channels containing vertical cylinders are described. The aim is to indicate methods for constructing solutions for particular cylinders and to describe a general existence proof for a class of vertical cylinders.

5.1. Two Types of Trapped Modes

In what follows we use notations that distinguish from those in the previous chapters. First, Cartesian axes are denoted (x, y, z), and they are chosen so that y is directed vertically upward and x and z are in the plane of the unperturbed free surface. Since problems under consideration are stated in one or other cross section of a water domain, we reserve the letter W for the corresponding cross section, and F, S, and B denote cross sections of the free surface, of immersed bodies or/and walls, and of the bottom, respectively. We write $\Phi(x, y, z, t)$ for a time-dependent velocity potential, which satisfies the Laplace equation

$$\Phi_{xx} + \Phi_{yy} + \Phi_{zz} = 0 \tag{5.1}$$

in the domain occupied by water. The linearized free surface condition has the form

$$\Phi_{tt} + g\Phi_y = 0 \tag{5.2}$$

on the corresponding part of $\{y = 0\}$. Since we are interested only in time-harmonic eigensolutions, we impose the homogeneous Neumann condition on rigid boundaries,

$$\partial\Phi/\partial n = 0, \tag{5.3}$$

where $\mathbf{n}$ is the unit normal directed into water.

Equations (5.1)–(5.3) form the basis for particular problems formulated in this subsection. They describe two distinct physical situations when the dependence upon z or y may be eliminated, leaving two-dimensional problems that may have trapped mode solutions.

First, we state problems for geometries allowing us to separate the z coordinate (see Subsection 5.1.1). These problems describe the edge waves (Section 5.2), trapped modes above submerged obstacles (Section 5.3), and oblique waves trapped by surface-piercing structures (Section 5.4). Second, we consider channels having vertical walls and containing vertical cylinders extending throughout the depth (see Subsection 5.1.2). In this situation the dependence on the y coordinate may be removed. Apparently, this evident classification has been proposed explicitly in the survey paper by Evans and Kuznetsov [75].

5.1.1. Modes Trapped by Surfaces with Horizontal Generators

Let water contain an obstruction having constant cross section in the z direction. This might be immersed cylinders or a bottom topography varying with x such as a sloping beach. Then solutions to (5.1)–(5.3), corresponding to waves of frequency ω and wavenumber k traveling along the obstruction can be sought in the form

$$\Phi(x, y, z, t) = \phi(x, y)\cos(kz - \omega t). \tag{5.4}$$

It follows by substituting (5.4) into (5.1)–(5.3) that the function $\phi(x, y)$, defined on a two-dimensional domain W that is a cross section of water orthogonal to the z axis, satisfies the following boundary value problem:

$$\phi_{xx} + \phi_{yy} = k^2\phi \quad \text{in } W, \tag{5.5}$$

$$\phi_y - \nu\phi = 0 \quad \text{on } F, \tag{5.6}$$

$$\partial\phi/\partial n = 0 \quad \text{on } S \cup B. \tag{5.7}$$

Here F, S, and B denote parts of the boundary ∂W, lying in the free surface, and in the solid surfaces of immersed bodies and of the bottom, respectively, and $\nu = \omega^2/g$.

The problem of (5.5)–(5.7) is a spectral problem, in the sense that one of the parameters involved (ν or k^2) should be treated as a *spectral parameter* (*eigenvalue*), which is to be found simultaneously with the corresponding nontrivial solution (*eigenfunction*). It makes no difference which of the parameters is considered to be the eigenvalue and is to be determined in terms of the other which is to be prescribed. Both approaches are equivalent and will be applied in what follows.

We say that a certain eigenvalue belongs to the *continuous spectrum* of the problem if the corresponding eigensolution ϕ has an infinite *energy norm*, so that the wave mode has infinite energy per unit length of the z axis. When the energy norm of ϕ is finite, the corresponding value of the spectral parameter is said to be a *point eigenvalue* or a point of the *point spectrum*. In this latter case the energy of the wave mode is finite per unit length of the z axis, and the mode or wave is said to be trapped. In application to water waves, notions of continuous spectrum and of point eigenvalues (they are well known in spectral operator theory) were introduced in the paper by Jones [127].

The same problem (5.5)–(5.7) arises when water waves are considered in a channel that is infinitely long in the x direction and has vertical walls $z = \pm b$. The latter are spanned by a solid obstruction with horizontal generators. In this case W denotes the longitudinal section of the channel parallel to the walls, and the velocity potential can be taken, for example, in the form

$$\Phi(x, y, z, t) = \text{Re}\{\phi(x, y)e^{i\omega t}\}\cos kz,$$

where $kb = n\pi (n = 1, 2, \ldots)$ to ensure that (5.7) is satisfied on the channel walls. A similar solution with a sine dependence upon z is also possible, provided $kb = (2n - 1)\pi/2$, $(n = 1, 2, \ldots)$.

Another physical situation described by (5.5)–(5.7) is as follows. Let infinitely long cylinders be immersed in water and let the crests of waves make a nonzero angle θ with the plane normal to the generators of the cylinders.

If waves have components of the wavenumber ν ($= 2\pi/\lambda$, where λ is the wavelength) given by k and l in directions parallel to and normal to the generators of the cylinders, respectively, then $k = \nu \sin\theta, l = \nu\cos\theta$, so that $l = (\nu^2 - k^2)^{1/2}$ with $k < \nu = \omega^2/g$. An example of application of such a problem is provided by Bolton and Ursell [29].

Finally, there is one more case leading to the same problem, namely the case of waves propagating without change of shape along a uniform channel of arbitrary cross section. This problem is not considered in the present chapter and we restrict ourselves to a couple of references: Rosenblat [300] and Groves [99]. In the latter paper, the author also considers channels of non-uniform cross section. It should be mentioned that Groves and Lesky [100] extend results from [99] and study the evolutionary problem of waves in a channel caused by time-harmonic surface pressure.

5.1.2. Modes Trapped by Vertical Cylinders in Channels

Let the obstruction be a vertical cylinder extending throughout the depth of water, and let S be the boundary of the cylinder's cross section. We denote by W a cross section of fluid orthogonal to the y axis, and we assume that W does not depend on y. Then the depth dependence may be removed from the problem by writing

$$\Phi(x, y, z, t) = \phi(x, z) \cosh k(y + d) \cos \omega t, \tag{5.8}$$

where in order to satisfy (5.2) k is chosen to be the real positive root of

$$\omega^2 = gk \tanh kd, \tag{5.9}$$

where d is the depth of the channel. Substituting (5.8) into (5.1) and (5.3), we obtain

$$\phi_{xx} + \phi_{zz} + k^2\phi = 0 \quad \text{in } W, \tag{5.10}$$

$$\partial\phi/\partial n = 0 \quad \text{on } S, \tag{5.11}$$

$$\phi_z = 0 \quad \text{on } z = \pm h. \tag{5.12}$$

Here the channel walls are described by

$$\{-\infty < x < +\infty, -d \leq y \leq 0, z = \pm h\}.$$

In considering possible trapped waves we shall be concerned with cylinders that are symmetric about the center line of the channel, and we shall seek solutions which are antisymmetric about the center line so that

$$\phi(x, z) = -\phi(x, -z), \quad (x, z) \in W,$$

from which it follows that

$$\phi = 0, \quad \text{on } z = 0 \text{ outside } S. \tag{5.13}$$

We shall seek solutions of (5.10)–(5.13) having finite energy in W.

The problem (5.10)–(5.13) also describe possible trapped acoustic modes in a two-dimensional waveguide. Examples of constructive methods for finding solutions of these equations for particular cylinders and the description of a general existence proof are given in Section 5.5.

5.2. Edge Waves

This section is devoted to a special geometry, when the free surface cross section is of the form

$$F = \{(x, y) : x > 0, y = 0\},$$

and W is a subdomain of the quadrant $\{(x, y) : x > 0, y < 0\}$. An important example of the bottom shape in this case is an explicitly given curve

$$S = \{(x, y) : x \geq 0, y = -d(x)\}, d(0) = 0, d(x) > 0 \quad \text{for } x > 0. \tag{5.14}$$

Since in this case the boundary of the water domain contains a straight coastline (the z axis), separating F and S, the trapped modes occurring for this geometry are usually termed *edge waves*.

A *sloping beach* with

$$d(x) = x \tan\alpha, \quad 0 < \alpha < \pi/2 \tag{5.15}$$

in (5.14) is the simplest case, for which edge waves exist. The first example of edge wave modes over a sloping beach was found as early as 1846 by Stokes. It took 105 years before other edge wave modes for this simple geometry were found. Thus in June 1951 Ursell announced some new edge wave modes at an NBS Symposium (see [324]). The following year he published a paper [325], which in addition to his analytical results presented experimental data confirming the existence of these new edge wave modes.

Now, almost 150 years after Stokes, the problem of edge waves over a sloping beach has got a final solution in the paper [144] by Komech, Merzon, and Zhevandrov. They have demonstrated that Stokes' and Ursell's edge waves are the only trapped modes over a sloping beach having finite energy (see also Merzon's brief presentation [243], where a certain additional restriction on smoothness of modes is made). These modes and a sketch of the considerations made by Komech et al. are given in Subsection 5.2.1, where a brief description of some other results is also given.

In this connection it should be mentioned that Lehman and Lewy [185] have demonstrated the uniqueness of edge waves in the class of functions bounded throughout water. More precisely, they proved that there is never more than one bounded solution, and such a solution cannot be zero at the beach. However, their result is restricted to the angles $\alpha \neq \pi/(2m)$ with integer m.

Recently, an extensive review of other geometries that permit the existence of edge waves was published by Bonnet–Ben Dhia and Joly [30]. They considered three types of bottom curve S, depending upon its behavior in a vicinity of the edge:

1. Cliff: $S = \{(x, y) : x = 0, -d_0 \leq y \leq 0\} \cup S_0$, where $d_0 > 0$ and $S_0 = \{(x, y) : x \geq 0, y = -d(x)\}$, $d(0) = d_0$, $d(x) > 0$ for $x \geq 0$.
2. Tangential bottom: $d'(0) = 0$ for h in (5.14), but $x^{2-\epsilon}/d(x) = O(1)$ as $x \to 0$, where $\epsilon > 0$.
3. Transversal bottom: d in (5.14) satisfies $d(x) \sim x \tan \alpha$ as $x \to 0$.

The function $d(x)$ is assumed to be constant at infinity; that is, $d(x) = d_\infty = \text{const}$ for $x \geq x_\infty > 0$. Precise restrictions on the smoothness of d are given by Bonnet–Ben Dhia and Joly [30], but for our purposes it is sufficient to assume that d is piecewise continuously differentiable.

For these geometries a necessary and sufficient condition is obtained for the existence of at least one edge wave mode for all values of frequency ω with the square of wavenumber k^2 being considered as the spectral parameter.

A review of these results by Bonnet–Ben Dhia and Joly [30] is given in Subsection 5.2.2, along with a description of earlier results obtained by Grimshaw [98], Evans and McIver [79], and McIver and Evans [238].

5.2.1. Edge Waves on a Sloping Beach

In his report on recent researches in hydrodynamics, Stokes [313] proposed the following very simple formula:

$$\phi_0(x, y) = \exp\{-k(x \cos \alpha - y \sin \alpha)\}, \tag{5.16}$$

which gives an eigensolution to (5.5)–(5.7) in W described by (5.14) and (5.15) (sloping beach domain), provided the corresponding point eigenvalue

ν_0 has the form

$$\nu_0 = k \sin \alpha. \tag{5.17}$$

Here k is considered as an arbitrary parameter, on which the spectral parameter ν and eigenfunction depend.

Since ϕ_0 decays exponentially at infinity, then it has a finite norm in the Sobolev space $H^1(W)$. The solution (5.16) and (5.17) has another remarkable property in that it satisfies the boundary condition (5.6) throughout the water domain.

In a breakthrough paper [325], Ursell writes down the Stokes' eigensolution (5.16) and (5.17), notes that values $\nu \geq k$ belong to the continuous spectrum (as was demonstrated by Peters [285] shortly before), and proposes the following formulae for a finite sequence of eigensolutions:

$$\nu_n = k \sin(2n+1)\alpha, \quad n = 1, 2, \ldots, \text{ such that } (2n+1)\alpha \leq \pi/2, \tag{5.18}$$

$$\begin{aligned} \phi_n(x, y) = \phi_0(x, y) & \\ + \sum_{m=1}^{n} A_{mn} (& \exp\{-k[x \cos(2m-1)\alpha + y \sin(2m-1)\alpha]\} \\ & + \exp\{-k[x \cos(2m+1)\alpha - y \sin(2m+1)\alpha]\}), \end{aligned} \tag{5.19}$$

where

$$A_{mn} = (-1)^m \prod_{r=1}^{n} \frac{\tan(n-r+1)\alpha}{\tan(n+r)\alpha}.$$

Ursell gives no derivation of (5.18) and (5.19), and only a brief discussion follows these formulae. He notes that the Stokes eigenvalue ν_0 decreases with α, and that when α reaches a critical angle $\pi/6$, the second point eigenvalue appears and also decreases with α. A new point eigenvalue appears at $\alpha = \pi/10$, and so on. The remainder of the paper is devoted to a detailed description of his experiments.

It was not until 1958 that a systematic approach to the sloping beach problem for arbitrary $0 < \alpha \leq \pi$ was carried out. It was Roseau who, in his extensive paper [298], obtained solutions in the form of integrals by solving certain functional difference equations. His integrals in the complex plane, resembling those obtained in the inversion of the Laplace transform, can be shown to include the solutions (5.16)–(5.19) as a particular case. Roseau also produced further edge wave solutions in the form of integrals, but these are all unbounded at the edge, and moreover, have infinite energy.

Subsequently, Whitham [359] simplified Roseau's method in order to determine the bounded solutions systematically. Evans [70] showed that the same method could be exploited for the sloping beach problem with the more general beach condition:

$$\partial\phi/\partial n + l\phi = 0 \quad \text{on } F, 0 < l < k.$$

He was able to reproduce the solution for this case, stated by Greenspan [97]. Shortly after that, Packham [274] showed how the same results could be obtained more economically by using the ideas of Williams [363]. In another paper, Evans [69] described two possible mechanisms for the excitation of edge waves over a sloping beach.

The question as to whether (5.16)–(5.19) give all eigensolutions to (5.5)–(5.7), having finite $H^1(W)$ norm, was recently finally resolved in a comprehensive paper by Komech et al. [144]. Their new approach to the problem enables them not only to derive the formulae (5.16)–(5.19) but also to prove uniqueness of these solutions in the class of functions having finite energy. What follows is a brief sketch of the work [144] (see also Merzon's paper [243], containing his talk given at the Third Conference on Wave Propagation in 1995, where a superfluous additional condition is imposed).

The first step is the following change of variables:

$$(x_1, x_2) = f(x, y), \quad \text{where } x_1 = k(x + y\cot\alpha), x_2 = -ky/\sin\alpha.$$

Hence, f is a one-to-one mapping of W onto the first quadrant

$$Q = \{(x_1, x_2) : x_1, x_2 > 0\}.$$

The problem (5.5)–(5.7) in the new variables takes the following form:

$$\frac{1}{\sin^2\alpha}\left(\frac{\partial^2}{\partial x_1^2} + \frac{\partial^2}{\partial x_2^2} - 2\cos\alpha\frac{\partial^2}{\partial x_1 \partial x_2}\right) u = u \quad \text{in } Q, \tag{5.20}$$

$$-\cot\alpha\frac{\partial u}{\partial x_1} + \frac{1}{\sin\alpha}\frac{\partial u}{\partial x_2} + \lambda u = 0 \quad \text{when } x_1 > 0, x_2 = 0, \tag{5.21}$$

$$\frac{1}{\sin\alpha}\frac{\partial u}{\partial x_1} - \cot\alpha\frac{\partial u}{\partial x_2} = 0 \quad \text{when } x_1 = 0, x_2 > 0. \tag{5.22}$$

The function $u(x_1, x_2) = \phi(f^{-1}(x_1, x_2))$ is the representation of the velocity potential in the new coordinates. The condition that $\phi \in H^1(W)$ is naturally equivalent to the requirement that $u \in H^1(Q)$.

The solutions (5.16)–(5.19) are equivalent to the following solutions of the spectral problem (5.20)–(5.22) with λ as the spectral parameter:

$$u_n(x_1, x_2) = \exp\{-(x_1 \cos\alpha + x_2)\}$$
$$+ \sum_{m=1}^{n} A_{mn}(\exp\{-[x_1 \cos(2m-1)\alpha + x_2 \cos 2m\alpha]\}$$
$$+ \exp\{-[x_1 \cos(2m+1)\alpha + x_2 \cos 2m\alpha]\}) \qquad (5.23)$$

with eigenvalue

$$\lambda_n = \sin(2n+1)\alpha, \quad n = 0, 1, 2, \ldots, \text{ such that } (2n+1)\alpha \leq \pi/2. \qquad (5.24)$$

Here A_{mn} $(1 \leq m \leq n)$ are the same coefficients as in (5.19).

The main result of Komech et al. is as follows:

All point eigenvalues of (5.20)–(5.22) *are given by* (5.24), *and* (5.23) *gives the corresponding eigenfunctions.*

Let us outline the proof's scheme. An application of the Fourier transform in the complex domain reduces (5.20)–(5.22) to an equivalent algebraic problem. This algebraic problem can be rewritten in the following equivalent form.

Find all $\lambda > 0$, *such that there exists a nontrivial function* $v(w)$ *with the following properties:*

1. $v(w)$ *is analytic in the strip* $0 < \operatorname{Im} w < \pi$.
2. $v(2i\pi - w) = v(w)$ *when* $-\alpha < \operatorname{Im} w < \pi + \alpha$.
3. $v(w)(\cosh w - \lambda) - v(w + 2i\alpha)[\cosh(w + 2i\alpha) + \lambda] = 0$ *when* $-\alpha < \operatorname{Im} w < \pi - \alpha$.

The latter property resembles the functional difference equations occurring in Roseau [298]. It can be shown that the last problem has nontrivial solutions if and only if λ is given by (5.24), and these solutions correspond to (5.23). However, the verification takes more than fifteen pages of hard analysis in [144], to which the interested reader is referred for details. It should also be mentioned that the technique developed in [144] has a much wider area of application (see Zhevandrov and Merzon [369] and references cited therein).

In conclusion of this subsection, it should be said that oblique waves over a plane beach were considered by Ehrenmark [61, 62].

5.2.2. Water Having Constant Depth at Infinity

The main part of this subsection is devoted to results obtained by Bonnet–Ben Dhia and Joly [30]. The method applied in this paper is rather general. It exploits the fact that water has constant depth at infinity, which allows the use of the separation of variables in this part of the water domain. The same idea was earlier used by Aranha [2], who investigated the trapped mode problem for a cylinder submerged in water of finite depth.

First, we recall that the problem (5.5)–(5.7) is to be solved in the water domain W bounded from below by the given bottom curve S:

$$y = -d(x), d(x) > 0 \quad \text{for } x > 0, \qquad d(x) = d_\infty \quad \text{for } x \geq x_\infty > 0.$$

The domain W is infinite in the direction of the positive x axis, but it is bounded in the opposite direction, since the free surface F and the bottom S meet at the edge placed at the origin. Let W_e be a semi-infinite strip

$$\{(x, y) : x > h, -d_\infty < y < 0\},$$

where $h > x_\infty$. By S' we denote the "artificial boundary"

$$\{(x, y) : x = h, -d_\infty < y < 0\},$$

and $W_h = W \setminus \bar{W}_e$. Hence, the boundary ∂W_h consists of three parts:

$$S', \quad F_h = F \cap \{x < h\}, \quad S_h = S \cap \{x < h\}.$$

Separation of variables in W_e allows us to reduce the original problem (5.5)–(5.7) to another problem in the bounded domain W_h. As a first step in this direction we introduce the following functions:

$$u_n(\nu, y) = \begin{cases} \alpha_0 \cosh k_0(y + d_\infty), & n = 0, \\ \alpha_n \cos k_n(y + d_\infty), & n = 1, 2, \ldots. \end{cases} \tag{5.25}$$

Here k_0 is the only positive root of the equation $k \tanh kd_\infty = \nu$, and k_n is the only root in the interval $(n\pi - \pi/2, n\pi + \pi/2)$ of the equation $k \tan kd_\infty = -\nu$ (cf. Subsection 1.1.2). The constants α_n $(n = 0, 1, \ldots)$ are chosen to normalize the functions $u_n(\nu, y)$.

It is a well-known result from the spectral theory of ordinary differential operators that the family (5.25) forms an orthonormal basis in $L^2(-d_\infty, 0)$. Now, with the help of this family, separation of variables in W_e can be carried out.

If $\phi \in H^1(W)$ is a solution to (5.5)–(5.7), then the following representation

$$\phi(x, y) = \sum_{n=N}^{\infty} a_n u_n(\nu, y) \exp\{-\beta_n(x - h)\} \tag{5.26}$$

holds for $(x, y) \in W_e$. Here $N = N(k^2, \nu)$ is the smallest nonnegative integer, such that $\mu_n > -k^2$, where $\mu_0 = -k_0^2$, and $\mu_n = k_n^2$ for $n \geq 1$. Furthermore,

$\beta_n = (\mu_n + k^2)^{1/2}$ and

$$a_n = \int_{-d_\infty}^{0} \phi(h, y) u_n(\nu, y)\, \mathrm{d}y. \tag{5.27}$$

It is clear that the series (5.26) and all its derivatives converge uniformly. Moreover, (5.26) gives an explicit representation of the solution in the exterior domain W_e in terms of coefficients a_n, which depend only upon the trace of ϕ on the artificial boundary S'.

An equivalent statement of the trapped modes problem can now be given, which only involves the restriction of the function ϕ to the bounded domain W_h and its boundary $S_h \cup F_h \cup S'$.

Find $\phi \in H^1(W_h)$, satisfying

$$\nabla^2 \phi - k^2 \phi = 0 \quad \text{in } W_h, \tag{5.28}$$

$$\phi_y - \nu\phi = 0 \quad \text{on } F_h, \tag{5.29}$$

$$\partial\phi/\partial n = 0 \quad \text{on } S_h, \tag{5.30}$$

$$\phi_x + T(k^2)\phi = 0 \quad \text{on } S'. \tag{5.31}$$

The last condition contains a nonlocal operator depending upon the spectral parameter k^2 (ν is here considered as an arbitrary parameter), and this operator is given as follows:

$$[T(k^2)\phi](y) = \sum_{n=N}^{\infty} \beta_n a_n u_n(\nu, y).$$

This operator depends on k^2 through $N(k^2, \nu)$ and β_n. Formula (5.27) for the coefficients a_n demonstrates the nonlocality of this operator.

Nevertheless, the usual tools of linear operator theory (see, for example, books by Birman and Solomyak [27] and Reed and Simon [296]) can be applied to investigate (5.28)–(5.31). A coercive symmetric bilinear form and a corresponding self-adjoint operator can be associated with this problem, and consequently, the min-max principle holds.

We begin with the statement of some general results, which follow from the application of this technique. After that, some more specific results for each of three geometries described above (cliff, tangential bottom, and transversal bottom) will be summarized.

General results are as follows.

1. *At least one trapped mode exists for any ν, if and only if the following inequality is true:*

$$\int_0^\infty [d(x) - d_\infty]\, \mathrm{d}x < 0.$$

2. *At least one edge wave exists for sufficiently large values of* ν, *when*

$$d^- \equiv \min\{d(x) : x \geq 0\} < d_\infty. \tag{5.32}$$

3. *For sufficiently small* ν, *at most one trapped mode exists.*
4. *The number of edge waves is finite for any* ν.
5. *The function* $\nu \mapsto k(\nu)$ *from the dispersion relation for trapped modes increases with* ν.
6. *If* $d^{(1)}(x) \leq d^{(2)}(x)$ *for* $x \geq 0$, *then the inequality*

$$k_m^{(1)}(\nu) \leq k_m^{(2)}(\nu)$$

holds for the eigenvalues of the same modes having the same frequency. Moreover, if $d_\infty^{(1)} = d_\infty^{(2)}$, *then for the total number of trapped modes the following inequality is true:*

$$N^{(1)}(\nu) \geq N^{(2)}(\nu) \quad \text{for all } \nu > 0.$$

7. *If* $k_1(\nu) > k_0$, *then* k_1^2 *is a simple eigenvalue and the corresponding eigenfunction* ϕ_1 *can be chosen to be nonnegative in* W.

Special results for different geometries are as follows.

1. Cliff. *For this geometry there exists only one edge wave for sufficiently large values of* ν, *and the fundamental eigenvalue* $k_1^2(\nu)$ *satisfies the inequalities*

$$0 \leq k_1^2(\nu) - k_0^2 \leq \frac{4\nu^2 \exp\{-4\nu d^-\}}{1 - 4\exp\{-2\nu d^-\}},$$

where d^- *is defined by* (5.32).
2. Tangential bottom. In this case, the function $d(x)$ is assumed to satisfy $x^{2-\epsilon}/d(x) = O(1)$ as $x \to 0$, where $0 < \epsilon < 1$. The exact result can be formulated as follows:
There exists an increasing sequence $\nu_1, \nu_2, \ldots, \nu_m, \ldots,$ *such that* $\nu_m \to \infty$ *as* $m \to \infty$, *and for* $\nu \geq \nu_m$ *the number of point eigenvalues is greater or equal to* m.
Hence, the total number of trapped modes is infinite. The sequence in the theorem can be chosen as follows:

$$\nu_m = [\gamma_m \tanh(\gamma_m d_\infty)]^2, \quad \text{where } \gamma_m = (C_1 + C_2 m^2)^{(2-\epsilon)/[2(1-\epsilon)]},$$

and C_1, C_2 are certain constants.
In his papers [325] and [338], Ursell also conjectured the above formulated behavior for the number of trapped edge modes in this case.

3. Transversal bottom. For this geometry the behavior of $d(x)$ is assumed to be in the form $d(x) \sim x \tan\alpha$ as $x \to 0$, and the following results are true: (i) *for sufficiently large values of* ν, *there exist at least* $M(\alpha)$ *trapped modes, where* $M(\alpha)$ *is the Ursell number, that is, the greatest integer* n, *such that* $(2n-1)\alpha < \pi/2$; *and* (ii) *if the depth profile* $d(x)$ *satisfies the inequality* $d(x) \geq \min(x, d_\infty)$, *then at most one edge wave exists for all* $\nu > 0$. *Hence, exactly one trapped mode exists for large* ν.

Merzon and Zhevandrov [244] extended the results formulated above by constructing asymptotic expansions for frequencies of modes trapped by a beach of nonconstant slope as the longshore wave number tends to infinity. For a beach of constant slope their formulae coincide with those of Ursell given earlier in Subsection 5.2.1. Another generalization of Ursell's result on edge waves over a beach of constant slope is concerned with modes trapped by the so-called gently sloping beach when $d(x) \ll 1$. The asymptotic formulae for the corresponding trapped-mode frequencies were heuristically derived by Shen, Meyer, and Keller [304] and justified by Miles [247] under an additional restriction excluding one particular case. Subsequently, Zhevandrov [368] not only proved the formulae from [304] but also obtained a uniform asymptotic expansion for eigenfunctions. Another approach to waves over a gently sloping beach was developed by Sun and Shen [315].

An interesting explicit solution for a transversal bottom has been given by Wehausen and Laitone [354] in the following form:

$$\phi(x, y) = e^{-kx\cos\beta}\frac{\cosh K(y+d)}{\cosh Kd}, \quad \text{where } K = k\sin\beta.$$

It satisfies (5.5), has finite energy, and satisfies (5.6) provided $\nu = K\tanh Kd$. In addition, (5.7) is satisfied on S to be determined from $\mathrm{d}y/\mathrm{d}x = -d'(x) = \phi_y/\phi_x$, which integrates to

$$\frac{\sinh K(y+d)}{\sinh Kd} = \exp\{-kx\sin\beta\tan\beta\}.$$

Thus, $d(x)$ has initial slope $\nu\sec\beta/k$ and increases monotonically to $d_\infty = d$. Note that by letting $d \to \infty$ we recover the Stokes edge wave (5.16) with $\beta = \alpha$.

Now we turn to a different approach developed by Grimshaw [98] that allows us to obtain a lower bound for eigenfrequencies of edge waves. Let us suppose that $d(x)$ describing the bottom profile satisfies the following conditions: (i) $d(x) \to 1$ as $x \to \infty$, *and* (ii) $d'(x) \geq M\,[1 - d(x)] \geq 0$, where M is a positive constant. The last condition implies that

$$1 - d(x) \geq \exp\{-Mx\}.$$

Then the following assertion holds.

If a trapped mode exists for such a geometry, the following inequality holds for k and ν in (5.5) *and* (5.6), *respectively:*

$$\nu > m \tanh m, \quad \text{where } m^2 = M\left[(k^2 + M^2/4)^{1/2} - M/2\right]. \tag{5.33}$$

The comparison method used to prove (5.33) is a powerful analytic tool (see, for example, the book [292] by Protter and Weinberger). Another interesting application of this method is given by McIver and Linton [233], who demonstrated with its help the nonexistence of trapped modes for some acoustic waveguides.

In his paper [98], Grimshaw begins derivation of (5.33) with the identity

$$|\nabla\phi|^2 + k^2\phi^2 = \psi^2|\nabla v|^2 + v^2\psi(k^2\psi - \nabla^2\psi) + \nabla\cdot(v^2\psi\nabla\psi),$$

where $\phi = \psi v$. Integrating this identity over $W_h = W \cap \{x < h\}$ and applying the divergence theorem, one arrives at

$$\int_{W_h} (|\nabla\phi|^2 + k^2\phi^2)\,\mathrm{d}x\mathrm{d}y - \nu\int_{F_h} \phi^2\,\mathrm{d}x$$
$$= \int_{F_h} v\psi(\psi_y - \nu\psi)\,\mathrm{d}x + \int_{W_h} [\psi^2|\nabla v|^2 + v^2\psi(k^2 - \nabla^2\psi)]\,\mathrm{d}x\mathrm{d}y$$
$$+ \int_{S'} \psi\psi_x v^2\,\mathrm{d}y - \int_{S_h} v^2\psi\frac{\partial\psi}{\partial n}\,\mathrm{d}s.$$

Let the comparison function ψ satisfy the conditions

$$\psi > 0 \quad \text{in } \bar{W} = W \cup F \cup S, \tag{5.34}$$

$$\nabla^2\psi - k^2\psi \leq 0 \quad \text{in } W, \tag{5.35}$$

$$\partial\psi/\partial n \leq 0 \quad \text{on } S, \tag{5.36}$$

$$\psi_y - \nu'\psi \geq 0 \quad \text{on } F, \tag{5.37}$$

$$\psi_x/\psi = O(1) \quad \text{as } x \to \infty. \tag{5.38}$$

Then, the last integral identity implies

$$\int_{W_h} (|\nabla\phi|^2 + k^2\phi^2)\,\mathrm{d}x\mathrm{d}y - \nu\int_{F_h} \phi^2\,\mathrm{d}x \geq \int_{S'} \frac{\psi_x}{\psi}\phi^2\,\mathrm{d}y$$
$$+ (\nu' - \nu)\int_{F_h} v^2\psi^2\,\mathrm{d}x. \tag{5.39}$$

Let ϕ be an edge wave; then the left-hand side of (5.39) tends to zero as $h \to \infty$. The first term on the right-hand side also tends to zero as $h \to \infty$, because of (5.38) and the finite energy property of ϕ. Hence,

$$\nu \geq \nu'. \tag{5.40}$$

In view of the properties of $d(x)$ let us choose

$$\psi(x, y) = \cosh m(y + 1) \exp\{-(k^2 - m^2)^{1/2} x\}$$

as a comparison function with m defined in (5.33). Then, ψ satisfies (5.34) and (5.38), and (5.35) and (5.37) have the form of equalities with

$$\nu' = m \tanh m. \tag{5.41}$$

Moreover, (5.36) holds from condition (ii), and the choice of m. So (5.33) follows from (5.40) and (5.41).

Another important property – the principle of monotony – was discovered by McIver and Evans [238] to hold for edge waves in the same way as it is true for sloshing problem (see, for example, Moiseyev's work [249] for the basic theory of sloshing). It can be formulated as follows:

Let edge waves exist for domains $W^{(i)}$ $(i = 1, 2)$, such that

$$W^{(1)} \subset W^{(2)}, \quad \text{but } F^{(1)} = F^{(2)} = \{(x, y) : x > 0, y = 0\}.$$

Let $\nu^{(i)}$ $(i = 1, 2)$ be the frequency of the fundamental mode with wavenumber k in the domain $W^{(i)}$; then $\nu^{(1)} \leq \nu^{(2)}$.

This assertion is in some sense the inverse of 5) and 6) in the list of general results obtained by Bonnet–Ben Dhia and Joly [30].

To conclude this section, we should note that the theoretical results are in good agreement with the numerical ones. For example, Evans and McIver [238] numerically examined the predictions for the number of edge waves over a rectangular shelf given by Jones' [127] theory. It proves possible to obtain a formula numerically for the number of trapped modes, which gives this number for all values of frequency; the theoretical formula only predicts to within one for certain ranges of frequency.

5.3. Trapped Modes Above Submerged Obstacles

The first paper demonstrating the existence of trapped modes other than edge waves was published by Ursell [323] in 1951. He demonstrated the possibility of trapped modes traveling along the top of a totally submerged horizontal circular cylinder of infinite extent, in infinitely deep water. In his proof, based on multipole expansions and infinite determinants, Ursell had to impose the restriction that the radius of the cylinder should be sufficiently small. This restriction was removed by Jones [127] in 1953. He considered cylinders of arbitrary cross section with symmetry about vertical axis. In his article, based on deep results in the theory of unbounded operators, the existence of trapped modes in water waves is given as an application of more general theorems concerning the spectrum of the Laplacian with certain boundary conditions on semi-infinite domains. The presentation in [127] is extremely condensed, which makes the paper difficult to follow.

The basic idea was to divide the semi-infinite domain into two parts: a finite domain with a purely point spectrum and a smaller semi-infinite domain, and to generalize to semi-infinite domains the theorem, which is well-known for finite domains, that the eigenvalues increase as the size of the region is diminished. Jones was able to obtain a number of interesting theorems on the spectrum of the Laplace operator in a fifteen-page article. At least half a dozen papers can be listed rediscovering his theorems.

In 1988, Aranha [2] simplified the approach of Jones by introducing artificial boundaries, which reduce the problem in an infinite domain to another problem in a finite domain. This technique was demonstrated in Subsection 5.2.2, where it was applied to the edge wave problem. It should be mentioned that Aranha was unaware of Jones' results, which are referred to only in a footnote.

Another method based on potential theory and comparison theorems was proposed by Ursell [338] in 1987. An outline of this method is given in Subsection 5.3.1 for the case of a horizontal cylinder of arbitrary cross section immersed in deep water. The advantage of this method is that it makes no difference between water of infinite and finite depth. However, the formulae are not so cumbersome in the case of infinite depth. In connection with trapped waves caused by a submerged cylinder, another work to be cited is that of Aranha [3] concerning a nonlinear mechanism for the excitation of trapped modes above submerged cylinders.

One more geometry for which trapped modes exist is where the water is of uniform depth apart from a protrusion. This geometry models an ocean of finite depth with an underwater mountain ridge. The existence of trapped modes in this case was first demonstrated by Jones [127] in 1953 for a protrusion symmetric about a vertical axis. Arbitrary protrusions were treated by Garipov in 1965, and an exposition of his results is given in §34 of Lavrentiev and Chabat's book [182]. Since the methods of Jones and Garipov are similar to the approach developed in Subsection 5.2.2 and based on the linear operator theory, we restrict our considerations to other results for water of finite depth with a locally uneven bottom.

The method used for this purpose involves a version of *the auxiliary integral identity* (see Subsection 2.2.3). This identity proves to be an appropriate analytic tool in many different problems arising in the linear theory of surface waves. Apart from those considered above and below, we mention two more. A lower bound for sloshing frequencies is obtained in Kuznetsov's paper [157]. In another work [162], he demonstrates the *absence* of trapped modes when a surface-piercing cylinder spans vertical walls of a channel and is contained within vertical lines drawn downward from the cylinder's intersection points with the free surface – the so-called John condition (see Subsection 3.2.1).

In Subsection 5.3.2, we give two applications of the auxiliary integral identity to the problem of trapped waves. First, the nonexistence of trapped modes is shown for channels with bottom obstructions (a by-product result in Vainberg and Maz'ya's paper [347], concerned mainly with the water-wave problem in three and two dimensions). Second, a lower bound for the eigenfrequencies of modes trapped by bottom protrusions is given, as announced by Kuznetsov [159].

A number of authors have numerically considered the question of trapped modes over a submerged obstacle. Thus McIver and Evans [238] used the Ursell formulation from [323] for the submerged horizontal circular cylinder and directly computed zeros of the infinite determinant corresponding to trapped modes frequencies. Their results suggested that for any size of cylinder and any depth of submergence there is always at least one trapped mode, and that the number of such modes increases as the top of the cylinder approaches the free surface. For a depth of submergence greater than about 1.07 times the radius, only a single mode exists. These results were extended by Porter and Evans [291] (see also [80]), who addressed the case of modes trapped by an arbitrary configuration of multiple submerged circular cylinders. They used a combination of multipole expansion methods and addition theorems for Bessel functions to derive an infinite algebraic system. Again, its vanishing determinant depending on geometric and wave parameters locates trapped modes frequencies. The following effect of multiple cylinders can be conjectured from numerical results. Let N identical cylinders, each supporting a total of Q trapped modes, be centered on a vertical plane and equally spaced. Then the number of trapped modes present for large enough cylinder separations is equal to NQ. Linton and Evans [203] used a matched eigenfunction expansion method to compute the trapped modes above a submerged horizontal thin plate. Their results suggested that trapped waves, which were antisymmetric about the midpoint of the plate, also exist, a result not previously given by any of the existence proofs.

Confirmation of this result was provided by Parsons and Martin [283], who showed how to compute the trapped waves over a symmetric curved convex-upward thin plate in infinitely deep water by using hypersingular integral equations. By considering different limiting cases they obtained agreement with both the submerged flat plate results of Linton and Evans [203] and the submerged cylinder results of McIver and Evans [238].

A rigorous proof of the existence of trapped modes antisymmetric with respect to the center of a submerged horizontal flat plate was recently provided in Fernyhough's thesis [85].

5.3.1. Trapped Waves Above Submerged Cylinders

Here we consider the problem (5.5)–(5.7) in a water domain W, which is a part of the half-plane $\{y < 0\}$, lying outside the closed smooth curve S placed below the free surface

$$F = \{(x, y) : -\infty < x < +\infty, y = 0\}.$$

We shall be concerned with solutions ϕ having the finite energy norm and so describing trapped modes.

Following the method proposed by Ursell [338], we reduce the spectral problem (5.5)–(5.7) with ν as spectral parameter to an operator spectral problem in the space $L^2(-\infty, +\infty)$. The corresponding operator is a bounded integral operator with a symmetric kernel. The first step is the construction of Green's function $g(x, y; \xi, 0)$ with the following properties:

$$\begin{gathered}(\nabla^2 - k^2)g = 0 \quad \text{in } W, \partial g/\partial n = 0 \text{ on } S,\\ g_y = 0 \quad \text{on } F \text{ except at } (\xi, 0),\\ g(x, y; \xi, 0) - K_0\{k[(x-\xi)^2 + y^2]^{1/2}\} \text{ is bounded as } (x, y) \to (\xi, 0),\\ g(x, y; \xi, 0) \text{ decays as } x^2 + y^2 \to \infty \text{ and } \xi \text{ is bounded.}\end{gathered}$$

Here $K_0(z)$ is the modified Bessel function having a logarithmic singularity when $z = 0$ and decaying exponentially at infinity.

Let $x = X(s)$, $y = Y(s)$ be parametric equations of S, where s is the arc length along the curve. By choosing Green's function in the form

$$\begin{aligned} g(x, y; \xi, 0) = {} & K_0\{k[(x-\xi)^2 + y^2]^{1/2}\} \\ & + \int_S m(s)\big[K_0\big(k\{[x - X(s)]^2 + [y - Y(s)]^2\}^{1/2}\big) \\ & + K_0\big(k\{[x - X(s)]^2 + [y + Y(s)]^2\}^{1/2}\big)\big]\, ds, \end{aligned} \tag{5.42}$$

Ursell arrives at the Fredholm integral equation for the unknown density $m(s)$:

$$-\pi m(s') + \int_S m(s) \frac{\partial}{\partial n} \big[K_0\big(k\{[x - X(s)]^2 + [y - Y(s)]^2\}^{1/2}\big)$$
$$+ K_0\big(k\{[x - X(s)]^2 + [y + Y(s)]^2\}^{1/2}\big)\big] \, \mathrm{d}s$$
$$= -\frac{\partial}{\partial n} K_0\big\{k[(x - \xi)^2 + y^2]^{1/2}\big\}.$$

This equation is uniquely solvable. Hence, Green's function $g(x, y; \xi, 0)$ exists and has the representation (5.42).

The second step is the reduction of (5.5)–(5.7) to a spectral operator problem in $L^2(-\infty, +\infty)$. Consider the following potential,

$$(V\mu)(x, y) = \frac{1}{\pi} \int_{-\infty}^{+\infty} \mu(\xi) \, g(x, y; \xi, 0) \, \mathrm{d}\xi,$$

where $\mu \in L^2(-\infty, +\infty)$. Then, $V\mu$ satisfies (5.5) and (5.7). By the usual property of a single-layer potential we have that

$$\partial(V\mu)/\partial y = \mu \quad \text{on } F.$$

From here and from (5.6), we see that $V\mu$ is a trapped-mode potential if and only if

$$\mu(x) = \frac{\nu}{\pi} \int_{-\infty}^{+\infty} \mu(\xi) \, g(x, 0; \xi, 0) \, \mathrm{d}\xi; \tag{5.43}$$

that is, μ must be an eigenfunction corresponding to the eigenvalue ν of the operator

$$(V_0\mu)(x) = \frac{1}{\pi} \int_{-\infty}^{+\infty} \mu(\xi) \, g(x, 0; \xi, 0) \, \mathrm{d}\xi.$$

This operator is bounded, symmetric, and positive. Consequently, its spectrum is real. Moreover, the eigenvalues corresponding to the point spectrum can lie only between $\|V_0\|^{-1}$ and k.

The main result is that at least one point eigenvalue ν exists below the cutoff k, that is, the lower bound of the continuous spectrum. The proof is based on the following comparison theorem (compare with the monotony principle for edge waves from McIver and Evans [238] mentioned earlier).

Suppose that the curve $S^{(1)}$ lies inside $S^{(2)}$, and the eigenvalue problem (5.5)–(5.7) *for $S^{(1)}$ or, equivalently,* (5.43) *for $S^{(1)}$, has p characteristic values $k > \nu_1^{(1)} \geq \cdots \geq \nu_p^{(1)}$. Then the same problem for $S^{(2)}$ has at least p characteristic values, and $\nu_s^{(1)} > \nu_s^{(2)}$ $(s = 1, 2, \ldots, p)$.*

Since a small circle can be placed inside the arbitrary closed curve S, then *there exists at least one trapped mode for* S, since the existence of such modes for small circles was proved earlier by Ursell [323]. Using an integral identity similar to that of Grimshaw [98], Motygin [254] obtained bounds for the trapped-mode frequencies.

5.3.2. Bottom Obstacles: Uniqueness and Trapped Waves

The aim of this subsection is to show how a version of the auxiliary integral identity applies to the investigation of the problem (5.5)–(5.7) in a strip W with locally uneven bottom B, which coincides with $\{y = -d_\infty\}$ at infinity. So, the free surface is

$$F = \{(x, y) : -\infty < x < +\infty, y = 0\}.$$

First, we briefly remind the reader of the derivation of the auxiliary integral identity. We restrict ourselves to a version of this identity to be used here (see Subsection 2.2.3 for the general case). Let $\mathbf{V}$ be a vector field in $\bar{W}$ with real components (V_1, V_2), which are uniformly Lipschitz functions. Furthermore, let

$$V_1 = -x, \qquad V_2 = 0 \quad \text{on } F. \tag{5.44}$$

We define a matrix $Q = [Q_{ij}]$ with elements

$$Q_{ij} = (1 + \nabla \cdot \mathbf{V})\delta_{ij} - \left(\frac{\partial V_i}{\partial x_j} + \frac{\partial V_j}{\partial x_i} \right), \quad 1 \leq i, j \leq 2,$$

where $x = x_1$, $y = x_2$, and δ_{ij} is the Kronecker delta. We assume that

$$(Qv) \cdot v \geq 0 \quad \text{in } \bar{W} \text{ for all } v \in \,]R^2, \tag{5.45}$$

so that Q is a nonnegative matrix on $\bar{W}$.

The following version of identity (see Subsection 2.2.3 for the general case),

$$\nabla \cdot [(2\mathbf{V} \cdot \nabla\phi - \phi)\nabla\phi] = (2\mathbf{V} \cdot \nabla\phi - \phi)\nabla^2\phi - (Q\nabla\phi) \cdot \nabla\phi + \nabla \cdot (|\nabla\phi|^2 \mathbf{V}),$$

can be easily verified by direct calculation. Integrating this over W and using the divergence theorem along with (5.5)–(5.7) and (5.44), one arrives at

$$\int_W (Q\nabla\phi) \cdot \nabla\phi \,\mathrm{d}x\mathrm{d}y + \int_S |\nabla\phi|^2\, \mathbf{V} \cdot \mathbf{n} \,\mathrm{d}s = k^2 \int_W (2\phi\mathbf{V} \cdot \nabla\phi - \phi^2) \,\mathrm{d}x\mathrm{d}y,$$

where $\mathbf{n}$ is the unit normal on S directed into W. Integrating by parts in the last integral and taking into account (5.44) again, we get

$$\int_W (Q\nabla\phi)\cdot\nabla\phi \,\mathrm{d}x\mathrm{d}y + \int_S (|\nabla\phi|^2 + k^2\phi^2)\,\mathbf{V}\cdot\mathbf{n}\,\mathrm{d}s$$
$$= -k^2 \int_W \phi^2(1+\nabla\cdot\mathbf{V})\,\mathrm{d}x\mathrm{d}y. \tag{5.46}$$

We note that, if $\mathbf{V}$ satisfies (5.44), (5.45), and

$$C(\mathbf{V}) = \min\{\mathbf{V}\cdot\mathbf{n} : (x, y) \in S\} > 0, \tag{5.47}$$

then it follows from (5.46) that ϕ is identically zero in $\bar{W}$; that is, a uniqueness theorem is true for (5.5)–(5.7), if the following quantity

$$A(\mathbf{V}) = \max\{-(1+\nabla\cdot\mathbf{V}) : (x, y) \in \bar{W}\} \tag{5.48}$$

is nonpositive. A very simple example of $\mathbf{V}$ and corresponding geometry with the uniqueness property was proposed by Vainberg and Maz'ya [347]. The vector field must have components (5.44) not only on F, but throughout $\bar{W}$. Then $A(\mathbf{V}) = 0$, and Q is a nonnegative matrix as is shown in Example 2, Subsection 3.2.3.1. For (5.47) to hold, the bottom topography must be as follows: (i) no part of B must lie above $\{y = -d_\infty\}$; (ii) for the appropriately chosen y axis the curve $B \cap \{x > 0, y < -d_\infty\}$ is a one-to-one image of an interval on the y axis, and the same is true for the curve $B \cap \{x < 0, y < -d_\infty\}$ and the same interval on the y axis.

The y axis positioning is essential in (ii), because the inequality (5.47) depends on it strongly for the vector field $\mathbf{V}$ chosen. The same conditions were shown to guarantee uniqueness in the two-dimensional water-wave problem (see Subsections 2.2.4 and 2.2.5).

In the rest of this section we are concerned with the following water cross-sectional domain:

$$W = \{(x, y) : -\infty < x < +\infty, -d(x) < y < 0\}.$$

Here $d(x)$ is an arbitrary smooth positive function, such that $d(x) = d_\infty =$ const for $|x| \geq x_\infty > 0$, and

$$0 < \min\{d(x) : -\infty < x < +\infty\} < d_\infty.$$

Hence, $S = \{(x, y) : -\infty < x < +\infty, y = -d(x)\}$.

The existence of trapped waves for such a geometry was investigated at the papers mentioned at the beginning of this chapter. In particular, Garipov demonstrated (see §34 in Lavrentiev and Chabat's book [182]) the existence of eigenfrequencies for sufficiently large values of k, which is regarded as an arbitrary parameter in his approach. Moreover, it follows from his considerations that if

$$0 < d(x) \leq d_\infty \quad \text{for all } -\infty < x < +\infty,$$

then for each $k > 0$ at least one trapped mode exists below the cut off; that is, the corresponding eigenvalue belongs to the interval $(0, k \tanh kd)$. Our aim here is to derive a lower bound for the eigenvalues under the assumption that $A(\mathbf{V})$ defined by (5.48) is a positive number. Then by (5.45) and (5.47), we obtain from (5.46) that

$$C(\mathbf{V}) \int_S (|\nabla\phi|^2 + k^2\phi^2)\, \mathrm{d}s \leq A(\mathbf{V}) k^2 \int_W \phi^2 \, \mathrm{d}x \mathrm{d}y.$$

Hence,

$$A(\mathbf{V}) \int_W |\nabla\phi|^2 \, \mathrm{d}x\mathrm{d}y + C(\mathbf{V}) k^2 \int_S \phi^2 \, \mathrm{d}s \leq A(\mathbf{V}) \int_W (|\nabla\phi|^2 + k^2\phi^2)\, \mathrm{d}x\mathrm{d}y.$$

Thus, according to (5.5)–(5.7) we obtain

$$\int_W |\nabla\phi|^2 \, \mathrm{d}x\mathrm{d}y + k^2 \frac{C(\mathbf{V})}{A(\mathbf{V})} \int_S \phi^2 \, \mathrm{d}s \leq \nu \int_F \phi^2 \, \mathrm{d}x.$$

From here it follows that

$$\min\left\{1, k^2 \frac{C(\mathbf{V})}{A(\mathbf{V})}\right\} \left[\int_W |\nabla\phi|^2 \, \mathrm{d}x\mathrm{d}y + d_m^{-1} \int_S \phi^2 \, \mathrm{d}s\right] \leq \nu \int_F \phi^2 \, \mathrm{d}x, \tag{5.49}$$

where $d_m = \max\{d(x) : -\infty < x < +\infty\}$.

Now let us write

$$\phi(x, 0) = \phi(x, -d(x)) + \int_{-d(x)}^0 \phi_y(x, y)\, \mathrm{d}y.$$

Then the Schwarz inequality yields

$$\begin{aligned} \phi^2(x, 0) &\leq 2\left\{\phi^2(x, -d(x)) + \left[\int_{-d(x)}^0 \phi_y(x, y)\, \mathrm{d}y\right]^2\right\} \\ &\leq 2\left\{\phi^2(x, -d(x)) + d_m \int_{-d(x)}^0 \phi_y^2(x, y)\, \mathrm{d}y\right\}. \end{aligned}$$

After integration over F we get

$$(2d_m)^{-1} \int_F \phi^2 \, \mathrm{d}x \leq \int_W |\nabla\phi|^2 \, \mathrm{d}x\mathrm{d}y + d_m^{-1} \int_S \phi^2 \, \mathrm{d}s.$$

Combining this inequality with (5.49) one finds that

$$2\nu d_m \geq \min\{1, k^2 d_m C(\mathbf{V})/A(\mathbf{V})\} \tag{5.50}$$

This inequality gives a lower bound for the eigenvalue ν corresponding to the fundamental trapped mode. For values below this bound the uniqueness theorem holds.

Consider the following example of a vector field satisfying (5.44) and (5.45) and having (5.48) positive for any domain W in the half-plane $y < 0$:

$$\mathbf{V} = (-x, -qy), \quad 0 < q \leq 2.$$

Then the following assertion is true:

Let $q \in (0, 2]$ exist, such that

$$C_q = \min\{-\partial(x^2 + qy^2)/\partial n : (x, y) \in S\} > 0. \tag{5.51}$$

Then, the first eigenvalue ν satisfies

$$\min\{2, k^2 d_m C_q q^{-1}\} \leq 4\nu d_m.$$

An example of S satisfying (5.51) can be constructed as follows. Let

$$d(x) = \begin{cases} ax^2 + h_\infty - b & \text{when } |x| < (b/a)^{1/2}, \\ h_\infty & \text{when } |x| \geq (b/a)^{1/2}, \end{cases}$$

where $a, b > 0$ and $d_\infty/b > 2/q$ with $q \in (0, 2]$. Then,

$$C_q = 2(qd_\infty - 2b)(1 + 4ab)^{-1/2} > 0.$$

In this case (5.50) takes the form

$$2\nu d_\infty \geq \min\left\{1, \frac{k^2 d_\infty (d_\infty q - 2b)}{q(1 + 4ab)^{1/2}}\right\}.$$

Some other examples of $\mathbf{V}$ and B, providing lower bounds for point eigenvalues, can easily be constructed.

5.4. Waves in the Presence of Surface-Piercing Structures

Here we consider the problem (5.5)–(5.7) as a spectral problem, in which ν is to be determined as an eigenvalue for a given value of parameter k, and the corresponding nontrivial solution ϕ must have a finite energy norm. The presentation follows the paper [173] by Kuznetsov et al. and considerations are mainly restricted to water domains of infinite depth, but some results for the finite-depth water are also formulated.

As early as 1929, it was shown by Havelock [107] that in the deep-water case the cutoff frequency existed. In other words, there is a positive lower bound of the continuous spectrum, which is equal to k. After it was discovered that trapped modes do exist below the cutoff frequency in the presence of totally submerged cylinders (see Subsection 5.3.1), it took forty years to recognize that such trapped waves could not take place, when a cylinder is partially immersed in water, at least if its cross section satisfies John's condition. Only in 1991, in an appendix to [235], did McIver give a proof of this proposition (see Subsection 5.4.1.1 in the following paragraphs), assuming that there is a single surface-piercing cylinder. The result was achieved by consideration of an integral, resembling the simple wave of order zero in Subsection 3.2.1.

In 1992, Simon [305] extended the uniqueness theorem for the below-cutoff case in two directions. He considered the finite-depth water, and he also applied a technique using nonvertical lines as in Simon and Ursell's work [307] on uniqueness in the two-dimensional problem (see Subsection 3.2.2.1).

Later on, it was realized (see Subsection 5.4.1.1), that the proof in [235] gave the result for multiple cylinders each satisfying John's condition. This result has also been extended to the above-cutoff case ($k < \nu$), generalizing, at the same time, the two-dimensional theorem in Subsection 4.2.2. The necessary amendments and supplements are given in Subsection 5.4.1.2. As in Subsection 4.2.2, a pair of cylinders, symmetric about the midplane between them, each satisfying John's condition, is considered. It occurs that solutions to symmetric and antisymmetric problems are unique, when the parameter lb belongs to complementary intervals. Here $2b$ denotes the distance between the innermost points of intersection of the bodies with the free surface (see Fig. 4.7), and $l = (\nu^2 - k^2)^{1/2}$.

We also extend the two-dimensional results, concerning waves trapped by a two-body surface-piercing structure. In Subsection 5.4.2.1, we outline the construction of the corresponding examples for $k < \nu$, based on utilizing an appropriate source/source or source/sink combination positioned in the x axis so as to cancel out the waves at infinity. The resulting wave field is computed, and it is shown that certain pairs of field lines intersect the x axis twice and enclose the sources. They can therefore be regarded as rigid cylindrical cross sections, and together with the resulting wave field they provide examples of trapping structures and waves.

In Subsection 5.4.3, a proof is given for the existence of such structures and waves for sufficiently small k/ν. For any value of k/ν, examples of trapping structures and waves are obtained by Motygin [252], who gives a proof of existence as well as numerical construction, which are based on using dipoles as in Subsection 4.2.2.3, where this method is applied to the two-dimensional water-wave problem.

5.4.1. Uniqueness Theorems

In this subsection we prove uniqueness theorems for both below and above the cutoff.

5.4.1.1. Uniqueness Below the Cutoff

First, we shall prove the following assertion.

When $\nu < k$, the problem (5.5)–(5.7) in deep water has only a trivial solution for any finite number of surface-piercing bodies, provided that they individually satisfy John's condition.

We recall that the latter condition means that each surface-piercing body is totally confined within a pair of vertical lines drawn downward from the points of intersection of the body with the free surface. The proof is a fairly straightforward extension of John's original proof for a single surface-piercing body in a strictly two-dimensional flow (see Subsection 3.2.1), and it was given by McIver [235], Appendix A.

Let $W_{ac} = W \cap \{-a < x < a, 0 < y < c\}$, where a, c are large enough to contain all wetted rigid contours within W_{ac}. Then, the divergence theorem gives

$$\int_{W_{ac}} \nabla \cdot (\bar{\phi}\, \nabla \phi)\, \mathrm{d}x\mathrm{d}y = \int_{\partial W_{ac}} \bar{\phi} \frac{\partial \phi}{\partial n}\, \mathrm{d}s,$$

where n denotes outward normal. From here, after use of (5.5), we get

$$\int_{W_{ac}} (|\nabla \phi|^2 + k^2 |\phi|^2)\, \mathrm{d}x\mathrm{d}y = \int_{\partial W_{ac}} \bar{\phi} \frac{\partial \phi}{\partial n}\, \mathrm{d}s. \tag{5.52}$$

It follows from Havelock's paper [107] that any solution satisfying (5.5) and (5.6) for $k > \nu$, for which $\partial \phi / \partial n$ is prescribed on any number of surface-piercing cylinders, is such that the sum of kinetic and potential energies per unit length of the z axis is finite; that is,

$$\int_W |\nabla \phi|^2\, \mathrm{d}x\mathrm{d}y + \nu \int_F |\phi|^2\, \mathrm{d}x < \infty.$$

Then taking the limit $a, c \to \infty$, and using (5.6) and (5.7), reduces (5.52) to

$$\int_W (|\nabla \phi|^2 + k^2 |\phi|^2)\, \mathrm{d}x\mathrm{d}y = \nu \int_F |\phi|^2\, \mathrm{d}x. \tag{5.53}$$

Now, for any $(x, 0) \in F$, we take the expression

$$\int_{-\infty}^{0} \phi_y(x, y)e^{\nu y}\,\mathrm{d}y, \tag{5.54}$$

which resembles the simple wave of order zero in Subsection 3.2.1 but contains the y derivative instead of the velocity potential itself. Integrating by parts, we obtain

$$\phi(x, 0) = \int_{-\infty}^{0} (\nu\phi + \phi_y)e^{\nu y}\,\mathrm{d}y,$$

whence

$$\begin{aligned}|\phi(x, 0)|^2 &\leq \int_{-\infty}^{0} (\nu\phi + \phi_y)^2\,\mathrm{d}y \int_{-\infty}^{0} e^{2\nu y}\,\mathrm{d}y \\ &= \frac{1}{2\nu}\left[\int_{-\infty}^{0} (\nu^2|\phi|^2 + |\phi_y|^2)\,\mathrm{d}y + \nu|\phi(x, 0)|^2\right],\end{aligned}$$

where the last term on the right-hand side has come from writing $2\phi\phi_y = \partial\phi^2/\partial y$. The above is simply

$$\nu|\phi(x, 0)|^2 \leq \int_{-\infty}^{0} (\nu^2|\phi|^2 + |\phi_y|^2)\,\mathrm{d}y,$$

which when substituted into (5.53) gives

$$\int_W (|\nabla\phi|^2 + k^2|\phi|^2)\,\mathrm{d}x\mathrm{d}y \leq \int_{W^c} (\nu^2|\phi|^2 + |\phi_y|^2)\,\mathrm{d}x\mathrm{d}y,$$

where W^c is the part of W covered by vertical lines from F downward. Then,

$$(k^2 - \nu^2)\int_W |\phi|^2\,\mathrm{d}x\mathrm{d}y \leq 0.$$

Since $k > \nu$, then $\phi \equiv 0$, and there are no trapped modes possible; that is, the uniqueness is proved.

The important point to notice here is that the proof is independent of the number of surface-piercing bodies and the spacing between them; the only restriction is that all bodies must satisfy John's condition. Also, this condition allows one to prove the following uniqueness theorem below the cutoff.

The problem (5.5)–(5.7) *in water of finite depth d has only a trivial solution below the cutoff,* that is, for $\nu < k \tanh kd$, *when there is any finite number of surface-piercing bodies, individually satisfying John's condition.*

This follows from Simon's considerations in [305], where

$$\int_{-d}^{0} \phi_y \sinh k(y+d)\, \mathrm{d}y$$

is used instead of (5.54) as a starting point.

We conclude this section with several remarks on the case of a *single* surface-piercing body. First, let the depth be infinite. Under this assumption, Simon [305] has established uniqueness for a single body confined between straight lines drawn at the angle β to the downward vertical from the points of intersection of the body with the free surface provided that $\nu < k \cos \beta$. This bound for the range of ν/k, where the uniqueness theorem holds for bodies having parts strictly below the free surface, depends on β and is optimal, since $\nu/k = \cos \beta$ for the Stokes edge wave (see Subsection 5.2.1, where $\alpha = \pi/2 - \beta$). For a single surface-piercing body satisfying John's condition, Kuznetsov [162] has given a unified uniqueness proof that is valid for both $\nu < k$ and $\nu > k$. It is based on using the extended auxiliary integral identity, which is derived in Subsection 5.3.2.

Second, the case of finite-depth water and a single body contained between straight lines described above has also been considered by Simon [305]. He obtained the following condition:

$$\nu \leq k \cos \beta \tanh(kd \sec \beta),$$

which guarantees the uniqueness of solution to (5.5)–(5.7).

5.4.1.2. Uniqueness Above the Cutoff

First we note that many results, proven for the two-dimensional water-wave problem, remain true for the problem (5.5)–(5.7) in the above cutoff case $\nu > k$. In particular, a solution to (5.5)–(5.7) decays at infinity, so that it has finite kinetic and potential energies. Furthermore, John's method for proving uniqueness is still applicable to (5.5)–(5.7), provided there is a single surface-piercing body satisfying his geometric condition. So, we will use these facts without repeating their proofs.

Here, we restrict ourselves to the case of deep water, containing a pair of bodies, that are symmetric about the y axis and that each satisfy John's condition (the corresponding geometry is that shown in Fig. 4.7, where one has to put $\beta = \pi/2$ and to drop out the horizontal bottom). The water region interior (exterior) to vertical lines, bounding the bodies from inside (outside), is denoted by W_0 (W_∞) as labeled in Fig. 4.7.

Our aim is to generalize the uniqueness result in Subsection 4.2.2. Since the geometry is symmetric and the problem is homogeneous, we may consider the symmetric and antisymmetric potentials separately by writing

$$\phi^{(\pm)}(x, y) = \pm\phi^{(\pm)}(-x, y), \tag{5.55}$$

where the superscript $+$ $(-)$ refers to symmetric (antisymmetric) mode. This clearly gives

$$\phi_x^{(+)}(0, y) = 0, \qquad \phi^{(-)}(0, y) = 0, \tag{5.56}$$

and we need only consider $x \geq 0$ with the extension to $x < 0$ coming from (5.55).

Following the pattern in Subsection 4.2.2, we introduce

$$a^{(\pm)}(x) = \int_{-\infty}^{0} \phi^{(\pm)}(x, y)e^{\nu y}\, \mathrm{d}y. \tag{5.57}$$

Then, $a_{xx}^{(\pm)} = (k^2 - \nu^2)a^{(\pm)} = -l^2 a^{(\pm)}$. The solution of this satisfying (5.55) is as follows:

$$a^{(\pm)} = C_{\pm} \cos\,(lx - \pi/4 \pm \pi/4) \quad \text{for } (x, 0) \in F_0, \tag{5.58}$$

and

$$a^{(\pm)}(x) = C_1 \cos lx + C_2 \sin lx, \quad \text{for } (x, 0) \in F_\infty$$

for some constants $C_{\pm}$, C_1, and C_2. However, as in the two-dimensional problem, we get

$$\lim_{|x| \to \infty} a^{(\pm)}(x) = 0,$$

and hence,

$$a^{(\pm)} \equiv 0 \quad \text{for } (x, 0) \in F_\infty.$$

Then John's method leads to the following inequality:

$$\nu \int_{F_\infty} \left|\phi^{(\pm)}(x, 0)\right|^2 \mathrm{d}x \leq \frac{1}{2} \int_{W_\infty} \left|\nabla\phi^{(\pm)}\right|^2 \mathrm{d}x\mathrm{d}y. \tag{5.59}$$

Now substituting (5.58) into (5.57) gives

$$C_{\pm} \cos\left(lx - \frac{\pi}{4} \pm \frac{\pi}{4}\right) = \int_{-\infty}^{0} \phi^{(\pm)}(x, y)e^{\nu y}\mathrm{d}y \quad \text{for } (x, 0) \in F_0. \tag{5.60}$$

In the same way as in Subsection 4.2.2, we get from here the following two inequalities:

$$\nu\left|\phi^{(\pm)}(x,0)\right|^2 \leq 2\nu^3 C_\pm^2 \cos^2\left(lx - \frac{\pi}{4} \pm \frac{\pi}{4}\right) + \int_{-\infty}^{0} \left|\phi_y^{(\pm)}(x,y)\right|^2 \mathrm{d}y, \tag{5.61}$$

and

$$2\nu l^2 C_\pm^2 \sin^2\left(lx - \frac{\pi}{4} \pm \frac{\pi}{4}\right) \leq \int_{-\infty}^{0} \left|\phi_x^{(\pm)}(x,y)\right|^2 \mathrm{d}y. \tag{5.62}$$

Also, (5.60) implies that

$$2\nu C_\pm^2 \cos^2\left(lx - \frac{\pi}{4} \pm \frac{\pi}{4}\right) \leq \int_{-\infty}^{0} \left|\phi^{(\pm)}(x,y)\right|^2 \mathrm{d}y,$$

which was not necessary when the two-dimensional case was considered in Subsection 4.2.2. After integrating (5.61) over F_0, we apply the last inequality and obtain that

$$\begin{aligned}\nu \int_{F_0} \left|\phi^{(\pm)}(x,0)\right|^2 \mathrm{d}x &\leq 2\nu(k^2 + l^2) C_\pm^2 \int_{F_0} \cos^2\left(lx - \frac{\pi}{4} \pm \frac{\pi}{4}\right) \mathrm{d}x \\ &+ \int_{W_0} \left|\phi_y^{(\pm)}\right|^2 \mathrm{d}x\mathrm{d}y \leq k^2 \int_{W_0} \left|\phi^{(\pm)}\right|^2 \mathrm{d}x\mathrm{d}y + \int_{W_0} \left|\phi_y^{(\pm)}\right|^2 \mathrm{d}x\mathrm{d}y \\ &+ 2\nu l^2 C_\pm^2 \int_{F_0} \cos^2\left(lx - \frac{\pi}{4} \pm \frac{\pi}{4}\right) \mathrm{d}x.\end{aligned}$$

The last integral can be estimated by using (5.62), provided the following inequality,

$$\int_0^b \cos^2\left(lx - \frac{\pi}{4} \pm \frac{\pi}{4}\right) \mathrm{d}x \leq \int_0^b \sin^2\left(lx - \frac{\pi}{4} \pm \frac{\pi}{4}\right) \mathrm{d}x, \tag{5.63}$$

holds. Thus we have

$$\nu \int_{F_0} \left|\phi^{(\pm)}(x,0)\right|^2 \mathrm{d}x \leq k^2 \int_{W_0} \left|\phi^{(\pm)}\right|^2 \mathrm{d}x\mathrm{d}y + \int_{W_0} \left|\nabla\phi^{(\pm)}\right|^2 \mathrm{d}x\mathrm{d}y.$$

This estimate and (5.59) combine to give

$$\nu \int_{F} \left|\phi^{(\pm)}(x,0)\right|^2 \mathrm{d}x \leq k^2 \int_{W_0} \left|\phi^{(\pm)}\right|^2 \mathrm{d}x\mathrm{d}y + \int_{W_0 \cup W_\infty} \left|\nabla\phi^{(\pm)}\right|^2 \mathrm{d}x\mathrm{d}y,$$

and this contradicts (5.53) unless $\phi^{(\pm)} \equiv 0$ in W. Thus (5.63) guarantees uniqueness. From (5.63) one immediately obtains the uniqueness intervals and hence the following theorem is proved, which is analogous to the uniqueness theorem in Subsection 4.2.2.2:

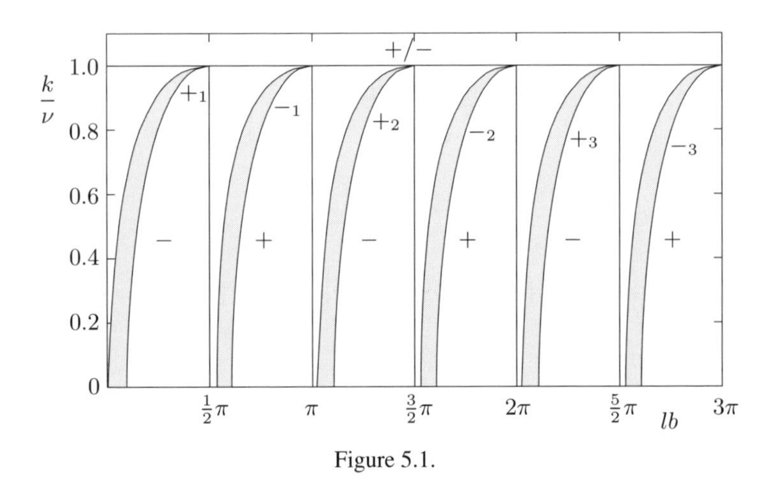

Figure 5.1.

Let W have infinite depth, and other geometric conditions formulated in the beginning of this subsection hold. If $\phi^{(\pm)}$ satisfies (5.5)–(5.7) *and* (5.56), *then it is trivial when*

$$\pi(m + 1/4 \pm 1/4) \leq lb \leq \pi(m + 3/4 \pm 1/4). \tag{5.64}$$

Both uniqueness theorems (just formulated, and that for the below-cut-off case in Subsection 5.4.1.1, to which the geometric assumptions of the present subsection must be added) are illustrated in Fig. 5.1. It shows regions of $(lb, k/\nu)$ parameter space where uniqueness is guaranteed. They are rectangles marked by plus (minus) signs for symmetric (antisymmetric) modes, and the quadrant above $k/\nu = 1$. On the vertical segments separating the rectangles, uniqueness is guaranteed for both modes. The shaded blades are related to examples of non-uniqueness to be considered in Subsection 5.4.2.

5.4.2. Examples of Trapped Waves

In this subsection we construct examples of trapping structures and the corresponding wave fields (Subsection 5.4.2.1), and we illustrate these examples numerically in Subsection 5.4.2.2. A brief discussion of the numerical procedure is also given there.

5.4.2.1. Construction of Non-Uniqueness Examples

The time-harmonic potential caused by a line source lying in the free surface at $(x, y) = (\xi, 0)$ and periodic in the z coordinate with wavenumber k

may be written in the form Re $\{G(x, y, \xi)e^{-i\omega t}e^{\pm ikz}\}$. Then G satisfies

$$(\nabla^2 - k^2)G = 0 \quad \text{for } y < 0, -\infty < x < +\infty, \tag{5.65}$$

$$G_y - \nu G = 0 \quad \text{for } y = 0, x \neq \xi, \tag{5.66}$$

$$G(x, y, \xi) \sim -\log \rho \quad \text{as } \rho^2 = (x - \xi)^2 + y^2 \to 0, \tag{5.67}$$

$$\nabla G(x, y, \xi) \to 0 \quad \text{as } y \to -\infty. \tag{5.68}$$

Green's function satisfying equations (5.65)–(5.68) plus a radiation condition is given by Ursell [329], and after an elementary change of variable it takes the form

$$G(x, y, \xi) = 2 \int_{\ell_-(|k|)} \frac{\mu e^{\mu y}}{(\mu^2 - k^2)^{1/2}(\mu - \nu)} \cos\left[(\mu^2 - k^2)^{1/2}(x - \xi)\right] \mathrm{d}\mu, \tag{5.69}$$

where the contour $\ell_-(|k|)$ runs along the real axis from $\mu = |k|$ and is indented from below at the pole $\mu = \nu$. It is clear that the two-dimensional Green's function for a source in the free surface is recovered by taking $k = 0$ (see Subsection 1.2.1). It is convenient to make a change of variable $l\tau = (\mu^2 - k^2)^{1/2}$, where $l^2 = \nu^2 - k^2$; whence

$$G(x, y, \xi) = 2 \int_\ell \frac{\nu + (k^2 + l^2\tau^2)^{1/2}}{l(\tau^2 - 1)} \exp\{y(k^2 + l^2\tau^2)^{1/2}\} \cos l(x - \xi)\tau \, \mathrm{d}\tau,$$

where ℓ denotes the contour coinciding with the positive half-axis indented below at $\tau = 1$. Putting $k/\nu = \sin\theta$ for $\theta \in [0, \pi/2)$, we get

$$G(x, y, \xi) = 2 \int_\ell \frac{\sec\theta + (\tan^2\theta + \tau^2)^{1/2}}{\tau^2 - 1} \exp\{ly(\tan^2\theta + \tau^2)^{1/2}\} \cos l(x - \xi)\tau \, \mathrm{d}\tau. \tag{5.70}$$

In the same way as in Subsection 1.2.1, it can be shown that

$$G(x, y, \xi) \sim 2\pi i \sec\theta e^{\nu y} e^{il|x-\xi|} \quad \text{as } |x - \xi| \to \infty. \tag{5.71}$$

So, by taking a combination of two sources or a source and a sink at certain separations, it is possible to cancel the waves at infinity. Let us write the potential as the combination

$$\phi_n^{(\pm)}(x, y) = 1/2\left\{G\left[x, y, a_n^{(\pm)}\right] \pm G\left[x, y, -a_n^{(\pm)}\right]\right\}, \quad n = 1, 2, \ldots, \tag{5.72}$$

where

$$la_n^{(\pm)} = (n - 1/4 \mp 1/4)\pi. \tag{5.73}$$

Here, the subscript n refers to the mode number, and the superscript + (−) refers to the symmetric (antisymmetric) mode. Thus, $\phi_1^{(+)}$ represents the first symmetric mode, $\phi_1^{(-)}$ the first antisymmetric mode and so on. Then, (5.71) and (5.73) ensure that no waves are radiated to infinity by $\phi_n^{(\pm)}$.

By (5.73) the contributions to (5.72) from two indentations in (5.70) cancel, and the velocity potential is real and is given by

$$\phi_n^{(\pm)}(x, y) = \int_0^\infty \frac{\sec\theta + (\tau^2 + \tan^2\theta)^{1/2}}{\tau^2 - 1} \exp\{ly(\tau^2 + \tan^2\theta)^{1/2}\}$$
$$\times \{\cos l[x - a_n^{(\pm)}]\tau \pm \cos l[x + a_n^{(\pm)}]\tau\} \, d\tau. \tag{5.74}$$

Here the integration is taken along the whole real τ axis in usual sense, because (5.73) guarantees that there is a removable singularity of the integrand at $\tau = 1$.

The expression (5.74) is convenient for the computation of the field lines that are tangential to the flow; that is, if $\mathbf{n}$ is a local normal to the field line, then it satisfies $\nabla\phi_n^{(\pm)} \cdot \mathbf{n} = 0$. A pair of field lines, each having both ends on the x axis so that they enclose the singularity points, can be interpreted as a rigid cross section of two cylinders. They form a structure providing a non-uniqueness example, because the wave field given by (5.74) is a nontrivial solution to the homogeneous boundary value problem outside of these cylinders (cf. Section 4.1, where the same technique is applied to the two-dimensional problem with $k = 0$). In Subsection 5.4.3, it is proved that this construction actually gives non-uniqueness examples, at least for sufficiently small k/ν.

5.4.2.2. Numerical Results

If a field line is given by $y = f(x)$, then it satisfies $f'(x) = \phi_y^{(\pm)}/\phi_x^{(\pm)}$. However, for numerical computations it is better to parameterize field lines by writing

$$\frac{dy}{ds} = \frac{\partial\phi_n^{(\pm)}(x, y)}{\partial y}, \qquad \frac{dx}{ds} = \frac{\partial\phi_n^{(\pm)}(x, y)}{\partial x}. \tag{5.75}$$

To gain a sufficient degree of accuracy when numerically solving this first-order system of differential equations, we find it sufficient to apply a fourth-order Runge–Kutta scheme with a step size of 0.01. The accuracy of the calculations can be checked in the following way. By starting a field line at a point on the x axis to the right of the source, we arrive at the x axis to the left of the source point, completing a particular field line. This process is reversed, with the field line started from the same point that the previous calculation had finished. This should give the same field line, and indeed the final point is accurate to within four decimal places of the original starting point in all cases considered in the next subsection. Only for starting points close to the source point more care has to be taken to account for rapidly varying gradients and high body curvatures.

Another important part of the numerical process is the computation of the integral in (5.74). For moderate or large values of $l|y|$, the exponential factor in the integrand allows the integral to be computed efficiently. For small values of $l|y|$, and indeed, for $ly = 0$, the integral is not so easily computed. For this, the upper limit of infinity is truncated to a value of T, say. The remainder is then approximated by the leading-order asymptotic form of the integrand, which is

$$\left(\frac{\sec\theta}{\tau^2}+\frac{1}{\tau}\right)e^{ly\tau}\left[\cos l\left(x-a_n^{(\pm)}\right)\tau \pm \cos l\left(x+a_n^{(\pm)}\right)\tau\right] \quad \text{as } \tau \to \infty.$$

The integral of this expression between T and $+\infty$ has an analytic form:

$$\frac{\sec\theta}{T}\operatorname{Re}\left(E_2\left\{l\left[|y|-\operatorname{Ci}\left(x-a_n^{(\pm)}\right)\right]T\right\}+E_2\left\{l\left[|y|-\operatorname{Ci}\left(x+a_n^{(\pm)}\right)\right]T\right\}\right)$$
$$+\operatorname{Re}\left(E_1\left\{l\left[|y|-\operatorname{Ci}\left(x-a_n^{(\pm)}\right)\right]T\right\}+E_1\left\{l\left[-y-\operatorname{Ci}\left(x+a_n^{(\pm)}\right)\right]T\right\}\right),$$

where $E_n(z)$ and $\operatorname{Ci}(z)$ are the exponential and cosine integrals, respectively. By carefully choosing T and combining the result of the truncated integral with the contribution from the leading-order asymptotic remainder, we find it possible to achieve an accuracy of at least six decimal places in the calculation of the potential, which is sufficient for the accuracy required in the computation of the field lines.

A further important check on the numerical scheme can be made by comparing the results obtained from this method when $k = 0$ with those obtained in Section 4.1 for the two-dimensional examples of non-uniqueness, where the conjugate stream function is defined and can be used to find streamlines directly. An agreement to within four decimal places is obtained. This gives us confidence in the results for $k > 0$.

Let us turn next to the results of the computations. Thus, Figs. 5.2–5.4 show field lines computed from (5.74) and (5.75) with $n = 1$ and the plus sign chosen for different values of k/ν. Namely, for the field lines of $\phi_1^{(+)}$ in Figs. 5.2, 5.3, and 5.4 this parameter is equal to 0, 1/2, and 0.99, respectively. In each case, the solid line is a limiting field line in the sense that it and all field lines interior to it enter the source position and cannot therefore be interpreted as delivering a non-uniqueness example.

Figure 5.2 is the special case of $k = 0$ considered in Section 4.1. Note that any of the field lines exterior to the limiting field line may be replaced by a rigid cylinder cross section, which increases in size as the innermost point of

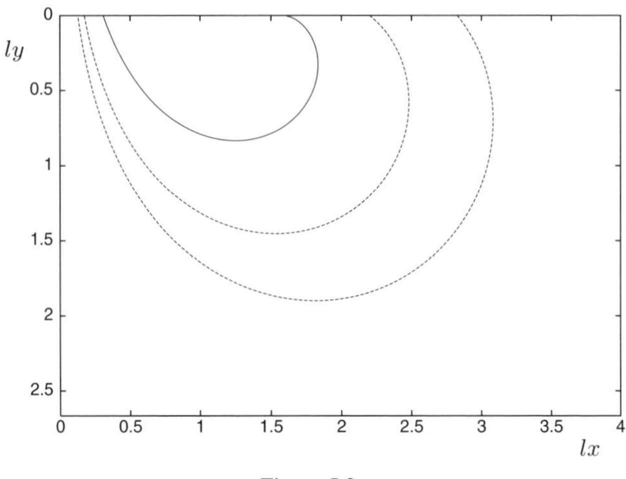

Figure 5.2.

intersection of the field line with the x axis approaches the origin. The situation is quite different for $k > 0$, as Figs. 5.3 and 5.4 illustrate. Here the field line shown dotted and labeled D is a dividing field line that tends to infinity. All field lines to the left of this also extend to infinity, whereas all field lines to the right, up to but excluding the solid field line, intersect the free surface again beyond the source point and may be replaced by a rigid cross section. It is now clear that in Fig. 5.2 (when $k = 0$), the streamline $x = 0$ is just a special case of the unique dividing field line, which arises in each case for $k > 0$.

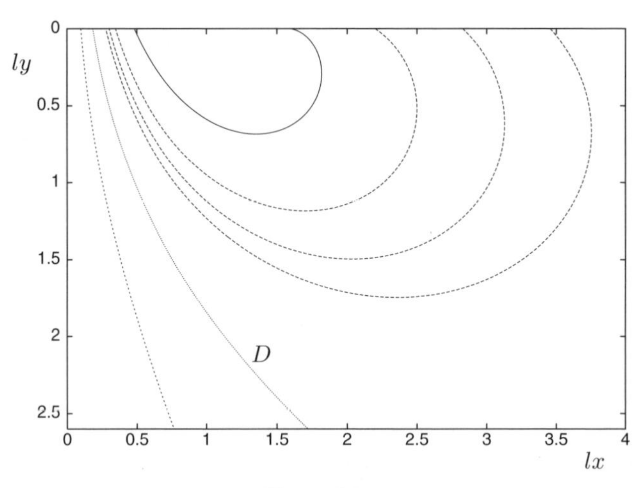

Figure 5.3.

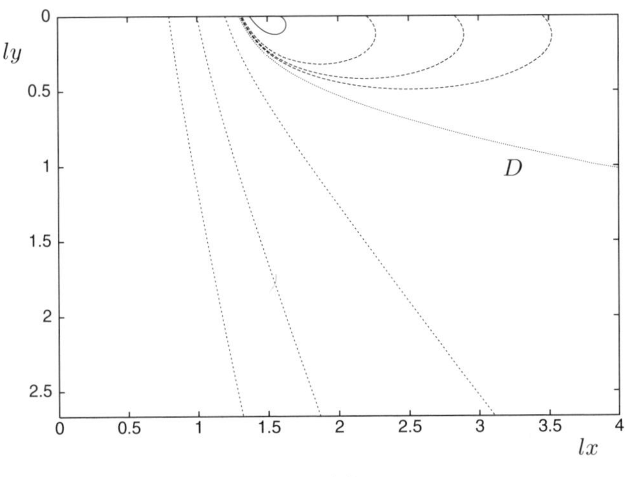

Figure 5.4.

An interesting interpretation of this is to regard any field line to the left of the dividing field line D as a rigid boundary and any of the field lines to the right, up to the limiting field line, as a rigid cylinder thereby producing trapped modes above the cutoff in the vicinity of a surface-piercing cylinder and a curved beach of particular shapes.

Of course, the simpler interpretation in which one of the field lines, intersecting the x axis twice and being exterior to the limiting solid field line, is regarded as a rigid cylindrical cross section, together with another one of the corresponding field lines from the mirror image set in $x < 0$, provides, just as in the case $k = 0$ studied in Section 4.1, an example of a trapped mode in the water domain exterior to both cylinders.

In Figs. 5.5 and 5.6, the spacing of the two sources is widened to the next value, which produces no waves; that is, we have $n = 2$ and the plus sign in (5.74) and (5.75). Figure 5.5 presents the strictly two-dimensional flow considered in Section 4.1. We see that the region $lx, ly \geq 0$ is separated by a dividing streamline labeled D, such that streamlines to its right starting at the x axis up to the limiting streamline all intersect the x axis again beyond the source position, and any one of them may be regarded as a cylinder cross section. Again, all streamlines to the left of D intersect the free surface twice and any one may be regarded as a cylinder cross section. Thus now we can have the existence of symmetric trapped modes in the vicinity of four surface-piercing cylinders, once a reflection in the y axis has been made. In contrast, choosing one streamline inside D as a cylinder can provide us with trapped modes near three surface-piercing cylinders. In Fig. 5.6, where

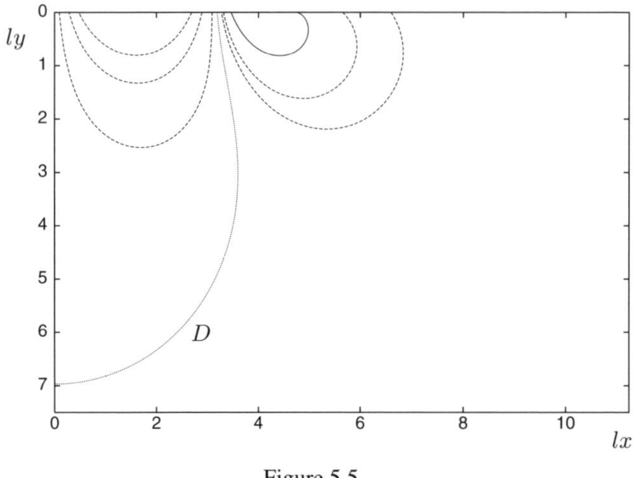

Figure 5.5.

$k/\nu = 1/2$, similar arguments apply, the only difference being that the single dividing streamline for $k = 0$ has split into two dividing field lines, between which all field lines go to infinity.

Figures 5.7 and 5.8 give examples of non-uniqueness involving the closest spacing of a source/sink combination, which produce cancellation of the wave field at infinity; that is, $n = 1$ and the minus sign is chosen. Here, in Fig. 5.7 for $k = 0$, we have for the first time a dividing streamline extending to infinity that is not the negative y axis. Streamlines to the right and exterior to the limiting streamline enter the x axis on either side of the source position, and

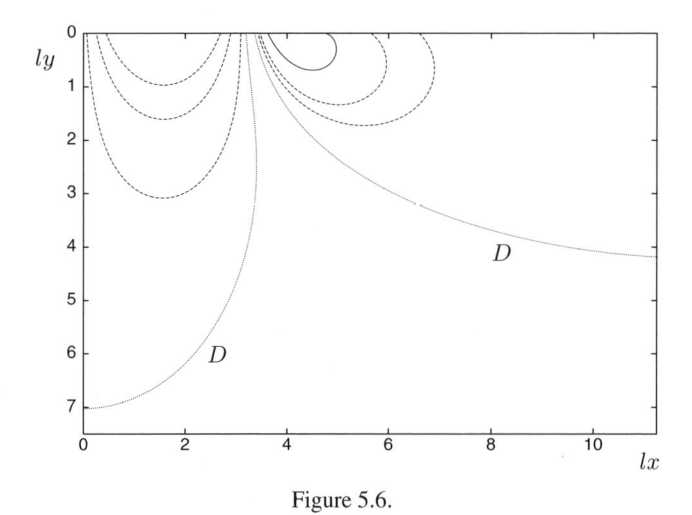

Figure 5.6.

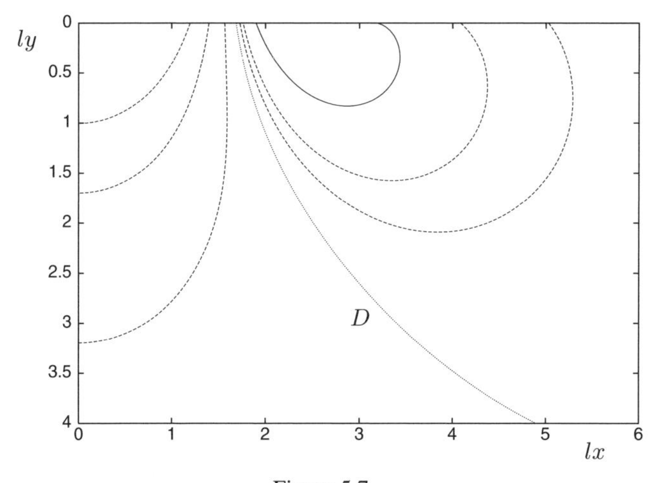

Figure 5.7.

they may be chosen as cylinder cross sections whereas all streamlines to the left enter the y axis normally and together with their image in the y axis can be chosen to represent a symmetric cylinder cross section. In Fig. 5.8 for the same n and sign, we have $k/\nu = 1/2$, and similar arguments apply except for the presence of a second dividing field line extending to infinity. All field lines between the two dividing field lines also go to infinity, and the case $k = 0$ may be regarded as a special case, when the two dividing field lines close up to form the single dividing streamline.

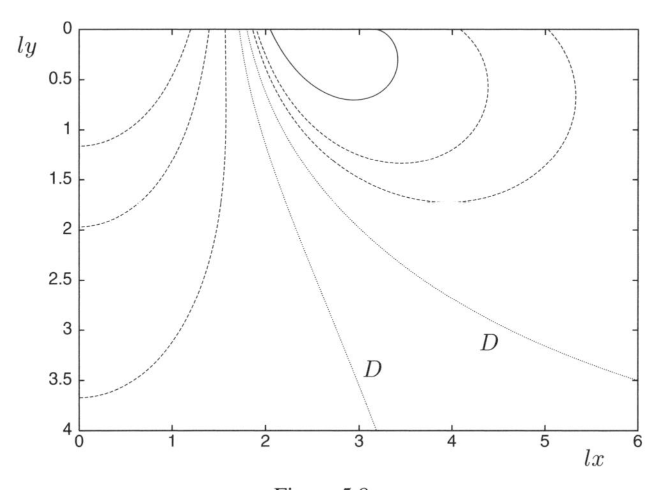

Figure 5.8.

It is possible, by considering larger values of n, to construct an increasing number of separate surface-piercing cylinders for which trapped modes occur. For a description of this procedure in the two-dimensional problem, see Section 4.1.

Finally, we shall bring together our main conclusions by returning to Fig. 5.1. We have already explained the rectangular regions in terms of the results in Subsection 5.4.1. We turn next to the narrow shaded blade-type regions. These form the boundaries of the parameter regions, within which non-uniqueness examples of various cylinder geometries can be constructed by using the technique of Subsection 5.4.2.1. To interpret the shaded regions, consider the first such region labeled $+_1$. Based on numerical solution of (5.75), and using (5.74) with $n = 1$ and the upper plus sign, examples of non-uniqueness in the form of symmetric trapped modes can be found for all k/ν and lb within the shaded region. This is consistent with the first rectangular region, in which no antisymmetric trapped modes exist. All subsequent shaded regions alternatively describing regions where symmetric and antisymmetric trapped modes can be constructed, using increased spacings of sources and sinks, are consistent with the rectangular regions of uniqueness. Notice how the shaded regions all approach the source position as $k/\nu \to 1$, a result suggested by Fig. 5.4, where $k/\nu = 0.99$.

5.4.3. Non-Uniqueness for Small k

In the previous section we gave various examples of non-uniqueness obtained by computing the field lines by using (5.73)–(5.75). Although these numerical results are convincing, it is desirable to provide a *proof* of non-uniqueness for $k > 0$. In Section 4.1 we were assisted in doing this in the case $k = 0$, because there was an explicit expression for the streamlines of the flow, unlike here where a proof for general $k > 0$ appears more difficult. We content ourselves to a proof for sufficiently small k using the function $\phi_1^{(+)}$. Specifically, we shall show that for each member of the family of cross sections in $x > 0$ with $k = 0$, which intersect the x axis twice and also enclose the source point, there exists a unique cross section having the same point of intersection with the x axis to the right of the source point, which also intersects the x axis in $x > 0$ and to the left of the source point provided k is sufficiently small. It is clear from Figs. 5.2 and 5.3 that it is desirable to choose the starting point on the right of the source point, since if we started from the left of it for $k > 0$, the existence of a dividing field line could prevent the curve from intersecting the free surface again if k is not small enough.

Thus, with $\phi_1^{(+)}$ given by (5.73) and (5.74), we seek a curve $v(x, y; k) =$ const, satisfying the equation

$$\nabla v \cdot \nabla \phi_1^{(+)} = 0 \quad \text{for } x, y > 0, \tag{5.76}$$

and having end points $(x_\pm, 0)$, such that

$$\pm\big[x_\pm - a_1^{(+)}(k)\big] > 0, \tag{5.77}$$

and we write $a_1^{(+)}(k)$ to emphasize that this value depends on k.

In Section 4.1, it was shown that for $k = 0$, there is a constant B belonging to $(0, \pi/2)$, such that for every $b > 0$, satisfying $\nu b < B$, a unique curve $v_b(x, y; 0) = C(b)$ exists and satisfies (5.76) and (5.77) with $x_- = b$, where $C(b)$ is a constant equal to $v_b(x_+, 0, 0)$. Furthermore, if $b_1 < b_2$, then $x_+^{(1)} > x_+^{(2)}$, where $(x_+^{(i)}, 0)$ is the right-hand end point of $v_{b_i}(x, y; 0) = C(b_i)$, $i = 1, 2$. The family $v_b(x, y; 0) = C(b)$ just consists of the streamlines of the harmonic function conjugate to $\phi_1^{(+)}$.

The ray $\{x > a_1^{(+)}(0), y = 0\}$ is not a characteristic of equation (5.76) or its equivalent system (5.75) for $k = 0$, where $n = 1$ and the plus sign is chosen because $y = 0$ is not a field line (streamline) for the harmonic conjugate to $\phi_1^{(+)}$ (see Subsection 4.1.1.2).

By (5.73) and the definition of l, the source point $a_1^{(+)}(k)$ approaches $a_1^{(+)}(0)$ from the right as $k \to 0$. Hence, for every choice of b $(= x_-)$ and the corresponding right end point $x_+(b)$, it is possible to ensure that $a_1^{(+)}(k) < x_+(b)$ for sufficiently small k.

Now, from (5.69), (5.72), and (5.73) it can be shown that $\phi_1^{(+)}$ satisfies a Lipschitz condition with respect to k. Since $|k|$ and $(\mu^2 - k^2)^{1/2}$ are Lipschitz functions, the only difficulty when demonstrating that the Green's function given by (5.69), and hence $\phi_1^{(+)}$, has the same property arises from the integration over an infinite interval. This can be overcome in a straightforward way by dividing $(0, +\infty)$ into two parts. The integral over a finite interval is certainly a Lipschitz function, and the dividing point can be chosen properly to estimate the integral over an infinite interval by $|k_1 - k_2|$. Thus we have that

$$\big|\nabla\phi_1^{(+)}(x_1, y_1; k_1) - \nabla\phi_1^{(+)}(x_2, y_2; k_2)\big| \le A(|k_1 - k_2| + |x_1 - x_2| + |y_1 - y_2|)$$

holds for a certain constant A.

Then, for sufficiently small k, the ray $\{x > a_1^{(+)}(k), y = 0\}$ is also not a characteristic of equation (5.76) or the system (5.75) for that value of k. In particular, the same is true for the ray $\{x > x_+(b), y = 0\}$.

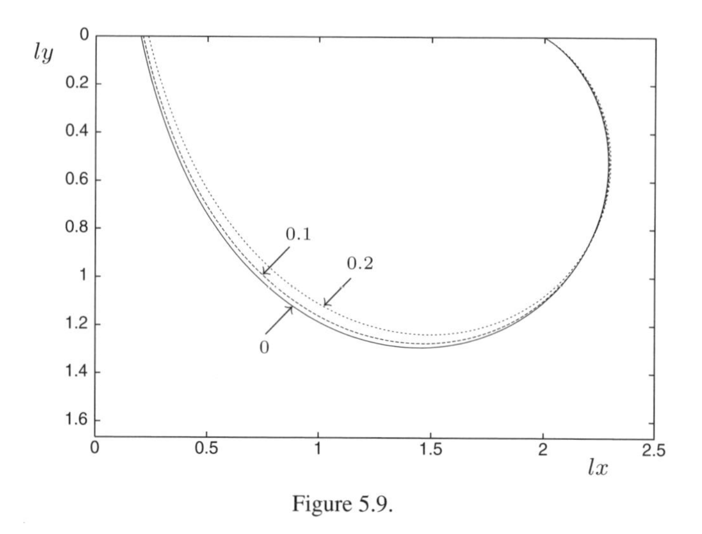

Figure 5.9.

We complement (5.76) with the following initial condition $v(x, 0; k) = C(b)$, where $x = x_+(b)$; that is, start both curves for $k = 0$ and $k > 0$ at the same right end point. The Cauchy problem for (5.76) has a solution, which is unique in a neighborhood of $(x_+(b), 0)$ (see, for example, Zachmanoglou and Thoe [366], Theorem 4.1). Furthermore, since (5.76) is equivalent to the autonomous system (5.75) with $n = 1$ and the plus sign chosen, the solution may be extended to the domain described by $0 < s < \infty$ since s is the arc length along the integral curve (see, for example, Petrovski [287], p. 164).

Finally, we apply a theorem on parameter dependence of solutions of differential equations (see Hille [114], Theorem 3.4.1) to show that

$$|v_b(x, y; 0) - v(x, y; k)| \leq k|e^{As} - 1|,$$

where A is the constant arising in the Lipschitz condition and s is the arc length. It follows, by fixing s, that the two curves can be made as close as we please by choosing k to be sufficiently small. An illustration of this is given in Fig. 5.9, where the field lines for $\phi_1^{(+)}$, each starting from a right end point $lx = 2$, are plotted for $k/\nu = 0$, 0.1, and 0.2. It can be seen how the perturbed curve for small k, starting from the same right end point, varies from the curve for $k = 0$.

5.5. Vertical Cylinders in Channels

It was shown in Subsection 5.1.2 that if the depth dependence can be removed, as is the case for a fixed vertical cylinder extending throughout the depth, the resulting two-dimensional equation to be satisfied is the Helmholtz equation (5.10), which also describes small acoustic perturbations. Although Ursell's proof of existence of trapped modes along the top of a completely submerged *horizontal* cylinder in deep water was well known within the water-wave community, it was some forty years before it was recognized that trapped modes also could exist in two-dimensional acoustic waveguides, since no uniqueness theorem was available for this situation.

In the case of a horizontal submerged cylinder, trapped modes are found only under the assumption that $\omega^2/g = \nu < k$, where k is the longshore wavenumber. The latter value is a cutoff frequency (corresponding to the boundary of the continuous spectrum), beyond which no waves can propagate away from the cylinder. By analogy, in the case of a vertical cylinder in a channel, the imposition of condition (5.13) ensures that here too a cutoff in frequency occurs. In this case k, playing a different role, is related to the frequency through (5.9), and no waves can propagate away from the cylinder provided $k < \pi/(2h)$.

A proof of existence for a vertical circular cylinder of sufficiently small radius can now be developed by precisely the method originally given by Ursell for the submerged horizontal circular cylinder.

Of course there is nothing special about the geometry of the circular cylinder. For example, the case of a vertical cylinder of rectangular cross section parallel to the channel walls can be considered by matching eigenfunction expansions in appropriate rectangular regions in much the same way as was done by Evans and McIver [79] in obtaining numerical results for edge waves over a continental shelf. Thus Evans and Linton [76] numerically show that there also existed at least one trapped mode, and that other modes appeared as the size of the rectangular cylinder increased in the direction parallel to the walls. Following this, Evans [71] proved that for a vertical thin plate positioned on the midplane and parallel to the channel walls, a trapped mode existed for sufficiently wide plates. A numerical scheme for determining trapped modes near a vertical cylinder of arbitrary cross section was developed by Linton and Evans [204] through the construction of an appropriate Green's function and the numerical solution of a homogeneous Fredholm integral equation of the second kind. The results confirmed the earlier numerical predictions for the special cases of circular or rectangular cross-section cylinders.

Because of the condition (5.13) imposed on $z = 0$ outside S, all of the above problems give rise to trapped modes having eigenfrequencies below the cutoff. However, in Evans, Linton, and Ursell [78], it was proved that trapped modes also occurred near an off-center plate and, since condition

(5.13) no longer applied, the trapped mode frequency was embedded in the continuous spectrum because k^2 is an eigenvalue for all $k > 0$ with corresponding eigenfunction $\exp\{\pm ikx\}$. A general existence proof for trapped modes near a cylinder of fairly general cross section was given in Evans, Levitin, and Vassiliev [74]. Further details of the methods used in some of these problems are presented in Subsection 5.5.1. Recently, another approach to this problem was developed by Kamotskii and Nazarov [129].

In the very recent past, interest in waves, trapped in waveguides by vertical obstacles, has been awakened because of their connection with the so-called Rayleigh–Bloch surface waves along periodic gratings. This interest grew up from the following problem, which is important in applications of the water-wave theory to ocean engineering: to calculate wave forces caused by incident waves on a large but finite number of vertical cylinders in the ocean. In particular, Maniar and Newman [214] demonstrated that at certain frequencies the exiting forces on those cylinders near the center of a large number of identical bottom-mounted circular cylinders in a linear array become extremely large, compared with the force on an isolated cylinder. These frequencies are very close to those of trapped modes around a single cylinder in the corresponding waveguide. Prompted by this, Evans and Porter in a series of papers obtained a substantial body of numerical results in this direction. In [80], they extended the ideas of Callan, Linton, and Evans [35] to consider trapped modes existing in a vicinity of any number of circular cylinders, having different radii and placed on the center line of a channel. It was found that in the general case as many modes existed as there were cylinders present. The second paper [82] by Evans and Porter gives convincing numerical evidence for the existence of an additional isolated point eigenvalue embedded in the continuous spectrum in the problem involving one circular cylinder on the center line. In the third paper [290], Porter and Evans numerically determined what types of cylinder cross sections can support Rayleigh–Bloch waves (similar, but less general results were obtained by Utsunomiya and Eatock Taylor [344]). In this connection, it should be mentioned that Davies and Parnovski [43] rigorously proved a number of conditions, under which trapped waves do and do not exist in the presence of two cylinders placed symmetrically about the center of the waveguide. Another approach to numerical construction of waves trapped by certain structures in waveguides and diffraction gratings was developed by McIver, Linton, and McIver [239]. Their method is similar to that described in Chapter 4 for obtaining two-dimensional trapping structures and structures having axial symmetry (also see Subsection 5.4.2). This technique is to take solutions of (5.10), satisfying the required conditions on the vertical walls and having the proper decay at infinity, and then to identify field lines, which can be interpreted as the boundary of a structure.

5.5.1. General Vertical Cylinders: Existence of Trapped Waves

Here we give an outline of the results obtained by Evans et al. [74] for fairly general-shaped vertical cylinders in channels with the antisymmetry condition (5.13) imposed. The method involves the use of an extension of the well-known Rayleigh quotient characterization for the lowest eigenvalue of differential operators on bounded domains where the spectrum is purely discrete, to differential operators on unbounded domains having combined discrete and continuous spectra. The proof is restricted to cross sections of cylinders that are piecewise smooth and symmetric about $z = 0$, and that intersect the x axis at $x = \pm a$, so that we need only consider the cross section W_+ of the water region orthogonal to the y axis, with $0 < z < h$.

We introduce the space C^∞ of infinitely differentiable functions at all the interior points of W_+, as well as up to the whole boundary of W_+, and we consider the subspace $\tilde{C}_0^\infty$ of functions belonging to C^∞, which also satisfy the boundary condition $\psi(x, 0) = 0$ for $|x| > a$, and $\psi(x, y) = 0$ uniformly over $0 \le z \le h$ for sufficiently large $|x|$. For any two functions $\psi_1, \psi_2, \in \tilde{C}_0^\infty$ we can define the following inner product:

$$(\psi_1, \psi_2) = \int_{W_+} \psi_1 \bar{\psi}_2 \, \mathrm{d}x\mathrm{d}z + \int_{W_+} \nabla\psi_1 \cdot \nabla\bar{\psi}_2 \, \mathrm{d}x\mathrm{d}z.$$

Hence, for any function $\psi \in \tilde{C}_0^\infty$ the corresponding norm is defined as follows:

$$\|\psi\| = \int_{W_+} |\psi|^2 \, \mathrm{d}x\mathrm{d}z + \int_{W_+} |\nabla\psi|^2 \, \mathrm{d}x\mathrm{d}z.$$

The closure of the space $\tilde{C}_0^\infty$ is the Hilbert space $\tilde{H}_0^1$, equipped with the same inner product and norm. Since $\tilde{C}_0^\infty$ is dense in $\tilde{H}_0^1$, it means that we can work in the more convenient smaller space $\tilde{C}_0^\infty$. Then we have the following fundamental result whose general form can be found in, for example, Birman and Solomyak's book [27].

Let

$$\lambda_0 = \inf \frac{\int_{W_+} |\nabla\psi|^2 \, \mathrm{d}x\mathrm{d}z}{\int_{W_+} |\psi|^2 \, \mathrm{d}x\mathrm{d}z},$$

where the infimum is taken over $\psi \in \tilde{C}_0^\infty$, such that ψ does not vanish identically (in fact, almost everywhere).

Then $\lambda_0 \leq \Lambda = \pi^2/4h^2$, where $2h$ is the width of the channel. Furthermore, if $\lambda_0 < \Lambda$, then $\lambda_0(= -k^2)$ is the smallest eigenvalue of the problem described by equations (5.10)–(5.13), *and if $\lambda_0 = \Lambda$ there are no eigenvalues of the problem below the continuous spectrum* $[\Lambda, +_\infty)$.

Notice that we seek the infimum among functions, which satisfy (5.13), but do not necessarily satisfy (5.11), since if $\lambda_0 < \Lambda$, then *all* of the equations (5.10)–(5.13) are automatically satisfied.

To show that $\lambda_0 < \Lambda$ we introduce the following function:

$$\Psi(x, z) = \chi\left(\frac{x}{A}\right) \sin \frac{\pi z}{2h},$$

where A is a certain positive constant and χ is a smooth cutoff function with the following properties:

$$\chi(x) = 1 \quad \text{for } |x| \leq 1, \qquad 0 < \chi(x) < 1 \quad \text{for } 1 < |x| < 2,$$
$$\chi(x) = 0 \quad \text{for } |x| \geq 2.$$

An example of χ is given in Evans et al. [74]. It can now be shown by elementary analysis that

$$\int_{W_+} |\nabla\Psi|^2 \, \mathrm{d}x\mathrm{d}z < \frac{\pi^2}{4h^2} \int_{W_+} |\Psi|^2 \, \mathrm{d}x\mathrm{d}z,$$

from which it follows immediately that a point eigenvalue $k = k_0 < \pi/(2h)$ exists. The proof is easily extended to the case of a thin vertical plate on the midplane, and also to the case in which the channel walls contain identical symmetric indentations.

All of the problems discussed in this section can be regarded as problems in the linear acoustic theory. The existence of trapped modes, termed *acoustic resonances* in the acoustics literature, is well known and appears to have been studied first by Parker [281] for the case of a thin plate, in a series of papers, both experimental and theoretical. For an excellent review of acoustic resonances, see Parker and Stoneman [282].

5.5.2. *Particular Geometries*

In this subsection we survey a number of particular geometries providing the existence of trapped modes.

5.5.2.1. A Vertical Circular Cylinder

In this case equations (5.10)–(5.13) hold with (5.11) replaced by

$$\partial\phi/\partial r = 0 \quad \text{on } r = (x^2 + z^2)^{1/2} = a, \tag{5.78}$$

where a is the cylinder's radius.

A fundamental solution, satisfying (5.10) and (5.13), is as follows:

$$H^{(1)}_{2n+1}(kr)\sin(2n+1)\theta, \quad \text{where } x = r\cos\theta,\ z = r\sin\theta.$$

This can be modified by adding an integral term chosen so that the resulting expression $\phi_{2n+1}(r,\theta)$ satisfies (5.12). A solution is now sought in the following form:

$$\phi = \sum_{n=0}^{\infty} a_n \phi_{2n+1}(r,\theta).$$

Then (5.78) leads to the homogeneous infinite system of equations

$$a_m + \sum_{n=0}^{\infty} B_{mn} a_n = 0, \quad m = 0, 1, \ldots,$$

for the unknown a_m. The matrix of this system has elements

$$B_{mn} = \frac{J'_{2m+1}(ka)}{Y'_{2n+1}(ka)} A_{mn},$$

where J_{2m+1} and Y_{2n+1} are Bessel functions, and

$$A_{mn} = -\frac{2}{\pi}(-1)^{m+n}\,\mathrm{Re}\int_{-\infty}^{\infty} \frac{e^{-\gamma h}\sinh(2n+1)\mu\,\sinh(2n+1)\mu\,\sinh\gamma h}{\gamma\cosh\gamma h}\,dt,$$

$$\gamma = \begin{cases} (t^2-k^2)^{1/2} & \text{for } |t| > k, \\ -i(k^2-t^2)^{1/2}, & \text{for } |t| < k, \end{cases} \quad t = k\cosh\mu.$$

It can be shown that

$$\sum_m \sum_n |B_{mn}| < \infty \quad \text{for } a/d < 1.$$

This ensures that Δ_N, the determinant of the truncated system, converges uniformly as $N \to \infty$ to Δ_∞, the determinant of the infinite system.

A careful analysis now shows that Δ_∞ vanishes if

$$a/d \to 0, \qquad kd \to \pi/2$$

simultaneously, so that

$$kd \sim \frac{\pi}{2}\left[1 - \frac{\pi^4}{8}\left(\frac{a}{d}\right)^4\right].$$

Numerical computation of Δ_N for increasing N suggests that in fact there exists a single trapped-mode frequency for *all* $a/d \le 1$.

5.5.2.2. A Thin Vertical Plate on the Midplane of the Channel

For problems involving vertical cylinders on the midplane of the channel having long middle sections parallel to the channel walls, it is possible to establish a simple approximate expression for the frequency of trapped modes. For such a cylinder symmetric about the origin, having length $2a \gg 1$, the simple solution

$$\phi = 2A \cos kx = A(e^{ikx} + e^{-ikx})$$

satisfying (5.10)–(5.12) holds away from the ends. However, the first wave term on the right-hand side, on reaching the end $x = a$, will be totally reflected [since $k < \pi/(2h)$] with a complex reflection coefficient R, satisfying $|R| = 1$. This coefficient is approximately equal to that corresponding to a wave, being reflected from the end $x = a$ of the semi-infinite cylinder obtained by extending the end $x = -a$ out to $x = -\infty$.

For consistency then we require

$$2A \cos kx = B\left[e^{ik(x-a)} + Re^{-ik(x-a)}\right],$$

from which it follows that $R = e^{-2ika}$ is the equation determining the trapped-mode frequency. For a general vertical cylinder of semi-infinite extent in x, the coefficient R is not known, but in the case of a thin semi-infinite barrier, the solution can be shown, for example, by using the Wiener-Hopf technique, to be $R = -e^{-2i\beta}$, where

$$\beta = \sum_{n=1}^{\infty}\left[\arctan\left(\frac{k}{\kappa_n}\right) - \arctan\left(\frac{k}{k_n}\right)\right],$$

and

$$\kappa_n = \left[\left(n - \frac{1}{2}\right)^2 \frac{\pi^2}{4} - k^2\right]^{1/2}, \qquad k_n = \left[\left(\frac{n\pi}{d}\right)^2 - k^2\right]^{1/2}.$$

Then, an approximation to the trapped modes is provided by solving the equation

$$\pi/2 - \beta = ka + n\pi,$$

where n is an integer. Since β is independent of a, it is easily shown that this equation has solutions, and that the number of solutions increases with increasing a/h.

This approximation forms the basis for a rigorous proof of the existence theorem for trapped modes, given by Evans [71] in the case of a thin plate of sufficiently large width $2a$. This proof also confirms the accuracy of the above approximation. The same idea was used by Linton and Evans [203] to obtain an estimate for the trapped-mode frequencies above a submerged horizontal plate as described in Section 5.3.

5.5.2.3. Truncated Cylinders

Trapped modes also occur when vertical cylinders do not extend throughout the water depth. Thus Linton and Evans [205] used the multipole method to consider both radiation and scattering problems for truncated cylinders in channels. They were primarily concerned with computations of hydrodynamic characteristics rather than trapped modes. However, it is clear that the existence of an antisymmetric trapped mode implies a non-uniqueness of the corresponding sway problems, in which the cylinder makes small horizontal oscillations in a direction normal to the channel walls. This non-uniqueness is manifested by a singularity in the added mass coefficient at the trapped-mode frequency in contrast to sharp (but finite) spikes in both the added mass and damping coefficients at frequencies corresponding to cross modes in the channel. Linton and Evans [205] found clear numerical evidence for trapped modes both for truncated cylinders intersecting the free surface and also for bottom-mounted completely submerged cylinders, for all drafts and depths of submergence. In the former case, as the draft increased and the clearance beneath the cylinder reduced, the trapped mode approached the expected value for a cylinder extending throughout the depth. However, this limit was not approached by reducing the clearance above the bottom-mounted cylinder, which is evidently a non-uniform limit on physical grounds. A fuller discussion of the implication of trapped-mode frequencies on hydrodynamic characteristics of bodies in channels is given in Linton and Evans [206].

5.5.2.4. Trapped Modes Near Vertical Periodic Coastlines

A novel example of a trapped mode is obtained by considering the possible wave field in the vicinity of a vertical coastline consisting of a periodic array of rectangular blocks. Once again the depth can be extracted so that the problem is identical to a diffraction grating problem in acoustics. The possibility of trapped modes is now equivalent to the question as to whether in the diffraction grating context Rayleigh–Bloch surface waves can exist in isolation in the absence of an incoming or reflected wave. According to Wilcox (see his book [362], pp. 11–12), no general criteria are known on the shape of the grating for such surface waves to exist. However, if the Neumann condition is replaced by the Dirichlet condition, then it appears that no surface waves or trapped modes exist. This result can be compared with recent results of McIver and Linton [233], who have considered a variety of problems for vertical cylinders in channels and showed the nonexistence of trapped modes when the Dirichlet condition is imposed on the cross section of the cylinders.

An approximation method based on an idea of Whitehead [358] was used by Hurd [121] to estimate the frequencies of trapped modes in terms of the geometric parameters in the special case, when the blocks reduced to a period array of thin barriers. This problem was considered afresh by Evans and Linton [77], who were able to *prove* the existence of trapped modes or edge waves traveling in the direction of the coastline, for sufficiently long barriers. The method used was based on the modified residue calculus technique, familiarized by Mittra and Lee [248], and used by Evans [71] for the vertical plate in the channel described in Subsection 5.5.2.2. Accurate numerical results for the periodic rectangular blocks were recently obtained by Evans and Fernyhough [73], using a Galerkin approximation to an integral equation representation of the solution. It is unlikely that there is anything special about the rectangular geometry, and one would expect that similar edge waves solutions exist for more general periodic coastlines. It is possible that the method used to derive the general existence proof for vertical cylinders in channels, described in Subsection 5.5.1, could be adapted to the array with a general period.

Part 2

Ship Waves on Calm Water

6
Green's Functions

As in the case of time-harmonic waves (see Chapter 1), we begin with the simplest model, replacing a ship by a point source in the uniform forward motion in calm water. The corresponding velocity potential is sometimes referred to as the Kelvin source, but to keep the terminology unified we call it the Green's function in what follows. Similar to the theory of time-harmonic waves developed in Part 1, the theory of ship waves presented here relies essentially on Green's functions. They are of importance not only for proving solvability theorems (see Chapters 7 and 8) but also for constructing examples of trapped waves (nontrivial solutions to homogeneous boundary value problems) in Section 8.4.

The three-dimensional Green's function of a point source in deep water is considered in detail in Sections 6.1 and 6.2. General facts about the three-dimensional Green's function are considered in Section 6.1 and the far-field expansions for Green's function and the corresponding elevation of the free surface are obtained in Section 6.2. Two-dimensional Green's functions are treated in Section 6.3, which we begin with the simpler case of deep water (Subsection 6.3.1). For water of finite depth, which will be referred to as shallow water, we consider Green's function in Subsection 6.3.2. There is an essential difference between Green's functions describing sources in deep and shallow water because in the latter case the velocity field behind the moving line source has two different regimes depending on whether or not the forward speed U exceeds the critical value $\sqrt{gd}$ (d is the depth of unperturbed water and g is the acceleration caused by gravity). As usual, bibliographical notes are collected in the last section, which is Section 6.4.

6.1. Three-Dimensional Problem of a Point Source in Deep Water

In this section, we consider Green's function G describing a point source in the uniform forward motion in calm water of infinite depth. In Subsection 6.1.1, we define G as a solution of two particular boundary value problems (for a submerged source and a source placed on the free surface), and we give an explicit integral representation for G. A heuristic derivation of the representation is given in Subsection 6.1.2. In Subsection 6.1.3, we prove that G solves the corresponding boundary value problems. The asymptotic behavior at infinity of the double integral involved in Green's function representation is investigated in Subsection 6.1.4. The detailed analysis of the asymptotic behavior of Green's function at infinity is given in Section 6.2. Subsection 6.1.5 is concerned with the question of uniqueness.

6.1.1. Statement of the Problem; Explicit Representations

6.1.1.1. Statement of the Problem

We use rectangular coordinates (x, y, z) having the origin in the free surface, the y axis directed downward and attached to the source, and so moving forward uniformly along the x axis at the speed U. Let the source be placed at

$$(0, y_0, 0) \in \overline{\mathbb{R}^3_-} = \{-\infty < x, z < +\infty,\ y \leq 0\}.$$

If $y_0 < 0$ (the case of submerged source), then Green's function $G(x, y, z; y_0)$ must satisfy

$$\begin{gathered}\nabla^2 G = -4\pi\,\delta_{(0,y_0,0)}\,(x, y, z) \quad \text{in } \mathbb{R}^3_-, \\ G_{xx} + \nu G_y = 0, \quad \text{when } y = 0,\end{gathered} \tag{6.1}$$

where $\nu = g/U^2$ and g is the acceleration caused by gravity. If the source is placed on the free surface; that is, $y_0 = 0$, then instead of (6.1) we have to solve

$$\begin{gathered}\nabla^2 G = 0 \quad \text{in } \mathbb{R}^3_-, \\ G_{xx} + \nu G_y = 4\pi\,\delta_{(0,0)}\,(x, z) \quad \text{when } y = 0.\end{gathered} \tag{6.2}$$

In either case, the following conditions at infinity should be imposed:

$$\sup_{x,z} |G| \to 0 \quad \text{as } y \to -\infty, \qquad \int_{-\infty}^{+\infty} |G(x, 0, z; y_0)|^2 \, dz \to 0 \quad \text{as } x \to +\infty. \tag{6.3}$$

These conditions mean that the velocity field decays with the depth and on each horizontal line in the free surface tending to infinity in the direction of the source motion. It will be shown in the following paragraphs that these conditions guarantee the uniqueness of Green's function.

From the hydrodynamic point of view, the velocity potential G solving (6.1) represents a sink, that is, a source of strength -1, but it is commonly accepted to call such solutions sources.

6.1.1.2. Integral Representation of Green's Function

The representation under consideration involves two integrals, and the first of them is a double integral of the following form:

$$I_2(x, y, z) = \frac{2\nu}{\pi} \mathrm{Re} \int_{-\infty}^{+\infty} \int_{\ell_\alpha} \frac{\exp\{y\sqrt{\zeta^2 + \tau^2} + i\,|x|\,\zeta + iz\tau\}}{\nu\sqrt{\zeta^2 + \tau^2} - \zeta^2}\, d\zeta\, d\tau. \tag{6.4}$$

Here ℓ_α is the ray $\zeta = \sigma e^{i\alpha}$ in the complex ζ plane, where $\sigma > 0$, $0 < \alpha \leq \pi/4$ and the branch of $\sqrt{\zeta^2 + \tau^2}$ is chosen so that

$$\mathrm{Re}\sqrt{\zeta^2 + \tau^2} \geq 0. \tag{6.5}$$

Below we will show that the integral (6.4) converges when $y < 0$, does not depend on α, and can be written in the following form:

$$I_2(x, y, z) = -\frac{2\nu}{\pi} \mathrm{Re} \lim_{\varepsilon \to +0} \int_{-\pi/2}^{\pi/2} \int_0^\infty \frac{\exp\{yk + i\,|x|\,k\cos\theta + izk\sin\theta\}}{k\cos^2\theta - \nu + i\varepsilon}\, dk d\theta. \tag{6.6}$$

In what follows, we understand $I_2(x, 0, z)$ as the limit of $I_2(x, y, z)$ given by (6.4) as $y \to -0$. The existence of this limit will be demonstrated in Subsection 6.1.4.

Representation of Green's function also involves a single integral

$$I_1(x, y, z) = -4\nu H(-x)\, \mathrm{Im} \int_{-\infty}^{+\infty} \exp\{\nu y(1 + t^2) + i\nu(x + tz)\sqrt{1 + t^2}\}\, dt, \tag{6.7}$$

where H is the Heaviside function defined as follows:

$$H(-x) = 1 \quad \text{for } x < 0, \qquad H(-x) = 0 \quad \text{for } x \geq 0.$$

In order to formulate the main result of the present section, we introduce the following notation. By $R = [x^2 + (y - y_0)^2 + z^2]^{1/2}$ we denote the distance between the source point and a point of observation $(x, y, z) \in \mathbb{R}^3_-$; also $R_0 = [x^2 + (y + y_0)^2 + z^2]^{1/2}$ denotes the distance between (x, y, z) and the mirror reflection of the source point in the free surface.

The boundary value problem (6.1) *and* (6.3) *has one and only one solution* G. *If* $y + y_0 < 0$, *then*

$$G(x, y, z; y_0) = R^{-1} - R_0^{-1} + I_2(x, y + y_0, z) + I_1(x, y + y_0, z). \quad (6.8)$$

It should be noted that if $y_0 < 0$, then (6.8) is valid in the whole $\overline{\mathbb{R}^3_-}$ and solves (6.1) and (6.3). If $y_0 = 0$, then (6.8) solves (6.2) and (6.3), where $G(x, 0, z; 0)$ and the corresponding derivatives on $y = 0$ are understood as the limits as $y \to -0$ and their existence is a consequence of the following assertion.

For Green's function $G(x, y, z; y_0)$ *solving* (6.1) *and* (6.3), *there exists the limit* $G(x, y, z; 0)$ *as* $y_0 \to -0$, *and the latter function solves* (6.2) *and* (6.3). *The convergence of* $G(x, y, z; y_0)$ *and its derivatives is uniform on any compact subset of* $\mathbb{R}^3_-$.

6.1.2. Formulae for Green's Function: Heuristic Derivation

6.1.2.1. Derivation of (6.8)

A solution of (6.1) and (6.3) is sought in the following form:

$$G(x, y, z; y_0) = R^{-1} - R_0^{-1} + v(x, y, z; y_0), \quad (6.9)$$

and so v must satisfy

$$\nabla^2 v = 0 \quad \text{in } \mathbb{R}^3_-,$$
$$v_{xx} + \nu v_y = -2\nu y_0 \left(x^2 + y_0^2 + z^2\right)^{-3/2} \quad \text{when } y = 0. \quad (6.10)$$

Let us consider the Fourier transform of v with respect to x and z:

$$\hat{v}(\zeta, y, \tau; y_0) = \int_{-\infty}^{+\infty} \int_{-\infty}^{+\infty} v(x, y, z; y_0) e^{-ix\zeta - iz\tau} \, dx dz. \quad (6.11)$$

From (6.10) we get

$$(-\zeta^2 - \tau^2)\hat{v} + \hat{v}_{yy} = 0, \quad y < 0,$$
$$-\zeta^2 \hat{v} + \nu \hat{v}_y = 4\pi \nu \exp\{y_0 \sqrt{\zeta^2 + \tau^2}\}, \quad y = 0, \quad (6.12)$$

where the right-hand-side term in the second equation is the Fourier transform of the corresponding term in the second equation of (6.10). In order to demonstrate this, let us find the inverse Fourier transform $F^{-1}h$ of

$$h = 4\pi(\zeta^2 + \tau^2)^{-1/2} \exp\{-\varepsilon\sqrt{\zeta^2 + \tau^2}\}.$$

By (k, θ) and (r, φ) we denote polar coordinates in (ζ, τ) plane and in (x, z) plane, respectively. Then

$$\begin{aligned} F^{-1}h &= \frac{1}{\pi}\int_{-\infty}^{+\infty}\int_{-\infty}^{+\infty}(\zeta^2+\tau^2)^{-1/2}\exp\{-\varepsilon\sqrt{\zeta^2+\tau^2}+ix\zeta+iz\tau\}\,d\zeta\,d\tau \\ &= \frac{1}{\pi}\int_0^{2\pi}\int_0^{+\infty}\exp\{-\varepsilon k+ikr\cos(\theta-\varphi)\}\,dk d\theta \\ &= \frac{1}{\pi}\int_0^{2\pi}\frac{d\theta}{\varepsilon-ir\cos(\theta-\varphi)} \\ &= \frac{1}{\pi}\int_0^{2\pi}\frac{d\theta}{\varepsilon-ir\cos\theta}=\frac{\varepsilon}{\pi}\int_0^{2\pi}\frac{d\theta}{\varepsilon^2+r^2\cos^2\theta}+\frac{ir}{\pi}\int_0^{2\pi}\frac{\cos\theta\,d\theta}{\varepsilon^2+r^2\cos^2\theta}. \end{aligned}$$

The last integral is equal to zero, and for evaluating the next to the last one the new variable $u=\tan\theta$ should be used. This leads to the equation giving $F^{-1}h$:

$$\begin{aligned} &\frac{1}{\pi}\int_{-\infty}^{+\infty}\int_{-\infty}^{+\infty}(\zeta^2+\tau^2)^{-1/2}\exp\{-\varepsilon\sqrt{\zeta^2+\tau^2}+ix\zeta+iz\tau\}\,d\zeta\,d\tau \\ &\quad = 2[r^2+\varepsilon^2]^{-1/2}. \end{aligned} \tag{6.13}$$

Differentiating this with respect to ε and putting $-\varepsilon = y_0$, we see that the right-hand-side term in the second equation (6.12) is the Fourier transform of the corresponding term in the second equation (6.10).

Now solving the first equation in (6.12), we obtain that

$$\hat{v} = C_1(\zeta,\tau,y_0)\exp\{y\sqrt{\zeta^2+\tau^2}\} + C_2(\zeta,\tau,y_0)\exp\{-y\sqrt{\zeta^2+\tau^2}\}.$$

From (6.9) and the first condition (6.3), it follows that $C_2 = 0$, and so

$$\hat{v} = C_1(\zeta,\tau,y_0)\exp\{y\sqrt{\zeta^2+\tau^2}\}. \tag{6.14}$$

Substituting this into the second equation (6.12), we get

$$(-\zeta^2+\nu\sqrt{\zeta^2+\tau^2})C_1(\zeta,\tau,y_0) = 4\pi\nu\exp\{y_0\sqrt{\zeta^2+\tau^2}\}.$$

Then $C_1(\zeta,\tau,y_0)$ can be obtained by dividing both sides of the last equality by $\zeta^2-\nu\sqrt{\zeta^2+\tau^2}$. Since this function vanishes at the origin and along a certain curve, such a division is impossible, but reasoning heuristically and doing this, we find that

$$C_1(\zeta,\tau,y_0) = \frac{4\pi\nu\exp\{y_0\sqrt{\zeta^2+\tau^2}\}}{\nu\sqrt{\zeta^2+\tau^2}-\zeta^2}.$$

Now (6.14) and (6.11) produce

$$v(x, y, z; y_0) = \frac{\nu}{\pi} \int_{-\infty}^{+\infty} \int_{-\infty}^{+\infty} \frac{\exp\{(y + y_0)\sqrt{\zeta^2 + \tau^2} + ix\zeta + iz\tau\}}{\nu\sqrt{\zeta^2 + \tau^2} - \zeta^2}\, d\zeta\, d\tau. \tag{6.15}$$

Let us represent this integral as a sum of two integrals over the half-planes $\zeta > 0$ and $\zeta < 0$. Changing variables $(\zeta, \tau) \mapsto (-\zeta, -\tau)$ in the second integral, we see that these two integrals are complex conjugate, and therefore,

$$v(x, y, z; y_0) = \frac{2\nu}{\pi} \operatorname{Re} \int_{-\infty}^{+\infty} \int_{0}^{+\infty} \frac{\exp\{(y + y_0)\sqrt{\zeta^2 + \tau^2} + ix\zeta + iz\tau\}}{\nu\sqrt{\zeta^2 + \tau^2} - \zeta^2}\, d\zeta\, d\tau. \tag{6.16}$$

The integrand is an analytic function of ζ branching at $\zeta = \pm i\tau$ and having poles at

$$\zeta = \pm\nu \left[\frac{1 \pm \sqrt{1 + 4\nu^{-2}\tau^2}}{2} \right]^{1/2}, \tag{6.17}$$

where $\zeta^2 - \nu\sqrt{\zeta^2 + \tau^2}$ vanishes. Hence the integrand is a single-valued function on the ζ plane cut along the imaginary axis from $\pm i\tau$ to infinity. In the half-plane $\operatorname{Re}\zeta > 0$, the integrand has only one pole at

$$\zeta_0 = \nu \left[\frac{1 + \sqrt{1 + 4\nu^{-2}\tau^2}}{2} \right]^{1/2},$$

and we understand (6.16) in the following sense:

$$v(x, y, z; y_0) = \frac{2\nu}{\pi} \operatorname{Re} \int_{-\infty}^{+\infty} \int_{\ell} \frac{\exp\{(y + y_0)\sqrt{\zeta^2 + \tau^2} + ix\zeta + iz\tau\}}{\sqrt{\zeta^2 + \tau^2} - \zeta^2}\, d\zeta\, d\tau, \tag{6.18}$$

where the contour ℓ goes along the half-axis $\zeta > 0$ indented above at $\zeta = \zeta_0$.

If $x \geq 0$, then ℓ can be replaced by ℓ_α because $e^{ix\zeta}$ is bounded on ℓ_α in this case. Hence, v coincides with $I_2(x, y + y_0, z)$, and (6.9) implies (6.8) for $x \geq 0$ because $I_1 = 0$ in this case.

If $x < 0$, then replacing ℓ in (6.18) by the complex conjugate path $\bar{\ell}$, we see that the residue theorem applied at $\zeta = \zeta_0$ gives that

$$\begin{aligned} v(x, y, z; y_0) = {} & \frac{2\nu}{\pi} \operatorname{Re} \int_{-\infty}^{+\infty} \int_{\bar{\ell}} \frac{\exp\{(y + y_0)\sqrt{\zeta^2 + \tau^2} - i\,|x|\,\zeta + iz\tau\}}{\nu\sqrt{\zeta^2 + \tau^2} - \zeta^2}\, d\zeta\, d\tau \\ & + 4\nu \operatorname{Im} \int_{-\infty}^{+\infty} \frac{\zeta_0}{\nu^2 - 2\zeta_0^2} \exp\left\{ (y + y_0)\sqrt{\zeta_0^2 + \tau^2} \right. \\ & \left. + ix\zeta_0 + iz\tau \right\} d\tau. \end{aligned} \tag{6.19}$$

Since there is only the real part of the double integral in (6.19), we replace this integral by its complex conjugate and change the variable as follows: $\tau \mapsto -\tau$ in the latter integral. Then we transform ℓ into ℓ_α and note that the double integral in (6.19) is equal to $I_2(x, y + y_0, z)$. Thus we have

$$v = I_2(x, y + y_0, z) + I_1(x, y + y_0, z), \quad \text{for } x < 0,$$

because the single integral in (6.19) reduces to $I_1(x, y + y_0, z)$ after the variable is changed: $\tau \mapsto \nu t\sqrt{1+t^2}$. Together with (6.9) this leads to (6.8) in the case $x < 0$.

Changing the paths of integration above is not justified, and a rigorous proof that the integrals in (6.4) and (6.6) are convergent will be given in Subsection 6.1.3 along with proving that (6.8) solves (6.1) and (6.3).

6.1.2.2. Equivalence of (6.4) and (6.6)

Considerations are still heuristic here and rigorous proofs will be given in the next subsection. We shall show that

$$f_\varepsilon = (1 - i\varepsilon)\nu\sqrt{\zeta^2 + \tau^2} - \zeta^2, \quad \text{where } \varepsilon > 0,$$

is not equal to zero for $0 \leq \arg\zeta \leq \pi/4$, and so we can write (6.4) in the form

$$I_2(x, y, z) = \frac{2\nu}{\pi}\operatorname{Re}\lim_{\varepsilon\to+0}\int_{-\infty}^{+\infty}\int_{\ell_\alpha}\frac{\exp\{y\sqrt{\zeta^2+\tau^2} + i\,|x|\,\zeta + iz\tau\}}{(1 - i\varepsilon)\nu\sqrt{\zeta^2+\tau^2} - \zeta^2}\,d\zeta\,d\tau.$$

If $\varepsilon > 0$, then we can replace ℓ_α by the real semiaxis ℓ_0 in the last integral. Then it remains for us to write the integral by using polar coordinates.

6.1.3. Justification of the Formulae for Green's Function

6.1.3.1. Convergence and Smoothness of Integrals I_2 and I_1

Let us show that (6.4) is a convergent integral and that it coincides with (6.6). Let us also demonstrate that (6.8) solves problems (6.1), (6.3) and (6.2), (6.3). We begin with the following assertion.

If $y < 0$, then (6.4) *is a convergent integral that does not depend on $\alpha \in (0, \pi/4]$. Moreover, for $y < 0$ formulae* (6.4) *and* (6.6) *define the same function I_2 and it is infinitely smooth for $x \neq 0$. At last, in order to calculate the derivatives of I_2 with respect to y and z, one may differentiate the integrand in* (6.4) *or* (6.6). *The same is true for the derivatives with respect to x if $x \neq 0$.*

To prove this assertion, we need some estimates for the integrand. First we note that

$$\left|e^{i|x|\zeta+iz\tau}\right| \le 1 \quad \text{when } 0 \le \arg\,\zeta \le \pi/4. \tag{6.20}$$

Also, for $0 \le \arg\,\zeta \le \pi/4$ we have $\operatorname{Re}\zeta^2 \ge 0$, and so

$$|\zeta^2 + \tau^2| \ge \max(|\zeta|^2, |\tau|^2) \ge 1/2(|\zeta|^2 + |\tau|^2) \tag{6.21}$$

for any real τ and $0 \le \arg\,\zeta \le \pi/4$. Therefore,

$$\begin{aligned}\operatorname{Re}\sqrt{\zeta^2+\tau^2} &= |\zeta^2+\tau^2|^{1/2}\cos\frac{\arg(\zeta^2+\tau^2)}{2} \ge \frac{1}{\sqrt{2}}\sqrt{|\zeta|^2+|\tau|^2}\cos\frac{\pi}{4}\\ &= \frac{\sqrt{|\zeta|^2+|\tau|^2}}{2} \quad \text{when } 0 \le \arg\,\zeta \le \frac{\pi}{4}.\end{aligned} \tag{6.22}$$

From here it follows that

$$\exp\{y\sqrt{\zeta^2+\tau^2} + i\,|x|\,\zeta + iz\tau\} \le \exp\{y\sqrt{|\zeta|^2+|\tau|^2}/2\} \tag{6.23}$$

for $0 \le \arg\,\zeta \le \pi/4$.

Let us now estimate the integrand's numerator in (6.4). If

$$\zeta = e^{i\alpha}\sigma, \qquad \sigma \ge 0, \qquad 0 \le \alpha \le \pi/4,$$

then we have that $0 \le \arg(\zeta^2+\tau^2) \le 2\alpha$, and so

$$0 \le \arg\sqrt{\zeta^2+\tau^2} \le \alpha, \qquad -2\alpha \le \arg(e^{-2i\alpha}\sqrt{\zeta^2+\tau^2}) \le -\alpha.$$

Combining these inequalities with (6.21), we get that

$$|\operatorname{Im}\{e^{-2i\alpha}\sqrt{\zeta^2+\tau^2}\}| \ge \frac{\sin\alpha}{2}\sqrt{|\zeta|^2+|\tau|^2}.$$

Hence,

$$\begin{aligned}|\nu\sqrt{\zeta^2+\tau^2} - \zeta^2| &= |\nu e^{-2i\alpha}\sqrt{\zeta^2+\tau^2} - \sigma^2| \ge |\operatorname{Im}\{\nu e^{-2i\alpha}\sqrt{\zeta^2+\tau^2}\} - \sigma^2|\\ &= \nu\,|\operatorname{Im}\{e^{-2i\alpha}\sqrt{\zeta^2+\tau^2}\}| \ge \frac{\nu\sin\alpha}{2}\sqrt{|\zeta|^2+|\tau|^2}\\ &\qquad \text{when } 0 \le \arg\,\zeta \le \frac{\pi}{4}.\end{aligned} \tag{6.24}$$

In the same way one obtains the following for $0 \le \arg\,\zeta \le \pi/4$:

$$|\zeta^2 - (1-i\varepsilon)\nu\sqrt{\zeta^2+\tau^2}| \ge \frac{\nu\sin(\alpha+\gamma)}{2}\sqrt{|\zeta|^2+|\tau|^2}, \tag{6.25}$$

where $\gamma = \arctan\varepsilon$.

Estimates (6.23) and (6.24) imply that (6.4) is a convergent integral and the fact that the derivatives of I_2 with respect to y, z and with respect to

x when $x \neq 0$ can be obtained by differentiating the integrand. Combining these estimates with the fact that the integrand is analytic in ζ, one proves that I_2 is independent of $\alpha \in (0, \pi/4]$. Estimates (6.23), (6.25) justify that (6.4) and (6.6) are equivalent (this was discussed in Subsection 6.1.2.2) and allow us to differentiate I_2 given by (6.6) simply differentiating the integrand.

It is obvious that (6.7) defines an infinitely differentiable function I_1 in the half-space $y < 0$ when $x \neq 0$. Derivatives of I_2 and I_1 may have jumps at $x = 0$, but $I_2 + I_1$ *is infinitely differentiable throughout the half-space* $y < 0$, *in particular on the plane* $x = 0$.

In order to prove this assertion we note that (6.23) and (6.24) imply that derivatives of I_2 have limits as $x \to \pm 0$, and for any n_1, n_2, and n_3 we have

$$\lim_{x\to\pm 0} \partial_x^{n_1}\partial_y^{n_2}\partial_z^{n_3} I_2$$

$$= \frac{2\nu}{\pi}\mathrm{Re}\int_{-\infty}^{+\infty}\int_{\ell_\alpha} \frac{(\pm i\zeta)^{n_1}(\sqrt{\zeta^2+\tau^2})^{n_2}(i\tau)^{n_3}\exp\{y\sqrt{\zeta^2+\tau^2}+iz\tau\}}{\nu\sqrt{\zeta^2+\tau^2}-\zeta^2}\,d\zeta\,d\tau. \tag{6.26}$$

Hence for even n_1 we have that $[\partial_x^{n_1}\partial_y^{n_2}\partial_z^{n_3} I_2] = 0$, where $[\cdot]$ denotes the jump at $x = 0$ of the function in the square brackets. Using (6.23), we find it is easy to verify that the same relation holds for I_1 as well.

Now let n_1 be odd. From (6.23) and (6.24) it follows that (6.26) can be considered as an iterated integral. Then we can replace ℓ_α in the inner integral by the contour ℓ used in (6.18), and after that the integral can be replaced by its complex conjugate because we consider only the real part of it in (6.26). Thus we get

$$\lim_{x\to -0} \partial_x^{n_1}\partial_y^{n_2}\partial_z^{n_3} I_2$$

$$= \frac{2\nu}{\pi}\mathrm{Re}\int_{-\infty}^{+\infty}\int_{\bar{\ell}} \frac{(i\zeta)^{n_1}(\sqrt{\zeta^2+\tau^2})^{n_2}(-i\tau)^{n_3}\exp\{y\sqrt{\zeta^2+\tau^2}-iz\tau\}}{\nu\sqrt{\zeta^2+\tau^2}-\zeta^2}\,d\zeta\,d\tau.$$

We change the variable $\tau \mapsto -\tau$ in the last integral and replace the integral over $\bar{\ell}$ by the integral over ℓ plus the difference of the integrals evaluated with the help of the residue theorem applied at $\zeta = \zeta_0$. Again replacing ℓ by ℓ_α, we obtain the following for odd n_1:

$$\left[\partial_x^{n_1}\partial_y^{n_2}\partial_z^{n_3} I_2\right]$$

$$= 4\,\mathrm{Im}\int_{-\infty}^{+\infty} (i\zeta_0)^{n_1}\left(\sqrt{\zeta_0^2+\tau^2}\right)^{n_2}(i\tau)^{n_3}\,\frac{\zeta_0\exp\left\{y\sqrt{\zeta_0^2+\tau^2}+iz\tau\right\}}{\nu^2-2\zeta_0^2}\,d\zeta\,d\tau.$$

It remains for us to change the variable $\tau \mapsto \nu t\sqrt{1+t^2}$ in the last integral in order to see that this integral and the jump at $x = 0$ of the corresponding derivative of I_1 defined by (6.7) have opposite signs.

6.1.3.2. Verification of the Laplace Equation and the Free Surface Boundary Condition

The considerations above make it easy to verify that for $y < 0$ we have

$$\nabla^2(I_1 + I_2) = 0, \tag{6.27}$$

$$\left(\partial_x^2 + \nu\partial_y\right)(I_1 + I_2) \\ = \frac{2\nu}{\pi} \operatorname{Re} \int_{-\infty}^{+\infty} \int_{\ell_\alpha} \exp\{y\sqrt{\zeta^2+\tau^2} + i\,|x|\,\zeta + iz\tau\}\,\mathrm{d}\zeta\,\mathrm{d}\tau. \tag{6.28}$$

Since the integrand in (6.28) has no singularities between ℓ_α and the half-axis $\zeta > 0$, we can replace ℓ_α by the latter half-axis. If $x < 0$, then we substitute the complex conjugate integral and change the variable $\tau \mapsto -\tau$. This leads to formula (6.28), where ℓ_α and $|x|$ are replaced by the half-axis $\zeta > 0$ and x, respectively. After that, one can apply the argument demonstrating the equivalence of (6.15) and (6.16), and so we have the following for $y < 0$:

$$\left(\partial_x^2 + \nu\partial_y\right)(I_1 + I_2) = \frac{\nu}{\pi} \int_{-\infty}^{+\infty} \int_{-\infty}^{+\infty} \exp\{y\sqrt{\zeta^2+\tau^2} + ix\zeta + iz\tau\}\,\mathrm{d}\zeta\,\mathrm{d}\tau. \tag{6.29}$$

For $y = y_0$ the right-hand side in (6.29) is the inverse Fourier transform of the right-hand-side term in the second equation of (6.12). It was shown in Subsection 6.1.2.1 that this transform is equal to the right-hand-side term in the second equation of (6.10), and so we get that for $y + y_0 < 0$,

$$\left(\partial_x^2 + \nu\partial_y\right)[I_1(x, y + y_0, z) + I_2(x, y + y_0, z)] \\ = -2\nu(y + y_0)[x^2 + (y + y_0)^2 + z^2]^{-3/2}. \tag{6.30}$$

Now we are in a position to show that G satisfies (6.1) and (6.2). Let $y_0 < 0$; then

$$\nabla^2 R^{-1} = -4\pi\,\delta_{(0,y_0,0)}(x, y, z), \qquad \nabla^2 R_0^{-1} = 0.$$

Therefore, the first equation of (6.1) follows from (6.8) and (6.27). The second equation of (6.1) is a direct consequence of (6.8) and (6.30). If $y_0 = 0$, then (6.8) implies that

$$G = I_2(x, y, z) + I_1(x, y, z),$$

and so the first equation of (6.2) coincides with (6.27) and the second one follows from (6.30) by virtue of a direct calculation using the theory of distributions.

6.1.3.3. Verification of Conditions at Infinity and Continuity of G as $y_0 \to 0$

In Subsection 6.1.4, we give an asymptotic expansion for I_2 at infinity. In particular, it implies that

$$|I_2(x, y + y_0, z)| < C/R \quad \text{for } R > 1,\ y + y_0 \leq 0. \tag{6.31}$$

From (6.7), it follows that

$$|I_1(x, y + y_0, z)| \leq 4\nu \int_{-\infty}^{+\infty} e^{\nu(y+y_0)(1+t^2)}\,dt = 4\sqrt{\nu\pi}\, e^{\nu(y+y_0)}\, |y + y_0|^{-1/2}. \tag{6.32}$$

Since I_1 contains $H(-x)$ as a factor, I_1 is equal to zero for $x > 0$, and so (6.3) follows from (6.8), (6.31), and (6.32). Thus, the fact that (6.8) solves (6.1)–(6.3) will be established after (6.31) is proved in the next subsection.

Since $I_1 + I_2$ is a smooth function, it follows from (6.8) that

$$G(x, y, z; y_0) \to G(x, y, z; 0) \quad \text{as } y_0 \to -0 \text{ when } y < 0.$$

Moreover, the convergence is uniform on any compact subset of $\mathbb{R}^3_-$ as well as the convergence of all derivatives.

6.1.4. Behavior of the Double Integral at Infinity

6.1.4.1. Statement of the Result

Here we are concerned with proving the following assertion.

Let $I_2(x, y, z)$ be given by (6.4) for $y < 0$, and let $I_2(x, 0, z)$ be the limit of $I_2(x, y, z)$ as $y \to -0$. Then the asymptotic expansion

$$I_2(x, y, z) \sim \sum_{j=1}^{\infty} \rho^{-j} f_j\left(\frac{x}{\rho}, \frac{y}{\rho}, \frac{z}{\rho}\right), \quad \text{where } f_1 = 2, \tag{6.33}$$

holds for $y \leq 0$ as $\rho = \sqrt{(x^2 + y^2 + z^2)} \to \infty$; f_j are infinitely smooth functions of their arguments.

Expansion (6.33) *is uniform; that is, for any* $N \geq 1$ *there exists* C_N *such that*

$$\left| I_2 - \sum_{j=1}^{N} \rho^{-j} f_j \right| \leq C_N \rho^{-(N+1)}, \quad \text{where } \rho > 1,\ y \leq 0. \qquad (6.34)$$

The asymptotic expansion (6.33) *is differentiable with respect to* x, y, *and* z *if* $x \neq 0$; *the expansions for derivatives are also uniform.*

Our proof is based on the following formula:

$$I_2(x, y, z) = \frac{2\nu}{\pi} \operatorname{Re} \int_{-\pi/2}^{\pi/2} \int_0^{+\infty} \frac{\exp\{S(\theta, \omega) k\rho + i\pi/4\}}{\nu f(\theta) - ik\cos^2\theta} \, dk d\theta, \qquad (6.35)$$

where $\omega = (\omega_1, \omega_2, \omega_3)$ is equal to $(|x|, y, z)/\rho$,

$$f(\theta) = \sqrt{i\cos^2\theta + \sin^2\theta}, \quad \operatorname{Re} f(\theta) \geq 0, \qquad (6.36)$$

and

$$S(\theta, \omega) = f(\theta)\omega_2 + \left(\frac{-1}{\sqrt{2}} + \frac{i}{\sqrt{2}} \right) \omega_1 \cos\theta + i\omega_3 \sin\theta. \qquad (6.37)$$

In order to obtain (6.35) one has to substitute $\alpha = \pi/4$ and $\zeta = \sigma e^{i\pi/4}$ into (6.4) and to use the polar coordinates (k, θ) in the plane (σ, τ).

Note that (6.22)–(6.24) imply that

$$\operatorname{Re} f(\theta) > \frac{1}{2}, \quad |\nu f(\theta) - ik\cos^2\theta| > \frac{\nu\sqrt{2}}{4}, \quad \operatorname{Re} S(\theta, \omega)\rho < \frac{y}{2} < 0. \qquad (6.38)$$

Let

$$Q = \frac{2\nu}{\pi} \left\{ \frac{1}{\nu f(\theta) - ik\cos^2\theta} - \frac{1}{\nu f(\theta)} \sum_{j=0}^{N} \left[\frac{ik\cos^2\theta}{\nu f(\theta)} \right]^j \right\}. \qquad (6.39)$$

Then (6.35) takes the following form:

$$I_2 = \sum_{j=1}^{N+1} J_j + \tilde{J}, \qquad (6.40)$$

where

$$J_j = \operatorname{Re} \int_{-\pi/2}^{+\pi/2} \int_0^{+\infty} \frac{2}{\pi f(\theta)} \left[\frac{ik\cos^2\theta}{\nu f(\theta)} \right]^{j-1} e^{S(\theta,\omega)k\rho + i\pi/4} \, dk d\theta, \qquad (6.41)$$

$$\tilde{J} = \operatorname{Re} \int_{-\pi/2}^{+\pi/2} \int_0^{+\infty} Q(\theta, \omega) e^{S(\theta,\omega)k\rho + i\pi/4} \, dk d\theta. \qquad (6.42)$$

Our aim is to show that J_j, where $j = 1, 2, \ldots$, give the terms in (6.33) and $\tilde{J}$ corresponds to the remainder.

6.1.4.2. Estimate for the Remainder

Let us demonstrate that $\tilde{J}$ *does not contribute* to the sum in (6.34). For achieving this we have to estimate $\tilde{J}$ in a specific way in different subregions of $\mathbb{R}^3_-$. In order to single out these subregions we fix $\varepsilon_0 > 0$ such that

$$|S(\theta, \omega)| > 1/10 \quad \text{when } |\sin\theta| \geq 1/3, \tag{6.43}$$

$$|S_\theta(\theta, \omega)| > 1/10 \quad \text{when } |\sin\theta| \leq 2/3 \tag{6.44}$$

for $|\omega_1| + |\omega_2| < \varepsilon_0$. The existence of ε_0 follows from the continuity of $|S|$ and $|S_\theta|$. We also have to take into account the fact that these functions tend to $|\sin\theta|$ and $|\cos\theta|$ respectively as $|\omega_1| + |\omega_2| \to 0$.

First, we assume that $|\omega_1| + |\omega_2| \geq \varepsilon_0$; then we integrate by parts with respect to k in (6.42) and do this $N + 1$ times. Since (6.39) and (6.38) imply that Q and $\partial_k^j Q, 1 \leq j \leq N$, do vanish for $k = 0$ and that the integrand decays exponentially as $k \to \infty$, we have the following after integration by parts:

$$\tilde{J} = \operatorname{Re} \int_{-\pi/2}^{+\pi/2} \int_0^{+\infty} [-\rho S(\theta, \omega)]^{-(N+1)} \frac{\partial^{N+1} Q}{\partial k^{N+1}} e^{S(\theta,\omega)k\rho + i\pi/4} \, dk d\theta. \tag{6.45}$$

Noting that the $(N + 1)$th derivative with respect to k vanishes for the second term on the right-hand side in (6.39), we write (6.45) in the following form:

$$\tilde{J} = C \operatorname{Re} \int_{-\pi/2}^{+\pi/2} \int_0^{+\infty} \left(\frac{i \cos^2\theta}{\rho S} \right)^{N+1} (\nu f - ik\cos^2\theta)^{-N-2} e^{Sk\rho + i\pi/4} \, dk d\theta. \tag{6.46}$$

Without loss of generality one may assume that $N \geq 3$. Then from (6.46) and (6.38) it follows that

$$|\tilde{J}| \leq C\rho^{-(N+1)} \int_{-\pi/2}^{+\pi/2} \int_0^{+\infty} \left| \frac{\cos\theta}{S} \right|^{N+1} \left| \frac{\cos^2\theta}{\nu f - ik\cos^2\theta} \right|^2 dk d\theta. \tag{6.47}$$

For estimating the integrand in (6.47), we note that $\omega_1 \geq 0$ and $\omega_2 \leq 0$, and so (6.37) and the first inequality (6.38) produce

$$\operatorname{Re} S(\theta, \omega) \leq -\left(\frac{1}{2} |\omega_2| + \frac{1}{\sqrt{2}} \omega_1 \cos\theta \right) \leq -\frac{1}{2} (|\omega_2| + \omega_1 \cos\theta)$$

for $|\theta| \le \pi/2$. Thus,

$$\left|\frac{S(\theta,\omega)}{\cos\theta}\right| \ge \frac{1}{2}\left(\frac{|\omega_2|}{\cos\theta}+\omega_1\right) \ge \frac{1}{2}(|\omega_2|+\omega_1) \ge \frac{\varepsilon_0}{2}. \tag{6.48}$$

Further, the second relation of (6.38) implies that

$$\left|\frac{\cos^2\theta}{\nu f(\theta)-ik\cos^2\theta}\right| = \frac{1}{k}\left|\frac{ik\cos^2\theta-\nu f+\nu f}{\nu f(\theta)-ik\cos^2\theta}\right|$$

$$= \frac{1}{k}\left|\frac{\nu f}{\nu f(\theta)-ik\cos^2\theta}-1\right| \le \frac{C}{k}.$$

Combining this with the first and second inequalities (6.38), we obtain that

$$\left|\frac{\cos^2\theta}{\nu f(\theta)-ik\cos^2\theta}\right| \le \frac{C}{1+k}. \tag{6.49}$$

From (6.47), (6.48), and (6.49) it follows that

$$|\tilde{J}| < C\rho^{-N-1} \quad \text{when } |\omega_1|+|\omega_2| \ge \varepsilon_0. \tag{6.50}$$

In order to estimate $\tilde{J}$ when $|\omega_1|+|\omega_2| \le \varepsilon_0$, let us introduce an infinitely differentiable cutoff function $\eta = \eta(\theta)$ such that $\eta = 1$ if $|\sin\theta| \le 1/3$ and $\eta = 0$ if $|\sin\theta| \ge 2/3$. Then we represent $\tilde{J}$ in the form $\tilde{J} = \tilde{J}_1 + \tilde{J}_2$, where $\tilde{J}_j$, $j = 1, 2$, are given by (6.42) having in the integrand the additional factors $\eta(\theta)$ and $1-\eta(\theta)$, respectively. Since (6.43) is valid on the support of $1-\eta$, $\tilde{J}_2$ can be estimated in the same way that $\tilde{J}$ was estimated above, and similarly to (6.50) we get

$$|\tilde{J}_2| \le C\rho^{-N-1} \quad \text{when } |\omega_1|+|\omega_2| \le \varepsilon_0. \tag{6.51}$$

Now we introduce another infinitely differentiable cutoff function $\xi = \xi(k)$ such that $\xi = 1$ for $k < 1$ and $\xi = 0$ for $k > 2$. Again, we represent $\tilde{J}_1$ in the form $\tilde{J}_{1,1} + \tilde{J}_{1,2}$, where $\tilde{J}_{1,1}$ and $\tilde{J}_{1,2}$ are given by (6.42) having in the integrand the additional factors $\eta(\theta)\xi(k)$ and $\eta(\theta)[1-\xi(k)]$, respectively. Integrating by parts with respect to θ, we have

$$\tilde{J}_{1,1} = \operatorname{Re}\int_{-\pi/2}^{+\pi/2}\int_0^{+\infty} \xi(k)\frac{\eta Q(\theta,\omega)}{S_\theta(\theta,\omega)}S_\theta(\theta,\omega)e^{Sk\rho+i\pi/4}\,\mathrm{d}k\mathrm{d}\theta$$

$$= \operatorname{Re}\int_{-\pi/2}^{+\pi/2}\int_0^{+\infty} \frac{\xi(k)}{-k\rho}\left(\frac{\eta Q}{S_\theta}\right)_\theta e^{Sk\rho+i\pi/4}\,\mathrm{d}k\mathrm{d}\theta,$$

because $\eta = 0$ when $\theta = \pm\pi/2$. Repeating this procedure $N+1$ times, we obtain the integral whose integrand does not exceed $C\rho^{-N-1}$ for $k \le 2$, because (6.44) estimates S_θ on the support of η and Q has zero of order $N+1$

at $k = 0$ [see (6.39)]. Since $\xi(k) = 0$ for $k > 2$, it follows that

$$|\tilde{J}_{1,1}| \leq C\rho^{-N-1} \quad \text{when } |\omega_1| + |\omega_2| \leq \varepsilon_0. \tag{6.52}$$

A similar argument is applicable to $\tilde{J}_{1,2}$, but one has to integrate by parts $N+2$ times. Since Q and its derivatives with respect to θ do not exceed Ck^N for $k > 1$ [see (6.39)], we get the integral whose integrand does not exceed $C\rho^{-N-2}k^{-2}$ for $k > 1$ and contains the factor $1 - \xi(k)$ vanishing for $k < 1$. Hence

$$|\tilde{J}_{1,2}| \leq C\rho^{-N-2} \quad \text{when } |\omega_1| + |\omega_2| \leq \varepsilon_0.$$

Combining this with (6.50)–(6.52), we arrive at the final estimate

$$|\tilde{J}| \leq C\rho^{-N-1}, \quad \text{where } \rho > 1,\ y < 0. \tag{6.53}$$

Note that the integrand in (6.39) depends smoothly on y, and therefore the estimates above justify the existence of the limit of $\tilde{J}$ as $y \to -0$ and the validity of (6.53) for that limit value.

6.1.4.3. Properties of Coefficients in the Asymptotic Expansion (6.33)

Our aim is to prove the following assertion.

Functions (6.41) *given for* $\{\rho > 0, y < 0\}$ *have infinitely differentiable extensions into* $\{\rho > 0, y \leq 0\}$. *These functions are homogeneous of the order* $-j$.

Since integral (6.41) converges in $\{\rho > 0, y < 0\}$ as a result of (6.38), the homogeneity of J_j follows after the change of variables $k \mapsto \rho^{-1}k_1$ in (6.41). In order to prove that J_j are smooth functions, we write (6.41) in the form

$$J_j = \int_{-\infty}^{+\infty}\int_{-\infty}^{+\infty} \frac{1}{\pi\sqrt{\zeta^2+\tau^2}} \left(\frac{\zeta^2}{\nu\sqrt{\zeta^2+\tau^2}}\right)^{j-1} \times \exp\{y\sqrt{\zeta^2+\tau^2} + ix\zeta + iz\tau\}\, d\zeta\, d\tau, \tag{6.54}$$

which is equivalent to (6.41). This can be verified in the same way as the equivalence of (6.28) and (6.29).

Let us prove the smoothness of functions (6.54) by induction. For J_1, this is a consequence of (6.13), which implies that $J_1 = 2\rho^{-1}$ and justifies the leading term in (6.33). Since (6.54) implies that

$$\nu\partial_y J_{j+1} = -\partial_x^2 J_j \quad \text{for } j \geq 1 \tag{6.55}$$

and J_j are homogeneous functions, then

$$J_{j+1} = -\frac{1}{\nu}\int_{-\infty}^{y} \partial_x^2 J_j \, \mathrm{d}y.$$

Combining this with the homogeneity of J_j, we get that J_{j+1} is an infinitely smooth function in $\{\rho \neq 0,\ y \leq 0\}$, if J_j has this property. Since the smoothness of J_1 was proved, all functions J_j are infinitely smooth.

We note that (6.34) holds because (6.40) can be written as

$$I_2 = \sum_{j=1}^{N} J_j + (J_{N+1} + \tilde{J}), \tag{6.56}$$

and the last two terms can be estimated by virtue of (6.53). Thus the first part of the theorem formulated in Subsection 6.1.4.1 is proved. In Subsection 6.1.3, it was demonstrated that if $y < 0$ and $x \neq 0$, then (6.35) has infinitely many derivatives with respect to x, y, and z; for their calculation it suffices to differentiate the integrand. One can verify that the asymptotic expansions for derivatives of I_2 can be obtained in the same way as those for I_2. Moreover, these expansions coincide with the corresponding derivatives of (6.33).

6.1.5. Uniqueness of Green's Function

For demonstrating the uniqueness of Green's function, it is sufficient to prove the following assertion.

The following boundary value problem,

$$\nabla^2 u = 0 \quad \text{in } \mathbb{R}^3_-, \tag{6.57}$$

$$u_{xx} + \nu u_y = 0 \quad \text{when } y = 0, \tag{6.58}$$

$$\sup_{x,z} |u(x, y, z)| \to 0 \quad \text{as } y \to -\infty,$$

$$\int_{-\infty}^{+\infty} |u(x, 0, z)|^2 \, \mathrm{d}z \to 0 \quad \text{as } x \to +\infty \tag{6.59}$$

has only a trivial solution.

Applying the Fourier transform with respect to x and z, we see that $\hat{u}(\zeta, y, \tau)$ must satisfy

$$\hat{u}_{yy} - (\zeta^2 + \tau^2)\hat{u} = 0 \quad \text{for } y < 0,$$

and so

$$\hat{u} = C_1(\zeta, \tau) e^{y\sqrt{\zeta^2+\tau^2}} + C_2(\zeta, \tau) e^{-y\sqrt{\zeta^2+\tau^2}} \quad \text{for } y \leq 0, \tag{6.60}$$

where C_1 and C_2 are distributions. Let us show that no term, increasing as y decreases, appears on the right-hand side in (6.60); that is, $C_2 = 0$. From (6.59), it follows that there exists $h < 0$ such that $|u| \leq 1$ for $y \leq h$. Let $\varphi = \varphi(x, z) = u(x, h, z)$. A bounded solution of the Laplace equation in the half-space $y \leq h$ is uniquely determined by the Dirichlet data, that is, by φ, and this solution has the form of convolution with respect to x and z of φ and $\phi = (2\pi)^{-1}(\partial/\partial y)[x^2 + (y-h)^2 + z^2]^{-1/2}$. Hence $\hat{u}$ is the product of the Fourier transforms of φ and ϕ. Since $\hat{\phi}$ can be found by using (6.13), we have

$$\hat{u} = (2\pi)^{-2}\hat{\varphi}(\zeta, \tau)e^{(y-h)\sqrt{\zeta^2+\tau^2}} \quad \text{for } y \leq h,$$

and so

$$\hat{u}_y - \sqrt{\zeta^2 + \tau^2}\hat{u} = 0 \quad \text{for } y \leq h. \tag{6.61}$$

Substituting (6.60) into (6.61), we get

$$\sqrt{\zeta^2 + \tau^2}e^{-y\sqrt{\zeta^2+\tau^2}}C_2(\zeta, \tau) = 0 \quad \text{for } y \leq h,$$

which implies that $C_2(\zeta, \tau) = C\delta_{(0,0)}(\zeta, \tau)$, where C is a constant. Then we can add $-C\delta_{(0,0)}(\zeta, \tau)$ to the coefficient C_1 in (6.60) and put $C_2 = 0$. Finally, (6.60) takes the form

$$\hat{u} = C_1(\zeta, \tau)e^{y\sqrt{\zeta^2+\tau^2}} \quad \text{for } y \leq 0. \tag{6.62}$$

From (6.58), it follows that

$$-\zeta^2\hat{u} + \nu\hat{u}_y = 0 \quad \text{when } y = 0.$$

Combining this with (6.62), we obtain the following equation:

$$(-\zeta^2 + \nu\sqrt{\zeta^2 + \tau^2})C_1(\zeta, \tau) = 0 \tag{6.63}$$

for determining $C_1(\zeta, \tau)$. Finding ζ from the equation $\zeta^2 = \nu\sqrt{\zeta^2 + \tau^2}$, we see that the set of real roots consists of the origin and two smooth curves $l_\pm$ given as follows:

$$\zeta = \pm\zeta(\tau), \quad \text{where } \zeta(\tau) = \nu\left[\frac{1 + \sqrt{1 + 4\nu^{-2}\tau^2}}{2}\right]^{1/2}.$$

Then (6.63) shows that $C_1(\zeta, \tau)$ is a linear combination of Dirac's measures placed at the origin and on $l_\pm$. More precisely, we write

$$C_1(\zeta, \tau) = \alpha\delta_{(0,0)}(\zeta, \tau) + \mu_+(\tau)\delta_{-\zeta(\tau)}(\zeta) + \mu_-(\tau)\delta_{+\zeta(\tau)}(\zeta), \tag{6.64}$$

where α is a constant and $\mu_\pm(\tau)$ are distributions on $l_\pm$.

Let $v(x,\tau)$ denote the Fourier transform of $u(x,0,z)$ with respect to z. From (6.62) and (6.64) we have that

$$v(x,\tau) = \frac{1}{\sqrt{2\pi}}\left[\alpha\delta(\tau) + \mu_+(\tau)e^{i\zeta(\tau)x} + \mu_-(\tau)e^{-i\zeta(\tau)x}\right], \tag{6.65}$$

and the second condition of (6.59) implies

$$\int_{-\infty}^{+\infty} |v(x,\tau)|^2\,\mathrm{d}\tau \to 0 \quad \text{as } x \to +\infty. \tag{6.66}$$

Let us demonstrate that

$$p = p(x,\tau) = \mu_+(\tau)e^{i\zeta(\tau)x} + \mu_-(\tau)e^{-i\zeta(\tau)x} \tag{6.67}$$

is equal to zero when $\tau \neq 0$. Let us fix an arbitrary $b > 0$; then (6.65) and (6.66) give that

$$\|p\|_{L^2(b^{-1}<|\tau|<b)} \to 0 \quad \text{as } x \to +\infty.$$

Hence for any $\varepsilon > 0$ there is T such that for any $x \geq T$ we have

$$\|p\|_{L^2(b^{-1}<|\tau|<b)} < \varepsilon \sin(\nu/M), \quad \text{where } M = \max_{|\tau|\leq b} \zeta(\tau). \tag{6.68}$$

Let us put $x = T$ and $x = T + M^{-1}$ in (6.67) and consider the resulting two equations as a system for $\mu_\pm$. The matrix A of the system has the form

$$A = \begin{bmatrix} \exp\{i\zeta(\tau)T\} & \exp\{-i\zeta(\tau)T\} \\ \exp\{i\zeta(\tau)(T+M^{-1})\} & \exp\{-i\zeta(\tau)(T+M^{-1})\} \end{bmatrix}.$$

The determinant D of this system is equal to $-\sin[\zeta(\tau)M^{-1}]$, and so $|D| \geq \sin(\nu/M)$ for $|\tau| \leq b$ because $\nu \leq \zeta(\tau) \leq M$ for such values of $|\tau|$. Therefore, the elements of the inverse matrix A^{-1} do not exceed $\sin^{-1}(\nu/M)$. Hence,

$$|\mu_\pm(\tau)| \leq \sin^{-1}(\nu/M)\{|p(T,\tau)| + |p(T+M^{-1},\tau)|\} \quad \text{for } |\tau| \leq b.$$

Combining this with (6.68), we obtain

$$\|\mu_\pm\|_{L^2(b^{-1}<|\tau|<b)} \leq 2\varepsilon.$$

Since b and ε are arbitrary, this means that $\mu_\pm(\tau) = 0$ for $\tau \neq 0$. Now (6.65) shows that v is a linear combination of $\delta(\tau)$ and its derivatives. The coefficients of this linear combination are polynomials of x. Then (6.66) implies that these coefficients are zeros, and so $\alpha = 0$ and $\mu_\pm(\tau) = 0$. Now we get from (6.64) that $C_1 = 0$, and this leads to $u = 0$ in view of (6.62), which completes the proof.

6.2. Far-Field Behavior of the Three-Dimensional Green's Function

In this section we study the asymptotic behavior of waves generated by the Kelvin wave source at infinity. For this purpose we use the velocity potential G obtained in the previous section and expressed by virtue of double and single integrals. It was demonstrated that the contribution of the double integral I_2 is negligibly small at infinity (see Subsection 6.1.4), and so waves are described by the single integral I_1 defined by (6.7) in Subsection 6.1.1.2. It is well known that Kelvin's source generates two systems of waves confined to Kelvin's angle between two vertical half-planes $\varphi = \pi \pm \arcsin 1/3$, where φ is the polar angle on the free surface in the coordinate system moving together with the source point. Amplitudes and frequencies of these waves depend on the angle formed with the direction of the forward motion. The frequencies of these two systems of waves coincide on the boundary of Kelvin's angle, and the velocity potential is described in terms of the Airy function there (see Ursell [326]).

Furthermore, Ursell noted in [326] and [328] that the far-field behavior of these two systems of waves is non-uniform in the neighborhood of a source's track. In fact, the amplitude of one of the wave systems and of the remainder tend to infinity as $\varphi \to \pi$ and $y = 0$ (that is, when both the source and observation point are on the free surface and the observation point approaches the source's track). At the same time, for $\varphi = \pi$ and $y < 0$ (that is, when the observation point is placed strictly below the source's track), one obtains that the amplitude of the same wave system vanishes and the remainder is bounded and decays at infinity. Ursell's description (see [326]) of the far-field behavior is based on the steepest descent method. It occurs that within Kelvin's angle there are two saddle points and each of them is responsible for its own wave system. When two saddle points merge the standard method fails to provide the result, and for describing asymptotic behavior near the boundary of Kelvin's angle uniformly with respect to φ the Airy function is required. When the observation point approaches the source's track, one of the saddle points goes to infinity, and so for obtaining a uniform asymptotic formula it is essential to determine the contribution of this infinitely remote saddle point. This requires the introduction of a special wave localized near the source's track. Such a wave was introduced in the work [229] by Maz'ya and Vainberg, and combining it with the two systems described above, one obtains a uniform far-field asymptotic representation with a remainder that is shown to be bounded and to decay uniformly at infinity.

Green's function of the Neumann–Kelvin problem has also a singularity on the source's track when the source is placed on the free surface (see Ursell's paper [339], where this singularity is described in the case when the distance between the observation point and the source is finite). Since we are interested in the asymptotic behavior in the far field, the results in the present section and those in [339] do not follow from each other but are complementary.

6.2.1. Formulation of the Results

For the sake of brevity we assume the parameter ν to be equal to one throughout the present section, but this does not lead to loss of generality. It fact, let $G^{(\nu)}$ denote Green's function in the case of arbitrary $\nu > 0$, and G we will use for Green's function of the Neumann–Kelvin problem in which $\nu = 1$, and so $G = G^{(1)}$. Then one immediately finds that

$$G^{(\nu)}(x, y, z; y_0) = \nu G(\nu x, \nu y, \nu z; \nu y_0),$$

which allows us to reformulate the asymptotic results obtained for G and its derivatives in terms of $G^{(\nu)}$ and the corresponding derivatives.

According to the first assertion in Subsection 6.1.1.2, we have the following for $y + y_0 < 0$:

$$G(x, y, z; y_0) = R^{-1} - R_0^{-1} + I_2(x, y + y_0, z) + I_1(x, y + y_0, z), \quad (6.69)$$

where

$$R = \sqrt{x^2 + (y - y_0)^2 + z^2}, \quad R_0 = \sqrt{x^2 + (y + y_0)^2 + z^2},$$

and

$$I_1(x, y, z) = -4H(-x)\mathrm{Im} \int_{-\infty}^{\infty} \exp\{y(1 + t^2) + i(x + tz)\sqrt{1 + t^2}\}\, dt. \quad (6.70)$$

Here $H(-x) = 1$ for $x < 0$ and $H(-x) = 0$ for $x \geq 0$. The double integral I_2 is given by (6.4) in Subsection 6.1.1.2, but the contribution of I_2 to the wave pattern is negligible (it was demonstrated in Subsection 6.1.4 that I_2 decays at infinity sufficiently fast), and so we are going to investigate only the wave integral I_1 here.

First, we recall that both I_1 and I_2 do not define smooth functions for $y + y_0 < 0$ and $x = 0$, but $I_1 + I_2$ is an infinitely differentiable function (see Subsection 6.1.3.1), and so $G - R^{-1}$ is infinitely differentiable when $y + y_0 < 0$. Furthermore, (6.69) is meaningless when $y + y_0 = 0$ because I_1 defined by (6.70) is a divergent integral if $y = 0$. Hence, for $y + y_0 = 0$, Green's function must be understood as the limit value of $G(x, y, z; y_0)$ as $y + y_0 \to -0$, and this is possible as follows from the first and second theorems formulated below. In fact, these theorems imply that $G - R^{-1}$ is a smooth function throughout the region, where $y + y_0 \le 0$, from which a half-line

$$l = \{y + y_0 = 0,\ \varphi = \pi\}$$

(that is, source's track) is removed.

Our main aim is to obtain an asymptotic expansion, which is uniform with respect to φ and y, for the wave integral

$$I_1(x, y + y_0, z) \quad \text{as } r = \sqrt{x^2 + z^2} \to \infty.$$

For this purpose we introduce functions describing frequencies and amplitudes of propagating waves. Let

$$\varphi_0 = \arctan \sqrt{2}/4 = \arcsin 1/3.$$

Then Kelvin's angle is defined by $|\varphi - \pi| < \varphi_0$, and for φ within this angle we introduce

$$t_\pm(\varphi) = -\frac{\cot\varphi}{4}\left(1 \pm \sqrt{1 - 8\tan^2\varphi}\right), \tag{6.71}$$

$$\tau_\pm(\varphi) = t_\pm^2(\varphi) + 1, \tag{6.72}$$

$$S_\pm(\varphi) = \sqrt{\tau_\pm(\varphi)}\,(\cos\varphi + t_\pm(\varphi)\sin\varphi)\,, \tag{6.73}$$

$$a_\pm(\varphi) = -4\sqrt{2\pi}\left[\frac{\tau_\pm(\varphi)}{1 - 9\sin^2\varphi}\right]^{1/4}. \tag{6.74}$$

It is obvious that t_-, τ_-, S_-, and a_- are infinitely differentiable functions when $|\varphi - \pi| < \varphi_0$. In contrast, t_+, τ_+, S_+, and a_+ are singular at $\varphi = \pi$, and these singularities can be characterized by noting that

$$t(\varphi) = t_+(\varphi)\sin\varphi, \qquad \tau(\varphi) = \tau_+(\varphi)\sin^2\varphi,$$
$$S(\varphi) = S_+(\varphi)\,|\sin\varphi|\,, \quad a(\varphi) = a_+(\varphi)\,|\sin\varphi|^{1/2} \tag{6.75}$$

are infinitely differentiable functions such that

$$t(\pi) = 1/2, \quad \tau(\pi) = 1/4, \quad S(\pi) = -1/4, \quad a(\pi) = -4\sqrt{\pi}.$$

Setting

$$A(\varphi) = \frac{S_+(\varphi) + S_-(\varphi)}{2}, \qquad B(\varphi) = \left\{\frac{3[S_+(\varphi) - S_-(\varphi)]}{4}\right\}^{2/3}, \tag{6.76}$$

we see that these functions defined for $|\varphi - \pi| < \varphi_0$ can be extended analytically to a certain neighborhood of $\varphi = \pi \pm \varphi_0$. This follows directly from (6.71)–(6.73), or it can be obtained from general results (see the second theorem formulated below in Subsection 6.2.3). When A and B are considered for $|\varphi - \pi| > \varphi_0$, they are understood as such extensions.

By v we denote the Airy function

$$v(t) = \frac{1}{2\sqrt{\pi}} \int_{-\infty}^{+\infty} \exp\left\{i\left(tx + \frac{x^3}{3}\right)\right\} dx, \tag{6.77}$$

and $O(R_0^{-\beta})$ denotes any function $u = u(x, y + y_0, z)$ defined in the half-space $y + y_0 \leq 0$ and such that

$$|u| \leq C R_0^{-\beta} \quad \text{for } R_0 \geq 1,$$

where C is a constant that does not depend on x, $y + y_0$, and z.

Let us formulate and discuss the main results of the present section.

Let $\delta > 0$ be sufficiently small and $\varepsilon > 0$. Then for any $y_0 \leq 0$ the following asymptotic formulae hold for G as $R_0 \to \infty$:

1. *If $|y + y_0| \geq \varepsilon r$ or $|\varphi - \pi| > \varphi_0 + \varepsilon$, then*

$$G = R^{-1} + R_0^{-1} + O(R_0^{-2}). \tag{6.78}$$

2. *If $|y + y_0| < r$ and $|\varphi - \pi| < \varphi_0 - \varepsilon$, then*

$$\begin{aligned} G &= R^{-1} + R_0^{-1} \\ &\quad + r^{-1/2} \sum_{\pm} a_\pm(\varphi) \exp\{(y + y_0)\tau_\pm(\varphi)\} \sin\left(S_\pm(\varphi) r \pm \frac{\pi}{4}\right) \\ &\quad + O(R_0^{-3/2}). \end{aligned} \tag{6.79}$$

3. *If $|y + y_0| < r$ and $|\varphi - \pi \pm \varphi_0| < \delta$, then*

$$\begin{aligned} G &= R^{-1} + R_0^{-1} \\ &\quad + 4\sqrt{2\pi} \left[\frac{B(\varphi)}{1 - 9\sin^2\varphi}\right]^{1/4} \operatorname{Re} \sum_{\pm} [\tau_\pm(\varphi)]^{1/4} \\ &\quad \times \exp\{(y + y_0)\tau_\pm(\varphi) + i A(\varphi) r\} \\ &\quad \times \left\{ i r^{-1/3} v\left[-B(\varphi) r^{2/3}\right] \pm r^{-2/3} \frac{v'\left[-B(\varphi) r^{2/3}\right]}{\sqrt{B(\varphi)}} \right\} + O(R_0^{-4/3}). \end{aligned} \tag{6.80}$$

Here $\tau_\pm$, $S_\pm$, $a_\pm$, A, and B are defined by (6.71)–(6.74) *and* (6.76).

The meaning of (6.80) should be explained. Since for $|\varphi - \pi| < \varphi_0$ we have that $B(\varphi) > 0$ and $\tau_\pm(\varphi) > 0$, the right-hand side in (6.80) is defined uniquely for $|\varphi - \pi| < \varphi_0$. If $|\varphi - \pi| \geq \varphi_0$, then the right-hand side in (6.80) is understood as an analytic extension with respect to φ. The existence of such an extension follows from three facts: (i) B is an analytic function of φ in the neighborhood of $\varphi = \pi \pm \varphi_0$; (ii) B has first-order zeros at these points; and

(iii) $\tau_\pm(\pi \pm \phi_0) = 3/2$ and $u(\lambda) = \lambda^{1/4} \exp\{(y + y_0)\lambda\}$ is an analytic function in the neighborhood of $\lambda = 3/2$. Therefore,

$$u[\tau_+(\varphi)] + u[\tau_-(\varphi)], \quad \{u[\tau_+(\varphi)] - u[\tau_-(\varphi)]\}/\sqrt{B(\varphi)}$$

are analytic functions of ϕ in a neighborhood of $\varphi = \pi \pm \varphi_0$. All above facts follow from formulae (6.71)–(6.73) and (6.76) or can be obtained from general considerations (see the second proposition in Subsection 6.2.3).

Let us discuss the theorem formulated above. First, it means that

$$G_1 = G - R^{-1} - R_0^{-1}$$

decreases rapidly with the depth (more precisely, as $y + y_0 \to -\infty$), and in the half-space upstream, that is, as $R_0 \to \infty$ for $x > 0$; see (6.78). Second, in the half-space downstream (for $x < 0$), waves are mainly localized within a dihedral angle (Kelvin's angle) formed by vertical planes intersecting the free surface along $\varphi = \pi \pm \varphi_0$. Outside of this angle, G_1 decreases rapidly with the distance from the source; see (6.78). Strictly inside Kelvin's angle, there are two wave systems whose frequencies depend on φ, and these waves decay rapidly with the depth; see (6.79). The frequencies of waves are $S_+(\varphi)/2\pi$ and $S_-(\varphi)/2\pi$, and they coincide at the boundary of Kelvin's angle.

Since S_+, a_+, and τ_+ are singular at $\varphi = \pi$, the amplitude and frequency of the corresponding wave tend to infinity as $\varphi \to \pi$ when $y + y_0 = 0$, but the amplitude of this wave vanishes for $\varphi = \pi$ and $y + y_0 < 0$, as was pointed out by Ursell as early as in 1960 (see [326]). The singularity of the wave under consideration in a neighborhood of the ray $l = \{y + y_0 = 0, \varphi = \pi\}$ (that is, in a neighborhood of the source's track) is characterized by (6.79) and by the fact that (6.75) determines smooth functions. From (6.79) and (6.75) one obtains that the wave's form in a neighborhood of l is as follows:

$$\frac{a(\varphi)}{\sqrt{|\sin\varphi|}} \exp\left\{\frac{(y + y_0)\tau(\varphi)}{\sin^2\varphi}\right\} \sin\left(\frac{S(\varphi)}{|\sin\varphi|} r + \frac{\pi}{4}\right),$$

where a, τ, and S are infinitely differentiable functions, and

$$a(\pi) = -4\sqrt{\pi}, \quad \tau(\pi) = -S(\pi) = 1/4.$$

We emphasize that the remainder in (6.79) decreases uniformly in all directions at infinity.

Formula (6.79) does not hold in a neighborhood of two half-planes $\varphi = \pi \pm \varphi_0$, and so the Airy function and its derivative are used in the asymptotics of G there [see (6.80)].

Formulae expressing the asymptotic behavior of G can be differentiated everywhere excluding a neighborhood of l because the derivatives of the remainder term are singular on l. However, these singularities are more weak than the singularities of the derivatives of terms in the asymptotic expansion. In the next two assertions, we give asymptotic expansions of ∇G and $G_x(x, 0, z; y_0)$; the latter is equal (up to a constant factor) to the elevation of the free surface. These asymptotic expansions for the region inside Kelvin's angle contain not only the derivatives of corresponding terms in the asymptotic expansion of G, but also additional terms. Outside a neighborhood of the source's track, these additional terms decay rapidly at infinity, and so they correspond to a wave propagating to infinity along the track. The remainder terms in these expansions are continuous and decay at infinity uniformly with respect to all variables.

In order to formulate the next theorem we need three functions defined for $\xi \leq 0$:

$$D_1(\xi) = \left. \frac{\mathrm{d}^2\left(\sigma e^{\xi\sigma^2}\right)}{\mathrm{d}\sigma^2} \right|_{\sigma=1}, \tag{6.81}$$

$$D_2(\xi) = \left. \frac{\mathrm{d}^2\left(\sigma^2 e^{\xi\sigma^2}\right)}{\mathrm{d}\sigma^2} \right|_{\sigma=1},$$

$$D_3(\xi) = \left. \frac{\mathrm{d}^4\left(\sigma e^{\xi\sigma^2}\right)}{\mathrm{d}\sigma^4} \right|_{\sigma=1}. \tag{6.82}$$

It is obvious that these functions are bounded and decay exponentially as $\xi \to -\infty$.

For any $y_0 \leq 0$ and $\varepsilon > 0$, we have the following as $R_0 \to \infty$:

1. *If $|y + y_0| \geq \varepsilon r$ or $|\varphi - \pi| > \varphi_0 + \varepsilon$, then*

$$\nabla G = \nabla\left(R^{-1} + R_0^{-1}\right) + O\left(R_0^{-3}\right). \tag{6.83}$$

2. *If $|y + y_0| < r$ and $|\varphi - \pi| < \varphi_0 - \varepsilon$, then*

$$\begin{aligned}
\nabla G = \nabla R^{-1} & \\
& + \operatorname{Im}\Bigg\{ r^{-1/2} \sum_{\pm} H_{\pm}(\varphi) a_{\pm}(\varphi) \\
& \times \exp\left\{ (y + y_0)\, \tau_{\pm}(\varphi) + i S_{\pm}(\varphi) r \pm \frac{i\pi}{4} \right\} \\
& + \frac{\sqrt{\pi}}{|z|^{3/2}} H(x, y + y_0, z) \\
& \times \exp\left\{ y + y_0 + i S_{+}(\varphi) r + \frac{i\pi}{4} \right\} \Bigg\} + O\left(R_0^{-3/2}\right).
\end{aligned}$$

3. *If $|y+y_0| < r$, $|\varphi - \pi \pm \varphi_0| < \delta$, and δ is sufficiently small, then*

$$\nabla G = \nabla R^{-1} + 4\sqrt{2\pi}\left[\frac{B(\varphi)}{1-9\sin^2\varphi}\right]^{1/4} \operatorname{Re}\sum_{\pm} H_{\pm}(\varphi) q_{\pm} + O\left(R_0^{-4/3}\right),$$

where by $q_{\pm}$ we denote the expression under the sign $\sum_{\pm}$ in (6.80).

In 2 and 3, $t_{\pm}$, $\tau_{\pm}$, $S_{\pm}$, and $a_{\pm}$ are given by (6.71)–(6.74), and we use the following vector functions:

$$H_{\pm}(\varphi) = [i\sqrt{\tau_{\pm}(\varphi)}, \tau_{\pm}(\varphi), it_{\pm}(\varphi)\sqrt{\tau_{\pm}(\varphi)}],$$

$$H(x, y+y_0, z) = \left(\frac{2|z|}{r} D_1, -iD_2 + \frac{|z|}{2r^2} D_3, D_2 \operatorname{sign} z + \frac{iz}{2r^2} D_3\right),$$

where $D_j = D_j((y+y_0)t_{\mp}^2(\varphi))$ and $D_j (j = 1, 2, 3)$ are defined in (6.81) and (6.82).

It is easy to verify that

$$H_{\pm}(\varphi)\exp\{(y+y_0)\,\tau_{\pm}(\varphi) + iS_{\pm}(\varphi)r\} = \nabla \exp\{(y+y_0)\,\tau_{\pm}(\varphi) + iS_{\pm}(\varphi)r\},$$

and that for $|z| > \varepsilon r$, the term containing $H(x, y+y_0, z)$ (see 2) has the same order as the remainder.

The asymptotic behavior of $G_x(x, 0, z; y_0)$, which is equal (up to a constant factor) to the elevation of the free surface, is an immediate corollary of the last theorem. It is convenient to put down the corresponding formulae, assuming that $\rho = \sqrt{x^2 + y_0^2 + z^2} \to \infty$, and so the source's depth of submergence $|y_0|$ can also tend to infinity.

Let $\delta > 0$ be sufficiently small and $\varepsilon > 0$; then we have the following as $\rho \to \infty$:

1. *If $|y_0| > \varepsilon r$ or $|\varphi - \pi| > \varphi_0 + \varepsilon$, then $G_x(x, 0, z; y_0) = O(\rho^{-3})$.*
2. *If $|y_0| < r$ and $|\varphi - \pi| < \varphi_0 - \varepsilon$, then*

$$G_x(x, 0, z; y_0) = r^{-1/2} \sum_{\pm} \sqrt{\tau_{\pm}(\varphi)} a_{\pm}(\varphi) e^{y_0 \tau_{\pm}(\varphi)} \cos\left(S_{\pm}(\varphi) r \pm \frac{\pi}{4}\right)$$
$$+ r^{-1} \frac{2\sqrt{\pi}}{|z|^{1/2}} D_1[y_0 t_{+}^2(\varphi)] e^{y_0} \sin\left[S_{+}(\varphi) r + \frac{\pi}{4}\right]$$
$$+ O\left(\rho^{-3/2}\right).$$

3. *If $|y_0| < r$ and $|\varphi - \pi \pm \varphi_0| < \delta$, then*

$$G_x(x, 0, z; y_0) = 4\sqrt{2\pi}\left[\frac{B(\varphi)}{1-9\sin^2\varphi}\right]^{1/4} \operatorname{Re}\sum_{\pm} \tau_{\pm}(\varphi) q_{\pm} + O\left(\rho^{-4/3}\right).$$

Here $t_\pm$, $\tau_\pm$, $S_\pm$, and $a_\pm$ are given by (6.71)–(6.74), *respectively; D_1 is defined in* (6.81)*; and by $q_\pm$ we denote the expression under the sign $\sum_\pm$ in* (6.80).

In the following subsections, we prove two theorems formulated above.

6.2.2. Reduction to an Oscillatory Integral

By $F = F_h(r, \varphi, y)$ (we recall that $x = r\cos\varphi$ and $z = r\sin\varphi$) we denote the following integral:

$$F = \int_{-\infty}^{\infty} f(t, y)\exp\{ir(\cos\varphi + t\sin\varphi)\sqrt{t^2+1}\}\,\mathrm{d}t, \tag{6.84}$$

where $f(t, y) = h(t)e^{y(t^2+1)}$ and $h \in C^\infty(\mathbb{R})$ is such that

$$h(t) = d^\pm|t|^\alpha + O(|t|^{\alpha-2}) \quad \text{as } t \to \pm\infty, \tag{6.85}$$

and the asymptotics can be differentiated. If $y = 0$ and α is large, then F is a divergent integral, and so $F|_{y=0}$ must be understood as the limit in (6.84) as $y \to -0$. The existence of this limit will be established in Subsections 6.2.4–6.2.7.

If $x \neq 0$, then the function defined by (6.70) and its gradient can be expressed in terms of the oscillatory integral (6.84) as follows:

$$I_1 = -4H(-x)\,\mathrm{Im}\,F|_{h=1}, \tag{6.86}$$

$$\nabla I_1 = -4H(-x)\,\mathrm{Im}\,F|_{h=(i\sqrt{t^2+1},\ t^2+1,\ it\sqrt{t^2+1})}. \tag{6.87}$$

These formulae combined with (6.69) and results in Subsection 6.1.4.1 reduce proving two theorems formulated in the previous subsection to the study of asymptotics as $r \to \infty$ for F when $x < 0$. It is obvious that the function h given by (6.86) and the components of the vector h given by (6.87) satisfy (6.85), and in Subsections 6.2.4–6.2.7 we are concerned with the asymptotic behavior of (6.84). For this purpose we apply the method of stationary phase described in Subsection 6.2.3, and the crucial notion of this method is the so-called point of the stationary phase.

Let S denote the phase function in (6.84); that is,

$$S = \sqrt{t^2+1}\cos\varphi + t\sqrt{t^2+1}\sin\varphi, \tag{6.88}$$

and so points of the stationary phase are determined by the following equation:

$$S_t = (t^2+1)^{-1/2}[t\cos\varphi + (1+2t^2)\sin\varphi] = 0. \tag{6.89}$$

We assume that $\cos\varphi < 0$ because the factor $H(-x)$ in (6.86) and (6.87) vanishes for $x < 0$. Equation (6.89) has no solutions when $|\varphi - \pi| > \varphi_0$ or $\cos\varphi < 0$. On the contrary, there are two points $t = t_{\pm}(\varphi)$ of stationary phase when φ is within Kelvin's angle $|\varphi - \pi| < \varphi_0$ and $\varphi \neq \pi$. These points are given by (6.71). The function t_- is infinitely smooth when $|\varphi - \pi| < \varphi_0$, and the same is true for t_+ when $\varphi \neq \pi$, but $t_+(\varphi) \to \mp\infty$ as $\varphi \to \pi \pm 0$ whereas $t_-(\varphi) \to 0$ as $\varphi \to \pi$. As $\varphi \to \pi + \varphi_0$, two points of the stationary phase merge at $t_{\pm}(\pi + \varphi_0) = -\sqrt{2}/2$, and the point of merging is $t_{\pm}(\pi - \varphi_0) = \sqrt{2}/2$ as $\varphi \to \pi - \varphi_0$.

By $S_{\pm} = S_{\pm}(\varphi)$ we denote the value of (6.88) at $t_{\pm}(\varphi)$; see (6.73), where these functions are described explicitly.

6.2.3. Stationary Phase Method

In this subsection we recall facts concerning the method of the stationary phase used for the investigation of (6.84). Many theoretical and applied problems require evaluation of the asymptotic behavior as $r \to \infty$ for functions defined as an oscillatory integral:

$$I(r) = \int_a^b f(t) e^{irS(t)}\,\mathrm{d}t, \quad \text{where } f, S \in C^\infty(\mathbb{R}),\ \operatorname{Im} S = 0. \tag{6.90}$$

In the case of superposed waves, f is the amplitude and S is the phase. What is important is not the phase's values (the more so because adding a multiple of 2π does not change the integrand), but the frequency $rS_t/2\pi$. If $S_t \neq 0$ and $r \to \infty$, then there are only high-frequency oscillations. The method of stationary phase allows us to find an asymptotic expansion for (6.90). Moreover, the method is applicable to multidimensional integrals as well. Below we give without proofs an account of the results related to the one-dimensional case required for our purposes in the present section. Proofs and multidimensional generalizations can be found elsewhere (see, for example, Fedoryuk [84] and Vainberg [345]).

We will assume that f vanishes near a and b. Also, let f as well as S depend smoothly on a parameter $\varphi \in \Delta$, where Δ is a closed segment, and so

$$I(r, \varphi) = \int_a^b f(t, \varphi) e^{irS(t,\varphi)}\,\mathrm{d}t, \quad \operatorname{Im} S = 0, \tag{6.91}$$

where $f,\ S \in C^\infty(\mathbb{R} \times \Delta)$, $f = 0$ when $t \notin [a + \varepsilon, b - \varepsilon]$ for a certain $\varepsilon > 0$.

Let us define some notions used in what follows. If $S_t(t, \varphi) = 0$ for $t = t(\varphi)$, then $t(\varphi)$ is called a point of stationary phasc, and such a point is called nondegenerate when $S_{tt}(t(\varphi), \varphi) \neq 0$. Let

$$Q(t, \varphi) = \frac{|S(t, \varphi) - S(t(\varphi), \varphi)|}{(t - t(\varphi))^2},$$

where $t(\varphi)$ is a point of stationary phase. By $\|f\|_k$ we denote the norm of f in the Banach space $C^k([a, b])$; that is,

$$\|f\|_k = \max_{0 \le j \le k} \max_{a \le t \le b} \left| f^{(j)}(t) \right|.$$

The first assertion is concerned with the case of nondegenerate points of stationary phase.

1. *Let S have only one stationary point $t = t(\varphi)$ in $[a, b]$ for each $\varphi \in \Delta$, and let this point be nondegenerate. Then we have the following for any N:*

$$I = \exp\{irS(t(\varphi), \varphi)\} \sum_{k=0}^{N} a_k(\varphi) r^{-k-1/2} + O\left(r^{-N-3/2}\right). \quad (6.92)$$

Here $a_k \in C^\infty(\Delta)$ $(k = 0, \ldots, N)$ have the following form:

$$a_k = \frac{\Gamma(k+1)}{(2k)!} \exp\left\{ i \frac{\pi(2k+1)}{4} \operatorname{sign} S_{tt}(t(\varphi), \varphi) \right\} \times \left[\frac{d^{2k}}{dt^{2k}} \left\{ f(t, \varphi) [Q(t, \varphi)]^{-k-1/2} \right\} \right]_{t=t(\varphi)}, \quad (6.93)$$

where $\Gamma(\cdot)$ is the gamma function. The reminder term in (6.92) *does not exceed*

$$C(a, b, S) \|f\|_{2N+2} r^{-N-3/2},$$

and this expansion admits differentiation with respect to r and φ.

2. *The one-term formula* (6.92) *is as follows:*

$$I = \exp\left\{ irS(t(\varphi), \varphi) + i\frac{\pi}{4} \operatorname{sign} S_{tt}(t(\varphi), \varphi) \right\} \times f(t(\varphi), \varphi) \sqrt{\frac{2\pi}{|S_{tt}(t(\varphi), \varphi)| r}} + O\left(r^{-3/2}\right).$$

3. *If for each $\varphi \in \Delta$ the phase S has several stationary points $t = t_j(\varphi)$, $j = 1, \ldots, j_0$, in $[a, b]$ and all of these points are nondegenerate, then the asymptotic expansion for I has the form of a sum whose terms are given in* (6.92), *but $t(\varphi) = t_j(\varphi)$.*

The next statement gives the asymptotic expansion for (6.91) when S has two nondegenerate stationary points merging at $t = t'$ as $\varphi \to \varphi'$.

Let S be an analytic function of (t, φ) in a neighborhood of (t', φ') and

$$S_t(t', \varphi') = S_{tt}(t', \varphi') = 0$$

whereas

$$\alpha_1 = S_{t\varphi}(t', \varphi') < 0, \quad \alpha_2 = S_{ttt}(t', \varphi') > 0.$$

Then there exist $\delta_1 > 0$ and $\delta > 0$ such that the following assertions hold.

1. *No stationary point exists in the interval $|\varphi - \varphi'| < \delta_1$ when $\varphi' - \delta < \varphi < \varphi'$. If $\varphi' < \varphi < \varphi' + \delta$, then S has exactly two stationary points $t = t_\pm(\varphi)$ in the interval $|t - t'| < \delta_1$, and $t_\pm$ are analytic functions of $\psi = \sqrt{\varphi - \varphi'}$ in a neighborhood of $\psi = 0$. Functions A and B defined for $\varphi \in (\varphi', \varphi' + \delta)$ by* (6.76), *where $S_\pm(\varphi) = S(t_\pm(\varphi), \varphi)$, have analytic continuations into the interval $|\varphi - \varphi'| < \delta$.*
2. *If $t' \in (a, b)$, where $b - a < \delta_1$ and $|\varphi - \varphi'| < \delta$, then the following asymptotic expansion holds for* (6.91) *as $r \to \infty$:*

$$I = e^{irA(\varphi)} \left\{ v\left(-B(\varphi) r^{2/3}\right) \sum_{j=0}^{N} b_{j1}(\varphi) r^{-1/3-j} \right.$$
$$\left. + v'\left(-B(\varphi) r^{2/3}\right) \sum_{j=0}^{N} b_{j2}(\varphi) r^{-2/3-j} \right\} + O\left(r^{-4/3-N}\right),$$

where v is the Airy function (6.77) *and b_{j1} and b_{j2} are analytic functions of φ. The expansion admits differentiation with respect to r and φ.*
3. *If t_+ is chosen so that $t_+(\varphi) > t_-(\varphi)$, then coefficients in the leading terms of expansions are as follows:*

$$b_{01}(\varphi) = \sqrt{2\pi}\, f\, B^{1/4} (S_{tt})^{-1/2} \Big|_{t=t_+(\varphi)}$$
$$+ \sqrt{2\pi}\, f\, B^{-1/4} (-S_{tt})^{-1/2} \Big|_{t=t_-(\varphi)},$$
$$b_{02}(\varphi) = -i\sqrt{2\pi}\, f\, B^{-1/4} (S_{tt})^{-1/2} \Big|_{t=t_+(\varphi)}$$
$$+ i\sqrt{2\pi}\, f\, B^{1/4} (-S_{tt})^{-1/2} \Big|_{t=t_-(\varphi)},$$

where the branches of all roots are chosen so that the roots are positive for $\varphi > \varphi'$

Note that $\pm S_{tt}|_{t=t_\pm(\varphi)} > 0$ when $\varphi > \varphi'$.

The propositions formulated in this subsection will be applied below for the investigation of (6.84). Also, the last proposition can easily be modified so that it is applicable when $\alpha_1 > 0$ or $\alpha_2 < 0$. For this purpose one has simply to replace φ by $-\varphi$ and to consider the complex conjugate integral.

6.2.4. Estimates Outside of Kelvin's Angle

In this subsection, we are concerned with proving the following lemma.

1. *If $y \leq -1$, then $|F| \leq Ce^y$.*
2. *Let $\varepsilon > 0$; then F is a continuous function when*

$$y \leq 0, \qquad x^2 + z^2 \neq 0, \qquad \varphi_0 + \varepsilon < |\varphi - \pi| \leq \pi/2, \tag{6.94}$$

and for any $N \geq 0$ there exists $C = C(\varepsilon, N)$ such that

$$|F| \leq C\,(1 + |y|)^{N/2}\, e^y r^{-N} \tag{6.95}$$

for $r > 1$ and φ, y specified in (6.94).

It is obvious that point 1 is true, and so we begin with proving that F is continuous when conditions (6.94) hold.

Let D be a set in $\mathbb{R}^3$ in which (y, φ, t) are the coordinates, and let ψ be an infinitely smooth function defined in D and such that

$$\left|\partial_t^j \psi\right| \leq C_j |t|^{s-j}$$

holds for any j, where C_j is independent of point $(y, \varphi, t) \in D$. By $M_s(D)$ we denote a space consisting of such functions, and we take

$$D = \{y \leq 0, \varphi_0 + \varepsilon \leq |\varphi - \pi| \leq \pi/2, |t| \geq 1\}.$$

Since the phase function (6.88) is such that $S_t \neq 0$ in D and $S_t \sim 2t \sin\varphi$ as $|t| \to \infty$ [see (6.89)], we see that $(S_t)^{-1} \in M_{-1}(D)$.

Let us demonstrate that

$$e^{yt^2} \in M_0(D). \tag{6.96}$$

Since

$$\partial_t^j e^{yt^2} = \sum_{0 \leq m \leq j} C_{m,j} y^m t^{2m-j} e^{yt^2}, \quad C_{m,j} = \text{const}, \tag{6.97}$$

and $|x|^m e^x$ is a bounded function when $x = yt^2 \leq 0$, we have that

$$\left|\partial_t^j e^{yt^2}\right| \leq C_j\, |t|^{-j} \quad \text{for } -\infty < t < +\infty, y \leq 0, \tag{6.98}$$

and so (6.96) is proved. Since $C_{m,j} = 0$ for $2m < j$, it follows from (6.97) that

$$\left|\partial_t^j e^{yt^2}\right| \leq C_j \, |y|^{j/2} \quad \text{for } -\infty < t < +\infty, \, y \leq 0 \tag{6.99}$$

(this will be used later on).

Let us split F into a sum $F_1 + F_2$, where F_j ($j = 1, 2$) are given by (6.84), but the corresponding integrands are multiplied by $\chi_j(t)$. Here $\chi_1 \in C^\infty$ is such that $\chi_1(t) = 1$ for $|t| \geq 2$ and $\chi_1(t) = 0$ for $|t| \leq 1$, and $\chi_2(t) = 1 - \chi_1(t)$. For estimating F_1 we multiply and divide the integrand by $iS_t r$, and integrating by parts we get

$$F_1 = \frac{1}{r} \int_{|t|>1} e^{y+iSr} \frac{\partial}{\partial t} \left(-\frac{\chi_1 h e^{yt^2}}{iS_t} \right) \mathrm{d}t, \quad y < 0.$$

Repeating this procedure N times, $N > (1+\alpha)/2$, we obtain

$$F_1 = r^{-N} \int_{|t|>1} e^{y+iSr} \frac{\partial}{\partial t} \left(\frac{-1}{iS_t} \frac{\partial}{\partial t} \right)^{N-1} \left(-\frac{\chi_1 h e^{yt^2}}{iS_t} \right) \mathrm{d}t, \quad y < 0. \tag{6.100}$$

Since $h \in M_\alpha(D)$ according to (6.85), we get from (6.96) that $\chi_1 h e^{yt^2} \in M_\alpha(D)$. Multiplication by $(S_t)^{-1}$ maps $M_s(D)$ into $M_{s-1}(D)$ because $(S_t)^{-1} \in M_{-1}$, and ∂_t has the same property. Hence the second factor in the integrand in (6.100) belongs to $M_{\alpha-2N}(D)$, and so it is majorized by $C|t|^\gamma$, where $\gamma = \alpha - 2N < -1$ because $N > (1+\alpha)/2$. Therefore, (6.100) is a converging integral and the convergence is uniform on (6.94). This implies that F_1 is a continuous function there and it satisfies (6.95) for $N > (1+\alpha)/2$. Moreover, there is no factor $(1+|y|)^{N/2}$ in the right-hand side in (6.95) for F_1. It is clear that this estimate also holds for smaller values of N.

Let us turn to estimating F_2. Since the domain of integration is finite in this case ($|t| \leq 2$), we see that in the procedure described above, the derivatives of $(S_t)^{-1}$ and $\chi_2 h$ are continuous and uniformly bounded for φ and y specified in the lemma's formulation. Also, the derivatives of e^{yt^2} satisfy (6.99). Thus F_2 is continuous, and integrating term by term, one arrives at (6.95) for F_2, which completes the proof.

6.2.5. Asymptotic Behavior Inside Kelvin's Angle

The following lemma provides the asymptotic behavior of (6.84) strictly inside Kelvin's angle, but outside of a neighborhood of the source's track.

Let $\varepsilon \geq 0$ and N be an arbitrary nonnegative integer; then for φ satisfying $\varepsilon \leq |\varphi - \pi| \leq \varphi_0 - \varepsilon$, we have that

$$F = \sum_{\pm} \exp\{y\tau_{\pm}(\varphi) + iS_{\pm}(\varphi)r\} \sum_{j=0}^{N} a_j^{\pm}(\varphi, y) r^{-1/2-j} + (1 + |y|)^{N+1} e^{y} O\left(r^{-3/2-N}\right) \tag{6.101}$$

holds for (6.84) *as $r \to \infty$. Here $a_j^{\pm}(\varphi, y)$ are polynomials in y whose orders are less than or equal to $2j$ and coefficients are infinitely differentiable functions of φ. Moreover,*

$$a_0^{\pm}(\varphi, y) = -1/4h(t_{\pm}(\varphi))a_{\pm}(\varphi)e^{\pm i\pi/4}. \tag{6.102}$$

The definitions of $t_{\pm}$, $\tau_{\pm}$, $S_{\pm}$, and $a_{\pm}$ are given by (6.71)–(6.74). In the proof of this lemma, we obtain a slightly better estimate for the remainder term in (6.101). Namely, one may replace $(1 + |y|)^{N+1}$ by $(1 + |y|)^{N/2+1}$, but we do not use this improvement in what follows. The formulated more rough estimate allows us to use formulae (6.101) and (6.102) for values of φ close to $\varphi = \pi$ (see the assertion formulated in Subsection 6.2.7).

We begin the proof by noting that for $\varepsilon > 0$ there exists $t_0 > 1$ such that the points of the stationary phase $t_{\pm}(\varphi)$ [see (6.71)] satisfy

$$2t_0^{-1} < t_{\pm}(\varphi) < t_0 \quad \text{for } \varepsilon \leq |\varphi - \pi| \leq \varphi_0 - \varepsilon. \tag{6.103}$$

Let us split F into a sum $F_1 + F_2 + F_3$ in which F_1, F_2, and F_3 are obtained by multiplying the integrand in (6.84) by $\chi_1(t/t_0)$, $\chi_2(t_0 t)$, and $\chi_3(t) = 1 - \chi_1(t/t_0) - \chi_2(t_0 t)$, respectively. Here $\chi_1(t)$ and $\chi_2(t)$ are the functions defined in the proof of the lemma in Subsection 6.2.4. It is obvious that for φ under consideration, $t_{\pm}(\varphi)$ lie outside of the supports of $\chi_1(t/t_0)$ and $\chi_2(t_0 t)$; that is, $S_t \neq 0$ on the supports of these functions. Let us change the definition of D given in the proof of the lemma in Subsection 6.2.4 and set

$$D = \{y \leq 0, \varepsilon \leq |\varphi - \pi| \leq \varphi_0 - \varepsilon, |t| \geq t_0\}.$$

Then $S_t \neq 0$ in D, and so F_1 and F_2 can be estimated in the same way as in the proof of the lemma in Subsection 6.2.4. Hence, $F_1 + F_2$ is defined for φ and y specified in the lemma's formulation. Moreover, $F_1 + F_2$ is continuous and (6.95) holds for it for these values of φ and y. Replacing N by $N + 2$ in that estimate, we find that $F_1 + F_2$ satisfies the estimate for the remainder on the right-hand side of (6.101), and so the contribution of $F_1 + F_2$ is negligible when one considers the asymptotics of F.

Let us turn to F_3, whose integrand has a compact support. Therefore, F_3 is defined and continuous for all $y \leq 0$, and the method of stationary phase is applicable. It was mentioned above that S has two stationary points $t = t_{\pm}(\varphi)$ defined by (6.71) for φ under consideration. At these stationary points we have

$$S_{tt}|_{t=t_{\pm}(\varphi)} = \frac{\pm\sqrt{1-9\sin^2\varphi}}{\tau_{\pm}(\varphi)} \neq 0, \quad \varepsilon \leq |\varphi - \pi| \leq \varphi_0 - \varepsilon. \tag{6.104}$$

That is, the points of the stationary phase are nondegenerate. Now the required assertion follows from the first proposition in Subsection 6.2.3, if one takes into account that f in F_3 has the form

$$f = \chi_3(t)h(t)e^{y(t^2+1)}, \tag{6.105}$$

and $\chi_3(t_{\pm}(\varphi)) = 1$, which follows from (6.103). Also, estimating the remainder term, one has to take into account that $\chi_3(t)h(t)$ in (6.105) depends on y belonging to the half-axis $y \leq 0$, which is not a compact set. However, the remainder arising when one applies the method of stationary phase can be estimated (see the first proposition in Subsection 6.2.3) by means of the derivatives of f with respect to t having orders less than or equal to $2N+2$ and by the derivatives of phase that does not depend on y. From (6.98), it follows that these derivatives are bounded by Ce^y on the support of χ_3. Hence F_3 satisfies (6.101), where the factor $(1+|y|)^{N+1}$ can be omitted in the remainder. It was shown above that $F_1 + F_2$ does not exceed the remainder in (6.101), which means that (6.101) is valid for $F_1 + F_2 + F_3$. Thus the proof is complete.

6.2.6. Asymptotics Near the Boundary of Kelvin's Angle

The aim of this subsection is to prove the following result.

There exists $\delta > 0$ such that for $|\varphi - \pi \pm \varphi_0| < \delta$, $y \leq 0$, and an arbitrary integer $N \geq 0$, we have the following as $r \to \infty$:

$$\begin{aligned} F = e^{irA(\varphi)} \Bigg[& v\left(-B(\varphi)r^{2/3}\right) \sum_{j=0}^{N} b_{j1}(\varphi, y) r^{-1/3-j} \\ & + v'\left(-B(\varphi)r^{2/3}\right) \sum_{j=0}^{N} b_{j2}(\varphi, y) r^{-2/3-j} \Bigg] \\ & + (1+|y|)^{1+N/2} e^y O\left(r^{-4/3-N}\right), \end{aligned} \tag{6.106}$$

where v is the Airy function (6.77), A and B are given by (6.76), and b_{j1} and b_{j2} are bounded infinitely differentiable functions.

The one-term formula is as follows:

$$F = -i\sqrt{2\pi}\left[\frac{B(\varphi)}{1-9\sin^2\varphi}\right]^{1/4}\sum_{\pm} h(t_\pm(\varphi))q_\pm(\varphi, y, r)$$
$$+(1+|y|)\,e^y O\left(r^{-4/3}\right), \tag{6.107}$$

where $q_\pm$ are the expressions under the sign $\sum_\pm$ in (6.80).

Since estimating (6.84) is similar for $|\varphi - \pi + \varphi_0| < \delta$ and $|\varphi - \pi - \varphi_0| < \delta$, we restrict ourselves to the case $|\varphi - \pi + \varphi_0| < \delta$. Let us introduce $\chi_i \in C^\infty(\mathbb{R})$, $i = 1, 2, 3$ as follows:

$$\chi_3(t) = 1 \text{ for } |t - \sqrt{2}/2| < \gamma \ll 1, \qquad \chi_3(t) = 0 \text{ for } |t - \sqrt{2}/2| > 2\gamma;$$
$$\chi_2(t) = 1 - \chi_3(|t|) \text{ for } |t| \le \sqrt{2}/2, \qquad \chi_2(t) = 0 \text{ for } |t| > \sqrt{2}/2;$$
$$\chi_1 = 1 - \chi_2 - \chi_3.$$

We split (6.84) into a sum $F = F_1 + F_2 + F_3$, where F_j is obtained by multiplying the integrand by $\chi_j(t)$, $j = 1, 2, 3$. We noted above that the points of stationary phase $t_\pm(\varphi)$ tend to $\sqrt{2}/2$ as $\varphi \to \pi - \varphi_0$. Taking $\delta > 0$ to be sufficiently small, we find that $t_\pm(\varphi)$ fall within a $\gamma/2$ neighborhood of $\sqrt{2}/2$ when $|\varphi - \pi + \varphi_0| < \delta$. Then $S_t \ne 0$ on supports of χ_1 and χ_2, and so F_1 and F_2 can be estimated in the same way as in the proof of the lemma in Subsection 6.2.4. Hence, these integrals are defined and continuous for all φ and y specified in the lemma's formulation and their contribution to the asymptotics of F is negligible.

Since one integrates over a finite segment in F_3, this integral can be treated by means of the method of stationary phase; one has to apply the version involving the phase $S(t, \varphi)$ that depends on the parameter φ and has two nondegenerate stationary points $t_\pm(\varphi)$ merging into $t_\pm(\pi - \varphi_0) = \sqrt{2}/2$ as $\varphi \to \pi - \varphi_0$. One can easily verify that

$$S_t = S_{tt} = 0,\ S_{\varphi t} < 0,\ S_{ttt} > 0, \quad \text{for } \ \varphi = \pi - \varphi_0, t = \sqrt{2}/2.$$

Noting that $\delta > 0$ can be chosen arbitrarily small, one obtains the required assertion as a direct consequence of the second proposition in Subsection 6.2.3. It remains for us to remark that the dependence on y in the integrand in F_3 creates no difficulties because the coefficients of the corresponding expansion and the estimate for the remainder can be written by using the derivatives in the same way as in the proof of the lemma in Subsection 6.2.5. It was noted there that by (6.98) these derivatives are bounded by Ce^y. The proof is complete.

6.2.7. Asymptotic Behavior Near the Source's Track

In this subsection we extend the result obtained in Subsection 6.2.5 to the case in which φ is close to π.

1. *The asymptotic expansions* (6.101) *and* (6.102) *hold for $y \leq 0$ and φ satisfying $|\varphi - \pi| \leq \varphi_0 - \varepsilon$, and remainders decay uniformly at infinity when $N \geq \alpha - 1/2$, where α is defined in* (6.85). *Furthermore, $a_j^-(\varphi, y)$ is a polynomial in y, its order is less than or equal to $2j$, and coefficients are infinitely smooth functions of φ;*

$$a_j^+(\varphi, y) = b_j(\varphi, yt_+^2(\varphi))\,|t_+(\varphi)|^{\alpha-j+1/2}, \tag{6.108}$$

where b_j is a polynomial of the second argument, its order is less than or equal to $2j$, and coefficients are bounded functions.

2. *If $\alpha \leq 2$, then for $y \leq 0$ and $|\varphi - \pi| < \varphi_0 - \varepsilon$ we have that*

$$\begin{aligned} F = &-\frac{r^{-1/2}}{4}\sum_{\pm} h(t_\pm(\varphi))\alpha_\pm(\varphi)\exp\left\{y\tau_\pm(\varphi) + iS_\pm(\varphi)r \pm \frac{i\pi}{4}\right\} \\ &+ d\frac{\sqrt{2\pi}}{2}r^{-3/2}\,|2\sin\varphi|^{1/2-\alpha}\exp\left\{y + iS_+(\varphi)r + \frac{i3\pi}{4}\right\} \\ &\times\left[\frac{\mathrm{d}^2}{\mathrm{d}\sigma^2}\left(\sigma^\alpha e^{yt_+^2(\varphi)\sigma^2}\right) + \frac{i|\sin\varphi|}{2r}\frac{\mathrm{d}^4}{\mathrm{d}\sigma^4}\left(\sigma^\alpha e^{yt_+^2(\varphi)\sigma^2}\right)\right]_{\sigma=1} \\ &+ (1+|y|)^3 e^y O\left(r^{-3/2}\right). \end{aligned} \tag{6.109}$$

Here $d = d^+$ if $\sin\varphi > 0$ and $d = d^-$ if $\sin\varphi < 0$, where $d^\pm$ are constants in (6.85), *and $t_\pm$, $\tau_\pm$, $S_\pm$, and $a_\pm$ are defined by* (6.71)–(6.74).

Before proving this assertion, we should make several remarks.

First, from (6.108) and (6.71), it follows that for $j < \alpha + 1/2$ the coefficients $a_j^+(\varphi, y)e^{yt_+^2(\varphi)}$ in (6.101) are, generally speaking, singular on the line $\{\varphi = \pi,\ y = 0\}$. Since $|x|^m e^{-x}$ is bounded when $x = yt_+^2(\varphi) \leq 0$, these coefficients are bounded for $j \geq \alpha + 1/2$ and tend to zero as $\varphi \to \pi$ for $j > \alpha + 1/2$.

Second, if $|\varphi - \pi| > \varepsilon$, then the middle term on the right-hand side in (6.109) (it contains $r^{-3/2}$) has the same estimate at infinity as the remainder, but it is impossible to omit this term because it tends to infinity as $\varphi \to \pi$. Despite the fact that this term and the leading one have singularities, the remainder decays uniformly at infinity.

Third, for $\alpha \leq 3/2$ the second term in the braces in (6.109) can be omitted because it does not exceed $Ce^y r^{-5/2}$ when combined with the factor preceding the braces. Here C does not depend on φ and y. The same reasoning shows that one can put $d = 0$ for $\alpha \leq 1/2$.

Turning to the proof, we note that according to the lemma proven in Subsection 6.2.5 it is sufficient to prove the present assertion under the assumption that $|\varphi - \pi| \leq \varepsilon$ for a certain $\varepsilon > 0$. Let us show that ε can be chosen so small that for $|\varphi - \pi| \leq \varepsilon$ we have

$$|t_+(\varphi)| > 8, \quad |t_-(\varphi)| < 1/2, \tag{6.110}$$

and for a certain $\gamma > 0$ the following inequality holds:

$$|S_t| > \gamma(1 + |t \sin\varphi|) \quad \text{when } 1 \leq |t| \leq 1/2\,|t_+(\varphi)| \text{ or } |t| > 2\,|t_+(\varphi)|. \tag{6.111}$$

Since $|t_+(\varphi)| \to \infty$ and $t_-(\varphi) \to 0$ as $\varphi \to \pi$, it is not difficult to satisfy (6.110). Further, (6.89) implies that for $|t| \geq 1$,

$$\begin{aligned} |S_t| &\geq \left|\cos\varphi + \frac{1+2t^2}{t}\sin\varphi\right| \geq |\cos\varphi| - \left|\frac{1+2t^2}{t}\sin\varphi\right| \\ &\geq |\cos\varphi| - |\sin\varphi| - 2\,|t\sin\varphi|\,, \end{aligned} \tag{6.112}$$

and similarly,

$$\begin{aligned} |S_t| &\geq \left|\frac{1+2t^2}{t}\sin\varphi\right| - |\cos\varphi| \geq 2\,|t\sin\varphi| - |\cos\varphi| \\ &\geq \frac{1}{3}(1 + |t\sin\varphi|) + \left(\frac{5}{3}|t\sin\varphi| - \frac{4}{3}\right). \end{aligned} \tag{6.113}$$

Let $1 \leq |t| \leq |t_+(\varphi)|/2$; then (6.71) implies that $|t_+(\varphi)| < |\cot\varphi|/2$. Hence, $1 \leq |t| \leq |\cot\varphi|/4$, and (6.112) yields

$$|S_t| > 1/2\,|\cos\varphi| - |\sin\varphi| \geq \gamma_1 \quad \text{for } 1 \leq |t| \leq |t_+(\varphi)|\,/2,$$

where $\gamma_1 > 0$ if ε is sufficiently small. On the other hand, we have the following for the same t:

$$1 + |t\sin\varphi| \leq 1 + 1/4\,|\cos\varphi| \leq 5/4,$$

and so (6.111) is valid for $1 \leq |t| \leq |t_+(\varphi)|\,/2$ (we may put $\gamma = 4\gamma_1/5$).

Now let $|t| > 2\,|t_+(\varphi)|$; then according to (6.71) we have that

$$|t_+(\varphi)\sin\varphi| \to 1/2 \quad \text{as } \varphi \to \pi.$$

If ε is sufficiently small to guarantee that $|t_+(\varphi)\sin\varphi| > 2/5$ holds for $|\varphi - \pi| < \varepsilon$, then $|t\sin\varphi| > 4/5$ and the second term is nonnegative on the right-hand side in (6.113). Hence, (6.111) is a consequence of (6.113) for $|t| \geq 2\,|t_+(\varphi)|$ (where $\gamma = 1/3$). Thus there exists $\varepsilon > 0$ such that (6.110) and (6.111) are valid for $|\varphi - \pi| \leq \varepsilon$.

Let $\chi_j \in C^\infty(\mathbb{R})$ $(j = 1, 2, 3)$ be such that

$$\chi_3(t) = 1 \quad \text{for } 1/2 \le t \le 2, \qquad \chi_3(t) = 0 \quad \text{for } t < 1/3 \quad \text{and for } t > 3;$$
$$\chi_2(t) = 1 \quad \text{for } |t| < 1, \qquad \chi_2(t) = 0 \quad \text{for } t > 2;$$
$$\chi_1(t, \varphi) = 1 - \chi_2(t) - \chi_3[t/t_+(\varphi)].$$

Splitting F into a sum $F_1 + F_2 + F_3$, where F_j has the form (6.84), but the integrand is multiplied by χ_j, let us investigate each F_j.

From (6.110) it follows that

$$\chi_1 = 0 \quad \text{for } |t| \le 1 \text{ and for } |t_+(\varphi)|/2 \le |t| \le 2\,|t_+(\varphi)|\,.$$

That is, (6.111) holds on the support of χ_1. Therefore, we can estimate F_1 by using the scheme applied for proving the lemma in Subsection 6.2.4. The only difference is that the following set,

$$\{y \le 0,\ |\varphi - \pi| \le \varepsilon,\ 1 \le |t| \le |t_+(\varphi)|/2 \text{ or } |t| \ge 2\,|t_+(\varphi)|\}\,,$$

must be used instead of D. Moreover, now $(S_t)^{-1}$ belongs to $M_0(D)$ instead of $M_{-1}(D)$ (this will be shown below). Consequently, for estimating F_1, we find it necessary to integrate by parts N times in F_1, where $N > \alpha + 1$. In order to complete the proof that the contribution of F_1 to the asymptotics of F is negligible, it remains for us to show that $(S_t)^{-1} \in M_0(D)$.

For demonstrating this we note that $\partial_t^j(1/S_t)$ is a linear combination of functions having the form

$$\prod_{p=1}^{k-1} \frac{(\partial_t^{\alpha_p} S)}{(S_t)^k}, \qquad \text{where } \alpha_p \ge 2, \sum_{p=1}^{k-1} \alpha_p - k = j - 1.$$

Besides (6.88) implies that for $|\varphi - \pi| \le \varepsilon$ and $|t| \ge 1$,

$$\left|\partial_t^n S\right| \le C_n(|t|^{1-n} + |t|^{2-n}|\sin\varphi|).$$

This and (6.111) give that on D,

$$\left|\partial_t^j \left(\frac{1}{S_t}\right)\right| \le C_j(1 + |t\sin\varphi|)^{-k} \prod_{p=1}^{k-1}(|t|^{1-\alpha_p} + |t|^{2-\alpha_p}|\sin\varphi|)$$
$$\le C_j|t|^{k-1-\Sigma\alpha_p} = C_j|t|^{-j},$$

and so $(S_t)^{-1} \in M_0(D)$.

Since one integrates over a finite interval ($|t| \le 2$) in F_2, the first proposition formulated in Subsection 6.2.3 can be applied to F_2. For $|\varphi - \pi| \le \varepsilon$, there is a single point of stationary phase strictly inside the integration interval. This point is nondegenerate, and so the asymptotics of this integral has the form (6.101), where the terms on the right-hand side containing t_-, S_-, and a_j^- must be omitted. This can be obtained in the same way as the asymptotic expansion (6.101) for F_3 in the proof of the lemma in Subsection 6.2.5, but the difference is that now the integration interval contains only one point of stationary phase instead of two such points. Moreover, a slightly different estimate arises for the remainder. We noted in the proof in Subsection 6.2.5 that the remainder is estimated by the t derivatives of f whose orders are less than or equal to $2N+2$. According to (6.99) these derivatives do not exceed $C(1+|y|)^{N+1}e^y$, which characterizes the dependence on y of the remainder in (6.101).

Substituting $t = \sigma t_+(\varphi)$ into F_3, one obtains the following form of this integral, which is usual in the stationary phase method:

$$e^{-y}F_3 = |t_+(\varphi)|^{\alpha+1}\int_l f\exp\{i\sqrt{\sigma^2+t_+^{-2}(\varphi)}[\cos\varphi+\sigma t_+(\varphi)\sin\varphi]k\}\,d\sigma. \tag{6.114}$$

Here $\varphi \neq \pi$, $l = \{1/3 \leq |\sigma| \leq 3\}$, $k = r\,|t_+(\varphi)| \to \infty$ as $r \to \infty$ (and also as $\varphi \to \pi$), and

$$f = \chi_3(\sigma)\,|t_+(\varphi)|^{-\alpha}\,h(\sigma t_+(\varphi))e^{\zeta\sigma^2}, \quad \zeta = yt_+^2(\varphi). \tag{6.115}$$

By virtue of (6.71) and (6.85) one obtains that

$$f = \chi_3(\sigma)\tilde{h}e^{\zeta\sigma^2} \quad \text{for } 1/3 \leq |\sigma| \leq 3 \text{ and } |\varphi-\pi| \leq \varepsilon,$$

where

$$\tilde{h} = |t_+(\varphi)|^{-\alpha}\,h(\sigma t_+(\varphi)) = \tilde{d}|\sigma|^{\alpha} + O(\sin^2\varphi) \quad \text{as } \varphi \to \pi, \tag{6.116}$$

and

$$\tilde{d} = d^+ \quad \text{for } \sigma\sin\varphi > 0, \qquad \tilde{d} = d^- \quad \text{for } \sigma\sin\varphi < 0.$$

Thus $\tilde{h}$ (and hence f) is infinitely differentiable when $1/3 \leq |\sigma| \leq 3$ and $\pi - \varepsilon \leq \varphi < \pi$, and it can be smoothly extended as $\varphi \to \pi - 0$. This function has the same properties when $1/3 \leq |\sigma| \leq 3$ and $\pi < \varphi \leq \pi + \varepsilon$, but, generally speaking, $\tilde{h}$ has different limiting values as $\varphi \to \pi - 0$ and $\varphi \to \pi + 0$. Taking into account these properties of $\tilde{h}$, we shall study the asymptotics of (6.114) separately for $\pi - \varepsilon \leq \varphi \leq \pi$ and for $\pi \leq \varphi \leq \pi + \varepsilon$.

According to (6.71), the phase function in (6.114) is infinitely smooth when $|\varphi - \pi| \leq \varepsilon$. In old variables, the point of stationary phase on the support of χ_3 is $t = t_+(\varphi)$, and so $\sigma = 1$ is such a point in new variables. One can easily verify that this point of stationary phase is nondegenerate. Then the first proposition formulated in Subsection 6.2.3 can be applied to (6.114) when either $\pi - \varepsilon \leq \varphi < \pi$ or $\pi < \varphi \leq \pi + \varepsilon$, and in the same way as (6.101) was derived for F_3 in Subsection 6.2.5 we obtain the following asymptotic expansion for (6.114) when $\varphi \neq \pi$:

$$e^{-y} F_3 = |t_+(\varphi)|^{\alpha+1} \Big\{ e^{\zeta + iS_+(\varphi) r} \sum_{j=0}^{N} b_j(\varphi, \zeta) \, [r \, |t_+(\varphi)|]^{-j-1/2} + O\big[r \, |t_+(\varphi)| \, \big]^{-N-3/2} \Big\}. \tag{6.117}$$

Here $\zeta = y t_+^2(\varphi)$ and b_j are polynomials in ζ having orders less than or equal to $2j$. The remainder in (6.117) is estimated uniformly in ζ and φ, which can be obtained in the same way as the similar assertion for F_3 in the proof of the lemma in Subsection 6.2.5 (the only difference is that we have ζ here instead of y in Subsection 6.2.5). Since f [see (6.115)] and the phase are smooth functions of φ, all b_j in (6.117) have limits as $\varphi \to \pi \pm 0$, and for the remainder $O(\lambda^{-N-3/2})$, where $\lambda = r \, |t_+(\varphi)|$, the following estimate

$$\left| O\left(\lambda^{-N-3/2}\right) \right| \leq C \lambda^{-N-3/2}, \quad \lambda > 1$$

holds uniformly in ζ and φ. Here C is independent of ζ and φ for $\zeta \leq 0$ and $|\varphi - \pi| \leq \varepsilon$. Since $|t_+(\varphi)| \to \infty$ as $\varphi \to \infty$, we have that

$$|t_+(\varphi)|^{\alpha+1} \left| O \left[r |t_+(\varphi)| \right]^{-N-3/2} \right| \leq C r^{-N-3/2}$$

if $N \geq \alpha - 1/2$, and therefore, the first assertion of the lemma formulated in this subsection is a corollary of (6.117) combined with the asymptotic expansion for F_2 and the estimate for F_1.

Let us turn to proving the second assertion formulated at the beginning of the present subsection. For finding the leading term in (6.101) when φ is close to π, one has to take into account that $|t_+(\varphi)| \to \infty$ as $\varphi \to \pi$. Let us determine the asymptotic behavior of b_k [see (6.117)] as $\varphi \to \pi \pm 0$. First, $e^{\zeta} b_k$ is given by (6.93), where f and S are those in (6.114). From (6.71) it follows that S is a smooth function of $\sin^2 \varphi$ and $\sigma = 1$ is the stationary point of S for all φ. Therefore, S has the form

$$S(\sigma, \varphi) = S(1, \varphi) + (\sigma - 1)^2 g(\sigma, \sin^2 \varphi),$$

where g is a smooth function and one can easily check that $g(\sigma, 0) = 1/2$. Hence, we have that

$$S(\sigma, \varphi) - S(1, \varphi) = \frac{(\sigma - 1)^2}{2} [1 + O(\sin^2 \varphi)] \quad \text{as } \sin \varphi \to 0, \tag{6.118}$$

where $O(\sin^2\varphi)$ depends also on σ. Since $\chi_3(\sigma) = 1$ in a neighborhood of $\sigma = 1$, (6.115), (6.118), and (6.93) imply that

$$e^{\zeta} b_k(\varphi, \zeta) = \frac{\Gamma(k+1/2)}{(2k)!} \exp\left\{\frac{i\pi(2k+1)}{4}\right\}$$
$$\times \left\{\frac{\mathrm{d}^{2k}}{\mathrm{d}\sigma^{2k}} \tilde{h}(\sigma,\varphi) e^{\zeta\sigma^2}\left[2^{k+1/2} + O(\sin^2\varphi)\right]\right\}_{\sigma=1} \quad \text{as } \varphi \to \pi, \tag{6.119}$$

where $\tilde{h}$ satisfies (6.116). From (6.116) it follows that

$$\tilde{h}(\sigma,\varphi) = d\sigma^{\alpha} + O(\sin^2\varphi) \quad \text{as } \varphi \to \pi,\ \sigma > 0,$$

where d is defined in the lemma's formulation. Since $e^{\zeta\sigma^2}$ and its derivatives with respect to σ are uniformly bounded for $\zeta \le 0$, (6.119) implies that as $\varphi \to \pi$,

$$e^{\zeta} b_1(\varphi,\zeta) = \frac{d\sqrt{2\pi}}{2} e^{i3\pi/4} \left[\frac{\mathrm{d}^2}{\mathrm{d}\sigma^2}\left(\sigma^{\alpha} e^{\zeta\sigma^2}\right)\right]_{\sigma=1} + O(\sin^2\varphi), \tag{6.120}$$
$$e^{\zeta} b_2(\varphi,\zeta) = \frac{d\sqrt{2\pi}}{8} e^{i5\pi/4} \left[\frac{\mathrm{d}^4}{\mathrm{d}\sigma^4}\left(\sigma^{\alpha} e^{\zeta\sigma^2}\right)\right]_{\sigma=1} + O(\sin^2\varphi), \tag{6.121}$$

where the remainders are estimated uniformly in ζ. From (6.71) we have that

$$|t_+(\varphi)| = \frac{1}{2\,|\sin\varphi|} + O(\sin\varphi) \quad \text{as } \varphi \to \pi. \tag{6.122}$$

In order to obtain the leading term in (6.101) for $|\varphi - \pi| < \varphi_0 - \varepsilon$ in such a form that the remainder decreases uniformly as $r \to \infty$, one has to proceed as follows: first, to use (6.101), where $N = 2$ (in this case the remainder decreases uniformly according to the just-proven part of the lemma); second, to move the following sum

$$\sum_{j=1}^{2} \exp\{y t_-^2(\varphi) + i S_-(\varphi) r\} a_j^- r^{-1/2-j}$$

arising in the expansion into the remainder (these terms decrease as $r \to \infty$ uniformly in φ and y because t_-, S_-, and a_j^- are smooth); and third, to replace $e^{\zeta} b_j$ and $|t_+|$ in

$$\sum_{j=1}^{2} e^{\zeta + iS_+(\varphi)r} a_j^+ r^{-1/2-j} = \sum_{j=1}^{2} e^{\zeta} b_j(\varphi,\zeta) e^{iS_+(\varphi)r}\, |t_+(\varphi)|^{\alpha-j+1/2} r^{-1/2-j}$$

by the leading terms in their asymptotics given by (6.120)–(6.122) (the arising error is estimated by $Cr^{-3/2}$, where C is independent of φ and $\zeta \leq 0$).

A result of this procedure is the leading term in the asymptotics for F given by (6.109). This completes the lemma's proof.

6.2.8. Concluding Remarks

Combining formulae (6.86)–(6.87) with lemmas proven in Subsections 6.2.4–6.2.7, one obtains asymptotic expansions at infinity downstream for I_1 and for its derivatives. These expansions, the result in Subsection 6.1.4 showing that I_2 decays at infinity, and (6.69) yield that the theorems formulated in Subsection 6.2.1 hold for $x \neq 0$. Then they are also valid for $x = 0$ because $G - 1/R \in C^\infty$ for $y + y_0 < 0$ (see the beginning of Subsection 6.2.1) and the asymptotic expansions obtained are uniform with respect to all horizontal directions and the depth.

6.3. Two-Dimensional Problems of Line Sources

In the present section we consider two-dimensional Green's functions. We recall (see the Conventions subsection in the Introduction) that two-dimensional problems describe wave motions parallel to a certain plane, that is, invariant with respect to translation in the direction orthogonal to that plane. Therefore, two-dimensional Green's functions correspond to straight horizontal source-lines. First, a line source in deep water is treated in Subsection 6.3.1; then, the case of a line source in shallow water is treated in Subsection 6.3.2.

6.3.1. Line Source in Deep Water

In this subsection, we consider in detail Green's function describing the uniform forward motion of a line source in deep water. In Subsection 6.3.1.1, the corresponding boundary value problem is given and a couple of equivalent representations are derived. Properties of Green's function are investigated in Subsection 6.3.1.2.

6.3.1.1. Boundary Value Problem and Derivation of Explicit Representations

We consider a two-dimensional stream of infinite depth about a source or, equivalently, the uniform forward motion of a submerged source at a constant speed U. Let the source be positioned at a point (ξ, η), $\eta < 0$, of the Cartesian coordinate system (x, y) moving together with the source. The induced steady-state velocity field is described by Green's function $G(x, y; \xi, \eta)$, which must be a solution of the following boundary value problem in $\mathbb{R}^2_- = \{(x, y) : -\infty < x < +\infty,\ y < 0\}$:

$$\nabla_z G = -2\pi \delta_\zeta(z) \quad \text{in } \mathbb{R}^2_-, \tag{6.123}$$

$$G_{xx} + \nu G_y = 0 \quad \text{when } y = 0, \tag{6.124}$$

$$\sup_{z \in \mathbb{R}^2_-} |\nabla_z (G(z, \zeta) + \log |z - \zeta|)| < \infty, \tag{6.125}$$

$$\lim_{x \to +\infty} |\nabla G(z, \zeta)| = 0. \tag{6.126}$$

Here $z = x + iy$ and $\zeta = \xi + i\eta$ are used for the sake of brevity. The main aim of the present subsection is to prove the following assertion.

The unique (up to an arbitrary constant term) solution of (6.123)–(6.126) *is given by*

$$G(z, \zeta) = -\Big[\log(\nu|z - \zeta|) + \log(\nu|z - \bar{\zeta}|) + 2 \int_0^\infty \frac{\cos k(x - \xi)}{k - \nu} e^{k(y+\eta)} \mathrm{d}k + 2\pi e^{\nu(y+\eta)} \sin \nu(x - \xi) \Big]. \tag{6.127}$$

Here the integral is understood as the Cauchy principal value. Another representation for G has the following form:

$$G(z, \zeta) = -\mathrm{Re} \left\{ \log(\nu[z - \zeta]) + \log(\nu[z - \bar{\zeta}]) + 2 \int_0^\infty \frac{e^{it} \mathrm{d}t}{t + \nu(z - \bar{\zeta})} \right\}. \tag{6.128}$$

First, let us show that (6.127) and (6.128) are equivalent. Using the path of integration ℓ_- going along the positive real k axis and indented below at $k = \nu$ (see Fig. 1.1), we have

$$\int_0^\infty \frac{\cos k(x - \xi)}{k - \nu} e^{k(y+\eta)} \mathrm{d}k + \pi e^{\nu(y+\eta)} \sin \nu(x - \xi) = \mathrm{Re} \int_{\ell_-} \frac{\exp\{-ik(z - \bar{\zeta})\}}{k - \nu} \mathrm{d}k.$$

In order to show that the last integral is equal to the integral in (6.128), we replace ℓ_- by the path ℓ emanating from $k = 0$ and passing through $k = -(z - \bar{\zeta})^{-1}$. This is possible because both points $z - \bar{\zeta}$ and $-(z - \bar{\zeta})^{-1}$ belong to the lower half-plane and the integrand has no poles there. Also, it is essential that $\exp\{-ik(z - \bar{\zeta})\}$ decays at infinity in the lower half-plane between ℓ_- and ℓ. Changing the integration variable $k \mapsto -t/(z - \bar{\zeta})$, we arrive at the required integral on the right-hand side in (6.128), which completes the proof of equivalence of (6.127) and (6.128).

In order to derive (6.128), a solution of (6.123)–(6.126) is sought in the form of the so-called complex potential w, which is a holomorphic function in $\{\mathrm{Im}\, z \leq 0\}\backslash\{\zeta\}$ and such that $G = \mathrm{Re}\, w$. Using the following representation,

$$w(z, \zeta) = -\{\log(z - \zeta) + H(z, \zeta)\},$$

we replace the original problem by the problem of finding $H(z, \zeta)$, which is holomorphic in the half-plane $\mathrm{Re}\, z < 0$. For choosing H so that $\mathrm{Re}\, w$ satisfies (6.124), we put

$$F(z) = i\,\frac{\mathrm{d}w}{\mathrm{d}z} - \nu w, \tag{6.129}$$

and we note that

$$F(z) = -\frac{i}{z - \zeta} + \nu \log(z - \zeta) + H_1(z, \zeta),$$

where H_1 is another holomorphic function in $\mathrm{Im}\, z < 0$. The reason for introducing F is that it satisfies a simpler boundary condition

$$\mathrm{Im}\, F = 0 \quad \text{when } y = 0. \tag{6.130}$$

In fact, (6.124) is equivalent to

$$\mathrm{Re}\left\{\frac{\mathrm{d}^2 w}{\mathrm{d}z^2} + i\nu\,\frac{\mathrm{d}w}{\mathrm{d}z}\right\} = 0 \quad \text{when } y = 0,$$

and this can be written as

$$\mathrm{Im}\, \mathrm{d}F/\mathrm{d}z = 0 \quad \text{or } \mathrm{Im}\, \mathrm{d}F/dx = 0 \quad \text{when } y = 0,$$

and so (6.124) and (6.130) are equivalent because w and F are defined up to an arbitrary constant term.

In view of (6.130), F can be analytically extended to the upper half-plane by the Schwarz reflection principle:

$$F(z) = -\frac{i}{z - \zeta} + \frac{i}{z - \bar{\zeta}} + \nu \log\{(z - \zeta)(z - \bar{\zeta})\} + H_2(z),$$

where H_2 is an entire function such that $\mathrm{Im}\, H_2 = 0$ when $y = 0$. Thus, if w is a solution of (6.129) with F just defined, then $\mathrm{Re}\, w$ satisfies (6.123) and (6.124). We choose H_2 to be equal to zero, thus guaranteeing (6.125) and (6.126) to hold (see the first assertion in Subsection 6.3.1.2). Thus w must be a solution to

$$i\,\frac{\mathrm{d}w}{\mathrm{d}z} - \nu w = -\frac{i}{z - \zeta} + \frac{i}{z - \bar{\zeta}} + \nu \log\{(z - \zeta)(z - \bar{\zeta})\} \tag{6.131}$$

in the whole z plane. The solution of (6.131) is as follows:

$$w(z,\zeta) = -\log(z-\zeta)(z-\bar{\zeta}) - 2\int_0^\infty \frac{e^{it}\,\mathrm{d}t}{t+\nu(x-\bar{\zeta})}. \tag{6.132}$$

In fact, the logarithmic term in (6.132) solves (6.131) with $i(z-\bar{\zeta})^{-1}$ replaced by $-i(z-\bar{\zeta})^{-1}$ on the right-hand side. Hence, it remains for us to verify that

$$w_1(z) = -2\int_0^\infty \frac{e^{it}\,\mathrm{d}t}{t+\nu(z-\bar{\zeta})}$$

is a solution to

$$i\,\frac{\mathrm{d}w_1}{\mathrm{d}z} - \nu w_1 = \frac{2i}{z-\bar{\zeta}}.$$

This follows from the fact that w_1 can be represented in the form of convolution of $2i(z-\bar{\zeta})^{-1}$ and a fundamental solution of the differential operator in (6.131). In order to obtain such a representation one has to change the variable $t \to -\nu t$ in the integral for w_1.

The uniqueness of Green's function is established in the following lemma. *Let $u \in C^2(\overline{\mathbb{R}^2_-})$ be a harmonic function in $\mathbb{R}^2_-$, and let it satisfy*

$$u_{xx} + \nu u_y = 0 \quad \text{when } y = 0, \ \sup_{y<0} |\nabla u(x,y)| < \infty.$$

Then

$$u(x,y) = k_1 e^{\nu(y+iz)} + k_2 e^{\nu(y-iz)} + k_3 x + k_4. \tag{6.133}$$

If additionally $|\nabla u| \to 0$ as $x \to +\infty$, then $u =$ const.

For the proof we consider w such that $u_{xx} + \nu u_y$ in $\overline{\mathbb{R}^2_-}$ and $w(x,y) = -w(x,-y)$ when $y > 0$. Since $w(x,0) = 0$, the Schwarz reflection principle yield that w is harmonic in $\mathbb{R}^2$. Then

$$\begin{aligned}
\pi|w(z)| &= \left|\iint_{|z_1-z|<1} w(z_1)\,\mathrm{d}x_1\mathrm{d}y_1\right| \\
&\le \left|\iint_{\substack{|z_1-z|<1\\ y_1<0}} \left(u_{x_1x_1} + \nu u_{y_1}\right)\mathrm{d}x_1\mathrm{d}y_1\right| \\
&\quad + \left|\iint_{\substack{|z_1-\bar{z}|<1\\ y_1<0}} \left(u_{x_1x_1} + \nu u_{y_1}\right)\mathrm{d}x_1\mathrm{d}y_1\right|.
\end{aligned}$$

Integrating $u_{x_1x_1}$ with respect to x_1, we find that $C\sup|\nabla u|$ is a bound for $|w|$, and so w is uniformly bounded in $\mathbb{R}^2$. Hence, $w =$ const, which vanishes in view of the boundary condition $w(x, 0) = 0$; that is,

$$u_{xx} + \nu u_y = 0 \quad \text{in } \mathbb{R}^2_-.$$

Taking into account that $\nabla^2 u = 0$ in $\mathbb{R}^2_-$, we get

$$u_{yy} - \nu u_y = 0 \quad \text{in } \mathbb{R}^2_-,$$

and so $u(x, y) = c_1(x) + c_2(x)e^{\nu y}$. Using the harmonicity of u again, we arrive at (6.133). The last assertion of the lemma immediately follows from (6.133). The proof is complete.

6.3.1.2. *Properties of Green's Function*

The following assertion contains an asymptotic representation of G at infinity, and thus it completes the proof of the theorem formulated in Subsection 6.3.1.1.

Let $|\zeta| <$ const; *then the following asymptotic formula holds as* $|z| \to \infty$:

$$G(z, \zeta) = -2\log(\nu|z|) - 4\pi H(-x)e^{\nu(y+\eta)} \sin \nu(x - \xi) + \varphi(x, y),$$

where H *is the Heaviside function. Also, the estimates*

$$\varphi = O(|z|^{-1}), \quad |\nabla\varphi| = O(|z|^{-2})$$

are true.

First we assume that $x > 0$, which implies that

$$2|t + \nu(z - \bar{\zeta})| \geq |t + \nu x - \nu\xi| + \nu|y + \eta| \\ \geq t + \nu x - \nu|\xi| + \nu|y| \geq t + \nu(|z| - |\zeta|).$$

This and the equality

$$\int_0^\infty \frac{e^{it}\,dt}{t + \nu(z - \bar{\zeta})} = \frac{i}{z - \bar{\zeta}} + i\int_0^\infty \frac{e^{it}\,dt}{[t + \nu(z - \bar{\zeta})]^2}$$

lead to the required asymptotic formulae for G and ∇G as $|z| \to \infty$ in the quadrant $\{x > 0,\ y < 0\}$.

Let us turn to the case $x < 0$. The same considerations lead to the required result after using

$$\int_0^\infty \frac{e^{it}\,dt}{t + \nu(z - \bar{\zeta})} = 2\pi i e^{-i\nu(z-\bar{\zeta})} - \int_{-\infty}^0 \frac{e^{it}\,dt}{t + \nu(z - \bar{\zeta})}.$$

Below we consider some properties of Green's function in the case in which the source point is placed on the free surface, that is, $\eta = 0$. One can easily see that the integrals in (6.127) and (6.128) are convergent for $\eta = 0$

and $(x-\xi)^2+y^2 \neq 0$, and so these representations for G remain valid. We begin with an expansion holding for $y, \eta \leq 0$.

For $y, \eta \leq 0$ *we have* $G(z,\zeta)=\log|z-\bar{\zeta}|-\log|z-\zeta|-g(z,\zeta)$, *where*

$$g(z,\zeta) = -2\,\mathrm{Re}\left\{ \log(\nu[z-\bar{\zeta}]) \sum_{m=1}^{\infty} \frac{[-i\nu(z-\bar{\zeta})]^m}{m!} \right.$$

$$\left. + \exp\{-i\nu(z-\bar{\zeta})\}\left(\gamma - \frac{\pi i}{2} + \sum_{m=1}^{\infty} \frac{[i\nu(z-\bar{\zeta})]^m}{m!m}\right)\right\}. \tag{6.134}$$

Here $\gamma = 0.5772\ldots$ *is Euler's constant.*

After changing the variable $t \mapsto \tau - \nu(z-\bar{\zeta})$, one obtains that the integral in (6.128) is equal to

$$-\exp\{-i\nu(z-\bar{\zeta})\}\int_{+\infty}^{\nu(z-\bar{\zeta})} e^{i\tau}\frac{\mathrm{d}\tau}{\tau}.$$

From 8.230 in Gradshteyn and Ryzhik [96] and Section 5.2 in Abramowitz and Stegun [1], it follows that the last integral is equal to

$$\mathrm{Ci}(\nu[z-\bar{\zeta}]) - i\,\mathrm{Si}(\nu[z-\bar{\zeta}]).$$

Replacing this combination of the sine and cosine integrals by an expansion into power series (see 8.232 in [96]), we get

$$G(z,\zeta) = -\mathrm{Re}\left\{ \log(\nu[z-\zeta]) + \log(\nu[z-\bar{\zeta}]) \right.$$

$$-2\exp\{-i\nu(x-\bar{\zeta})\}\left[\log(\nu[z-\bar{\zeta}]) + \gamma - \frac{\pi i}{2}\right.$$

$$\left.\left. + \sum_{m=1}^{\infty}(-1)^m\left(\frac{[\nu(z-\bar{\zeta})]^{2m}}{2m(2m)!} - i\frac{[\nu(z-\bar{\zeta})]^{2m-1}}{(2m-1)(2m-1)!}\right)\right]\right\}.$$

This can be written as follows:

$$G(z,\zeta) = -\mathrm{Re}\left\{ \log(\nu[z-\zeta]) - \log(\nu[z-\bar{\zeta}]) \right.$$

$$+2\,(1-\exp\{i\nu(z-\bar{\zeta})\})\log(\nu[z-\bar{\zeta}])$$

$$\left. -2\exp\{-i\nu(z-\bar{\zeta})\}\left[\gamma - \frac{\pi i}{2} + \sum_{m=1}^{\infty}\frac{[i\nu(z-\bar{\zeta})]^m}{m!m}\right]\right\}.$$

Substituting the series of the first exponential function, we arrive at (6.134) for $g(z,\zeta)$.

Three consequences of the last assertion are as follows. First, the equality

$$G(x, y; \xi, 0) = g(x, y; \xi, 0)$$

holds and so Green's function has no singularity when the source point is placed on the free surface. However, the gradient of Green's function has a logarithmic singularity as the following formula shows:

$$|\nabla g| = O[\nu \log(\nu|z - \bar{\zeta}|)] \quad \text{as } \nu|z - \bar{\zeta}| \to 0. \tag{6.135}$$

Second, when one substitutes $\eta = 0$ into (6.134) and then differentiates the result with respect to x, the following formula arises:

$$G_x(x, y; \xi, 0) = 2\nu \sum_{m=0}^{\infty} \frac{(-\nu r)^m}{m!} \left\{ \left(\varphi - \frac{\pi}{2}\right) \cos m \left(\varphi + \frac{\pi}{2}\right) \right.$$
$$\left. + \left[\log(\nu r) - \frac{\Gamma'(m+1)}{\Gamma(m+1)}\right] \sin m \left(\varphi + \frac{\pi}{2}\right) \right\}, \tag{6.136}$$

where $re^{i\varphi} = (x - \xi) + iy$, $\varphi \in [-\pi, 0]$. Finally, from here we obtain the jump formula for Green's function on the free surface:

$$\lim_{x \to \xi \pm 0} G_x(x, 0, \xi, 0) = \pi\nu(-2 \pm 1). \tag{6.137}$$

The next lemma, resulting from integration by parts, is essential for estimating the gradient of Green's function as $\nu \to \infty$.

The asymptotic formula

$$\mathrm{Re} \int_0^\infty \frac{e^{it}\, \mathrm{d}t}{t + \nu(z - \bar{\zeta})} = \mathrm{Re} \sum_{k=1}^{N-1} (k-1)! \left[\frac{i}{\nu(z - \bar{\zeta})}\right]^k + O[(\nu|z - \bar{\zeta}|)^{-N}]$$

holds as $\nu|z - \bar{\zeta}| \to \infty$.

6.3.2. Line Source in Shallow Water

Unlike the water-wave problem for which wave patterns of sources are similar in deep and shallow water (only quantitative differences occur in the behavior of the corresponding Green's functions, as is demonstrated in Subsection 1.2.1), the Neumann–Kelvin problem demonstrates more complicated behavior of Green's function. It depends essentially on the parameter $\nu d = gd/U^2$ related to the so-called Froude number, which is a similarity parameter in the theory of wave-making resistance (see, for example, Newman [262]).

6.3.2.1. Boundary Value Problem and Its Explicit Solution

Let L be a strip $\{(x, y) \in \mathbb{R}^2 : d < y < 0\}$, and for $\eta \in (-h, 0)$ let (ξ, η) be the projection of a line source on the (x, y) plane. The corresponding velocity potential $G(z, \zeta)$ usually referred to as Green's function must satisfy the following boundary value problem:

$$\nabla_z^2 G = -2\pi\delta_\zeta(z) \quad \text{in } L, \tag{6.138}$$

$$G_{xx} + \nu G_y = 0 \quad \text{when } y = 0, \tag{6.139}$$

$$G_y = 0 \quad \text{when } y = -d, \tag{6.140}$$

$$\limsup_{|x| \to -\infty} |\nabla G| < \infty, \qquad \lim_{x \to +\infty} |\nabla G| = 0. \tag{6.141}$$

In what follows, λ_0 denotes the unique positive root of $\nu \tanh \lambda d = \lambda$ existing only when $\nu d > 1$. This wavenumber characterizes waves behind the source as the following theorem shows.

For $\nu d \neq 1$ the unique (up to an arbitrary constant term) solution to (6.138)–(6.141) *is as follows*:

$$\begin{aligned}
G(z, \zeta) = &- \log|z - \zeta| - \log|z + 2id - \bar{\zeta}| + \frac{\pi\nu(x - \xi)}{\nu d - 1} \\
&+ H(\nu d - 1)\frac{2\pi\nu \cosh \lambda_0(y + d) \cosh \lambda_0(\eta + d)}{\lambda_0(\nu d - \cosh^2 \lambda_0 d)} \sin \lambda_0(x - \xi) \\
&+ 2\int_0^\infty \left[\frac{k + \nu}{k} e^{-kd} \frac{\cos k(x - \xi) \cosh k(y + d) \cosh k(\eta + d)}{\nu \sinh kd - k \cosh kd}\right. \\
&\left. + \frac{e^{-kd}}{k} - \frac{\nu}{(\nu d - 1)k^2}\right] \mathrm{d}k.
\end{aligned} \tag{6.142}$$

Here $-d \leq y \leq 0$, $-d < \eta < 0$, H is the Heaviside function, and the integral is understood as the Cauchy principal value at $k = \lambda_0$. Moreover, the integrand is regularized so that it has a finite limit as $k \to 0$.

Relations (6.138)–(6.140) can easily be verified by direct calculation for G given by (6.142). The validity of (6.141) follows from the first assertion in Subsection 6.3.2.2 and the uniqueness of G is a consequence of the following lemma.

Let $u \in C^2(L)$ be a harmonic function in L satisfying the following boundary conditions:

$$u_{xx} + \nu u_y = 0 \quad \text{when } y = 0, \qquad u_y = 0 \quad \text{when } y = -d,$$

and such that $\sup_L |\nabla u| < \infty$. *If* $\nu d \neq 1$, *then*

$$u(x, y) = c_1 + c_2 x + H(\nu d - 1)\cosh \lambda_0 (y + d)[c_3 e^{i\lambda_0 x} + c_4 e^{-i\lambda_0 x}],$$

where H *is the Heaviside function.*

Let us consider $u(\cdot, y)$ as a distribution belonging to $S'(\mathbb{R}_x)$ and depending smoothly on y. Then the Fourier transform $\hat{u}(\sigma, y)$ is in $S'(\mathbb{R}_\sigma)$ for $-d \leq y \leq 0$ and satisfies the following boundary value problem:

$$\begin{aligned} \hat{u}_{yy} - \sigma^2 \hat{u} &= 0 \quad \text{for } -d < y < 0, \\ \nu \hat{u}_y - \sigma^2 \hat{u} &= 0 \quad \text{when } y = 0, \\ \hat{u}_y &= 0 \quad \text{when } y = -d. \end{aligned}$$

Solving the differential equation in a wider class of distributions $D'(\mathbb{R}_\sigma)$, we get

$$\hat{u} = C_1(\sigma) e^{\sigma y} + C_2(\sigma) e^{-\sigma y}, \quad \text{where } C_1, C_2 \in D'(\mathbb{R}_\sigma).$$

From the last boundary condition we obtain that

$$\hat{u} = A(\sigma)\cosh \sigma(y + d), \quad A \in D'(\mathbb{R}_\sigma). \tag{6.143}$$

Now the first boundary condition gives

$$A(-\sigma^2 \cosh \sigma d + \nu\sigma \sinh \sigma d) = 0.$$

Here the second factor is an even function of σ having a zero of the second order at $\sigma = 0$. Moreover, this function does not vanish on the half-axis $\sigma > 0$ when $\nu d < 1$; if $\nu d > 1$, then this function has only one simple zero at $\sigma = \lambda_0$. These facts imply that A has the following form:

$$A(\sigma) = C_1 \delta(\sigma) + C_2 \delta'(\sigma) + H(\nu d - 1)\left[C_3 \delta(\sigma - \lambda_0) + C_4 \delta(\sigma + \lambda_0)\right],$$

where δ is Dirac's measure on the σ axis. The last expression combined with (6.143) proves the lemma.

Results obtained here suggest that two cases of the Neumann–Kelvin problem must be distinguished for shallow water. We will speak about a *subcritical flow* (problem) when $\nu d > 1$ and there exists the wavenumber λ_0 corresponding to propagating free waves on the surface of stream having a constant depth. When the opposite inequality $\nu d < 1$ holds, there are no propagating waves and a flow (and the Neumann–Kelvin problem) is called *supercritical.*

According to this definition the infinite-depth flow is always subcritical.

6.3.2.2. Properties of Green's Function

As in Subsection 6.3.1.2, we first formulate the lemma concerning the asymptotic behavior of G at infinity.

Let $|\xi| <$ const *and* $|x| \to \infty$. *Then*

$$G(x, y; \xi, \eta) = 2\pi H(-x) \left[\frac{\nu(x-\xi)}{\nu d - 1} + H(\nu d - 1) \frac{2\nu \cosh \lambda_0(y+d) \cosh \lambda_0(\eta + d)}{\lambda_0(\nu d - \cosh^2 \lambda_0 d)} \times \sin \lambda_0(x - \xi) \right] + \varphi(x, y)$$

holds, where $\varphi = O(|x|^{-1})$ *and* $|\nabla \varphi| = O(|x|^{-2})$.

This lemma is a consequence of results in Sections 2, 5, and 8 of Bochner's book [28]; also, the following assertion should be taken into account.

For $-d \leq y, \eta \leq 0$, *another representation of Green's function is true:*

$$G(z, \zeta) = -\log |x - \zeta| + \log |z - \bar{\zeta}| + g(z, \zeta),$$

where

$$g(z, \zeta) = \frac{\pi \nu(x-\xi)}{\nu d - 1} + 2I(x, y; \xi, \eta) + H(\nu d - 1) \frac{2\pi \nu \cosh \lambda_0(y+d) \cosh \lambda_0(\eta+d)}{\lambda_0(\nu d - \cosh^2 \lambda_0 d)} \sin \lambda_0(x - \xi)$$

and the function

$$I(x, y; \xi, \eta) = \int_0^\infty \left\{ \cos k(x - \xi) \frac{e^{-kd}}{k} \left[\cosh k(y + \eta + d) + \frac{(k+\nu)\cosh k(y+d) \cosh k(\eta + d)]}{\nu \sinh kd - k \cosh kd} \right] - \frac{\nu}{(\nu d - 1)k^2} \right\} dk$$

is continuous in $\bar{L} \times \bar{L}$.

In order to obtain I from (6.142), it is sufficient to notice that

$$\log |z - \bar{\zeta}| + \log |z + 2id - \bar{\zeta}| = 2 \int_0^\infty \frac{e^{-kd}}{d} [1 - \cos k(x - \xi) \cosh k(y + \eta + d)] \, dk,$$

which is a consequence of 3.951.8 in Gradshteyn and Ryzhik [96] (see also integral representations of logarithms in Subsection 1.2.1). Since the integrand in I is bounded by $C(d)k^{-2}$ as $k \to \infty$, this integral converges uniformly and is a continuous function.

The function $G(x, 0; \xi, 0)$ *is continuous for* $-\infty < x, \xi < +\infty$; *also,*

$$\lim_{x \to \xi \pm 0} \frac{\partial G}{\partial x}(x, 0; \xi, 0) = \pi \nu \left[\frac{1}{\nu d - 1} \pm 1 + H(\nu d - 1) \frac{2 \cosh^2 \lambda_0 d}{\nu d - \cosh^2 \lambda_0 d} \right]. \tag{6.144}$$

The first assertion follows immediately from the representation obtained in the previous lemma. The same representation gives the following after simple manipulation:

$$\frac{\partial G}{\partial x}(x, 0; \xi, 0) = \frac{\pi \nu}{\nu d - 1} + 2 \frac{\partial I}{\partial x}(x, 0; \xi, 0) + H(\nu d - 1) \frac{2\pi \nu \cosh^2 \lambda_0 d}{\nu d - \cosh^2 \lambda_0 d} \cos \lambda_0 (x - \xi),$$

where

$$\frac{\partial I}{\partial x}(x, 0; \xi, 0) = -\nu \int_0^\infty \frac{\sin k(x - \xi)}{\nu \tanh kd - k} \, dk.$$

Changing the variable of integration, we get that for $\pm(x - \xi) > 0$,

$$\int_0^\infty \frac{\sin k(x - \xi)}{\nu \tanh kd - k} \, dk = \int_0^{\pm\infty} \frac{\sin k \, dk}{\nu (x - \xi) \tanh [kd/(x - \xi)] - k}.$$

Letting $x - \xi \to \pm 0$, and using the well-known formula (see 3.721.1 in Gradshteyn and Ryzhik [96])

$$\int_0^{\pm\infty} \frac{\sin k}{k} \, dk = \pm \frac{\pi}{2},$$

we obtain

$$\lim_{x \to \xi \pm 0} (\partial I / \partial \xi)(x, 0; \xi, 0) = \pm \pi \nu / 2,$$

which implies (6.144). It is obvious that (6.144) becomes (6.137) as d tends to infinity.

6.4. Bibliographical Notes

There are numerous papers treating Green's functions for sources in the uniform motion in three and two dimensions. Here we mention some of them (mainly more recent works), and other references can be found in the survey papers by Wehausen and Laitone [354] and Wehausen [353].

6.1.1. Various representations of Kelvin's source in deep water are given by many authors; see the classic survey by Wehausen and Laitone [354] and the later works on this topic by Bessho [23], Eggers, Sharma, and Ward [63], Noblesse [268, 270], Euvrard [64], and Newman [264].

The velocity potential for a source point moving in a layer of finite depth can be found in Kostyukov [147] and in the works mentioned above ([353, 354]).

6.1.2. The derivation of Green's function for the three-dimensional ship waves in deep water is given in Havelock [108], Kochin [141], Peters and Stoker [286], Kostyukov [147], and elsewhere.

6.1.3. The justification of formulae for Green's function presented here was not published earlier. Ursell [336] justified Bessho's representation proposed in [23].

6.1.4. Asymptotic behavior of the double integral was found by Newman [264].

6.1.5. To our knowledge, the question of uniqueness has not been considered earlier for Kelvin's source.

Other works. Some questions concerning the wave pattern caused by Kelvin's source and line distributions of such sources were considered by Bauer [20], Noblesse [269], Tuck, [320], and Ursell [340], respectively.

6.2. The asymptotic results presented in this section are borrowed from the work [229] by Maz'ya and Vainberg.

Investigations in this field were initiated in the famous lecture by Kelvin [134] published in 1887. In this lecture and in a later paper [135], Kelvin considered the V pattern for waves produced by a moving concentrated pressure applied to the free surface. Further results in this direction were obtained by Hogner [115, 116], Peters [284], Ursell [339], and Euvrard [64]. These papers contain the far-field asymptotic expansions uniform in some regions, in particular in a neighborhood of the boundary of Kelvin's angle. Paper [229] concluded this series of works, and the expansions obtained in it are uniform with respect to the location of the observation point as well as the depth of submergence of the source. This was achieved by virtue of an additional wave discovered by the authors and concentrated near the source's track.

Numerical results. In conclusion of the bibliographical notes on Kelvin's source in deep water, we mention several papers treating computational aspects of the problem. Newman [264] developed a numerical procedure greatly facilitating the computation of the double integral. Three other papers (Newman [265], Clarisse and Newman [38], and Baar and Price [17]) are concerned with various methods of computation of the single integral.

6.3.1. For deriving Green's function of the line source in deep water, we apply the method proposed by Keldysh [131] (see also Keldysh and Lavrentiev [132], Kochin [141], and Kostyukov [147]). Among other methods used for this purpose, we mention application of the Fourier transform by Lenoir [187], who derived the two-dimensional Green's function for both deep and shallow water. The proofs of uniqueness and of the asymptotic formula for Green's function were given by Vainberg and Maz'ya [346]. The near-field expansion in Subsection 6.3.1.2 is similar to that obtained by Ursell [335].

6.3.2. The material in this subsection is borrowed from Vainberg and Maz'ya's paper [346].

7
The Neumann–Kelvin Problem for a Submerged Body

As for the water-wave problem investigated in Part 1, it is natural to solve the Neumann–Kelvin problem by applying integral equation techniques, since Green's function is constructed. However, in the theory of ship waves this approach is less straightforward than in the theory of time-harmonic waves. First of all, well-posed statements of the two-dimensional Neumann–Kelvin problem are different for totally submerged and surface-piercing bodies because certain supplementary conditions should be imposed in the latter case. Another essential point distinguishes the Neumann–Kelvin problem for a subcritical flow from the water-wave problem. In fact, any solution to the homogeneous water-wave problem has a finite energy, but for solutions of the homogeneous Neumann–Kelvin problem the unconditional validity of this property is still an open question.

So, using integral equations, we have to rely on the method that does not involve an a priori knowledge of uniqueness in the boundary value problem. Such a method was applied to the water-wave problem in Chapters 2 and 3. Its main features are related to the analyticity of integral operators as functions of the parameter ν and to the properties of these operators in limiting cases.

As in Part 1, we treat the simplest problem first, and this is the two-dimensional problem of a body totally submerged in water of infinite depth (see Section 7.1). This is the only problem for which the most complete result (the unique solvability for all $\nu > 0$) is obtained for a particular geometry (when the body is a circular cylinder). In Section 7.2, we turn to the case of shallow water and begin with results for a subcritical regime that are similar to those derived for deep water. In Subsection 7.2.2, the supercritical regime is considered in detail. In the short section, Section 7.3, formulae for the wave resistance of a submerged cylinder are derived for deep and shallow water. The three-dimensional problem of a submerged body is treated in Section 7.4, where the solvability theorem is proved for all $\nu > 0$ except possibly for a finite number of values. Bibliographical notes are collected in Section 7.5.

7.1. Cylinder in Deep Water

We consider a uniform stream of infinite depth about an infinitely long, horizontal, submerged cylinder with generators orthogonal to the stream direction, and so the arising water motion is two dimensional. Let the cylinder's cross section be a bounded, simply connected domain $D \subset \mathbb{R}^2_-$ such that $\bar{D} \subset \mathbb{R}^2_-$. By S we denote a C^3 curve bounding D, and $W = \mathbb{R}^2_- \setminus \bar{D}$ is the cross section of the water domain. For convenience of reference we recall that a potential $u(x, y)$ describing the steady-state velocity field in W must satisfy the following boundary value problem (see the Linear Ship Waves section in the Introduction):

$$\nabla^2 u = 0 \quad \text{in } W, \tag{7.1}$$

$$u_{xx} + \nu u_y = 0 \quad \text{when } y = 0, \tag{7.2}$$

$$\partial u / \partial n = f \quad \text{on } S, \tag{7.3}$$

$$\sup_W |\nabla u(x, y)| < \infty, \qquad \lim_{x \to +\infty} |\nabla u| = 0. \tag{7.4}$$

We also recall that an arbitrary constant term added to u does not violate the validity of conditions (7.1)–(7.4).

We begin by establishing the asymptotic behavior at infinity for solutions of this problem (Subsection 7.1.1). In Subsection 7.1.2, we prove that (7.1)–(7.4) is solvable for all $\nu > 0$ with a possible exception of a finite set of values. The uniqueness theorem is proved in Subsection 7.1.3, and the same restriction on ν is imposed (however, it is possible that a greater set of values should be excluded). Subsection 7.1.4 is concerned with an auxiliary problem, and a particular solution of this problem is used in Subsection 7.1.5 for deriving necessary and sufficient conditions guaranteeing the unique solvability of the Neumann–Kelvin problem (7.1)–(7.4).

7.1.1. Asymptotic Behavior of Solutions at Infinity

In order to apply a lemma describing the asymptotics of Green's function, we need the following assertion on Green's representation.

Let $u \in C^2(\bar{W})$ be a solution of (7.1)–(7.4). *Then, for $z \in W$,*

$$u(z) = \frac{1}{2\pi} \int_S \left[u(\zeta) \frac{\partial G}{\partial n_\zeta}(z, \zeta) - \frac{\partial u}{\partial n_\zeta} G(z, \zeta) \right] ds + c,$$

where $c =$ const.

Let u_0 be a function in $C^2(\mathbb{R}^2_-)$ obtained by an arbitrary extension of u into D. Then $-\nabla^2 u_0 = f$ in $\mathbb{R}^2_-$, where supp $f \subset \bar{D}$, and (7.2) and (7.4) hold for u_0. The function

$$u_1(z) = \frac{1}{2\pi} \int_D G(z, \zeta) f(\zeta) \, d\xi \, d\eta$$

has the same properties as u_0, and so applying the second assertion in Subsection 6.3.1.1 to $u_0 - u_1$, we get

$$u_0 - u_1 = k_1 e^{\nu(y+ix)} + k_2 e^{\nu(y-ix)} + k_3 x + k_4.$$

Since (7.4) implies $k_1 = k_2 = k_3 = 0$, it follows that

$$u(z) = \frac{1}{2\pi} \int_D G(z, \zeta) f(\zeta) \, d\xi \, d\eta + c, \quad \text{for } z \in W.$$

Applying Green's theorem to the last integral, we arrive at the required integral representation.

The following theorem is a direct consequence of the lemma just proven and the asymptotic formula obtained for Green's function in Subsection 6.3.1.2.

Let $u \in C^2(\bar{W})$ be a solution of (7.1)–(7.4). Then the following asymptotic formula holds as $|z| \to \infty$:

$$u(z) = Q \log(\nu|z|) + c + \psi(x, y) + H(-x)(\mathcal{A} \sin \nu x + \mathcal{B} \cos \nu x) e^{\nu y},$$

where $\psi = O(|z|^{-1})$, $|\nabla \psi| = O(|z|^{-2})$, and Q, $\mathcal{A}$, and $\mathcal{B}$ are constants given by the formulae

$$Q = \pi^{-1} \int_S \frac{\partial u}{\partial n} \, ds,$$

$$\mathcal{A} = -2 \int_S \left[u \frac{\partial}{\partial n} (e^{\nu y} \cos \nu x) - \frac{\partial u}{\partial n} e^{\nu y} \cos \nu x \right] ds,$$

$$\mathcal{B} = 2 \int_S \left[u \frac{\partial}{\partial n} (e^{\nu y} \sin \nu x) - \frac{\partial u}{\partial n} e^{\nu y} \sin \nu x \right] ds.$$

The last two constants are proportional to amplitudes of sine and cosine waves, respectively, representing the wave pattern at infinity downstream. Furthermore, $Q\pi/2$ is equal to the mean value of the extra rate of flow at infinity downstream that is due to the presence of a cylinder.

7.1.2. Kochin's Solvability Theorem

Let us seek a solution to (7.1)–(7.4) in the following form:

$$u(z) = \frac{1}{\pi} \int_S G(z, \zeta) \mu(\zeta) \, ds + c, \tag{7.5}$$

where $\mu \in C^{1,\alpha}(S)$ is an unknown density. In what follows, we use properties of the single-layer potential (7.5) without further reference (these properties and other related material from potential theory are given in Subsection 2.1.1). We have that $u \in C^2(\bar{W})$ and satisfies (7.1), (7.2), and (7.4). Also, (7.3) leads to the Fredholm integral equation:

$$-\mu(z) + (T_\nu \mu)(z) = f(z) \quad z \in S, \tag{7.6}$$

where the integral operator T_ν is defined as follows:

$$(T_\nu \mu)(z) = \frac{1}{\pi} \int_S \mu(\zeta) \frac{\partial G}{\partial n_z}(z, \zeta) \, \mathrm{d}s.$$

Since (see the second assertion in Subsection 6.3.1.2)

$$G(z, \zeta) = -[\log|z - \zeta| - \log|z - \bar{\zeta}| + g(z, \zeta)],$$

where by virtue of (6.127)

$$g(z, \zeta) = 2\left\{\pi e^{\nu(y+\eta)} \sin \nu(x - \xi)\right.$$

$$\left. + \int_0^\infty \left[\frac{\cos k(x - \xi)}{k - \nu} e^{\mu(y+\eta)} + \frac{1 - e^{k(y+\eta)} \cos k(x - \xi)}{k}\right] \mathrm{d}k\right\},$$

a straightforward but lengthy calculation gives that

$$\frac{\partial G}{\partial n_z}(z, \zeta) = \frac{\cos(n_z, r)}{r} - \frac{\cos(n_z, r_0)}{r_0}$$

$$+ 2\nu \int_x^{+\infty} \sin[\nu(t - x) - (r_t, n_z)] \frac{\mathrm{d}t}{r_t}. \tag{7.7}$$

Here $r = |z - \zeta|$, $r_0 = |z - \bar{\zeta}|$, and $r_t = [(t - \xi)^2 + (y + \eta)^2]^{1/2}$; $\mathbf{r}$ and $\mathbf{r}_0$ are vectors directed to z from ζ and $\bar{\zeta}$, respectively, and $\mathbf{r}_t$ is directed from $\bar{\zeta}$ to $t + iy$; at last, the angle (r_t, n_z) is measured from $\mathbf{n}_z$ to $\mathbf{r}_t$ counterclockwise. It is known (see, for example, Petrovskii [288]) that the kernel (7.7) belongs to $C(S \times S)$. Also, we notice that (7.7) is an analytic function of ν having its values in $C(S \times S)$.

Aiming to apply the *invertibility theorem* formulated in Subsection 2.1.2.1, we have to investigate the integral equation (7.6) as ν tends to zero and $+\infty$. Thus we begin with the following lemma concerning small values of ν.

For all sufficiently small values $\nu > 0$, the integral equation (7.6) *is uniquely solvable in* $C(S)$.

Let us consider (7.6) for $\nu = 0$:

$$-\mu(z) + (T_0 \mu)(z) = f_0(z), \quad z \in S, \tag{7.8}$$

where by virtue of (7.7) we have

$$(T_0\mu)(z) = \frac{1}{\pi}\int_S \mu(\zeta)\frac{\partial}{\partial n_z}(\log|z-\zeta| - \log|z-\bar{\zeta}|)\,ds.$$

We extend μ and f_0 to $S' = \{(x, -y) : (x, y) \in S\}$ as odd functions of y. Then (7.8) takes the form of the integral equation corresponding to the Neumann problem in the domain outside the union of S and S'.

In order to demonstrate that the latter equation has a unique solution, we consider a solution μ_0 of the corresponding homogeneous equation:

$$-\mu_0(z) + \frac{1}{\pi}\int_{S\cup S'} \mu_0(\zeta)\,\frac{\cos(n_z, r)}{r}\,ds = 0.$$

Then the single-layer potential

$$\frac{-1}{\pi}\int_{S\cup S'} \mu_0(\zeta)\log|z-\zeta|\,ds \tag{7.9}$$

is a solution of the homogeneous Neumann problem in the domain outside of $S \cup S'$. Hence, (7.9) is equal to a constant in this domain and also on $S \cup S'$. By the uniqueness theorem for the Dirichlet problem, (7.9) is equal to the same constant inside the union of S and S'. Now, the jump formula for the normal derivative [see (2.4) in Subsection 2.1.1] implies that $\mu_0 = 0$ on $S \cup S'$. Thus, the unique solvability of (7.8) is a consequence of the Fredholm alternative.

To complete the proof it is sufficient to show that $T - T_0$ has a small norm in $C(S)$ for values of ν close to zero. According to the second lemma in Subsection 6.3.1.2, the kernel of $T - T_0$ is equal to $-\pi^{-1}(\partial g/\partial n_z)(z, \zeta)$. Then the required assertion is a consequence of the estimate

$$(\partial g/\partial n_z)(z, \zeta) = O(\nu \log(\nu|z-\bar{\zeta}|)) \quad \text{as } \nu \to 0$$

which follows from (6.135).

The next lemma establishes a similar result for another limiting case.

Equation (7.6) *is uniquely solvable in* $C(S)$ *for all sufficiently large positive* ν.

Instead of (7.8), we now consider the integral equation

$$-\mu(z) + (T_\infty\mu)(z) = f_\infty(z), \quad z \in S, \tag{7.10}$$

where

$$(T_\infty\mu)(z) = \frac{-1}{\pi}\int_S \mu(\zeta)\frac{\partial}{\partial n_z}(\log|z-\zeta| + \log|z-\bar{\zeta}|)\,ds.$$

We extend μ and f_0 to $S' = \{(x, -y) : (x, y) \in S\}$ as even functions of y (cf. with the previous lemma, where the odd extension has been used). Then (7.10) takes the form of the integral equation corresponding to the Neumann problem in the domain outside the union of S and S', and so this equation has a unique solution as was demonstrated above.

Now we have to show that the norm of $T - T_\infty$ is small in $C(S)$ when ν is sufficiently large. It follows from (6.128) that the kernel of $T - T_\infty$ is equal to

$$\frac{-1}{\pi} \mathrm{Re} \frac{\partial}{\partial n_z} \int_0^\infty \frac{e^{it} \mathrm{d}t}{t + \nu(z - \bar{\zeta})}.$$

This formula yields the required assertion that is a consequence of the last lemma in Subsection 6.3.1.2.

The two proven lemmas put us in the position to derive the main theorem about the integral equation (7.6).

Equation (7.6) *has a unique solution in* $C(S)$ *for all* $\nu > 0$ *with a possible exception for a finite number of values.*

It was noted above that (7.7) is an analytic function of ν having values in $C(S \times S)$, and so the corresponding operator T_ν analytically depends on ν. Therefore, we can apply the invertibility theorem formulated in Subsection 2.1.2. This theorem guarantees that if there is a single value of ν such that $I - T_\nu$ has an inverse operator, then $I - T_\nu$ is invertible for all $\nu > 0$ except possibly for a sequence tending to infinity. In the first lemma in the present subsection, it is demonstrated that $I - T_\nu$ has an inverse for all sufficiently small $\nu > 0$ and we can use the invertibility theorem. Furthermore, the sequence of exceptional values, for which $I - T_\nu$ might not be invertible, cannot be infinite because the second lemma means that $(I - T_\nu)^{-1}$ does exist for sufficiently large $\nu > 0$. Thus, the solvability of (7.6) can be lost for a finite number of values ν at most.

Let us turn to the solvability of the Neumann–Kelvin problem. The last theorem and facts from the potential theory formulated in Subsection 2.1.1 lead to the following corollary.

The Neumann–Kelvin problem is solvable for all $\nu > 0$ *with a possible exception for a finite number of values and for every* $f \in C(S)$. *The normal derivative on the left-hand side in* (7.3) *should be understood as the regular normal derivative.*

In conclusion of the present subsection, we note that the more regular right-hand-side term f in (7.3) provides more regularity of a solution. In particular, the assertion on the continuity of solutions of integral equations formulated in Subsection 2.1.1 guarantees that $\mu \in C^{1,\alpha}(S)$ when f is in the same class. Substituting such a μ into (7.5), we get $u \in C^2(\bar{W})$.

7.1.3. On Uniqueness in the Neumann–Kelvin Problem

Let us consider the problem of the stream running about the cylinder D along the x axis; that is, the direction of the relative cylinder's motion is opposite to that in the original Neumann–Kelvin problem. A solution $u'(x, y)$ of the new problem must satisfy (7.1)–(7.3) and the first condition (7.4). The second condition (7.4) should be replaced by the following one:

$$\lim_{x \to -\infty} |\nabla u'| = 0.$$

According to the first lemma in Subsection 6.3.1.2, Green's function for the new problem has the following form:

$$G'(z, \zeta) = G(z, \zeta) + 4\pi e^{\nu(y+\eta)} \sin \nu(x - \xi).$$

If $|\zeta| <$ const, then the following asymptotic formula holds as $|z| \to \infty$:

$$G'(z, \zeta) = -2 \log(\nu|z|) + \varphi'(x, y) + 4\pi H(x) e^{\nu(y+\eta)} \sin \nu(x - \xi),$$

where $\varphi' = O(|z|^{-1})$, $|\nabla \varphi'| = O(|z|^{-2})$.

In the same way as in Subsection 7.1.1, one proves that

$$u'(z) = Q' \log(\nu|z|) + c' + \psi'(x, y) + H(x)(\mathcal{A}' \sin \nu x + \mathcal{B}' \cos \nu x) e^{\nu y} \tag{7.11}$$

as $|z| \to \infty$. Here $\psi' = O(|z|^{-1})$, $|\nabla \psi'| = O(|z|^{-2})$, and for the constants Q', $\mathcal{A}'$, and $\mathcal{B}'$ we have the following formulae (cf. similar formulae in Subsection 7.1.1):

$$Q' = \pi^{-1} \int_S \frac{\partial u'}{\partial n} \, ds, \tag{7.12}$$

$$\mathcal{A}' = 2 \int_S \left[u' \frac{\partial}{\partial n} (e^{\nu y} \cos \nu x) - \frac{\partial u'}{\partial n} e^{\nu y} \cos \nu x \right] ds,$$

$$\mathcal{B}' = -2 \int \left[u' \frac{\partial}{\partial n} (e^{\nu y} \sin \nu x) - \frac{\partial u'}{\partial n} e^{\nu y} \sin \nu x \right] ds.$$

In the next theorem, presenting Green's formula for solutions of two Neumann–Kelvin problems, we rely on the fact that the expressions for $\mathcal{A}'$ and $\mathcal{B}'$ are similar to those for $\mathcal{A}$ and $\mathcal{B}$ but have opposite signs.

Let u be a solution of (7.1)–(7.4) *and let u' be a solution defined in the present subsection. If*

$$\int_S f \, ds = \int_S f' \, ds = 0$$

for the right-hand-side terms in (7.3) *for u and u', then*

$$\int_W (u'\nabla^2 u - u\nabla^2 u')\,\mathrm{d}x\mathrm{d}y = \int_S \left(u\frac{\partial u'}{\partial n} - u'\frac{\partial u}{\partial n}\right)\mathrm{d}s. \tag{7.13}$$

Let $R_{\alpha\beta}$ be a rectangle $\{|x| < \alpha, -\beta < y < 0\}$, containing $\bar{D}$, and $W_{\alpha\beta} = R_{\alpha\beta}\backslash\bar{D}$. By $p_{\alpha 0}$, $p_{\alpha\beta}$, and $q_{\pm\alpha\beta}$ we denote the top, bottom, and right and left sides, respectively, of $R_{\alpha\beta}$. Green's formula for $W_{\alpha\beta}$ is as follows:

$$\int_{W_{\alpha\beta}} (u'\nabla^2 u - u\nabla^2 u')\,\mathrm{d}x\mathrm{d}y = \int_{\partial W_{\alpha\beta}} \left(u\frac{\partial u'}{\partial n} - u'\frac{\partial u}{\partial n}\right)\mathrm{d}s, \tag{7.14}$$

where $\mathbf{n}$ is directed into $W_{\alpha\beta}$. Let us consider the integral

$$\int_{\partial R_{\alpha\beta}} \left(u\frac{\partial u'}{\partial n} - u'\frac{\partial u}{\partial n}\right)\mathrm{d}s.$$

According to (7.12) and to the assumption made, we have $Q' = 0$ in (7.11); the same is true for u as well. This implies that the integral over $p_{\alpha\beta}$ vanishes as $\beta \to \infty$.

Using the asymptotic formulae for u and u' in

$$\int_{q_{+\alpha\infty}} \left(u\frac{\partial u'}{\partial n} - u'\frac{\partial u}{\partial n}\right)\mathrm{d}s = \int_{-\infty}^{0} [u'u_x - uu'_x]_{x=+\alpha}\,\mathrm{d}y$$

under assumption that $\alpha \gg 1$, we obtain that the last integral is equal to

$$-c(\mathcal{A}'\cos\nu\alpha - \mathcal{B}'\sin\nu\alpha) + O(\alpha^{-1}).$$

Similarly, the integral over $p_{-\alpha\infty}$ is equal to

$$-c'(\mathcal{A}\cos\nu\alpha + \mathcal{B}\sin\nu\alpha) + O(\alpha^{-1}).$$

Finally, we have

$$\begin{aligned}\int_{p_{\alpha 0}} \left(u\frac{\partial u'}{\partial n} - u'\frac{\partial u}{\partial n}\right)\mathrm{d}s &= \nu^{-1}\int_{-\alpha}^{+\alpha} [uu'_{xx} - u'u_{xx}]_{y=0}\,\mathrm{d}x \\ &= \nu^{-1}[u(x,0)u'_x(x,0) - u'(x,0)u_x(x,0)]_{x=-\alpha}^{x=+\alpha},\end{aligned}$$

where the boundary condition (7.2) is applied. Again, the asymptotic formulae for u and u' give that the last expression is equal to

$$c(\mathcal{A}'\cos\nu\alpha - \mathcal{B}'\sin\nu\alpha) + c'(\mathcal{A}\cos\nu\alpha + \mathcal{B}\sin\nu\alpha) + O(\alpha^{-1}).$$

After substituting the obtained expressions into (7.14), we let $\beta \to \infty$ first and then tend α to infinity. The result of this limiting procedure is the required formula (7.13).

Let us turn to the question of existence of u'. Since Green's functions G and G' and their derivatives have the same singularities, all of the results, proved in Subsection 7.1.2 for the original Neumann–Kelvin problem, remain valid for the new statement. Thus u' does exist for all $\nu > 0$ with a possible exception of a finite set of values (of course, exceptional sets could be not the same for two problems). The fact of solvability of the problem, which may be considered as "adjoint" to (7.1)–(7.4), in the sense of Green's formula (7.13), allows us to prove the following uniqueness theorem.

For any $\nu > 0$ with a possible exception of a finite set of values, there is at most one (up to a constant term) solution of (7.1)–(7.4).

Let us fix ν such that (7.1)–(7.4) is solvable for this value and also u' does exist for this ν. By u we denote a solution to the homogeneous problem (7.1)–(7.4). Substituting u into (7.13), we get

$$\int_S u \frac{\partial u'}{\partial n} \, ds = 0.$$

Since $\partial u'/\partial n$ in this orthogonality condition can be taken arbitrarily from the set of functions satisfying

$$\int_S \frac{\partial u'}{\partial n} \, ds = 0$$

(see assumptions in the previous lemma), we find that $u = \text{const}$ on S. Taking into account that $\partial u/\partial n = 0$ on S, we can apply the theorem that the Cauchy problem for the Laplace equation has a unique solution. Hence, $u = \text{const}$ in W, which completes the proof.

7.1.4. Auxiliary Problem of the Scattering Type

In order to investigate the unique solvability of the Neumann–Kelvin problem (7.1)–(7.4) (for the sake of brevity it will be referred to as Problem I) for a given $\nu > 0$, we first consider another boundary value problem of the scattering type (we shall call it Problem II).

7.1.4.1. Statement of the Problem; Green's Function

We say that $w(z)$ is a solution of Problem II if (7.1)–(7.3) and the following representation,

$$w(z) = Q \log(\nu|z|) + w_0(z), \tag{7.15}$$

hold for it. Here Q is a certain constant and w_0 satisfies

$$\sup_W |\nabla w_0(z)| < \infty \qquad \text{and} \quad \partial w_0/\partial|x| - i\nu w_0 = o(1) \quad \text{as } |x| \to \infty. \tag{7.16}$$

Considerations in Subsection 6.3.1 yield that

$$E(z, \zeta) = G(z, \zeta) - 2\pi i \exp\{\nu[y + \eta + i(x - \xi)]\} \tag{7.17}$$

is Green's function for Problem II; also, for $|\zeta| \leq \text{const}$ we have

$$E(z, \zeta) = -2 \log(\nu|z|) + 2\pi i \exp[\nu(y + \nu + i|x - \xi|)] + \varphi(z, \zeta) \tag{7.18}$$

as $|z| \to \infty$, where $\varphi = O(|z|^{-1})$ and $|\nabla\varphi| = O(|z|^{-2})$.

7.1.4.2. Asymptotics at Infinity

In the same way as in Subsection 7.1.1, one obtains the integral representation for solutions of Problem II:

$$w(z) = \frac{1}{2\pi} \int_S \left[w(\zeta) \frac{\partial E}{\partial n_\zeta}(z, \zeta) - \frac{\partial w}{\partial n_\zeta} E(z, \zeta) \right] ds.$$

This result and the asymptotic formula (7.18) lead to the following theorem. *Let w be a solution of Problem II. Then, as $z \to \infty$,*

$$w(x, y) = Q \log(\nu|z|) + \mathcal{D}_\pm e^{\nu(y \pm ix)} + \psi_\pm(x, y) \quad \text{for } \pm x > 0.$$

Here $\psi_\pm = O(|z|^{-1})$, $|\nabla\psi_\pm| = O(|z|^{-2})$, and the constants are given as follows:

$$Q = \pi^{-1} \int_S \frac{\partial w}{\partial n} ds$$

$$\mathcal{D}_\pm = i \int_S \left[\frac{\partial w}{\partial n} e^{\nu(y \mp ix)} - w \frac{\partial}{\partial n} e^{\nu(y \pm ix)} \right] ds.$$

7.1.4.3. On Uniqueness in Problem II

Let us begin with a simple technical assertion.

Let $w \in C^1(\bar{W})$ satisfy the finite energy condition:

$$\int_{y=0} |w_x|^2 dx + \int_W |\nabla w|^2 dxdy < \infty. \tag{7.19}$$

Also, we assume that

$$\int_{-\infty}^0 |w(a, y)|^2 dy < \infty \tag{7.20}$$

for all sufficiently large $|a|$. Then we have the following:

$$\lim_{x\to\infty} x^{-1}|w(x,0)|^2 = 0, \tag{7.21}$$

$$\lim_{x\to\infty} x^{-1}\int_{-\infty}^{0} |w(x,y)|^2\,dy = 0, \tag{7.22}$$

$$\lim_{y\to-\infty} y^{-1}\int_{-\alpha}^{+\alpha} |w(x,y)|^2\,dx = 0 \quad \text{for } \alpha > 0, \tag{7.23}$$

$$\liminf_{x\to\infty} |x|\left[|w_x(x,0)|^2 + \int_{-\infty}^{0} |\nabla w(x,y)|^2\,dy\right] = 0, \tag{7.24}$$

$$\liminf_{y\to-\infty} |y|\int_{-\infty}^{+\infty} |\nabla w(x,y)|^2\,dx = 0. \tag{7.25}$$

Let a be a real number having a sufficiently large absolute value. For $y \leq 0$ and $x \geq a$ we immediately have

$$|w(x,y)| \leq 2\left\{|w(a,y)|^2 + (x-a)\int_a^x |w_x(x,y)|^2\,dx\right\}. \tag{7.26}$$

This implies

$$\limsup_{x\to+\infty} x^{-1}|w(x,0)|^2 \leq 2\int_a^{\infty} |w_x(x,0)|^2\,dx,$$

and so $x^{-1}|w(x,0)|^2 \to 0$ as $x \to +\infty$. In the same way we obtain the result as $x \to -\infty$, which completes the proof of (7.21).

Let us integrate with respect to y the inequality (7.26) divided by x:

$$\limsup_{x\to+\infty} x^{-1}\int_{-\infty}^{0} |w(x,y)|^2\,dy \leq 2\int_{-\infty}^{0}\int_a^{\infty} |\nabla w|^2\,dx dy.$$

In order to prove (7.22), we must notice that the right-hand side tends to zero as $a \to \infty$. In the same manner (7.23) can be obtained. This completes the proof because (7.24) and (7.25) are obvious.

Now we consider an integral identity that will provide the uniqueness theorem.

Let $w \in C^2(\bar{W})$ satisfy (7.1), (7.2), (7.15), and (7.16). Then we have

$$2\int_W |w_y|^2\,dx dy + \int_S \mathbf{x}\cdot\mathbf{n}\,|\nabla w|^2\,ds = \mathrm{Re}\int_S (2x\bar{w}_x - \bar{w})\,\frac{\partial w}{\partial n}\,ds, \tag{7.27}$$

where $\mathbf{x} = (x, 0)$.

One can directly verify that

$$\mathrm{Re}\{(2x\bar{w}_x - \bar{w})\nabla^2 w\} = 2|w_y|^2 - (x|\nabla w|^2)_x - \mathrm{Re}\,\nabla\cdot[(2x\bar{w}_x - \bar{w})\nabla w]. \tag{7.28}$$

Let $R_{\alpha\beta}$ be the rectangle defined in Subsection 7.1.3. Integrating (7.28) over $W_{\alpha\beta} = R_{\alpha\beta}\backslash\bar{D}$, we get

$$\begin{aligned} &2\int_{W_{\alpha\beta}} |w_y|^2\,\mathrm{d}x\mathrm{d}y + \int_S \mathbf{x}\cdot\mathbf{n}|\nabla w|^2\,\mathrm{d}s + \left\{\int_{q_{-\alpha\beta}} - \int_{q_{\alpha\beta}}\right\} x|\nabla w|^2\,\mathrm{d}y \\ &+ \mathrm{Re}\left\{\int_{q_{\alpha\beta}} - \int_{q_{-\alpha\beta}}\right\}(2x\bar{w}_x - \bar{w})w_x\,\mathrm{d}y - \mathrm{Re}\int_{p_{\alpha\beta}}(2x\bar{w}_x - \bar{w})w_y\,\mathrm{d}x \\ &+ \mathrm{Re}\int_{p_{\alpha 0}}(2x\bar{w}_x - \bar{w})w_y\,\mathrm{d}x = \mathrm{Re}\int_S (2x\bar{w}_x - \bar{w})\frac{\partial w}{\partial n}\,\mathrm{d}s. \end{aligned}$$

The last term on the left-hand side can be transformed by using the boundary condition (7.2):

$$\begin{aligned} \mathrm{Re}\int_{p_{\alpha 0}}(2x\bar{w}_x - \bar{w})w_y\,\mathrm{d}x &= -\nu^{-1}\mathrm{Re}\int_{p_{\alpha 0}}(2x\bar{w}_x - \bar{w})w_{xx}\,\mathrm{d}x \\ &= -\nu^{-1}[x|w_x(x,0)|^2 - \mathrm{Re}\,\bar{w}(x,0)w_x(x,0)]_{x=-\alpha}^{x=+\alpha}. \end{aligned} \tag{7.29}$$

From (7.23) and (7.25), we have the following for a fixed $\alpha < \infty$ and a certain sequence $\beta_k \to +\infty$:

$$\lim_{k\to\infty}\mathrm{Re}\int_{p_{\alpha\beta_k}}(2x\bar{w}_x - \bar{w})w_y\,\mathrm{d}x = 0.$$

The integrals over $q_{\pm\alpha\beta}$, where $\beta = \infty$, and the quantity (7.29) tend to zero for some choice of the sequence $\alpha_k \to \infty$. Letting $\beta_k \to \infty$ first, and then $\alpha_k \to \infty$, we arrive at (7.27).

An immediate corollary of (7.27) is the following uniqueness theorem.

Let $\mathbf{x}\cdot\mathbf{n} \geq 0$ *on* S, *and let* $w \in C^2(\bar{W})$ *satisfy* (7.1), (7.2), *the homogeneous condition* (7.3), *and the following conditions:*

$$\sup_W |\nabla w(x,y)| < \infty, \qquad \lim_{|x|\to\infty}|\nabla w(x,y)| = 0.$$

Then $w = \mathrm{const}$ *in* $\bar{W}$.

The assumptions made show that Q, $\mathcal{A}$, and $\mathcal{B}$ vanish in the asymptotic formula for w (see Subsection 7.1.4.2), and so $w = w_0 + \mathrm{const}$, where w_0 satisfies the integral identity (7.27). Since $\partial w_0/\partial n = 0$ on S, we get from this identity that $w_0(x,y) = w_0(x)$. Now (7.20) implies that $w_0 = 0$ in $\bar{W}$.

It should be noted that a similar argument provides the uniqueness of a solution in the problem describing a flow about a submerged cylinder in a channel. Let in a channel having infinite depth the sidewalls $z = \pm b$ be spanned by a submerged cylinder with generators parallel to the z axis. Taking the velocity potential in the form $u(x, y) \cos kz$, we find that the homogeneous Neumann condition on the walls is satisfied provided $k = \pi l (l = 1, 2, \ldots)$, and u must satisfy the two-dimensional boundary value problem:

$$\nabla^2 u = k^2 u \text{ in } W, \quad u_{xx} + \nu u_y = 0 \text{ when } y = 0, \quad \partial u / \partial n = f \text{ on } S,$$

where W and S are the same as above. In this case the integral identity takes the following form:

$$2 \int_W (|u_y|^2 + k^2 |u|^2) \, \mathrm{d}x \mathrm{d}y + \int_S \mathbf{x} \cdot \mathbf{n} (|\nabla u|^2 + k^2 |u|^2) \, \mathrm{d}s$$
$$= \mathrm{Re} \int_S (2x \bar{u}_x - \bar{u}) \frac{\partial u}{\partial n} \, \mathrm{d}s.$$

This equality implies the uniqueness of solution provided $\mathbf{x} \cdot \mathbf{n} \geq 0$ on S.

Turning to Problem II again, let us show that a solution of the homogeneous Problem II satisfies the assumptions of the previous theorem.

Let w be a solution of Problem II such that $\int_S \partial w / \partial n \, \mathrm{d}s = 0$. Then

$$|\mathcal{D}_+|^2 + |\mathcal{D}_-|^2 = 2 \, \mathrm{Im} \int_S w \frac{\partial \bar{w}}{\partial n} \, \mathrm{d}s, \tag{7.30}$$

where $\mathcal{D}_\pm$ are coefficients in the asymptotic formula in Subsection 7.1.4.2.

Let us use the notation applied in the proof of (7.27). From Green's formula, it follows

$$\int_{\partial R_{\alpha\beta}} \left(\bar{w} \frac{\partial w}{\partial n} - w \frac{\partial \bar{w}}{\partial n} \right) \mathrm{d}s = - \int_S \left(\bar{w} \frac{\partial w}{\partial n} - w \frac{\partial \bar{w}}{\partial n} \right) \mathrm{d}s$$
$$= 2i \, \mathrm{Im} \int_S w \frac{\partial \bar{w}}{\partial n} \, \mathrm{d}s, \tag{7.31}$$

where $\mathbf{n}$ is directed into $W_{\alpha\beta}$. We are going to let $\beta \to \infty$ first and then $\alpha \to \infty$. The asymptotic formula from Subsection 7.1.4.2 gives

$$\lim_{\beta \to \infty} \int_{p_{\alpha\beta}} \left(\bar{w} \frac{\partial w}{\partial n} - w \frac{\partial \bar{w}}{\partial n} \right) \mathrm{d}s = 0.$$

In the same way, we find that for $\pm x > 0$,

$$\bar{w} \frac{\partial w}{\partial |x|} - w \frac{\partial \bar{w}}{\partial |x|} = 2i \nu |\mathcal{D}_\pm|^2 e^{2\nu y} + h_\pm(x, y),$$

where $h_{\pm} = O(|z|^{-2} + |z|^{-1}e^{\nu y})$ as $|z| \to \infty$. Hence

$$\lim_{\alpha\to\infty} \int_{q_{\pm\alpha,\infty}} \left(\bar{w}\frac{\partial w}{\partial n} - w\frac{\partial \bar{w}}{\partial n} \right) \mathrm{d}s = -i|\mathcal{D}_{\pm}|^2,$$

$$\lim_{\alpha\to\infty} [\bar{w}(x,0)w_0(x,0) - w(x,0)\bar{w}_x(x,0)]_{x=-\alpha}^{x=+\alpha} = 2i\nu(|\mathcal{D}_+|^2 + |\mathcal{D}_-|^2).$$

The last equality and the boundary condition (7.2) give that

$$\int_{p_{\alpha 0}} \left(\bar{w}\frac{\partial w}{\partial n} - w\frac{\partial \bar{w}}{\partial n} \right) \mathrm{d}s = \nu^{-1} \int_{p_{\alpha 0}} (\bar{w}w_{xx} - w\bar{w}_{xx})\,\mathrm{d}x$$
$$= \nu^{-1}[\bar{w}(x,0)w_x(x,0) - w(x,0)\bar{w}_x(x,0)]_{x=-\alpha}^{x=+\alpha} \to 2i(|\mathcal{D}_+|^2 + |\mathcal{D}_-|^2)$$

as $\alpha \to \infty$. This and (7.31)–(7.32) prove (7.30).

The asymptotic formula in Subsection 7.1.4.2 and the last assertion imply the following theorem.

Let only a trivial solution of the homogeneous Problem II satisfy (7.19). *Then the homogeneous Problem II has only a trivial solution.*

An immediate consequence of this result and the previous uniqueness theorem is as follows.

If $\mathbf{x} \cdot \mathbf{n} \geq 0$ *on S, then Problem II can have one solution at most.*

7.1.4.4. Existence Theorem for Problem II

Our aim is to prove the following theorem.

If Problem II has no more than one solution, then for any $f \in C^{1,\alpha}(S)$ *there exists a unique solution* $w \in C^2(\bar{W})$ *of Problem II.*

Let us seek a solution in the following form:

$$w(z) = \frac{1}{\pi} \int_S E(z,\zeta)\mu(\zeta)\,\mathrm{d}s, \tag{7.32}$$

with an unknown density $\mu \in C^{1,\alpha}(S)$. Then $w \in C^2(\bar{W})$ and satisfies (7.1), (7.2), and (7.16). Using the following formula for the normal derivative of the single-layer potential (7.32) on the side of S directed to W (see Subsection 2.1.1),

$$\frac{\partial w}{\partial n}(z) = -\mu(z) + \frac{1}{\pi} \int_S \frac{\partial E}{\partial n_z}(z,\zeta)\mu(\zeta)\,\mathrm{d}s,$$

we get from (7.3) the Fredholm integral equation:

$$-\mu + \mathcal{T}_\nu \mu = f, \qquad (\mathcal{T}_\nu \mu)(z) = \frac{1}{\pi} \int_S \frac{\partial E}{\partial n_z}(z,\zeta)\mu(\zeta)\,\mathrm{d}s. \tag{7.33}$$

Let μ_0 be a solution of the homogeneous equation (7.33). Substituting μ_0 into (7.32), one obtains w_0, which vanishes in $\bar{W}$ by the theorem's assumption. Moreover, w_0 is a continuous function in $\overline{\mathbb{R}^2_-}$ and w_0 is harmonic in D. Hence this function vanishes throughout $\overline{\mathbb{R}^2_-}$. Then the jump formula for the normal derivative, see (2.4) in Subsection 2.1.1, implies that $\mu_0 = 0$ on S, and so the homogeneous equation (7.33) has only a trivial solution. Therefore, the Fredholm alternative yields that (7.33) is uniquely solvable for any $f \in C^{1,\alpha}(S)$. Then (7.32) gives a unique solution of Problem II.

7.1.5. Unique Solvability of the Neumann–Kelvin Problem

The plan of this subsubsection is as follows. First, we investigate the null space of Problem I (see Subsection 7.1.5.1). Second, we obtain an integral equation equivalent to Problem I in Subsection 7.1.5.2. At last, a list of equivalent necessary and sufficient conditions providing the existence of a unique solution of Problem I is given in Subsection 7.1.5.3.

7.1.5.1. On Nontrivial Solutions of the Homogeneous Problem I

The following theorem provides a condition guaranteeing that the null space of Problem I is one dimensional.

Let only a trivial solution of the homogeneous Problem II satisfy (7.19) *(for example, this is true when* $\mathbf{x} \cdot \mathbf{n} \geq 0$ *on* S*). Then the homogeneous Problem I can have at most one nontrivial solution (up to a constant factor and a constant term).*

Let u_1 and u_2 be two solutions of the homogeneous Problem I. The theorem in Subsection 7.1.1 shows that the following asymptotic formula holds as $|z| \to \infty$:

$$u_i(x, y) = \varphi_i(x, y) + H(-x)e^{\nu y}(\mathcal{A}_i \sin \nu x + \mathcal{B}_i \cos \nu x), \qquad (7.34)$$

where $\varphi_i = O(|z|^{-1})$ and $|\nabla \varphi_i| = O(|z|^{-2})$. Let us introduce two vectors $(\mathcal{A}_1, \mathcal{B}_1)$ and $(\mathcal{A}_2, \mathcal{B}_2)$ that must be linearly independent. In fact, if one assumes the opposite, then a linear combination of u_1 and u_2 can be found giving a solution of the homogeneous Problem II and satisfying (7.19). Since this combination must be trivial by the theorem's assumption, we arrive at a contradiction.

Therefore, a certain linear combination of u_1 and u_2 must have the asymptotic behavior (7.34) with $\mathcal{A} = -i$ and $\mathcal{B} = 1$, and so this behavior coincides with that which a solution of Problem II has in the case when $Q = 0, \mathcal{D}_+ = 0$, and $\mathcal{D}_- = 1$ (see the assertion in Subsection 7.1.4.2). Hence, a nontrivial solution of the homogeneous Problem II is obtained that contradicts the theorem's assumption.

7.1.5.2. Equivalence of Problem I to an Integral Equation

We recall (see Subsection 7.1.2) that, seeking a solution of Problem I in the form (7.5), we arrive at the Fredholm integral equation (7.6) for the unknown density $\mu \in C^{1,\alpha}(S)$, and the kernel of the corresponding integral operator is $\pi^{-1}(\partial G/\partial n_z)(z, \zeta)$. Our aim is to prove the theorem demonstrating equivalence of this equation and Problem I.

Let Problem II be uniquely solvable (for example, when $\mathbf{x} \cdot \mathbf{n} \geq 0$ *on* S*). Then Problem I and the integral equation* (7.6) *are equivalent in the following sense. Any solution of Problem I has a unique representation in the form of* (7.5)*, where* μ *satisfies* (7.6)*. Conversely, substituting any solution* μ *of* (7.6) *into* (7.5)*, one obtains a solution of Problem I.*

Let u be a solution of Problem I corresponding to f in the Neumann condition (7.3). Using the asymptotic formula for u obtained in Subsection 7.1.1, we see that

$$w = u - c - e^{\nu(y+ix)}(\mathcal{B} - i\mathcal{A})/2$$

solves Problem II (for another boundary function in the Neumann condition). It was shown in Subsection 7.1.4.4 that this solution has the representation

$$w(z) = \frac{1}{\pi}\int_S \mu(\zeta)E(z, \zeta)\, ds.$$

This formula and (7.17) combined produce

$$\begin{aligned} u - c - e^{\nu(y+ix)}\frac{(\mathcal{B} - i\mathcal{A})}{2} \\ = \int_S \mu(\zeta)G(z, \zeta)\, ds - i\int_S \mu(\zeta)\exp\{\nu[y + \eta + i(x - \xi)]\}\, ds, \end{aligned}$$

and so

$$u(x, y) = \int_S \mu(\zeta)G(z, \zeta)\, ds + c_1 e^{\nu(y+ix)} + c.$$

Now, substituting the asymptotic formula for G as $x \to +\infty$ (see Subsection 6.3.1.2), we get that $c_1 = 0$, because otherwise there is a contradiction with the asymptotic result in Subsection 7.1.1. Thus we arrive at the required representation (7.5) for u solving Problem I.

Since

$$\int_S \mu(\zeta) G(z, \zeta)\, \mathrm{d}s \neq \text{const} \quad \text{in } W \text{ for } \mu \neq 0$$

(see the proof in Subsection 7.1.4.4), we have that (7.5) is unique. The first part of the theorem is proved and the converse assertion is obvious.

7.1.5.3. Necessary and Sufficient Conditions for the Unique Solvability of Problem I

Six conditions that are equivalent to the unique solvability and that use various auxiliary functions are given in the following theorem.

If Problem II is uniquely solvable (for example, when $\mathbf{x} \cdot \mathbf{n} \geq 0$ on S), then the following assertions are equivalent:

1. *Problem I is uniquely solvable (up to a constant term) for any $f \in C^{1,\alpha}(S)$.*
2. *We have that $\mathcal{D}_+ \neq 1$, where $\mathcal{D}_+$ is the constant in the asymptotic formula at infinity for $\mathcal{W}$ solving Problem II in which $f = \partial e^{\nu(y+ix)}/\partial n$ on S in* (7.3).
3. *For $\mathcal{W}$ defined in 2, we have*

$$\int_S \mathcal{W} \frac{\partial}{\partial n} e^{\nu(y-ix)}\, \mathrm{d}s \neq 2\nu^2 \int_D e^{2\nu y}\, \mathrm{d}x\mathrm{d}y + i.$$

4. *For $\mathcal{W}$ defined in 2, we have that $|\mathcal{D}_-| \neq 1$.*
5. *For $\mathcal{W}$ defined in 2, we have*

$$\left| \int_S \mathcal{W} \frac{\partial}{\partial n} e^{\nu(y+ix)}\, \mathrm{d}s \right| \neq 1.$$

6. *Let μ_+ be a solution of*

$$-\mu_+(z) + (\mathcal{T}\mu_+)(z) = \partial e^{\nu(y+ix)}/\partial n, \quad z \in S, \qquad (7.35)$$

where $\mathcal{T}$ is defined in (7.33); *then*

$$i \int_S \mu_+ e^{\nu(y-ix)}\, \mathrm{d}s \neq -1.$$

7. *For μ_+ solving* (7.35) *we have*

$$\left| \int_S \mu_+ e^{\nu(y+ix)}\, \mathrm{d}s \right| \neq 1.$$

$1 \Leftrightarrow 2$. Let $\mathcal{D}_+ = 1$; then the asymptotic formula for $\mathcal{W}$ (see Subsection 7.1.4.2) guarantees that $\mathcal{W} - e^{\nu(y+ix)}$ is a nontrivial solution of the homogeneous Problem I.

Conversely, let u_0 be a nontrivial solution of the homogeneous Problem I. The assertion in Subsection 7.1.5.1 shows that u_0 can be assumed to be real. The leading terms in the asymptotic representation of u_0 as $x \to -\infty$ have the form

$$1/2(\mathcal{B} - i\mathcal{A})e^{\nu(y+ix)} + 1/2(\mathcal{B} + i\mathcal{A})e^{\nu(y-ix)} + c,$$

where $\mathcal{A}$ and $\mathcal{B}$ are real constants. According to the first uniqueness theorem in Subsection 7.1.4.3, these constants do not vanish simultaneously. Putting

$$w = \frac{2}{\mathcal{B} - i\mathcal{A}}(c - u_0) + e^{\nu(y+ix)},$$

we see that

$$\frac{\partial w}{\partial n} = \frac{\partial}{\partial n} e^{\nu(y+ix)} \quad \text{on } S.$$

It follows from the asymptotic formula in Subsection 7.1.4.2 that w is a solution of Problem II. Also, $\mathcal{D}_+ = 1$ for w, and so $w = \mathcal{W}$ because Problem II has a unique solution.

2 ⇔ 3. From the asymptotic formula in Subsection 7.1.4.2, we have

$$\mathcal{D}_+ = -i \int_S \mathcal{W} \frac{\partial}{\partial n} e^{\nu(y-ix)} \, ds + i \int_S e^{\nu(y-ix)} \frac{\partial}{\partial n} e^{\nu(y+ix)} \, ds. \tag{7.36}$$

In order to complete the proof of this item it is sufficient to note that

$$\int_S e^{\nu(y-ix)} \frac{\partial}{\partial n} e^{\nu(y+ix)} \, ds = \int_D |\nabla e^{\nu(y+ix)}|^2 \, dx dy = 2\nu^2 \int_D e^{2\nu y} \, dx dy. \tag{7.37}$$

2 ⇔ 4. From (7.36) and (7.37), we get that

$$\operatorname{Re} \mathcal{D}_+ = \operatorname{Im} \int_S \mathcal{W} \frac{\partial}{\partial n} e^{\nu(y-ix)} \, ds.$$

Combining this with (7.30), where $w = \mathcal{W}$, we get

$$|\mathcal{D}_+|^2 + |\mathcal{D}_-|^2 = 2 \operatorname{Re} \mathcal{D}_+, \tag{7.38}$$

and so $\mathcal{D}_+ = 1$ if and only if $|\mathcal{D}_-| = 1$.

4 ⇔ 5. Applying to $\mathcal{W}$ the assertion obtained in Subsection 7.1.4.2, and using

$$\int_S e^{\nu(y+ix)} \frac{\partial}{\partial n} e^{\nu(y+ix)} \, ds = \frac{1}{2} \int_S \frac{\partial}{\partial n} e^{2\nu(y+ix)} \, ds = 0, \tag{7.39}$$

we get that

$$\mathcal{D}_- = -i \int_S \mathcal{W} \frac{\partial}{\partial n} e^{\nu(y+ix)}\, ds,$$

which proves the equivalence of 4 and 5.

$1 \Leftrightarrow 6$. The theorem obtained in Subsection 7.1.5.2 shows that the unique solvability of Problem I is equivalent to the fact that the integral equation

$$-\mu_0 + T\mu_0 = 0$$

has only a trivial solution. The last equation can be written as follows:

$$-\mu_0 + \mathcal{T}\mu_0 = -2i \left[\frac{\partial}{\partial n} e^{\nu(y+ix)} \right] \int_S e^{\nu(\eta - i\xi)} \mu_0 \, ds. \tag{7.40}$$

Let μ_0 be a nontrivial solution of (7.40) (in Subsection 7.1.4.4, it is demonstrated that this equation is uniquely solvable). Then $-\mu + \mathcal{T}\mu = 0$ has only a trivial solution, and so

$$K = -i \int_S e^{\nu(\eta - i\xi)} \mu_0 \, ds \neq 0.$$

Therefore, $K^{-1}\mu_0$ satisfies (7.35); that is, $K^{-1}\mu_0 = \mu_+$. Hence,

$$i \int_S \mu_+ e^{\nu(y-ix)}\, ds = ik^{-1} \int_S \mu_0 e^{\nu(\eta - i\xi)}\, ds = -1.$$

Conversely, if

$$i \int_S \mu_+ e^{\nu(y-ix)}\, ds = -1,$$

then (7.35) can be written in the following form:

$$-\mu_+ + \mathcal{T}\mu_+ = -2i \left[\frac{\partial}{\partial n} e^{\nu(y+ix)} \right] \int_S \mu_+ e^{\nu(\eta - i\xi)}\, ds,$$

which means that $-\mu_+ + T\mu_+ = 0$. Thus, (7.40) has a nontrivial solution, and so the unique solvability of Problem I is violated.

$5 \Leftrightarrow 7$. Let μ_+ be a solution of (7.35); then

$$(V\mu_+)(z) = \int_S \mu_+(\zeta) E(z, \zeta)\, ds$$

satisfies (7.3), where $f = \partial e^{\nu(y+ix)}/\partial n$. Hence, the theorem's assumption implies that this potential is equal to $\mathcal{W}$ defined in 2. According to the jump formula for the normal derivative (see Subsection 2.1.1), we have

$$2\mu_+ = \frac{\partial(V\mu_+)}{\partial n_-} - \frac{\partial(V\mu_+)}{\partial n_+},$$

and so

$$2\int_S \mu_+ e^{\nu(y+ix)}\,ds = \int_S e^{\nu(y+ix)} \frac{\partial(V\mu_+)}{\partial n_-}\,ds - \int_S e^{\nu(y+ix)} \frac{\partial}{\partial n}\, e^{\nu(y+ix)}\,ds. \tag{7.41}$$

Since $V\mu_+$ is a harmonic function in D and $V\mu_+ = \mathcal{W}$ on S, Green's theorem yields

$$\int_S e^{\nu(y+ix)} \frac{\partial(V\mu_+)}{\partial n_-}\,ds = \int_S \mathcal{W} \frac{\partial}{\partial n} e^{\nu(y+ix)}\,ds.$$

This combined with (7.39) and (7.41) produce

$$\int_S \mu_+ e^{\nu(y+ix)}\,ds = \int_S \mathcal{W} \frac{\partial}{\partial n} e^{\nu(y+ix)}\,ds,$$

which completes the proof.

The last proof also demonstrates that

$$i\mathcal{D}_- = \int_S \mathcal{W} \frac{\partial}{\partial n} e^{\nu(y+ix)}\,ds = \int_S \mu_+ e^{\nu(y+ix)}\,ds. \tag{7.42}$$

From (7.38), we see that the absolute value of this number cannot be greater than one for any contour S. Hence, if for a certain curve S, Problem I has more than one solution, then the absolute value of (7.42) attains its global maximum for S.

7.1.6. Unconditional Theorem for a Circular Cylinder

In this subsection, we prove the unique solvability of the Neumann–Kelvin problem for all $\nu > 0$ in the case of a submerged circular cylinder. The crucial point is the following lemma.

Let $D = \{z = x + iy : |z + ia| < R\}$, where a and R are positive constants such that $a > R$. Then a solution u of the homogeneous problem (7.1)–(7.4) *satisfies* (7.19) *and* (7.20).

From Subsection 7.1.1, it follows that

$$u(z) = H(-x)e^{\nu y}(\mathcal{A}\sin \nu x + \mathcal{B}\cos \nu x) + c + \psi(x, y) \quad \text{as } |z| \to \infty,$$

where $\psi = O(|z|^{-1})$ and $|\nabla \psi| = O(|z|^{-2})$. Therefore, for proving the assertion it is sufficient to verify that

$$|\nabla u| = o(1) \quad \text{as } x \to -\infty. \tag{7.43}$$

For this purpose we consider a stream function v, that is, a harmonic in W function that is conjugate to u. Since

$$\int_S \frac{\partial u}{\partial n} \, \mathrm{d}s = 0,$$

v is determined uniquely up to an arbitrary constant term, despite the fact that W is not a simply connected domain.

Using the Cauchy–Riemann equations, one can readily find that v can be chosen to satisfy the boundary value problem:

$$\nabla^2 v = 0 \quad \text{in } W, \tag{7.44}$$

$$v_y - \nu v = 0 \quad \text{when } y = 0, \tag{7.45}$$

$$v = \text{const} \quad \text{on } S = \{|z + ia| = R\}, \tag{7.46}$$

$$\sup_W |\nabla v(x, y)| < \infty, \qquad \lim_{x \to +\infty} |\nabla v| = 0. \tag{7.47}$$

Now we shall prove that

$$|\nabla v| = o(1) \quad \text{as } x \to -\infty, \tag{7.48}$$

which is equivalent to (7.43). The conformal mapping

$$z \mapsto \zeta = (z + ib)/(z - ib), \quad \text{where } b = (a^2 - R^2)^{1/2},$$

transforms W into an annulus $\{h < |\zeta| < 1\}$, where

$$h = \frac{(a^2 + R^2)^{1/2} - b}{(a^2 + R^2)^{1/2} + b}.$$

Putting $w(\zeta) = v(z(\zeta))$, and conversely, $v(z) = w(\zeta(z))$, we see that w is a harmonic function in the annulus, and from (7.46), we get that

$$w = \text{const} \quad \text{when } \rho = h, \tag{7.49}$$

where $\zeta = \rho e^{i\theta}$. Let us determine the boundary condition for w on $\{\rho = 1\}$ that is the image of the x axis. It is clear that

$$v_y(x, 0) = |\zeta'(x)| w_\rho(\zeta(x)).$$

Since $\zeta'(z) = -2bi(z - ib)^{-2}$, we have

$$v_y(x, 0) = \frac{2b}{|x - ib|^2} \, w_\rho(\zeta(x)).$$

Now from $x = ib(1+e^{i\theta})/(1-e^{i\theta})$, we obtain

$$|x - ib|^2 = \frac{4b^2}{|e^{i\theta}-1|^2},$$

and so we arrive at the following boundary condition:

$$|e^{i\theta}-1|^2 w_\rho - 2b\nu w = 0 \quad \text{when } \rho = 1.$$

It is convenient to write it in the following form:

$$(e^{i\theta} - 2 + e^{-i\theta}) w_\rho + \lambda w = 0 \quad \text{when } \rho = 1, \tag{7.50}$$

where $\lambda = 2b\nu$.

Seeking w in the form

$$w = \text{const} + \sum_{n=1}^{\infty} (w_n e^{in\theta} + w_{-n} e^{-in\theta}) \left(\rho^n - \frac{h^{2n}}{\rho^n} \right),$$

we immediately satisfy the Laplace equation and the boundary condition (7.49). Substituting

$$w_\rho = \sum_{n=1}^{\infty} n(w_n e^{in\theta} + w_{-n} e^{-in\theta})(1 + h^{2n}), \quad \text{where } \rho = 1,$$

into (7.50), we get

$$(e^{i\theta} - 2 + e^{-i\theta}) \sum_{n=1}^{\infty} n(w_n e^{in\theta} + w_{-n} e^{-in\theta})(1 + h^{2n})$$
$$+ \lambda\, \text{const} + \lambda \sum_{n=1}^{\infty} (w_n e^{in\theta} + w_{-n} e^{-in\theta})(1 - h^{2n}) = 0.$$

Using $W_n = w_n |n|(1 + h^{2|n|})$ for $n = \pm 1, \pm 2, \ldots$, we write the last equation in the following form:

$$(e^{i\theta} - 2 + e^{-i\theta}) \sum_{n \neq 0} W_n e^{in\theta} + \lambda\, \text{const} + \lambda \sum_{n \neq 0} W_n |n|^{-1} \frac{1 - h^{2n}}{1 + h^{2n}} e^{in\theta} = 0,$$

which is equivalent to

$$0 = \sum_{n \neq 0, \pm 1} \left[(W_{n+1} - 2W_n + W_{n-1}) + \frac{\lambda\big(1 - h^{2|n|}\big)}{|n|\big(1 + h^{2|n|}\big)} W_n \right] e^{in\theta}$$
$$+ W_1 + W_{-1} + \lambda\, \text{const}$$
$$+ \left[\left(\lambda \frac{1 - h^2}{1 + h^2} - 2 \right) W_1 + W_2 \right] e^{i\theta} + \left[\left(\lambda \frac{1 - h^2}{1 + h^2} - 2 \right) W_{-1} + W_{-2} \right] e^{-i\theta}.$$

This leads to the algebraic system for W_n:

$$W_1 + W_{-1} + \lambda \text{ const} = 0,$$
$$\{\lambda[(1-h^2)/(1+h^2)] - 2\}W_{\pm 1} + W_{\pm 2} = 0,$$
$$W_{n+1} - 2W_n + W_{n-1} + \left\{\left[\lambda\left(1-h^{2|n|}\right)\right]/\left[|n|\left(1+h^{2|n|}\right)\right]\right\}W_n = 0,$$
$$n = \pm 2, \pm 3, \ldots .$$

Let $\theta_\pm = \arg W_{\pm 1}$; then we find from the system that both sequences

$$\{e^{-i\theta_+}W_n\}_{n\geq 1}, \qquad \{e^{-i\theta_-}W_n\}_{n\leq -1}$$

are real. Therefore,

$$w = \sum_{n=1}^{\infty} W_n \zeta^n + \sum_{n=1}^{\infty} W_{-n}\bar{\zeta}^n + w_0(\zeta) = w_+(\zeta) + w_-(\zeta) + w_0(\zeta),$$

where w_0 is smooth in a neighborhood of $\zeta = 1$. Furthermore, w_+ is holomorphic and w_- is antiholomorphic, and also we have

$$w_\pm(\bar{\zeta}) = e^{2i\theta_\pm}\overline{w_\pm(\zeta)}.$$

The decomposition of w induces the corresponding decomposition of

$$v(z) = v_+(z) + v_-(z) + v_0(z) = w_+(\zeta(z)) + w_-(\zeta(z)) + w_0(\zeta(z)).$$

Since $\zeta(-\bar{z}) = \overline{\zeta(z)}$ and $\zeta(\infty) = 1$, we get that

$$v_\pm(-\bar{z}) = e^{2i\theta_\pm}\overline{v_\pm(z)}. \tag{7.51}$$

and $|\nabla v_0| = o(1)$ as $|z| \to \infty$. This and the second condition (7.47) imply that

$$|\nabla(v_+ + v_-)|^2 = |\nabla(v - v_0)|^2 = o(1) \quad \text{as } x \to +\infty.$$

Now, from the identities

$$\begin{aligned}|\nabla v_+|^2 + |\nabla v_-|^2 &= 2\left(\left|\frac{\partial v_+}{\partial z}\right|^2 + \left|\frac{\partial v_-}{\partial \bar{z}}\right|^2\right)\\ &= 2\left(\left|\frac{\partial}{\partial z}(v_+ + v_-)\right|^2 + \left|\frac{\partial}{\partial \bar{z}}(v_+ + v_-)\right|^2\right) = |\nabla(v_+ + v_-)|^2,\end{aligned}$$

we obtain that

$$|\nabla v_\pm| = o(1) \quad \text{as } x \to +\infty.$$

Combining this with (7.51), we arrive at (7.48), which completes the proof.

An immediate consequence of the lemma proven is as follows.

Let S be a circle and let ν be an arbitrary positive number. Then the boundary value problem (7.1)–(7.4) *has one and only one solution (up to a constant term) for any f in* (7.3).

Since (7.19) and (7.29) hold for a solution of the homogeneous Problem I as is demonstrated above, the uniqueness of solution (up to a constant term) follows from Subsection 7.1.4.3 because $\mathbf{x} \cdot \mathbf{n} \geq 0$ on S. For the proof of existence we refer to the integral equation method developed in Subsection 7.1.2. The solvability of the integral equation (7.6) is a consequence of the uniqueness of its solution. The last fact follows from the uniqueness in the boundary value problem.

7.2. Cylinder in Shallow Water

When the fluid depth d is finite, two different cases of the Neumann–Kelvin problem require separate treatments. We recall that the stream is called subcritical if $\nu d = gdU^{-2} > 1$. The results in this case are similar to those in Section 7.1 and are considered briefly in Subsection 7.2.1. The case of the supercritical stream when $\nu d < 1$ differs essentially from the subcritical one because there are no waves in both directions from the body. In Subsection 7.2.2, we consider the problem of the supercritical stream in detail.

7.2.1. Submerged Body in the Subcritical Stream

7.2.1.1. Statement of the Problem, Asymptotics at Infinity, and Uniqueness

Let the cylinder's cross section be a bounded simply connected domain D such that $\bar{D} \subset L = \{-\infty < x < +\infty, -d < y < 0\}$. As in Section 7.1, we suppose that $\partial D = S$ is a C^3 curve. By W we denote the water domain $L \backslash \bar{D}$. In the present case, the statement of the Neumann–Kelvin problem (7.1)–(7.4) should be complemented by the no-flow condition on the horizontal bottom:

$$u_y = 0 \quad \text{when } y = -d. \tag{7.52}$$

Also, we assume that $\nu d > 1$, and therefore, *for $u \in C^2(\bar{W})$ solving the Neumann–Kelvin problem, the asymptotic formula of the same type as in Subsection 7.1.1,*

$$u(x, y) = c_{\pm} + \psi_{\pm}(x, y) + H(-x)[Qx + \cosh \lambda_0(y + d)(\mathcal{A} \sin \lambda_0 x + \mathcal{B} \cos \lambda_0 x)], \quad \pm x > 0, \tag{7.53}$$

holds as $|x| \to \infty$. *Here* $\psi_\pm = O(|x|^{-1})$, $|\nabla \psi_\pm| = O(|x|^{-2})$, λ_0 *is the unique positive root of* $\nu \tanh \lambda d = \lambda$, *and* H *is the Heaviside function. Furthermore,*

$$(c_+ - c_-)(1 - \nu d) = \nu \int_S \left[x \frac{\partial u}{\partial n} - u \cos(n, x) \right] ds$$

and the constants Q, $\mathcal{A}$, *and* $\mathcal{B}$ *are as follows:*

$$Q(1 - \nu d) = \nu \int_S \frac{\partial u}{\partial n} ds,$$

$$\mathcal{A}\lambda_0(\nu d - \cosh \lambda_0 d)$$
$$= 2\nu \int_S \left\{ u \frac{\partial}{\partial n} [\cosh \lambda_0 (y + d) \cos \lambda_0 x] - \frac{\partial u}{\partial n} \cosh \lambda_0 (y + d) \cos \lambda_0 x \right\} ds,$$
$$\mathcal{B}\lambda_0(\nu d - \cosh \lambda_0 d)$$
$$= -2\nu \int_S \left\{ u \frac{\partial}{\partial n} [\cosh \lambda_0 (y + d) \sin \lambda_0 x] - \frac{\partial u}{\partial n} \cosh \lambda_0 (y + d) \sin \lambda_0 x \right\} ds.$$

This theorem can be proved in the same way as the asymptotic formula in Subsection 7.1.1. Turning to the question of uniqueness, we have the following analogs of results and methods in Subsection 7.1.4.3.

Let $\mathbf{x} \cdot \mathbf{n} \geq 0$ *on* S. *If* $u \in C^2(\bar{W})$ *is a solution of the homogeneous Neumann–Kelvin problem and*

$$\int_{y=0} |u_x|^2 dx + \int_W |\nabla u|^2 dx dy < \infty, \tag{7.54}$$

then $u = \text{const}$ *in* W.

Again, starting from (7.28) and using the free surface boundary condition and the homogeneous Neumann condition on S, one derives the integral identity

$$2 \int_W |u_y|^2 dx dy + \int_S \mathbf{x} \cdot \mathbf{n} |\nabla u|^2 ds = 0.$$

Now the geometric assumption yields the assertion.

7.2.1.2. Problem II

A solution w to the auxiliary problem of the scattering type (referred to as Problem II; the name Problem I is reserved for the Neumann–Kelvin problem) must satisfy (7.1)–(7.3), (7.52), and the first condition (7.4), and it must admit the following representation:

$$w(x, y) = Q|x| + c \operatorname{sign} x + w_0(x, y), \tag{7.55}$$

where Q and c are certain constants and the radiation condition

$$\lim_{|x|\to\infty}\left(\frac{\partial w_0}{\partial|x|} - i\lambda_0 w_0\right) = 0$$

holds.

Using the asymptotic formula for Green's function of Problem I (see Subsection 6.3.2.2), we find that Green's function of Problem II has the following form:

$$E(z,\zeta) = G(z,\zeta) + \frac{\pi\nu(x-\xi)}{1-\nu d} + 2\pi i\,\frac{\nu\cosh\lambda_0(y+d)\cosh\lambda_0(\eta+d)}{\lambda_0(\nu d - \cosh^2\lambda_0 d)}\exp\{i\lambda_0(x-\xi)\},$$

and the asymptotics at infinity for $E(z,\zeta)$ follows from the asymptotic formula for $G(z,\zeta)$ (see the first assertion in Subsection 6.3.2.2). This allows us to prove the following theorem in the same way as the corresponding proposition in Subsection 7.1.4.2.

Let w be a solution of Problem II, and let w_0 be defined in (7.55). *Then we have the following as $|x|\to\infty$:*

$$w_0(x,y) = \mathcal{D}_\pm\cosh\lambda_0(y+d)e^{\pm i\lambda_0 x} + \psi_\pm(x,y), \quad \pm x > 0, \tag{7.56}$$

where $\psi_\pm = O(|x|^{-1})$ and $|\nabla\psi_\pm| = O(x^{-2})$. The constants Q, c, and $\mathcal{D}_\pm$ are given by

$$2Q(\nu d - 1) = \nu\int_S \frac{\partial w}{\partial n}\,ds,$$

$$2c(1-\nu d) = \nu\int_S\left[x\frac{\partial w}{\partial n} - w\cos(n,x)\right]ds,$$

$$\mathcal{D}_\pm\lambda_0(\nu d - \cosh^2\nu d) = i\nu\int_S\left\{w\frac{\partial}{\partial n}[\cosh\lambda_0(y+d)e^{\mp i\lambda_0 x}] - \frac{\partial w}{\partial n}\cosh\lambda_0(y+d)e^{\mp i\lambda_0 x}\right\}ds.$$

The proof of formula analogous to (7.30) differs from that in Subsection 7.1.4.3 because of the second term on the left-hand side in (7.55), and so we prove the following theorem at length.

Let w be a solution of Problem II such that $\int_S \partial w/\partial n\,ds = 0$. Then

$$\lambda_0(\cosh^2\lambda_0 d - \nu d)(|\mathcal{D}_+|^2 + |\mathcal{D}_-|^2) = 2\nu\,\mathrm{Im}\int_S w\frac{\partial\bar{w}}{\partial n}\,ds. \tag{7.57}$$

Green's identity gives

$$\int_{\partial W_\alpha} \left(\bar{w} \frac{\partial w}{\partial n} - w \frac{\partial \bar{w}}{\partial n} \right) \mathrm{d}s = 0,$$

where $W_\alpha = W \cap \{|x| > \alpha\}$ and α is sufficiently large. Using the boundary conditions on the free surface and bottom, we get

$$\begin{aligned} 2i \,\mathrm{Im} \int_S w \frac{\partial \bar{w}}{\partial n} \,\mathrm{d}s = \nu^{-1} \int_{-\alpha}^{+\alpha} [\bar{w} w_{xx} - w \bar{w}_{xx}]_{y=0} \,\mathrm{d}x \\ - \sum_{\pm} \int_{W \cap \{\pm x = \alpha\}} \left(\bar{w} \frac{\partial w}{\partial |x|} - w \frac{\partial \bar{w}}{\partial |x|} \right) \mathrm{d}y. \end{aligned} \quad (7.58)$$

The theorem's assumption and the previous assertion show that $Q = 0$ in (7.55). Then from (7.56), it follows that

$$\begin{aligned} \int_{-\alpha}^{+\alpha} [\bar{w} w_{xx} - w \bar{w}_{xx}]_{y=0} \,\mathrm{d}x &= [\, \overline{w(x,0)} w_x(x,0) - w(x,0) \overline{w_x(x,0)}\,]_{x=-\alpha}^{x=+\alpha} \\ &= 2i\lambda_0 \cosh^2 \lambda_0 d (|\mathcal{D}_+|^2 + |\mathcal{D}_-|^2) \\ &\quad + ic\lambda_0 \cosh \lambda_0 d [(\mathcal{D}_+ - \mathcal{D}_-) e^{i\lambda_0 \alpha} \\ &\quad + (\bar{\mathcal{D}}_+ - \bar{\mathcal{D}}_-) e^{-i\lambda_0 \alpha}] + O(\alpha^{-1}) \end{aligned}$$

and

$$\begin{aligned} & - \sum_{\pm} \int_{W \cap \{\pm x = \alpha\}} \left(\bar{w} \frac{\partial w}{\partial |x|} - w \frac{\partial \bar{w}}{\partial |x|} \right) \mathrm{d}y \\ &= ic \sinh \lambda_0 d [(\mathcal{D}_+ - \mathcal{D}_-) e^{i\lambda_0 \alpha} + (\bar{\mathcal{D}}_+ - \mathcal{D}_-) e^{-i\lambda_0 \alpha}] \\ &\quad - 2i\lambda_0 (|\mathcal{D}_+|^2 + |\mathcal{D}_-|^2) \int_{-d}^{0} \cosh^2 \lambda_0 (y + d) \,\mathrm{d}y + O(\alpha^{-1}) \\ &= ic \sinh \lambda_0 d [(\mathcal{D}_+ - \mathcal{D}_-) e^{i\lambda_0 \alpha} + (\bar{\mathcal{D}}_+ - \bar{\mathcal{D}}_-) e^{-i\lambda_0 \alpha}] \\ &\quad - i(\lambda_0 d - \sinh \lambda_0 d \cosh \lambda_0 d)(|\mathcal{D}_+|^2 + |\mathcal{D}_-|^2) + O(\alpha^{-1}). \end{aligned}$$

Substituting these expressions into (7.58), we obtain

$$\begin{aligned} & i(2\lambda_0 \nu^{-1} \cosh^2 \lambda_0 d - \lambda_0 d - \sinh \lambda_0 d \cosh \lambda_0 d)(|\mathcal{D}_+|^2 + |\mathcal{D}_-|^2) \\ &= 2i \,\mathrm{Im} \int_S w \frac{\partial \bar{w}}{\partial n} \,\mathrm{d}s + O(\alpha^{-1}), \end{aligned}$$

because the term containing c vanishes in view of the definition of λ_0. Tending $\alpha \to \infty$ and using the definition of λ_0 again, we arrive at (7.57).

The last two assertions imply that a solution w of the homogeneous Problem II satisfies (7.54). Then, the last theorem in Subsection 7.2.1.1 gives that $w = \text{const}$ in W, and taking into account (7.55), we arrive at the following theorem.

Let $\mathbf{x} \cdot \mathbf{n} \geq 0$ *on* S*; then Problem II has a unique solution and it has the form*

$$w(z) = \frac{1}{\pi} \int_S \mu(\zeta) E(z, \zeta) \, \mathrm{d}s,$$

where μ *must be determined from the uniquely solvable Fredholm equation*

$$-\mu(z) + (\mathcal{T}\mu)(z) = f(z), \qquad (\mathcal{T}\mu)(z) = \frac{1}{\pi} \int_S \mu(\zeta) \frac{\partial E}{\partial n_z}(z, \zeta) \, \mathrm{d}s.$$

7.2.1.3. On the Unique Solvability of Problem I

The theorem in Subsection 7.1.5.1 remains true for the subcritical stream of finite depth. Replacing $\exp\{\nu(y + ix)\}$ by $e^{i\lambda_0 x} \cosh \lambda_0(y + d)$ in the considerations in Subsections 7.1.5.2 and 7.1.5.3, we arrive at the following two theorems. The first of them establishes the equivalence of Problem II and a boundary integral equation.

Let Problem II be uniquely solvable (for example, let $\mathbf{x} \cdot \mathbf{n} \geq 0$ *on* S*). Then*

$$u(z) = \frac{1}{2\pi} \int_S \mu(\zeta) G(z, \zeta) \, \mathrm{d}s + c$$

gives a one-to-one correspondence between solutions of Problem I and pairs (μ, c)*, where* $c = \text{const}$ *and* μ *solves the Fredholm integral equation*

$$-\mu + T\mu = f, \qquad (T\mu)(z) = \frac{1}{\pi} \int_S \mu(\zeta) \frac{\partial G}{\partial n_z}(z, \zeta) \, \mathrm{d}s.$$

The second theorem provides several necessary and sufficient conditions guaranteeing the unique solvability of Problem I.

Let Problem II be uniquely solvable; then the following assertions are equivalent.

1. *Problem I has one and only one (up to a constant term) solution for* $f \in C^{1,\alpha}(S)$.
2. *For* $\mathcal{W}$ *solving Problem II, where*

$$f = \frac{\partial}{\partial n} [e^{i\lambda_0 x} \cosh \lambda_0(y + d)]$$

in the Neumann condition on S, we have $\mathcal{D}_+(\mathcal{W})$ in (7.56) not equal to one.

3. *The inequality*

$$\int_S \mathcal{W} \frac{\partial}{\partial n}[e^{-i\lambda_0 x} \cosh \lambda_0(y+d)] \, \mathrm{d}s$$
$$\neq \lambda_0^2 \int_D \cosh 2\lambda_0(y+d) \, \mathrm{d}x\mathrm{d}y - i\lambda_0 \nu^{-1}(\nu d - \cosh^2 \lambda_0 d)$$

holds for $\mathcal{W}$ defined in 2.

4. *We have $|\mathcal{D}_-(\mathcal{W})| \neq 1$ for $\mathcal{W}$ defined in 2.*
5. *The inequality*

$$\left| \int_S \mathcal{W} \frac{\partial}{\partial n}[e^{i\lambda_0 x} \cosh \lambda_0(y+d)] \, \mathrm{d}s + \lambda_0^2 \int_D e^{2i\lambda_0 x} \, \mathrm{d}x\mathrm{d}y \right|$$
$$\neq \lambda_0 \nu^{-1}(\cosh^2 \lambda_0 d - \nu d)$$

holds for $\mathcal{W}$ defined in 2.

6. *Let μ_+ be a solution of*

$$-\mu_+(z) + (\mathcal{T}\mu_+)(z) = \frac{\partial}{\partial n}[e^{i\lambda_0 x} \cosh \lambda_0(y+d)], \quad z \in S, \quad (7.59)$$

where $\mathcal{T}$ is defined in the last theorem in Subsection 7.2.1.2; then

$$i\nu \int_S \mu_+ e^{i\lambda_0 x} \cosh \lambda_0(y+d) \, \mathrm{d}s \neq \lambda_0(\nu d - \cosh^2 \lambda_0 d).$$

7. *For μ_+ solving (7.59) we have*

$$\nu \left| \int_S \mu_+ e^{i\lambda_0 x} \cosh \lambda_0(y+d) \, \mathrm{d}s \right| \neq \lambda_0(\cosh^2 \lambda_0 d - \nu d).$$

In the proof of 1 ⇔ 6, the following addition should be made in the case of shallow water. From the homogeneous integral equation

$$-\mu_0 + T\mu_0 = 0, \quad \text{where } (T\mu_0)(z) = \frac{1}{\pi} \int_S \frac{\partial G}{\partial n_z}(z, \zeta)\mu_0(\zeta) \, \mathrm{d}s, \quad (7.60)$$

the equivalent equation

$$-\mu_0(z) + (\mathcal{T}\mu_0)(z) = -\frac{\nu \cos(n, x)}{1 - \nu d} \int_S \mu \, \mathrm{d}s$$
$$- 2i \left[\frac{\partial}{\partial n} \frac{\nu \cosh \lambda_0(y+d) e^{i\lambda_0 x}}{\lambda_0(\cosh^2 \lambda_0 d - \nu d)} \right]$$
$$\times \int_S e^{-i\lambda_0 \xi} \cosh \lambda_0(\eta + d)\mu_0(\zeta) \, \mathrm{d}s \quad (7.61)$$

arises after G is replaced by E. However, there is an extra term in the latter equation when it is compared with (7.40) (it is the first term on the right-hand side). Since $G(z, \zeta) + \log|z - \zeta|$ is a harmonic function, we have that

$$\int_S \frac{\partial G}{\partial n_z}(z, \zeta)\, \mathrm{d}s_z = -\int_S \log|z - \zeta|\, \mathrm{d}s_z = -\pi \quad \text{for } \zeta \in S.$$

Therefore, integrating (7.60) over S, one gets that $\int_S \mu_0\, \mathrm{d}s = 0$. Now, the integral equation (7.61) takes the same form as (7.40) and the argument from Subsection 7.1.5.3 applies.

7.2.2. Submerged Body in the Supercritical Stream

Here we consider the same problem as in Subsection 7.2.1, but we assume that $0 < \nu d < 1$. The same method as in Subsection 7.1.1 leads to the following theorem demonstrating that there is no wave pattern behind the body.

Let $u \in C^2(\bar{W})$ be a solution of the Neumann–Kelvin problem; then we have the following as $|x| \to \infty$:

$$u(x, y) = QxH(-x) + c_\pm + \psi_\pm(x, y), \quad \pm x > 0, \tag{7.62}$$

where $\psi_\pm = O(|x|^{-1})$, $|\nabla\psi_\pm| = O(|x|^{-2})$, and H is the Heaviside function. Furthermore,

$$(c_+ - c_-)(1 - \nu d) = \nu \int_S \left[x\frac{\partial u}{\partial n} - u\cos(n, x) \right] \mathrm{d}s,$$

$$Q(1 - \nu d) = \nu \int_S \frac{\partial u}{\partial n}\, \mathrm{d}s.$$

Since there is no term in (7.62) describing waves at infinity downstream, the uniqueness theorem differs from that in the subcritical case because no condition should be imposed on S.

Let $u \in C^2(\bar{W})$ be a solution of the homogeneous Neumann–Kelvin problem; then $u =$ const in W.

First, u satisfies (7.54) according to the previous theorem. Let us consider a stream function v, that is, a conjugate to u harmonic function in W. The homogeneous Neumann condition on S implies that v is determined uniquely up to a constant term. According to (7.52), this constant can be chosen so that

$$v = 0 \quad \text{when } y = -d. \tag{7.63}$$

Using the Cauchy–Riemann equations, one readily finds (cf. Subsection 7.1.6) that v must satisfy the following boundary conditions:

$$v = \text{const} \quad \text{on } S, \tag{7.64}$$

$$v_y - \nu v = 0 \quad \text{when } y = 0. \tag{7.65}$$

When deriving the last relation, one has to take into account the second condition (7.4), (7.63), and the fact that

$$v(x, 0) = \int_{-h}^{0} v_y(x, y)\, \mathrm{d}y, \quad \text{if } x \notin \{x \in \mathbb{R} : (x, y) \in S\}. \tag{7.66}$$

In view of (7.64), let us extend v into D as a constant. Now v is defined throughout the strip $\bar{L}$ and (7.66) holds for all $x \in \mathbb{R}$. Therefore, the Schwarz inequality gives that

$$|v(x, 0)|^2 \leq d \int_{-d}^{0} |v_y(x, y)|^2\, \mathrm{d}y, \quad -\infty < x < +\infty,$$

and so

$$\int_{y=0} |v|^2\, \mathrm{d}x \leq d \int_W |\nabla v|^2\, \mathrm{d}x\mathrm{d}y \tag{7.67}$$

because ∇v vanishes outside $\bar{W}$.

On the other hand, Green's identity yields

$$\int_W |\nabla v|^2\, \mathrm{d}x\mathrm{d}y = \int_{\partial W} v(\partial v/\partial n)\, \mathrm{d}s = \nu \int_{y=0} |v|^2\, \mathrm{d}x. \tag{7.68}$$

Here (7.63) and (7.65) are taken into account, and it is noted that

$$\int_S v \frac{\partial v}{\partial n}\, \mathrm{d}s = \text{const} \int_S \frac{\partial u}{\partial s}\, \mathrm{d}s = 0$$

because S is a closed curve. Comparing (7.67) with (7.68), we obtain

$$(1 - \nu d) \int_{y=0} |v|^2\, \mathrm{d}x\mathrm{d}y \leq 0.$$

Since $0 < \nu d < 1$, we have that $v = 0$ when $y = 0$. Now (7.68), (7.65), and the uniqueness theorem for the Cauchy problem for the Laplace equation imply that $v = 0$ in $\bar{W}$. Therefore, $u = \text{const}$ in $\bar{W}$, and the proof is complete.

In order to prove the existence theorem we apply the same integral equation method as in Subsection 7.1.5.2. A solution is sought in the following form:

$$u(z) = \int_S \mu(\zeta) G(z, \zeta)\, \mathrm{d}s + c, \tag{7.69}$$

where μ is an unknown density. The kernel of a single-layer potential is Green's function G for shallow water (see Subsection 6.3.2). Then, as in Subsection 7.1.5.2 we arrive at the Fredholm equation

$$-\mu + T\mu = f, \quad \text{where } (T\mu)(z) = \frac{1}{\pi} \int_S \mu(\zeta) \frac{\partial G}{\partial n_z}(z, \zeta)\, \mathrm{d}s. \tag{7.70}$$

The literal repeating of the proof in Subsection 7.1.4.4 with the reference to the previous uniqueness theorem in the present subsection gives the following result.

For all $\nu \in (0, d^{-1})$ the Neumann–Kelvin problem has one and only one solution (up to a constant term) for any $f \in C^{1,\alpha}(S)$. The solution $u \in C^2(\bar{W})$ can be found in the form of (7.69), *where μ satisfies* (7.70).

7.3. Wave Resistance

We conclude the study of the two-dimensional problem of a submerged body in the uniform forward motion in calm water with formulae for the wave resistance that is the horizontal component $\mathcal{R}$ of the force acting on the body, and so we can use the following general formula,

$$\mathcal{R} = -\int_S p \cos(n, x)\, \mathrm{d}s, \tag{7.71}$$

where p is the pressure in water. Since we consider the steady-state waves caused by a cylinder beneath the free surface, the pressure can be found from Bernoulli's integral for a two-dimensional flow (see the Equations of Motion section in the Introduction):

$$p = \text{const} + \rho g y - \rho v^2/2. \tag{7.72}$$

Here ρ is the density of water, g is the acceleration due to gravity, and

$$v^2 = (u_x - U)^2 + u_y^2, \tag{7.73}$$

where U is the body's forward speed and (u_x, u_y) is the velocity field caused by the potential u. Since

$$\int_S \cos(n, x)\, \mathrm{d}s = \int_S \frac{\partial y}{\partial s}\, \mathrm{d}s = 0, \qquad \int_S y \cos(n, x)\, \mathrm{d}s = \int_S y \frac{\partial y}{\partial s}\, \mathrm{d}s = 0$$

for any closed contour in $\mathbb{R}^2$, we get the following from (7.71)–(7.73):

$$\mathcal{R} = \frac{\rho}{2}\int_S \left[(u_x - U)^2 + u_y^2\right]\cos(n,x)\,\mathrm{d}s$$
$$= \frac{\rho}{2}\int_S (|\nabla u|^2 - 2Uu_x)\cos(n,x)\,\mathrm{d}s, \tag{7.74}$$

which is convenient for use in what follows.

7.3.1. Cylinder in Deep Water

The aim of the present subsection is to prove the following assertion.

Let u satisfy the boundary value problem (7.1)–(7.4), *where*

$$\partial u/\partial n = U\cos(n,x) \quad \text{on } S. \tag{7.75}$$

If $\rho = 1$, *then we have*

$$\mathcal{R} = -\nu(\mathcal{A}^2 + \mathcal{B}^2)/4, \tag{7.76}$$

where $\mathcal{A}$ *and* $\mathcal{B}$ *are the coefficients in the asymptotic formula obtained in Subsection* 7.1.1.

Let us begin by demonstrating that

$$\mathcal{R} = \int_{\partial R_{\alpha\beta}} \left[u_x\frac{\partial u}{\partial n} - \frac{1}{2}|\nabla u|^2\cos(n,x)\right]\mathrm{d}s, \tag{7.77}$$

where $R_{\alpha\beta}$ is the rectangle $\{|x| < \alpha, -\beta < y < 0\}$ such that $\alpha > \max_S |x|$ and $\beta > \max_S -y$. Using the notation from Subsection 7.1.3, by $p_{\alpha 0}$, $p_{\alpha\beta}$, and $q_{\pm\alpha\beta}$ we denote the top, bottom, right and left sides of $R_{\alpha\beta}$, respectively; also, we assume the normals to these sides to be directed into $R_{\alpha\beta}$. From Green's identity

$$0 = \int_{W\cap R_{\alpha\beta}} u_x\nabla^2 u\,\mathrm{d}x\mathrm{d}y = -\int_{W\cap R_{\alpha\beta}} \nabla u\cdot\nabla u_x\,\mathrm{d}x\mathrm{d}y - \int_{S\cup\partial R_{\alpha\beta}} u_x\frac{\partial u}{\partial n}\,\mathrm{d}s,$$

we obtain

$$-\int_S Uu_x\cos(n,x)\,\mathrm{d}s = -\int_S u_x\frac{\partial u}{\partial n}\,\mathrm{d}s$$
$$= \int_{\partial R_{\alpha\beta}} u_x\frac{\partial u}{\partial n}\,\mathrm{d}s + \frac{1}{2}\int_{W\cap R_{\alpha\beta}} (|\nabla u|^2)_x\,\mathrm{d}x\mathrm{d}y$$
$$= \int_{\partial R_{\alpha\beta}} u_x\frac{\partial u}{\partial n}\,\mathrm{d}s - \frac{1}{2}\int_{S\cup\partial R_{\alpha\beta}} |\nabla u|^2\cos(n,x)\,\mathrm{d}s,$$

where (7.75) and the divergence theorem are applied. Comparing the last formula with (7.74), we arrive at (7.77) because $\rho = 1$.

Let us calculate the integrals in (7.77) over $p_{\alpha 0}$, $p_{\alpha\beta}$, and $q_{\pm\alpha\beta}$ by using the asymptotic formula proved in Subsection 7.1.1. It should be taken into account that

$$Q = \pi^{-1} \int_S \frac{\partial u}{\partial n} \, \mathrm{d}s = U \pi^{-1} \int_S \cos(n, x) \, \mathrm{d}s = 0.$$

It is clear that the integral over $p_{\alpha\beta}$ in (7.77) tends to zero as $\beta \to \infty$. Also, the second term in the integral over $p_{\alpha 0}$ vanishes because $\cos(n, x) = 0$. According to (7.2), we have

$$\int_{p_{\alpha 0}} u_x \frac{\partial u}{\partial n} \, \mathrm{d}x = \nu^{-1} \int_{-\alpha}^{+\alpha} u_x u_{xx} \, \mathrm{d}x = (2\nu)^{-1} \left[u_x^2(x, 0) \right]_{x=-\alpha}^{x=+\alpha}.$$

The asymptotic formula gives that

$$\left[u_x^2(x, 0) \right]_{x=-\alpha}^{x=+\alpha} = -\nu^2 (\mathcal{A} \cos \alpha\nu + \mathcal{B} \sin \alpha\nu)^2 + O(\alpha^{-1}) \tag{7.78}$$

as $\alpha \to \infty$. The integral over $\mathbb{R}^2_- \cap \{x = +\alpha\}$ can be written in the form

$$\frac{1}{2} \int_{-\infty}^{0} \left(u_y^2 - u_x^2 \right) \big|_{x=+\alpha} \, \mathrm{d}y,$$

and it is equal to $O(\alpha^{-1})$ as $\alpha \to \infty$. Hence the contribution of the last integral into the right-hand side of (7.77) can be included into (7.78). The integral over $\mathbb{R}^2_- \cap \{x = -\alpha\}$ is equal to

$$\frac{1}{2} \int_{-\infty}^{0} \left(u_x^2 - u_y^2 \right) \big|_{x=-\alpha} \, \mathrm{d}y,$$

which gives, in view of the asymptotic formula,

$$\frac{\nu^2}{2} [(\mathcal{A} \cos \alpha\nu + \mathcal{B} \sin \alpha\nu)^2 - (\mathcal{B} \cos \alpha\nu - \mathcal{A} \sin \alpha\nu)^2] \int_{-\infty}^{0} e^{2\nu y} \, \mathrm{d}y + O(\alpha^{-1}). \tag{7.79}$$

Summing up (7.78) and (7.79) and tending α to infinity, we arrive at (7.76), which completes the proof.

Formula (7.77) shows that the resistance of a two-dimensional rigid body totally submerged in deep water is purely wave making.

7.3.2. Cylinder in the Subcritical Stream

In the present subsection, our aim is to prove the analog of (7.76) for the wave resistance of a two-dimensional body submerged in a subcritical stream; that is, $\nu d > 1$. Again, we use the no-flow condition (7.75) together with the bottom condition

$$u_y = 0 \quad \text{when } y = -d,$$

and we assume that $\rho = 1$.

Using (7.74) and applying Green's formula in the same way as in Subsection 7.3.1, we get that

$$\mathcal{R} = \int_{\partial R_{\alpha d}} \left[u_x \frac{\partial u}{\partial n} - \frac{1}{2} |\nabla u|^2 \cos(n, x) \right] \mathrm{d}s. \tag{7.80}$$

Here $R_{\alpha d} = \{|x| < \alpha, -d < y < 0\}$ and $\alpha > \max_S |x|$. The integral over $\{|x| < \alpha, y = -d\}$ vanishes because of the bottom condition and the equality $\cos(n, x) = 0$ holding there. The integrals over $p_{\alpha 0}$ and $q_{\pm\alpha 1}$ in (7.80) must be treated by using the asymptotic formula (7.53), where

$$Q = \frac{\nu}{1 - \nu d} \int_S \frac{\partial u}{\partial n} \, \mathrm{d}s = \frac{\nu U}{1 - \nu d} \int_S \cos(n, x) \, \mathrm{d}s = 0.$$

The integral over $p_{\alpha 0}$ can be calculated in the same way as in Subsection 7.3.1, but here we have

$$\left[u_x^2(x, 0) \right]_{x=-\alpha}^{x=+\alpha} = -\lambda_0^2 \cosh^2 \lambda_0 d (\mathcal{A} \cos \lambda_0 \alpha + \mathcal{B} \sin \lambda_0 \alpha)^2 + O(\alpha^{-1})$$

instead of (7.78). The integral over $q_{\alpha d}$ is equal to $O(\alpha^{-1})$. The last integral in (7.80) over $q_{\alpha d}$ can be written as

$$\frac{1}{2} \int_{-d}^{0} \left(u_x^2 - u_y^2 \right) \big|_{x=-\alpha} \mathrm{d}y.$$

Substituting the asymptotic formula (7.53), where $Q = 0$, into this integral, we arrive at

$$\frac{\lambda_0^2}{4} \big[(\mathcal{A} \cos \lambda_0 \alpha - \mathcal{B} \sin \lambda_0 \alpha)^2 \left(\lambda_0^{-1} \sinh \lambda_0 d \cosh \lambda_0 d + d \right)$$
$$- (\mathcal{A} \cos \lambda_0 \alpha + \mathcal{B} \cos \lambda_0 \alpha)^2 \left(\lambda_0^{-1} \sinh \lambda_0 d \cosh \lambda_0 d - d \right) + O(\alpha^{-1}) \big].$$

Summing up the expressions obtained for the integrals over $p_{\alpha 0}$ and $q_{\pm\alpha d}$, tending α to infinity, and taking into account the definition of λ_0, we find that

$$\mathcal{R} = -\frac{\lambda_0^2}{4\nu} (\cosh^2 \lambda_0 d - \nu d)(\mathcal{A}^2 + \mathcal{B}^2).$$

This formula is similar to (7.77), and it shows that the resistance of a totally submerged two-dimensional rigid body is purely wave making when the body has a subcritical speed.

7.4. Three-Dimensional Body in Deep Water

7.4.1. Statement of the Problem and Auxiliary Results

First we recall the problem's statement from the Linear Ship Waves section of the Introduction. We consider the uniform forward motion of a body submerged in water of infinite depth. Let $\mathbb{R}^3_- = \{(x, y, z) : y < 0\}$ be the undisturbed water domain and let D be a bounded domain occupied by a rigid body. We assume that $\bar{D} \subset \mathbb{R}^3_-$ and that the body's surface $S = \partial D$ is sufficiently smooth. By $W = \mathbb{R}^3_- \setminus \bar{D}$ we denote the disturbed water domain. Let the body moves at the speed U in the direction of the x axis; then the induced velocity field can be described by a velocity potential $u(x, y, z)$ in the coordinate system attached to the body, and u must satisfy

$$\begin{aligned} \nabla^2 u &= 0 \quad \text{in } W, \\ u_{xx} + \nu u_y &= 0 \quad \text{when } y = 0, \\ \partial u/\partial n &= -U \cos(n, x) \quad \text{on } S. \end{aligned} \tag{7.81}$$

Here $\nu = g/U^2$ and n is the unit normal to S directed into W. For discussing the existence and uniqueness for the problem (7.81) we have to prescribe certain conditions at infinity. It is convenient to formulate these conditions by using the results obtained in Section 6.2, and so instead of conditions at infinity (see the Conditions at Infinity Upstream and Downstream section in the Introduction) we require u to admit the following representation:

$$u(x, y, z) = \int_S G(x - x_0, y, z - z_0; y_0)\psi(x_0, y_0, z_0)\, dS, \tag{7.82}$$

where G is Green's function constructed in Subsection 6.1.2 for a point source moving forward in deep water and ψ is a continuous function on S.

In what follows we will use two auxiliary assertions, where we apply multi-index notation for partial derivatives. Let $\beta = (\beta_1, \beta_2, \beta_3)$ be a multi-index, that is, a vector with integer nonnegative components. By ∂^β we denote

$$\frac{\partial^{|\beta|}}{\partial x^{\beta_1} \partial y^{\beta_2} \partial z^{\beta_3}}, \quad \text{where } |\beta| = \sum \beta_i.$$

Let Green's function be written in the form

$$G(x, y, z; y_0) = R^{-1} - R_0^{-1} + I(x, y + y_0, z).$$

Then for any β and d there exists $C = C(\beta, d)$ such that for $-d < \nu y < 0$ we have

$$|\partial^\beta I(x, y, z)| \leq C\nu^{1/2} |y|^{-|\beta|-1/2}. \tag{7.83}$$

Before proving this assertion we note that $Q = R^{-1} - R_0^{-1}$ satisfies

$$\nabla^2 Q = -4\pi \delta(x, y - y_0, z) \quad \text{in } \mathbb{R}^3_-, \quad Q|_{y=0} = 0,$$

and so Q is Green's function of the Dirichlet problem in $\mathbb{R}^3_-$. Thus the assertion formulated means that G tends to Q as $\nu \to 0$ and $y < 0$.

In order to prove (7.83) we note that (6.8) in Subsection 6.1.1.2 gives

$$I(x, y, z) = I_2(x, y, z) + I_1(x, y, z), \tag{7.84}$$

where I_2 and I_1 are defined by formulae (6.4) and (6.7), respectively. It remains for us to prove (7.83) for $I = I_2 + I_1$. We showed in Subsection 6.1.3 that $I_2 + I_1$ is an infinitely differentiable function in $\mathbb{R}^3_-$, and so it is sufficient for us to prove (7.83) for each function I_2 and I_1 separately and only when $x \neq 0$.

If $x \neq 0$ and $-d < \nu y < 0$, then formula (6.7) in Subsection 6.1.1 implies that

$$\begin{aligned}
|\partial^\beta I_1| &\leq 4\nu^{|\beta|+1} \int_{-\infty}^{\infty} (1 + t^2)^{|\beta|} e^{\nu y t^2} \mathrm{d}t \\
&\leq C(\beta)\nu^{|\beta|+1} \left(\int_{-\infty}^{\infty} e^{\nu y t^2} \mathrm{d}t + \int_{-\infty}^{\infty} t^{2|\beta|} e^{\nu y t^2} \mathrm{d}t \right) \\
&\leq C(\beta)\nu^{|\beta|+1} \left(\frac{c_1}{|\nu y|^{1/2}} + \frac{c_2}{|\nu y|^{|\beta|+1/2}} \right) \leq C(\beta, d)\nu^{1/2} |y|^{-|\beta|-1/2}.
\end{aligned} \tag{7.85}$$

In order to prove a similar estimate for $\partial^\beta I_2$, we note that for differentiating I_2 when $x \neq 0$, one has simply to differentiate the integrand in (6.4); see Subsection 6.1.3 for the proof. Hence $\partial^\beta I_2$ is equal to the integral (6.4), where the integrand is multiplied by

$$f_\beta = (i\zeta \ \mathrm{sign}\, x)^{\beta_1} (\zeta^2 + \tau^2)^{\beta_2/2} (i\tau)^{\beta_3},$$

and it is obvious that

$$|f_\beta| \leq (|\zeta|^2 + \tau^2)^{|\beta|/2}.$$

Putting $\alpha = \pi/4$ and $\zeta = \sigma e^{i\pi/4}$ and using polar coordinates (k, θ) in the plane (ζ, τ), we obtain an integral representation for $\partial^\beta I_2$, and this representation is obtained from (6.35) in Subsection 6.1.4 by inserting an additional factor f_β in the integrand. This implies that for $x \neq 0$ we have

$$|\partial^\alpha I_2| \leq \frac{2\nu}{\pi}\int_{-\pi/2}^{\pi/2}\int_0^\infty \frac{k^{|\beta|}\,|\exp\{S(\theta,\omega)k\rho\}|}{|\nu f(\theta) - ik\cos^2\theta|}\,dk d\theta$$
$$\leq \frac{2\nu}{\pi}\int_{-\pi/2}^{\pi/2}\int_0^\infty \frac{k^{|\beta|}\exp\{yk/2\}}{|\nu f(\theta) - ik\cos^2\theta|}\,dk d\theta. \quad (7.86)$$

The last estimate follows from (6.36) in Subsection 6.1.4.

Now, let us improve estimate (6.38) obtained in Subsection 6.1.4 for the denominator in (7.86). From (6.36), it follows that

$$2(\operatorname{Re} f)(\operatorname{Im} f) = \cos^2\theta.$$

Since $\operatorname{Re} f \geq 1/2$, see (6.38), the last equality implies that $|\operatorname{Im} f| \leq \cos^2\theta$. Therefore,

$$|\nu f(\theta) - ik\cos^2\theta| \geq (\nu\,|\operatorname{Re} f| + |\nu \operatorname{Im} f - k\cos^2\theta|)/2$$
$$\geq (\nu + 2|\nu \operatorname{Im} f - k\cos^2\theta|)/4$$
$$\geq \begin{cases} \nu/4 & \text{for } k \leq 2\nu \\ (\nu + k\cos^2\theta)/4 & \text{for } k > 2\nu \end{cases}.$$

Hence, we have the following for $x \neq 0$:

$$|\partial^\beta I_2| \leq \frac{8}{\pi}\int_{-\pi/2}^{\pi/2}\int_0^{2\nu} k^{|\beta|}\exp\{yk/2\}\,dk d\theta$$
$$+ \frac{8\nu}{\pi}\int_{-\pi/2}^{\pi/2}\int_{2\nu}^\infty \frac{k^{|\beta|}\exp\{yk/2\}}{\nu + k\cos^2\theta}\,dk d\theta$$
$$\leq C\nu^{|\beta|+1} + \frac{8\nu}{\pi}\int_{2\nu}^\infty\int_{-\pi/2}^{\pi/2} \frac{k^{|\beta|}\exp\{yk/2\}}{\nu + k\cos^2\theta}\,d\theta dk.$$

Evaluating the inner integral, we get the following for $x \neq 0$ and $-d < \nu y < 0$:

$$|\partial^\beta I_2| \leq C\nu^{|\beta|+1} + 8\sqrt{\nu}\int_{2\nu}^\infty \frac{k^{|\beta|}\exp\{yk/2\}}{\sqrt{\nu + k}}\,dk$$
$$\leq C\nu^{|\beta|+1} + \frac{8\sqrt{\nu}}{\pi}\int_0^\infty k^{|\beta|-1/2}\exp\{yk/2\}\,dk$$
$$= C\nu^{|\alpha|+1} + C'\nu^{1/2}\,|y|^{-|\beta|-1/2} \leq C(\beta,d)\nu^{1/2}\,|y|^{-|\beta|-1/2}.$$

This and (7.85) imply (7.83), which completes the proof.

The following assertion follows from (6.78) and (6.83) in Subsection 6.2.1.

Let Green's function be written in the form

$$G(x, y, z; y_0) = R^{-1} + R_0^{-1} + F(x, y + y_0, z).$$

Then for any $\varepsilon > 0$ *there exists* $C = C(\varepsilon)$ *such that*

$$|F(x, y + y_0, z)| + |\nabla F(x, y + y_0, z)| \leq \frac{C}{\nu\, |y + y_0|^2}$$

when $\nu\, |y + y_0| \geq 1$ *and* $|y + y_0| \geq \varepsilon\sqrt{x^2 + z^2}$.

We note that $Q_+ = R^{-1} + R_0^{-1}$ satisfies

$$\nabla^2 Q_+ = -4\pi\,\delta(x, y - y_0, z) \quad \text{in } \mathbb{R}^3_-, \quad \partial Q_+/\partial y|_{y=0} = 0,$$

and so Q_+ is Green's function of the Neumann problem in $\mathbb{R}^3_-$. Thus the second assertion means that G tends to Q_+ as $\nu \to \infty$ and $y < 0$.

7.4.2. Solvability of the Problem

The aim of this subsection is to prove the following theorem.

For all $\nu > 0$ *except possibly for a finite number of values, there exists a unique solution* u *of* (7.81) *such that* u *has the form of* (7.82) *for a certain* $\psi \in C(S)$.

Seeking u in the form of (7.82) with unknown $\psi \in C(S)$, we see that u satisfies the Laplace equation and the free surface boundary condition because of the properties of Green's function. Since standard results about single-layer potentials (see Subsection 2.1.1.1) are applicable to (7.82), the Neumann condition in (7.81) holds if and only if ψ satisfies the following equation:

$$-\psi(x, y, z) + (T_\nu \psi)(x, y, z) = -(2\pi)^{-1} U \cos(n, x), \qquad (7.87)$$

where

$$(T_\nu \psi)(x, y, z) = \frac{1}{2\pi} \int_S \frac{\partial G}{\partial n_{(x,y,z)}}(x - x_0, y, z - z_0; y_0)\, \psi(x_0, y_0, z_0)\, dS.$$

Now we have to show that (7.87) is uniquely solvable in $C(S)$ for all $\nu > 0$ except at most a finite number of values.

From (6.8) in Subsection 6.1.1, it follows that

$$I = G - \left(R^{-1} - R_0^{-1}\right) = I_2(x, y + y_0, z) + I_1(x, y + y_0, z), \quad (7.88)$$

and so I is a function of $(x, y + y_0, z)$. Since for $y, y_0 < 0$ we have

$$\nabla^2 \left(R^{-1} - R_0^{-1}\right) = -4\pi\,\delta(x, y - y_0, z)$$

and the same equation holds for G, we see that I is a harmonic function in $\mathbb{R}^3_-$. Consequently, $I(x, y + y_0, z)$ is a real analytic function of x, $y + y_0$, and z for $-\infty < x, z < +\infty$, and $y + y_0 < 0$.

For emphasizing the dependence on ν, we will write $I^{(\nu)}$, $I_2^{(\nu)}$, and $I_1^{(\nu)}$ instead of I, I_2, and I_1, respectively. From (6.4) and (6.7) in Subsection 6.1.1, it follows that for any $\nu > 0$ we have

$$I_j^{(\nu)}(x, y + y_0, z) = \nu I_j^{(1)}[\nu x, \nu(y + y_0), \nu z], \quad j = 1, 2,$$

and so we get the following for $\nu > 0$ and $y + y_0 < 0$:

$$I^{(\nu)}(x, y + y_0, z) = \nu I^{(1)}(\nu x, \nu y + \nu y_0, \nu z).$$

This and the fact that $I^{(\nu)}$ is an analytic function of x, $y + y_0$, and z imply that

$$\left[T_1^{(\nu)}\psi\right](x, y, z) = \frac{1}{2\pi}\int_S \frac{\partial I^{(\nu)}}{\partial n_{(x,y,z)}}(x - x_0, y + y_0, z - z_0)\, \psi(x_0, z_0, y_0)\, \mathrm{d}S$$

is a compact operator in $C(S)$ depending analytically on $\nu > 0$.

Let us introduce two kernels:

$$\frac{\partial}{\partial n_{(x,y,z)}} \frac{1}{2\pi\sqrt{(x - x_0)^2 + (y \mp y_0)^2 + (z - z_0)^2}}.$$

By T_2 (T_3) we denote the integral operator in $C(S)$ having the kernel, where the minus (plus) sign is taken. Hence T_2 is the integral operator arising in the integral equations for the interior and exterior Neumann problems for the Laplace equation. Since S is smooth, the kernel of T_2 is a continuous function on $S \times S$, which implies that T_2 is a compact operator (see, for example, Subsection 2.1.1.1, where more detailed references to books by Kellogg [136], Mihlin [246], Petrovskii [288], and Vladimirov [348] are given). Since the kernel of T_3 is a smooth function when $y + y_0 < 0$, this operator is also compact in $C(S)$. Therefore, T_ν is a compact operator in $C(S)$ because $T_\nu = T_1^{(\nu)} + T_2 - T_3$ [see (7.88)] and T_2 and T_3 do not depend on ν. Moreover, T_ν depends analytically on $\nu > 0$ because $T_1^{(\nu)}$ has this property.

The first assertion in Subsection 7.4.1 implies that T_ν tends to $T_2 - T_3$ as $\nu \to 0$. Let us show that the equation

$$-\psi + (T_2 - T_3)\psi = f \tag{7.89}$$

is uniquely solvable in $C(S)$ for any $f \in C(S)$. Assuming that the corresponding homogeneous equation has a solution $\psi = \psi_0$, that is,

$$-\psi_0 + (T_2 - T_3)\psi_0 = 0, \quad (x, y, z) \in S \tag{7.90}$$

holds, let us modify this equation. By S' we denote the mirror image of S in the plane $y = 0$. Let $\tilde{S} = S \cup S'$ and $\tilde{\psi}$ be the odd in y extension of ψ_0 to S'. Then (7.90) implies that for $(x, y, z) \in S$ we have

$$-\tilde{\psi}(x, y, z) + \frac{1}{2\pi} \int_{\tilde{S}} \frac{\partial}{\partial n_{(x,y,z)}} \frac{1}{\sqrt{(x - x_0)^2 + (y - y_0)^2 + (z - z_0)^2}} \times \tilde{\psi}(x_0, y_0, z_0) \, \mathrm{d}S = 0,$$

which is the homogeneous integral equation for the Laplacian Neumann problem in the domain exterior to $\tilde{S}$. The latter equation is known to have only a trivial solution (see references given above), and so $\tilde{\psi}_0 = 0$ on S. Since T_2 and T_3 are compact operators, Fredholm's alternative guarantees that (7.89) has a unique solution. Now the fact that

$$T_\nu \to T_2 - T_3 \quad \text{as } \nu \to +0$$

implies that $-I + T_\nu$ (here I denotes the identity operator) is an invertible operator for sufficiently small $\nu > 0$. Since T_ν is a compact operator depending analytically on $\nu > 0$, the invertibility of $-I + T\nu$ for a certain value of $\nu > 0$ implies that $(-I + T_\nu)^{-1}$ is a meromorphic function of ν for $\nu > 0$ (see Subsection 2.1.2.1 for the formulation of the theorem and references). Thus (7.87) is uniquely solvable for all $\nu > 0$ except possibly for a discrete sequence $\{\nu_j\}$ tending to infinity.

For completing the proof it is sufficient to demonstrate that $-I + T_\nu$ is invertible for sufficiently large values of $\nu > 0$. The second assertion in Subsection 7.4.1 shows that T_ν tends to $T_2 + T_3$ as $\nu \to \infty$, and so we have to prove that the equation

$$-\psi + (T_2 + T_3)\psi = f$$

is uniquely solvable in $C(S)$. Since T_2 and T_3 are compact operators, verifying that the corresponding homogeneous equation has only a trivial solution provides the unique solvability. The homogeneous equation can be treated in the same way as (7.90), but ψ_0 must be extended to $\tilde{S}$ as an even function of y. The proof is complete.

Note that the same argument based on the second assertion in Subsection 7.4.1 leads to the following proposition.

Let $\nu > 0$ be fixed in the Neumann–Kelvin problem; then this problem is uniquely solvable for sufficiently large values of the distance from the body of fixed shape to the free surface.

7.5. Bibliographical Notes

The main reference for Sections 7.1 and 7.2 is the paper [346] by Vainberg and Maz'ya, and so we provide comments only for subsections using other sources.

7.1.2. The solvability theorems for the integral equation (7.6) guaranteeing the existence of a solution for small and large values of ν were proved by Kochin [141]. In [346], Vainberg and Maz'ya noted that combining Kochin's theorems with the invertibility theorem (see, for example, Gohberg and Krein [95]) and Subsection 2.1.2), one obtains the solvability of (7.6), and therefore, of the Neumann–Kelvin problem for all ν except possibly a finite set of values. These results were extended by Motygin [253] to the problem of a subcritical stream of finite depth about a totally submerged body.

7.1.3. The method applied here for proving the uniqueness of solution was first used by Kuznetsov and Maz'ya [165] in the case of a surface-piercing body. For a totally submerged body this result was not published earlier.

7.1.6. The material of this subsection is borrowed from the paper [211] by Livshits and Maz'ya.

7.2.2. The uniqueness theorem for a cylinder in a supercritical stream is from Lahalle [178].

7.3.1. The expression for the wave resistance of a cylinder in deep water equivalent to the formula derived here was obtained by Kochin [141], who introduced for this purpose the so-called H-function (see Wehausen and Laitone [354] and Kostyukov [147]).

7.3.2. Haskind [105] extended the definition of Kochin's H-function to the case of shallow water and obtained an expression of wave resistance equivalent to that derived here.

7.4.1. A condition at infinity in the form of (7.82) was proposed by Gutmann [101]. The first proposition describing the behavior of Green's function as $\nu \to 0$ was proved by Maz'ya and Vainberg [229]. The second result concerning the case of large values of ν was established in the same paper, but Kochin [141] investigated how the kernel of the integral operator T_ν introduced in Subsection 7.4.2 behaves for such values of ν.

7.4.2. The material in this subsection is borrowed from [229], but the solvability of the integral equation for large values of ν was demonstrated earlier in [141].

Other works. Another approach to the Neumann–Kelvin problem in the case of a submerged body (two or three dimensional) was developed by Dern [47]. Quenez [294] demonstrated that the Neumann–Kelvin problem has no more than one solution in the case of a subcritical stream over a flat bottom having a small protrusion. Pagani and Pierotti [276] investigated the problem of a horizontal plane wing (rigid plate) in a supercritical stream of finite depth. They demonstrated that there exists a one-parameter family of solutions decaying at infinity upstream and downstream, and a particular solution can be chosen when the value of circulation is prescribed. The results obtained are applied by the authors [277] for consideration of the nonlinear problem in the case of a slender submerged cylinder.

Wave resistance of a submerged three-dimensional body has been a topic of extensive studies by many authors. A comprehensive theory of this question was developed by Kochin [141] as early as in the 1930s. His results can be found in the book [147] by Kostyukov (also see survey papers by Wehausen and Laitone [354] and Wehausen [353] containing voluminous lists of references).

8

Two-Dimensional Problem for a Surface-Piercing Body

This chapter is concerned with various statements of the two-dimensional Neumann–Kelvin problem for a surface-piercing body. We speak about *various* statements because the Neumann–Kelvin problem as it is formulated in Chapter 7 for totally submerged bodies proves to be underdefinite when a body is surface piercing. It took several decades to realize that this underdefiniteness occurs and to develop several well-posed formulations of the problem (see a brief consideration of the question's history in Section 8.6).

The plan of this chapter is as follows. The problem augmented by general linear supplementary conditions is considered in Section 8.1. The question of total resistance to the forward motion for a surface-piercing cylinder is considered in a short Section 8.2, where we present formulae generalizing those in Section 7.3. A number of other statements of the Neumann–Kelvin problem are reviewed in Section 8.3. Among them, there are statements leading to the so-called least singular and wave-free solutions. Also, a statement of the Neumann–Kelvin problem for a tandem of surface-piercing cylinders is considered. This statement involves a set of four supplementary conditions canceling both the wave resistance and the spray resistance and providing a well-posed statement of the problem. This means that a unique solution exists for all values of the forward speed U except for a sequence tending to zero.

At the same time, for the exceptional values of U, examples of non-uniqueness are constructed in Section 8.4. The construction of these examples is based on the inverse procedure considered in Section 4.1 for a non-uniqueness example in the water-wave problem. Thus, there are two different kinds of non-uniqueness in the Neumann–Kelvin problem. The non-uniqueness of the first kind can be removed by imposing proper supplementary conditions. The non-uniqueness of the second kind is intrinsic to the problem with some supplementary conditions, but it is absent in the problem with other ones.

The problem of a supercritical stream about a two-dimensional body is considered in Section 8.5, where the equation, boundary conditions, and conditions at infinity are augmented by the same general linear supplementary conditions as in Section 8.1. Bibliographical notes and historical remarks are collected in Section 8.6.

8.1. General Linear Supplementary Conditions at the Bow and Stern Points

In this section, we are concerned with the problem describing the uniform forward motion of a cylinder in deep water. The plan of this section is as follows. A statement of the problem for the velocity potential is given in Subsection 8.1.1. Subsection 8.1.2 is concerned with asymptotics at infinity for the potential. In Subsection 8.1.3, the problem including supplementary conditions is reformulated in terms of the stream function, and both formulations are used in Subsection 8.1.6.2 for proving that the problem has at most one solution having the finite kinetic energy in the whole water domain. A uniqueness theorem of another kind (it is similar to that proven in Subsection 7.1.3 for a totally submerged body) is established in Subsection 8.1.5, but before it in Subsection 8.1.4, we prove that the problem with supplementary conditions is solvable for all positive values of ν except possibly for a sequence tending to infinity. The last two subsections, Subsections 8.1.6 and 8.1.7, are concerned with results analogous to those in Subsections 7.1.4 and 7.1.5 and treat the auxiliary problem of scattering type and its application for obtaining necessary and sufficient conditions of the unique solvability for the Neumann–Kelvin problem with supplementary conditions.

8.1.1. Statement of the Problem

We begin this subsection with a formulation of the main set of conditions involved in the statement of the Neumann–Kelvin problem for a surface-piercing two-dimensional body (see Subsection 8.1.1.1). This set includes the Laplace equation, boundary conditions, conditions at infinity, and the condition of local finiteness of the kinetic energy. The last condition implies the form of the local asymptotics for a velocity potential near the bow and stern points, which are the corner points of the water domain (see Subsection 8.1.1.2). In Subsection 8.1.1.3, two supplementary conditions are formulated. These conditions involve the limiting values of the x derivative at the corner points, and the existence of these values is a consequence of results in Subsection 8.1.1.2.

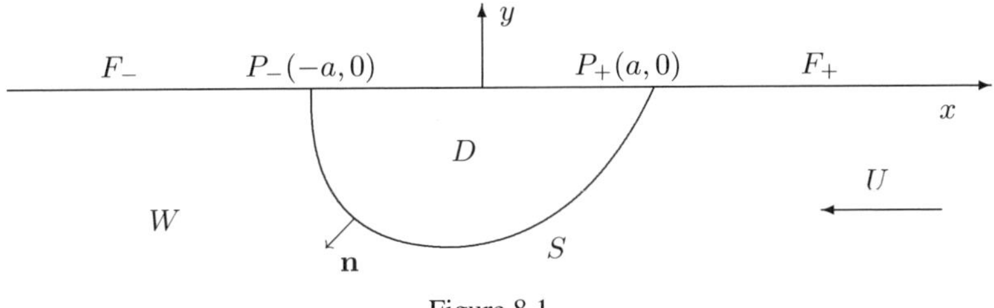

Figure 8.1.

8.1.1.1. The Main Set of Conditions

Let the cross section of a moving cylinder be a bounded, simply connected domain D in $\mathbb{R}^2_-$ such that ∂D consists of a segment $\{x \in [-a, a], y = 0\}$ ($a > 0$), and of a simple closed C^2 arc S having end points $P_\pm = (\pm a, 0)$; P_+ (P_-) will be referred to as the bow (stern) point. Let $W = \mathbb{R}^2_- \setminus \bar{D}$ be a region occupied by water, and let $F_\pm = \{a < \pm x < +\infty, y = 0\}$ be two parts of the free surface (see Fig. 8.1). We assume that the unilateral tangent to S at $P_\pm$ forms an angle $\beta_\pm \neq 0, \pi$ with the vector $\pm\mathbf{i}$ (by $\mathbf{i}$ we denote the unit vector directed along the x axis). Furthermore, $\mathbf{n}$ denotes the unit normal directed into W.

The velocity potential u describing the velocity field must satisfy the following boundary value problem:

$$\nabla^2 u = 0 \quad \text{in } W, \tag{8.1}$$

$$u_{xx} + \nu u_y = 0 \quad \text{on } F = F_+ \cup F_-, \tag{8.2}$$

$$\partial u/\partial n = U \cos(n, x) \quad \text{on int } S = S \setminus \{P_+, P_-\}, \tag{8.3}$$

$$\lim_{x \to +\infty} |\nabla u| = 0, \tag{8.4}$$

$$\sup\{|\nabla u| : (x, y) \in W \setminus E\} < \infty, \tag{8.5}$$

$$\int_{W \cap E} |\nabla u|^2 \, dx dy < \infty. \tag{8.6}$$

Here $\nu = gU^{-2}$; by E we denote an arbitrary compact set in $\overline{\mathbb{R}^2_-}$ such that $\bar{D} \subset E$ and $F_\pm \cap E \neq \emptyset$, and so the last condition (8.6) means that the kinetic energy must be locally finite near the bow and stern points despite the fact that it might have singularities there whereas away from $P_\pm$ the velocity field is bounded according to (8.5). It is obvious that u is defined up to an arbitrary constant term.

The right-hand-side term in (8.3) expresses the no-flow condition on the rigid cylinder's surface, but in what follows we do not use the specific form of this term and replace (8.3) by the general Neumann condition

$$\partial u/\partial n = f \quad \text{on int}\, S \tag{8.7}$$

with an arbitrary f from the Hölder space $C^{0,\alpha}(S)$, $0 < \alpha < 1$.

8.1.1.2. Asymptotic Formulae for a Potential Near the Bow and Stern Points

In order to formulate supplementary conditions, we have to establish that the quantities involved in these conditions do exist because this fact is not obvious beforehand. For this purpose we formulate the result describing the behavior of any solution to (8.1), (8.2), and (8.4)–(8.7) near $P_\pm$. Let $(\rho_\pm, \theta_\pm)$ be the polar coordinates with the pole at $P_\pm$ and having the half-line $F_\pm$ as the polar axis; θ_+ and θ_- are measured clockwise and anticlockwise, respectively; $0 \leq \theta_\pm \leq \beta_\pm < \pi$. Then any solution under consideration has the following asymptotic expansion as $\rho_\pm \to 0$:

$$u = \begin{cases} C^{(1)} + C^{(2)} \rho^{\pi/2\beta} \sin \pi\theta/2\beta \\ +C^{(3)} \rho \cos(\theta - \alpha) + O(\rho^{1+\delta}) & \text{for } \beta > \pi/2, \\ C^{(1)} + C^{(2)} [\rho \log \rho \sin \theta - \rho(\theta - \pi/2) \cos \theta] \\ +C^{(3)} \rho \cos(\theta - \alpha) + O(\rho^{1+\delta}) & \text{for } \beta = \pi/2, \\ C^{(1)} + C^{(3)} \rho \cos(\theta - \alpha) + O(\rho^{1+\delta}) & \text{for } \beta < \pi/2. \end{cases} \tag{8.8}$$

It can be formally obtained by separation of variables and justified by reference to the work of Kondratyev [145]. The subscripts $\pm$ indicating that $C^{(1)}$, $C^{(2)}$, $C^{(3)}$, ρ, θ, α, and δ depend on $P_\pm$ are omitted in (8.8) for the sake of simplicity. Moreover, $\delta > 0$, which allows us to differentiate (I.7) at least once. Hence, two limits along the free surface

$$\lim_{x \to \pm a \pm 0} u_x(x, 0) \tag{8.9}$$

do exist, although for $C^{(2)} \neq 0$ the velocity vector ∇u is singular when approaching $P_\pm$ along all nonhorizontal directions. Below, we denote limits (8.9) as $u_x(P_\pm)$.

8.1.1.3. Supplementary Conditions

The existence of the limiting values (8.9) allows us to introduce the following supplementary conditions:

$$A u_x(P_-) + B[u(P_+) - u(P_-)] = C, \tag{8.10}$$

$$u_x(P_+) - u_x(P_-) = K, \tag{8.11}$$

where A, B, C, and K are given real numbers such that $A^2 + B^2 \neq 0$. Thus, (8.10) prescribes a linear relation between the free surface elevation at P_- [$u_x(P_-)$ is proportional to this quantity] and the velocity circulation along S. The second supplementary condition (8.11) prescribes the value of the extra rate of flow at infinity caused by the presence of the immersed cylinder. This is a consequence of the asymptotic formula (8.12) and relation (8.13) established in Subsection 8.1.2.

8.1.2. Asymptotic Behavior of a Potential at Infinity

The aim of this subsection is to establish the following result similar to the second assertion in Subsection 7.1.1. As usual, we write $z = x + iy$ for the sake of brevity.

Any solution of (8.1)–(8.7) *has the following asymptotics as* $|z| \to \infty$:

$$u(x, y) = c + Q \log |z| + H(-x) e^{\nu y} (\mathcal{A} \sin \nu x + \mathcal{B} \cos \nu x) + \psi(x, y). \tag{8.12}$$

Here c is an arbitrary constant, H is the Heaviside function, and the estimates $\psi = O(|z|^{-1})$ and $|\nabla \psi| = O(|z|^{-2})$ hold. The constants Q, $\mathcal{A}$, and $\mathcal{B}$ are determined as follows:

$$\pi \nu Q + u_x(P_+) - u_x(P_-) = \nu \int_S \frac{\partial u}{\partial n} \, ds. \tag{8.13}$$

$$\mathcal{A} = -2 \left\{ \int_S \left[u \frac{\partial}{\partial n} (e^{\nu y} \cos \nu x) - \frac{\partial u}{\partial n} e^{\nu y} \cos \nu x \right] ds \right.$$
$$\left. + \, \nu^{-1} \cos \nu a [u_x(P_+) - u_x(P_-)] + \sin \nu a [u(P_+) + u(P_-)] \right\}, \tag{8.14}$$

$$\mathcal{B} = 2 \left\{ \int_S \left[u \frac{\partial}{\partial n} (e^{\nu y} \sin \nu x) - \frac{\partial u}{\partial n} e^{\nu y} \sin \nu x \right] ds \right.$$
$$\left. + \, \nu^{-1} \sin \nu a [u_x(P_+) + u_x(P_-)] - \cos \nu a [u(P_+) - u(P_-)] \right\}. \tag{8.15}$$

The coefficients in (8.12) have a clear physical meaning: $\mathcal{A}$ and $\mathcal{B}$ are proportional to the amplitudes of sine and cosine waves at infinity downstream; $-Q\pi/2$ is equal to the extra rate of flow at infinity that is due to the presence of cylinder. If S is rigid, then by (8.3) the right-hand side in (8.13) vanishes. In this case Q is defined by the difference $u_x(P_+) - u_x(P_-)$, and so Q is related to sprays that can occur at $P_\pm$.

In order to prove (8.12) one has to use the method applied in Subsection 2.2.1 and based on the use of a cutoff function. This results in representing u at infinity as a volume potential having Green's function as the kernel and integration spreading over a bounded subregion of $\mathbb{R}^2_-$. Then the asymptotic formula for Green's function obtained in Subsection 6.3.1.2 leads to (8.12). Now it remains to prove formulae (8.13), (8.14), and (8.15).

Let the rectangle $R_{\alpha\beta} = \{|x| < \alpha, -\beta < y < 0\}$ be such that $D \subset R_{\alpha-1\beta-1}$. By $p_{\alpha 0}$, $p_{\alpha\beta}$, and $q_{\pm\alpha\beta}$ we denote the upper, lower, and right and left sides of $R_{\alpha\beta}$, and the normals to these sides are assumed to be directed into the rectangle. Let us write Green's formula

$$0 = \int_S \frac{\partial u}{\partial n}\,\mathrm{d}s - \int_{(\partial R_{\alpha\beta})\setminus\bar{D}} \frac{\partial u}{\partial n}\,\mathrm{d}s. \tag{8.16}$$

In fact, Green's formula can be written for $R_{\alpha\beta}\setminus(\cup_\pm B_\epsilon(P_\pm) \cup \bar{D})$, $\epsilon > 0$; by $B_\epsilon(P_\pm)$ we denote an open disk of radius ϵ centered at $P_\pm$. The local asymptotics (8.8) near $P_\pm$ shows that the integrals over the circular arcs tend to zero as $\epsilon \to 0$, and so (8.16) is true.

The second integral in (8.16) is a sum of integrals over segments belonging to $\partial R_{\alpha\beta}$, and we are going to express these integrals by using (8.12) and assuming that $\alpha \to \infty$. First, we have

$$\int_{F_+\cap p_{\alpha 0}} u_y\,\mathrm{d}x = -\nu^{-1}\int_{+a}^{\alpha} u_{xx}\,\mathrm{d}x = \nu^{-1}u_x(+a,0) + O(\alpha^{-1}),$$

$$\int_{F_-\cap p_{\alpha 0}} u_y\,\mathrm{d}x = -\nu^{-1}u_x(-a,0) + \mathcal{A}\cos\nu\alpha + \mathcal{B}\sin\nu\alpha + O(\alpha^{-1}). \tag{8.17}$$

Similarly, we get

$$-\int_{p_{\alpha\beta}} u_y\,\mathrm{d}x = -Q\int_{-\alpha}^{+\alpha} \frac{\beta}{x^2+\beta^2}\,\mathrm{d}x - \nu e^{-\nu\beta}\int_{-\alpha}^{+\alpha} [\mathcal{A}\sin\nu x + \mathcal{B}\cos\nu x]\,\mathrm{d}x + O(\beta^{-1}), \tag{8.18}$$

and so for any fixed α the right-hand-side terms tend to zero as $\beta \to \infty$. Further, we find that

$$-\int_{q_{-\alpha\beta}} u_x\,\mathrm{d}y = Q\arctan\left(\frac{\beta}{\alpha}\right) - (1 - e^{-\nu\beta})(\mathcal{A}\cos\nu\alpha + \mathcal{B}\sin\nu\alpha) + O(\alpha^{-1}). \tag{8.19}$$

As $\beta \to \infty$ for fixed α, the expression on the right-hand side in (8.19) is equivalent to $Q\pi/2 - \mathcal{A}\cos\nu\alpha - \mathcal{B}\sin\nu\alpha + O(\alpha^{-1})$. Finally, analogous computations show that for fixed α

$$\lim_{\beta\to\infty}\int_{q_{\alpha\beta}} u_x\,\mathrm{d}y = \frac{Q\pi}{2} + O(\alpha^{-1}). \tag{8.20}$$

Substituting (8.17)–(8.20) into (8.16), we arrive at (8.13).

For finding $\mathcal{A}$ we use Green's formula

$$\begin{aligned} 0 = &\int_S \left[u\frac{\partial}{\partial n}(e^{\nu y}\cos \nu x) - \frac{\partial u}{\partial n} e^{\nu y}\cos \nu x \right] \mathrm{d}s \\ &- \int_{(\partial R_{\alpha\beta})\setminus \bar{D}} \left[u\frac{\partial}{\partial n}(e^{\nu y}\cos \nu x) - \frac{\partial u}{\partial n} e^{\nu y}\cos \nu x \right] \mathrm{d}s, \end{aligned} \tag{8.21}$$

which must be derived in the same way as (8.16). Using the free surface boundary condition and integrating by parts, we get

$$\begin{aligned} &\int_{F\cap p_{\alpha 0}} [u\nu \cos \nu x - u_y \cos \nu x]\,\mathrm{d}x \\ &\quad = [\nu^{-1}u_x(x,0)\cos \nu x + u(x,0)\sin \nu x]_{x=-a}^{x=+a} \\ &\qquad - [\nu^{-1}u_x(x,0)\cos \nu x + u(x,0)\sin \nu x]_{x=-\alpha}^{x=+\alpha}. \end{aligned} \tag{8.22}$$

By virtue of (8.12), we see that the last line in (8.22) takes the following form:

$$2[Q\log(\nu\alpha) + c]\sin \nu\alpha - \mathcal{A} + O(\alpha^{-1}).$$

Averaging this expression with respect to α over $(a, 2a)$, and then letting $a \to \infty$, we obtain $-\mathcal{A}$ because

$$\begin{aligned} &\frac{1}{a}\int_a^{2a} \log(\nu\alpha)\sin \nu\alpha\,\mathrm{d}\alpha \\ &\quad = \frac{-1}{a\nu}\left\{ [\log(\nu\alpha)\cos \nu\alpha]_{\alpha=a}^{\alpha=2a} - \int_a^{2a} \frac{\cos \nu\alpha}{\alpha}\,\mathrm{d}\alpha \right\} \to 0. \end{aligned}$$

Applying the same calculations to the remaining integrals over the segments forming $(\partial R_{\alpha\beta})\setminus\bar{D}$, we get the following results. The contribution of the integral over $q_{-\alpha\beta}$ is equal to $\mathcal{A}/2$, whereas $p_{\alpha\beta}$ and $q_{\alpha\beta}$ make no contribution. After averaging with respect to α, the values of the first integral on the right-hand side of (8.21) and of the first double substitution on the right-hand side of (8.22) remain the same. Hence, averaging with respect to α and letting $\beta \to \infty$ and $a \to \infty$, we obtain (8.14). In order to derive (8.15), one has to apply the same procedure.

8.1.3. Statement of the Problem for the Stream Function

As in Chapter 7, it is convenient to use a short name for the problem with supplementary conditions formulated in Subsection 8.1.1.

We say that u solves Problem I, if it satisfies (8.1), (8.2), (8.4)–(8.7), *and* (8.10) *and* (8.11).

Let v be a harmonic function in W conjugate to u. Defined up to a constant term, v is referred to as the stream function. Along with Problem I we need the equivalent boundary value problem for v, and it is derived in the following theorem.

Conditions (8.1), (8.2), *and* (8.4)–(8.7) *hold for u if and only if the stream function v is a solution of the boundary value problem:*

$$\nabla^2 v = 0 \quad \text{in } W, \tag{8.23}$$

$$v_y - \nu v = 0 \quad \text{on } F_+, \tag{8.24}$$

$$v_y - \nu v = C_- \quad \text{on } F_-, \tag{8.25}$$

$$v = C_0 + \int_{S_\ell} f \, \mathrm{d}s \quad \text{on } S, \tag{8.26}$$

$$\lim_{x \to +\infty} v = 0, \tag{8.27}$$

$$\sup_W |v(x, y)| < \infty, \tag{8.28}$$

$$\int_{W \cap E} |\nabla v|^2 \, \mathrm{d}x\mathrm{d}y < \infty, \tag{8.29}$$

where a certain value of additive constant is chosen. In (8.29), $E \subset \overline{\mathbb{R}^2_-}$ *is an arbitrary compact set of the type described after* (8.6). *Moreover, two equalities:*

$$C_0 = \nu^{-1} u_x(P_-) - \pi Q, \tag{8.30}$$

$$C_- = \pi \nu Q, \tag{8.31}$$

hold. Here Q is the constant in (8.12) *for u, and S_ℓ denotes the subarc of S having length ℓ, and the initial point at P_-.*

Let u satisfy (8.1), (8.2), and (8.4)–(8.7). Then (8.29) obviously follows from (8.6). Using the Cauchy–Riemann equations, we can write (8.2) in the form

$$u_{xx} - \nu v_x = 0 \quad \text{on } F_+ \cup F_-.$$

Integrating this and applying the Cauchy–Riemann equations again, we get

$$v_y - \nu v = C_\pm \quad \text{on } F_\pm,$$

where C_+ and C_- are real constants. An arbitrary additive constant can be specified for v so that (8.24) holds. It is clear that (8.7) becomes (8.26) after integration.

Let us prove (8.30) and (8.31). On one hand, from (8.26) we get that

$$v(P_+) = C_0 + \int_0^{|S|} f \, \mathrm{d}s,$$

where $|S|$ is the length of S. On the other hand, (8.24) implies

$$v(P_+) = \nu^{-1} v_y(P_+) = \nu^{-1} u_x(P_+).$$

This and (8.11) give (8.30). Substituting $v(P_-) = C_0$ into (8.25) and using (8.30), we find that

$$C_- = v_y(P_-) - \nu[\nu^{-1} u_x(P_-) - \pi Q] = \pi \nu Q.$$

Now, it remains to derive (8.27) and (8.28). Since $u_x(P_\pm)$ are finite, (8.24)–(8.26) and (8.4) and (8.5) imply that v is bounded on ∂W. An arbitrary point $x + iy = z \in W$ can be connected by a vertical segment with $x + iy_1 = z_1 \in \partial W$. Hence,

$$v(z) = v(z_1) - \int_y^{y_1} v_\eta(x, \eta) \, \mathrm{d}\eta = v(z_1) - \int_y^{y_1} u_x(x, \eta) \, \mathrm{d}\eta. \qquad (8.32)$$

The last integral is bounded because of (8.8) and (8.12), which proves (8.28). According to (8.24), (8.4), and the Cauchy–Riemann equations, (8.27) is true on $y = 0$. For $y < 0$, (8.27) follows from (8.32) and (8.12).

It is shown that (8.23) to (8.31) are consequences of (8.1), (8.2), and (8.4)–(8.7). Using the converse argument, one completes the proof of equivalence.

There are two immediate consequences of the theorem proven and, we formulate them as corollaries.

Problem I is equivalent to the boundary value problem (8.23) *to* (8.29) *where C_- is prescribed in* (8.25), *complemented by*

$$A v_y(P_-) - B \int_S \frac{\partial v}{\partial n} \, \mathrm{d}s = C. \qquad (8.33)$$

Also, $C_0 = \nu^{-1}[v_y(P_-) - C_-]$ *in* (8.26).

In order to prove this, we first get from (8.13) that (8.11) is equivalent to (8.25), where C_- is prescribed. Then, the formula for C_0 follows by virtue of the Cauchy–Riemann equations from (8.30) and (8.31). Applying the Cauchy–Riemann equations again, one obtains (8.33) from (8.10).

Let u satisfy (8.1), (8.2), *and* (8.4) *to* (8.7). *Then u is bounded at infinity if and only if v satisfies*

$$v_y - \nu v = 0 \quad \text{on } F, \tag{8.34}$$

$$v = \nu^{-1} v_y(P_-) + \int_{S_\ell} f \, \mathrm{d}s \quad \text{on } S. \tag{8.35}$$

Also, if u is bounded, then

$$\int_{-\infty}^{0} u_x(x, y) \, \mathrm{d}y = \nu^{-1} u_x(x, 0) \tag{8.36}$$

holds for sufficiently large values of $|x|$.

Since u is bounded at infinity if and only if $Q = 0$, (8.34) and (8.35) follow from (8.30), (8.31), and the Cauchy–Riemann equations. Also, if u is bounded at infinity, then $v(x, y) \to 0$ as $y \to -\infty$, and so for sufficiently large $|x|$ we have

$$\int_{-\infty}^{0} u_x(x, y) \, \mathrm{d}y = \int_{-\infty}^{0} v_y(x, y) \, \mathrm{d}y = v(x, 0).$$

Hence, (8.34) yields (8.36).

We note that the free surface elevation η is equal to $U g^{-1} u_x(x, 0)$, and so the rate of flow caused by the induced velocity field is proportional to η according to (8.36). Hence, the boundedness of u at infinity is equivalent to the fact that the mean level of the disturbed free surface coincides with the surface of the undisturbed stream far downstream.

8.1.4. The Integral Equation Method for Problem I

As in Subsection 7.1.2, it is convenient for us to seek a solution to Problem I by using the integral equation method.

8.1.4.1. Reduction of Problem I to an Integroalgebraic System

We begin with necessary properties of a single-layer potential:

$$(\mathcal{U}\mu)(z) = \frac{1}{\pi} \int_S \mu(\zeta) G(z, \zeta) \, \mathrm{d}s_\zeta, \quad z \in \overline{\mathbb{R}^2_-}.$$

Here $G(z, \zeta)$ is Green's function described in Subsection 6.3.1, and we assume that μ belongs to $C^{0,\alpha}(\text{int } S)$, $0 < \alpha < 1$, and to a Banach space $C_\kappa(S)$. The latter consists of continuous on int S functions and is supplied with the following norm:

$$\|\mu\|_\kappa = \sup\{|y|^{1-\kappa} |\mu(z)| : z \in \text{int } S\}, \quad 0 < \kappa < 1.$$

From the second assertion in Subsection 6.3.1.1, we have that

$$-\pi(\mathcal{U}\mu)(z) = \int_{S\cup S'} \mu(\zeta)\log|z-\zeta|\,\mathrm{d}s + \int_S \mu(\zeta)g(z,\zeta)\,\mathrm{d}s, \quad (8.37)$$

where $S' = \{z \in \overline{\mathbb{R}^2_+} : \bar{z} \in S\}$, and μ is extended to S' as an odd in y function. The second integral in (8.37) belongs to $C^{1,\beta}(\overline{\mathbb{R}^2_-})$ for every $\beta \in (0,\kappa)$ (see Theorem 4 in Chapter 11, §3 in Kantorovich and Akilov [130]). Since $\mathcal{U}\mu$ coincides with the second integral in (1.1) when $y = 0$, we have that $\mathcal{U}\mu \in C^{1,\beta}(\partial\mathbb{R}^2_-)$ for every $\beta \in (0,\kappa)$. Results in Chapter 11, §3 in Kantorovich and Akilov [130] provide more properties of $\mathcal{U}\mu$ when $\mu \in C_\kappa(S)$. First, $\mathcal{U}\mu$ is bounded in a rectangle $R_{db} = \{|x| < d, -b < y < 0\}$ for every finite $b, d > 0$. Second,

$$\int_{R_{db}} |\nabla\mathcal{U}\mu|^2\,\mathrm{d}x\mathrm{d}y < \infty.$$

Some of the properties of (8.37) that will be used in what follows are described in Subsection 2.1.1 (also see Subsection 7.1.2). In particular, for the normal derivative of $\mathcal{U}\mu$ we have

$$\begin{aligned} \partial\mathcal{U}\mu/\partial n_z &= -\mu(z) + (T\mu)(z), \quad z \in \operatorname{int} S, \\ (T\mu)(z) &= \pi^{-1}\int_S \mu(\zeta)\frac{\partial G}{\partial n_z}(z,\zeta)\,\mathrm{d}s, \end{aligned} \quad (8.38)$$

and the kernel of T has the form given by (7.7).

Other properties required here are the consequences of considerations in Subsection 2.1.3. We note that T is not a compact operator in $C_\kappa(S)$, but for some values $\kappa \in (0,1)$ Fredholm's alternative holds for $I - T$ in this space because $|T| < 1$ in $C_\kappa(S)$; by $|T|$ we denote the essential norm of T. It follows from results in Subsection 2.1.3 that

$$|T| < \max_{\pm} \frac{\sin\kappa|\pi - 2\beta_\pm|}{\sin\kappa\pi}.$$

Hence, if

$$\kappa < \min_{\pm}\left[1 + \left|1 - \frac{2\beta_\pm}{\pi}\right|\right]^{-1}, \quad (8.39)$$

then the required estimate $|T| < 1$ is valid.

For reducing Problem I to an integroalgebraic system, we use the following representation:

$$u(z) = (\mathcal{U}\mu)(z) + \sum_{\pm}\mu_\pm G(z,\pm a). \quad (8.40)$$

Here $\sum_\pm$ denotes the summation of two terms, $\mu_\pm$ are real unknown numbers, and $\mathcal{U}\mu$ is a potential with an unknown density μ. For any $X = (\mu, \mu_+, \mu_-)^t$, u given by (8.40) satisfies (8.1), (8.2), and (8.4)–(8.6). This function is a solution to Problem I when X is found so that u satisfies (8.7)–(8.11) as well. From (8.38) and (8.7) we obtain

$$-\mu(z) + (T\mu)(z) + \sum_\pm \mu_\pm \frac{\partial G}{\partial n_z}(z, \pm a) = f(z), \quad z \in \operatorname{int} S. \tag{8.41}$$

Then, (6.136) and (8.11) produce

$$T_+\mu - T_-\mu + \sum_\pm \mu_\pm[G_x(a, \pm a) - G_x(-a, \pm a)] = K,$$
$$T_\pm\mu = \pi^{-1}\int_S \mu(\zeta) G_x(\pm a, \zeta)\, \mathrm{d}s. \tag{8.42}$$

Eventually, (8.10) yields

$$A\left[T_-\mu + \sum_\pm \mu_\pm G_x(-a, \pm a)\right]$$
$$+ B\left\{L\mu + \sum_\pm \mu_\pm[G(a, \pm a) - G(-a, \pm a)]\right\} = C, \tag{8.43}$$

where

$$L\mu = \pi^{-1}\int_S \mu(\zeta)[G(a, \zeta) - G(-a, \zeta)]\, \mathrm{d}s.$$

Equations (8.41)–(8.43) constitute an integroalgebraic system for X. In the space $C_\kappa(S) \times \mathbb{R}^2$, having $\max\{\|\nu\|_\kappa, |\mu_+|, |\mu_-|\}$ as the norm, this system can be written as follows:

$$(-\mathcal{I} + \mathcal{P})X = V, \tag{8.44}$$

where $V = (f, K, C)^t$, $\mathcal{I}$ is the unit matrix operator, and $\mathcal{P}$ has the form

$$\begin{bmatrix} T & N^{(+)} & N^{(-)} \\ T_+ - T_- & a_{22} & a_{23} \\ AT_- + BL & a_{32} & a_{33} \end{bmatrix}. \tag{8.45}$$

Here $N^\pm$ is the operator of multiplication by $(\partial G/\partial n_z)(x, \pm a)$, and elements of the second-order matrix in the lower right-hand-side corner are as follows:

$$a_{22} = 1 - \pi\nu - G_x(-a, a), \quad a_{23} = G_x(a, -a) + 3\pi\nu,$$
$$a_{32} = AG_x(-a, a) + B[2\gamma - G(-a, a)],$$
$$a_{33} = 1 - 3A\pi\nu + B[G(a, -a) - 2\gamma].$$

Here we applied (6.134)–(6.137).

Now, let us prove the theorem establishing that Fredholm's alternative is valid for (8.44).

If κ *satisfies* (8.39), *then* $\mathcal{P} - \mathcal{I}$ *is a Fredholm operator in* $C_\kappa(S) \times \mathbb{R}^2$, *and its index vanishes [thus, Fredholm's alternative holds for* (8.44)*].*

As was mentioned above, Fredholm's alternative is true for $\mathcal{P}$ when $|\mathcal{P}| < 1$, and $|\mathcal{P}|$ is the essential norm of $\mathcal{P}$ in $C_\kappa(S) \times \mathbb{R}^2$. In order to verify the last inequality under condition (1.3), let us split (8.45) into a sum

$$\begin{bmatrix} T & 0 & 0 \\ 0 & 0 & 0 \\ 0 & 0 & 0 \end{bmatrix} + \begin{bmatrix} 0 & N^{(+)} & N^{(-)} \\ T_+ - T_- & a_{22} & a_{23} \\ AT_- + BL & a_{32} & a_{33} \end{bmatrix}.$$

Here the first operator has the same essential norm in $C_\kappa \times \mathbb{R}^2$ as T in $C_\kappa(S)$, and it is strictly less than one when (8.39) holds. Now, it is sufficient to show that the second operator is a finite-dimensional operator in $C_\kappa(S) \times \mathbb{R}^2$. This follows from the two facts: $T_\pm$ and L are continuous functionals in $C_\kappa(S)$, and $(\partial G/\partial n_z)(z, \pm a)$ belongs to $C_\kappa(S)$, which is a consequence of (6.134) and (6.136).

8.1.4.2. The Solvability of Problem I

As in Subsection 7.1.2, we are going to apply *the invertibility theorem*, and here we have to use the version of this theorem formulated in Subsection 2.1.3.1. Therefore, we have to demonstrate that (8.44) is solvable for sufficiently small ν, and so we begin with the following lemma.

For $a > 0$ *and sufficiently small values of* $\nu > 0$, *numbers* $\mu_\pm$ *can be found from* (8.42) *and* (8.43). *Moreover, if* $B = 0$, *then this can be done for any value* $\nu > 0$.

In order to prove the first assertion, we find it sufficient to verify that for small $\nu > 0$

$$\Delta = \begin{vmatrix} a_{22} - 1 & a_{23} \\ a_{32} & a_{33} - 1 \end{vmatrix} \neq 0.$$

From (6.134) and (6.136) we have

$$\begin{aligned} a_{22} - 1 = a_{23} &= \pi\nu[1 - 2a\nu \log \nu + O(\nu)], \\ a_{32} &= \nu[-3\pi(2Ba + A) + 4a(A + Ba)\nu \log \nu + O(\nu)], \\ a_{33} - 1 &= \nu[-\pi(2Ba + 3A) - B4a^2\nu \log \nu + O(\nu)]. \end{aligned}$$

Hence,

$$\begin{aligned} \Delta &= 4\pi a\nu^2[1 - 2a\nu \log \nu + O(\nu)][\pi B - (A + 2aB)\nu \log \nu + O(\nu)] \\ &= 4\pi a\nu^2\{\pi B - [A + 2a(1 + \pi)B]\nu \log \nu + O(\nu)\}. \qquad (8.46) \end{aligned}$$

Since $A^2 + B^2 \neq 0$, we get that $\Delta \neq 0$ for sufficiently small $\nu > 0$.

When $B = 0$, then (8.42) and (8.43) take the following form:

$$T_+\mu + \sum_{\pm} G_x(a, \pm a) = K + C, \quad T_-\mu + \sum_{\pm} G_x(-a, \pm a) = C,$$

and so

$$\Delta = \begin{vmatrix} G_x(+a, +a) & G_x(+a, -a) \\ G_x(-a, +a) & G_x(-a, -a) \end{vmatrix}.$$

From (6.137), (6.127), and formula 3.893.1 in Gradshteyn and Ryzhik [96], we obtain that

$$\Delta = 3\pi^2\nu^2 + \left[2\nu \int_0^\infty \frac{\sin 2ak \, \mathrm{d}k}{k - \nu} \right]^2 - 4\pi^2\nu^2 \cos^2 2a\nu.$$

According to 3.722.5 and 3.354.1 in [96] we have

$$\int_0^\infty \frac{\sin 2ak \, \mathrm{d}k}{k - \nu} = \pi \cos 2a\nu - \int_0^\infty \frac{e^{-2a\nu k} \, \mathrm{d}k}{1 + k^2}.$$

Let us denote the last integral by $J(2a\nu) = J(\nu_*)$; then

$$\Delta = \nu^2\{3\pi^2 + [2J(\nu_*)]^2 - 8\pi J(\nu_*) \cos \nu_*\}. \tag{8.47}$$

Differentiating with respect to ν_*, we find that

$$\Delta'/(8\nu^2) = J'(\nu_*) \, [J(\nu_*) - \pi \cos \nu_*] + \pi J(\nu_*) \sin \nu_*,$$

where

$$J'(\nu_*) = -\int_0^\infty \frac{k e^{-\nu_* k} \, \mathrm{d}k}{1 + k^2} < 0.$$

Since $0 < J(\nu_*) \leq J(0) = \pi/2$, it follows that $\Delta'/(8\nu^2)$ is positive for $0 < \nu_* \leq \pi/3$ and tends to $+\infty$ as $\nu_* \to +0$. Then (8.47) implies that $\Delta > 0$ for $0 < \nu_* \leq \pi/3$.

If $\nu_* \geq \pi/3$, then $J(\nu_*) < \nu_*^{-1} < 3/\pi$, and the same is true for $|J(\nu_*) \times \cos \nu_*|$. This and (8.47) yield that $\Delta > 0$ for $\nu_* \geq \pi/3$, which completes the proof.

Direct computation gives

$$\mu_\pm = \Delta^{-1} \left[\int_S \mu(\zeta) \Delta_\pm(\zeta) \, \mathrm{d}x + d_\pm(f, C, K) \right],$$

where

$$\Delta_\pm(\zeta) = \pm \begin{vmatrix} b_{11}^{(\pm)} & b_{12}(\zeta) \\ b_{21}^{(\pm)} & b_{22}(\zeta) \end{vmatrix}, \quad d_\pm(f, C, K) = \pm \begin{vmatrix} b_{11}^{(\pm)} & K \\ b_{21}^{(\pm)} & C \end{vmatrix},$$

and

$$
\begin{aligned}
b_{11}^{(\pm)} &= G_x(a, \mp a) - G_x(-a, \mp a), \\
b_{21}^{(\pm)} &= AG_x(-a, \mp a) + B[G(a, \mp a) - G(-a, \mp a)], \\
b_{12}(\zeta) &= G_x(a, \zeta) - G_x(-a, \zeta), \\
b_{22}(\zeta) &= AG_x(-a, \zeta) + B[G(a, \zeta) - G(-a, \zeta)].
\end{aligned}
$$

Thus, we have established the following lemma.

For sufficiently small $\nu > 0$, (8.44) is equivalent to

$$
-\mu(z) + (T_\nu \mu)(z) = f_\nu(z), \quad z \in \operatorname{int} S, \tag{8.48}
$$

where

$$
(T_\nu \mu)(z) = (T\mu)(z) + \frac{1}{\pi \Delta} \int_S \mu(\zeta) \sum_{\pm} \Delta_\pm(\zeta) \frac{\partial G}{\partial n_z}(z, \pm a)\, ds,
$$

and

$$
f_\nu(z) = f(z) + (\pi \Delta)^{-1} \sum_{\pm} d_\pm(f, C, K) \frac{\partial G}{\partial n_z}(z, \pm a).
$$

If $B = 0$, then (8.44) *is equivalent to* (8.48) *for all $\nu > 0$.*

From the second assertion in Subsection 6.3.1.2, we have that $T_\nu - T$ is a finite-dimensional operator in $C_\kappa(S)$. Therefore, $|T_\nu| = |T|$, and Fredholm's alternative holds for (8.48) when κ satisfies (8.39). Moreover, $f_\nu \in C_\kappa(S)$ if $f \in C_\kappa(S)$.

Let κ satisfy (8.39)*; then* (8.48) *is uniquely solvable in $C_\kappa(S)$, if $\nu > 0$ is sufficiently small.*

As in Subsection 7.1.2, we consider the equation

$$
-\mu(z) + (T_0 \mu)(z) = f_0(z), \quad z \in \operatorname{int} S, \tag{8.49}
$$

where

$$
(T_0 \mu)(z) = \pi^{-1} \int_S \mu(\zeta) \frac{\partial}{\partial n_z} (\log |z - \bar{\zeta}| - \log |z - \zeta|)\, ds.
$$

After extending μ and f_0 to $S' = \{z \in \overline{\mathbb{R}^2_+} : \bar{z} \in S\}$ as odd in y functions, (8.49) coincides with the integral equation for the Neumann problem in the domain exterior to $S \cup S'$. Carleman [36] established the unique solvability of this equation in $C_\kappa(S)$ when (8.39) holds. Therefore, for proving the theorem it is sufficient to show that $T_\nu - T_0$ has a small norm in this space for positive ν close to zero.

According to the previous lemma and the second assertion in Subsection 6.3.1.2, the kernel of $T_\nu - T_0$ is equal to

$$-\frac{1}{\pi}\left[\frac{\partial g}{\partial n_z}(z,\zeta) + \frac{1}{\Delta}\sum_{\pm}\Delta_\pm(\zeta)\frac{\partial g}{\partial n_z}(z,\pm a)\right], \tag{8.50}$$

and for all $\zeta \in S$ the following asymptotic formula is true:

$$(\partial g/\partial n_z)(z,\zeta) = -2\nu \log \nu \cos(n_z, y) + O(\nu) \quad \text{as } \nu \to 0.$$

Here (6.136) and a similar formula for G_y are also applied. Then assuming that ν is small, we can write (8.50) as follows:

$$[2\pi^{-1}\cos(n_z, y)\nu \log \nu + O(\nu)]\left[1 + \Delta^{-1}\sum_{\pm}\Delta_\pm(\zeta)\right]. \tag{8.51}$$

Let us analyze the behavior of the second factor in (8.51) for small ν. Since Δ has the asymptotic expansion (8.46), we have to consider the asymptotics of $\sum_\pm \Delta_\pm(\zeta)$. Again applying the properties of Green's function mentioned above, we obtain

$$b_{11}^{(\pm)} = \nu\{2\pi - 4a\nu\log\nu + \pi[3 - 2\gamma - 2\log(2a)]\nu + O(\nu^2\log\nu)\},$$

$$b_{21}^{(\pm)} = \nu\{-3\pi A - 2\pi a B(2 \mp 1) + 2a[A(\mp 1 + 1) \mp Ba]\nu\log\nu + O(\nu)\}.$$

From (6.127) we have that

$$G(z,\zeta) = G(\zeta, z) - 4\pi e^{\nu(y+\eta)}\sin\nu(x - \xi).$$

Hence, we get from (6.134) that $b_{22}(\zeta) = O(\nu\log\nu)$ for any A and B. Furthermore, the same formulae yield

$$b_{12}(\zeta) = 2\nu(\varphi_- - \varphi_+) + O(\nu^2\log\nu),$$

where $\varphi_\pm$ denotes the angle between the x axis and the vector directed from $P_\pm$ to ζ.

Now, we can write

$$\sum_{\pm}\Delta_\pm(\zeta)$$

$$= \begin{vmatrix} b_{11}^{(+)} - b_{11}^{(-)} & b_{12}(\zeta) \\ b_{21}^{(+)} - b_{21}^{(-)} & b_{22}(\zeta) \end{vmatrix}$$

$$= \begin{vmatrix} O(\nu^3\log\nu) & 2\nu(\varphi_- - \varphi_+) + O(\nu^2\log\nu) \\ 4a\nu[\pi B - (A + Ba)\nu\log\nu + O(\nu)] & O(\nu\log\nu) \end{vmatrix}.$$

In what follows, it is convenient to treat cases $B \neq 0$ and $B = 0$ separately. If $B \neq 0$, then the last formula and (8.46) imply

$$\begin{aligned}\Delta^{-1} \sum_{\pm} \Delta_{\pm}(\zeta) &= \frac{-8\pi a B(\varphi_- - \varphi_+)\nu^2 + O(\nu^3 \log \nu)}{4\pi^2 a B \nu^2 + O(\nu^3 \log \nu)} \\ &= 2\pi^{-1}(\varphi_+ - \varphi_-) + O(\nu \log \nu).\end{aligned}$$

For $B = 0$ we find

$$\begin{aligned}\Delta^{-1} \sum_{\pm} \Delta_{\pm}(\zeta) &= \frac{8a A(\varphi_- - \varphi_+)\nu^3 \log \nu + O(\nu^3)}{-4\pi a A \nu^3 \log \nu + O(\nu^3)} \\ &= 2\pi^{-1}(\varphi_+ - \varphi_-) + O(\nu \log \nu).\end{aligned}$$

Thus, the asymptotic representation for $\Delta^{-1} \sum_{\pm} \Delta_{\pm}(\zeta)$ is the same in both cases. This leads to the following asymptotic formula for the kernel (8.51):

$$2\pi^{-1} \cos(n_z, y)[1 + 2\pi^{-1}(\varphi_+ - \varphi_-)]\nu \log \nu + O(\nu),$$

and so the kernel tends to zero as $\nu \to 0$. The proof is complete.

Combining the theorem proven with the previous lemma, establishing the equivalence of (8.44) and (8.48) for small ν, we obtain the following corollary.

Let κ *satisfy* (8.39). *If* $\nu > 0$ *is sufficiently small, then* (8.44) *is uniquely solvable in* $C_\kappa(s) \times \mathbb{R}^2$.

Now we are in a position to formulate the main result concerning the unique solvability of the integroalgebraic system.

For all $\nu > 0$ *except possibly for a sequence tending to infinity,* (8.44) *is uniquely solvable in* $C_\kappa(S) \times \mathbb{R}^2$ *when* κ *satisfies* (8.39).

From the second assertion in Subsection 6.3.1.2 we deduce that the kernel of T as well as all other elements of (8.45) depend analytically on ν in a neighborhood of the half-line $\nu > 0$. Since $\mathcal{P} - \mathcal{I}$ is invertible for sufficiently small values of $\nu > 0$ by the previous corollary, the result follows from the invertibility theorem formulated in Subsection 2.1.3.2.

Let us turn to the solvability of Problem I and deduce the following corollary.

Problem I has a solution for any $(f, C, K) \in C^{0,\alpha}(S) \times \mathbb{R}^2$ *and for all values of* $\nu > 0$, *except possibly for a sequence tending to infinity.*

From the previous theorem we deduce that (8.44) is solvable for all $\nu > 0$ except for a discrete sequence. After substituting X that satisfies (8.44) into (8.40), we get u, which solves Problem I as we are going to demonstrate now. First, all conditions of Problem I, except (8.7), (8.10), and (8.11), are met by u (properties of Green's function yield this; see Subsection 6.3.1.1). The supplementary conditions (8.11) and (8.10) are satisfied because (8.42) and (8.43) hold. Thus we have to verify that $\mu \in C^{0,\alpha}(\operatorname{int} S)$, because in this case the potential $\mathcal{U}\mu$ has the normal derivative on $\operatorname{int} S$, and (8.7) follows from (8.41). In fact, (6.136) implies that $(\partial G/\partial n_z)(z, \pm a)$ belongs to $C^{0,\alpha}(\operatorname{int} S)$. Rewriting (8.42) in the form

$$-\mu(z) + (T\mu)(z) = f(z) - \sum_{\pm} \mu_{\pm} \frac{\partial G}{\partial n_z}(z, \pm a),$$

we see that the right-hand-side term is in $C^{0,\alpha}(\operatorname{int} S)$ for any $\mu_{\pm}$. Then properties of T guarantee that μ is in the same class (see Subsection 2.1.1).

8.1.5. On Uniqueness in Problem I

In this subsection, we prove the uniqueness theorem of the same kind as that in Subsection 7.1.3, where the case of a totally submerged body was considered. However, we have to modify the proof in order to handle the supplementary conditions (8.10) and (8.11).

Let us introduce an auxiliary problem (Problem I′ for brevity), which differs from Problem I by conditions at infinity because the new problem describes the forward motion of the same cylinder in the opposite direction. Namely, a solution u' of Problem I′ must be bounded at infinity, and we require the following condition,

$$\lim_{x \to -\infty} |\nabla u'| = 0,$$

to hold instead of (8.4). Also, we have to change the supplementary conditions. We omit (8.11) (the condition of boundedness at infinity is introduced instead), and we replace (8.10) by

$$A u'_x(P_+) + B[u'(P_+) - u'(P_-)] = 0. \tag{8.52}$$

The following asymptotic formula holds for u' as $|z| \to \infty$:

$$u'(x, y) = c' + H(x) e^{\nu y} (\mathcal{A}' \sin \nu x + \mathcal{B}' \cos \nu x) + \psi'(x, y),$$

where $\psi' = O(|z|^{-1})$ and $|\nabla \psi'| = O(|z|^{-2})$. For obtaining the asymptotics for u', the same method as in Subsection 8.1.2 should be applied. However, since u' is bounded at infinity, (8.13) must hold for u' with $Q = 0$, and one obtains formulae for $\mathcal{A}'$ and $\mathcal{B}'$ simply by changing sign on the right-hand side in (8.14) and (8.15).

In the next theorem we prove an analog of (7.13).

Let u and u' be solutions of Problems I and I′, respectively, and let $Q = 0$ in (8.12). *Then*

$$\int_W (u'\nabla^2 u - u\nabla^2 u')\,\mathrm{d}x\mathrm{d}y = \int_S \left(u\frac{\partial u'}{\partial n} - u'\frac{\partial u}{\partial n}\right)\mathrm{d}s + \nu^{-1}[u'(x,0)u_x(x,0) - u(x,0)u'_x(x,0)]_{x=-a}^{x=+a}. \tag{8.53}$$

Let R_{db} be a rectangle containing D, $W_{db} = R_{db}\backslash\bar{D}$, and let p_{d0}, p_{db}, and $q_{\pm db}$ be upper, lower, and right and left sides of R_{db}, respectively. Green's formula

$$\int_{W_{db}} (u'\nabla^2 u - u\nabla^2 u')\,\mathrm{d}x\mathrm{d}y = \int_{\partial W_{db}} \left(u\frac{\partial u'}{\partial n} - u'\frac{\partial u}{\partial n}\right)\mathrm{d}s, \tag{8.54}$$

where $\mathbf{n}$ is directed into W_{db}, can be justified in the same way as in Subsection 8.1.2. Let us consider integrals over straight segments on the right-hand side in (8.54). By means of (8.12), where $Q = 0$, and a similar formula for u' (see above), the integral over p_{db} tends to zero as $b \to \infty$.

Now, let us turn to

$$\int_{q_{+d\infty}} \left(u\frac{\partial u'}{\partial n} - u'\frac{\partial u}{\partial n}\right)\mathrm{d}s = \int_{-\infty}^{0} [u'u_x - uu'_x]_{x=d}\,\mathrm{d}y,$$

where d is large. Using the asymptotics at infinity for u and u', one immediately obtains that the last integral is equal to

$$-c(\mathcal{A}'\cos\nu d - \mathcal{B}'\sin\nu d) + O(d^{-1}).$$

Similarly, the integral over $p_{-d\infty}$ is equal to

$$-c'(\mathcal{A}\cos\nu d + \mathcal{B}\sin\nu d) + O(d^{-1}).$$

At last, according to (8.2) we have

$$\begin{aligned}\nu\int_{p_{d_0}\cap\partial W} \left(u\frac{\partial u'}{\partial n} - u'\frac{\partial u}{\partial n}\right)\mathrm{d}s &= \left(\int_{-d}^{-a} + \int_a^d\right)[uu'_{xx} - u'u_{xx}]_{y=0}\,\mathrm{d}x \\ &= [u(x,0)u'_x(x,0) - u'(x,0)u_x(x,0)]_{x=-d}^{x=+d} \\ &\quad - [u(x,0)u'_x(x,0) - u'(x,0)u_x(x,0)]_{x=-a}^{x=+a}.\end{aligned}$$

Taking into account the asymptotics of u and u' at infinity, we get that the first term on the right-hand side is equal to

$$c(\mathcal{A}' \cos \nu d - \mathcal{B}' \sin \nu d) + c'(\mathcal{A} \cos \nu d + \mathcal{B} \sin \nu d) + O(d^{-1}).$$

Letting $b \to \infty$ first, and $d \to \infty$ after that, we arrive at (8.53).

Green's function G' for Problem I′ is given in Subsection 7.1.3, and using G' instead of G in the integral equation method developed in Subsection 8.1.4, one finds that all results on the solvability hold for Problem I′ and its integroalgebraic system. In particular, Problem I′ has a solution for all values $\nu > 0$, except possibly for a sequence tending to infinity.

Now, we are able to show that Problem I has only one solution, when Problem I′ is solvable. Hence, the main result can be formulated as follows.

For all $\nu > 0$ *except possibly for a discrete sequence of values, Problem I has a unique (up to an additive constant) solution for any* $(f, C, K) \in C^{0,\alpha}(S) \times \mathbb{R}^2$.

The last assertion in Subsection 8.1.4 guarantees that Problem I has a solution of the form of (8.40) for all $\nu > 0$ except possibly for a discrete sequence. Previous remark says that the same is true for Problem I′.

In order to prove uniqueness we choose ν so that Problems I and I′ are both solvable. Let u be a solution of the homogeneous Problem I; that is, $f = 0$ in (8.7), $C = 0$ in (8.10), and $K = 0$ in (8.11). Then, (8.13) implies that $Q = 0$ in (8.12). Let u' be a solution of Problem I′ having the Neumann data on int S such that $\int_S (\partial u'/\partial n)\, ds = 0$. Then we have

$$u_x(P_+) = u_x(P_-), \qquad u'_x(P_+) = u'_x(P_-),$$

and the second term on the right-hand side of (8.53) can be written as follows:

$$\nu^{-1}\{u_x(P_-)[u'(P_+) - u'(P_-)] - u'_x(P_+)\,[u(P_+) - u(P_-)]\}.$$

From the homogeneous conditions (8.10) and (8.52), we get that the last expression is equal to zero.

Since u and u' are harmonic function, and the homogeneous Neumann condition holds for u, (8.53) takes the form

$$\int_S u \frac{\partial u'}{\partial n}\, ds = 0.$$

We can take $\partial u'/\partial n$ arbitrarily, and so $u = \text{const}$ on S. As $\partial u/\partial n = 0$ on S, we can apply the uniqueness theorem for the Cauchy problem for the Laplace equation which gives that $u = \text{const}$ in W.

8.1.6. Auxiliary Problem of the Scattering Type

As in Chapter 7, we have to consider an auxiliary problem of the scattering type (Problem II) for investigating the unique solvability of Problem I for all $\nu > 0$.

8.1.6.1. Statement of Problem II; Asymptotics at Infinity

We say that w is a solution of Problem II if it satisfies (8.1), (8.2), (8.5)–(8.7), and (8.10), and it admits the following representation:

$$w(z) = Q \log(\nu|z|) + w_0(z). \tag{8.55}$$

Here Q is a given constant, and w_0 must satisfy the radiation condition

$$\lim_{|x|\to\infty} \left(\frac{\partial w_0}{\partial |x|} - i\nu w_0 \right) = 0.$$

Green's function of Problem II has the following form [see (7.17)]:

$$E(x, y; \xi, \eta) = G(x, y; \xi, \eta) - i \exp\{\nu(y + \eta - i[x - \xi])\}.$$

The behavior of a solution to Problem II at infinity can be investigated as for Problem I, and the following analog of the result in Subsection 8.1.2 is true.

The following identity holds:

$$\pi\nu Q + w_x(P_+) - w_x(P_-) = \nu \int_S \frac{\partial w}{\partial n} \,\mathrm{d}s, \tag{8.56}$$

and for w_0 we have, as $|z| \to \infty$,

$$w_0(x, y) = \mathcal{D}_\pm e^{\nu(y \pm ix)} + \psi_\pm(x, y), \quad \pm x > 0, \tag{8.57}$$

where $\psi_\pm = O(|z|^{-1})$, $|\nabla\psi_\pm| = O(|z|^{-2})$, and

$$-i\mathcal{D}_\pm = \int_S \left[e^{\nu(y \mp ix)} \frac{\partial w}{\partial n} - w \frac{\partial}{\partial n} e^{\nu(y \mp ix)} \right] \mathrm{d}s \\ - [e^{\mp i\nu x}\{\nu^{-1} w_x(x, 0) \pm i w(x, 0)\}]_{x=-a}^{x=+a}.$$

8.1.6.2. On Uniqueness in Problem II

First we consider the question of uniqueness in Problem II under the following additional assumption:

$$\int_W |\nabla w|^2 \,\mathrm{d}x\mathrm{d}y < \infty, \tag{8.58}$$

in which case Problem II is indistinguishable from Problem I. The following assertion extends the uniqueness result from Subsection 7.1.4.3 to surface-piercing bodies.

Let w be a solution to the homogeneous Problem II; that is, $Q = 0$ in (8.55), *$f = 0$ in* (8.7), *and $C = 0$ in* (8.10). *Let w also satisfy* (8.58), *and $\mathbf{x} \cdot \mathbf{n} \geq 0$ on S. If one of the following two assumptions holds, (i) $B = 0$, (ii) $2a + AB^{-1} \geq 0$, then $w = 0$ in W.*

As Q and $\partial w / \partial n$ vanish, (8.56) takes the form

$$w_x(P_+) - w_x(P_-) = 0. \tag{8.59}$$

From (8.58) we see that coefficients $\mathcal{D}_\pm$ vanish in (8.57) for w. Following the scheme used in Subsection 7.1.4.3 and based on the identity

$$\operatorname{Re}[(2x\bar{w}_x - \bar{w})\,\nabla^2 w] = 2|w_y|^2 - (x|\nabla w|)^2_x + \operatorname{Re}\nabla\,[(2x\bar{w}_x - \bar{w})\,\nabla w],$$

we obtain

$$\begin{aligned} &\nu^{-1}(2a|w_x(P_-)|^2 - \operatorname{Re}\{w_x(P_-)[\bar{w}(P_+) - \bar{w}(P_-)]\}) \\ &+ 2\int_W |w_y|^2\,\mathrm{d}x\mathrm{d}y + \int_S \mathbf{x} \cdot \mathbf{n}\,|\nabla w|^2\,\mathrm{d}s = \operatorname{Re}\int_S (2x\bar{w}_x - \bar{w})\,\frac{\partial w}{\partial n}\,\mathrm{d}s = 0. \end{aligned} \tag{8.60}$$

Here (8.59) is also taken into account.

If $B = 0$, then the homogeneous condition (8.10) gives $w_x(P_-) = 0$, and this cancels terms outside of integrals on the left-hand side. Since the remaining terms on the left-hand side are strictly positive for a nontrivial w, (8.60) implies that $w(x, y) = w(x)$ in W. Now the result follows from (8.2) and (8.57).

If $B \neq 0$ and (ii) holds, then we have $\nu^{-1}[2a + AB^{-1}|w_x(P_-)|^2]$ in the first line in (8.60), which is nonnegative. So in the same way as above (8.60) establishes the result.

Since Problem II is invariant with respect to a horizontal shift of the origin, it is clear that it is sufficient if $\mathbf{x} \cdot \mathbf{n} \geq 0$ holds on S for a certain choice of the origin.

There are other sufficient geometric conditions that ensure the uniqueness of a solution with the finite Dirichlet integral. To this end we use a method proposed by Simon and Ursell [307] and presented in Subsection 3.2.2.1 for the water-wave problem.

Let D be contained in a truncated angle whose boundary is formed by $\{-a \leq x \leq +a,\, y = 0\}$, and two half-lines emanating symmetrically from $P_\pm$ at the angle $\pi/4$ to the vertical. If one of the following assumptions holds, (i) $B = 0$, (ii) $A/B \geq 0$, then $w = 0$ in W, where w is a solution of the homogeneous Problem II satisfying (8.58).

Without loss of generality we assume that w is real. Let v be a harmonic function in W that is conjugate to w. According to the results in Subsection 8.1.3, v satisfies

$$v_y - \nu v = 0 \quad \text{on } F, \tag{8.61}$$

$$v = \nu^{-1} v_y(P_-) \quad \text{on } S, \tag{8.62}$$

$$A v_y(P_-) - B \int_S \frac{\partial v}{\partial n} \, \mathrm{d}s = 0. \tag{8.63}$$

Furthermore, (8.58) holds for v. Then we can write Green's formula as

$$\int_W |\nabla v|^2 \, \mathrm{d}x\mathrm{d}y = \nu \int_F v^2 \, \mathrm{d}x - \nu^{-1} v_y(P_-) \int_S \frac{\partial v}{\partial n} \, \mathrm{d}s. \tag{8.64}$$

The condition (8.63) yields

$$\nu^{-1} v_y(P_-) \int_S \frac{\partial v}{\partial n} \, \mathrm{d}s = \begin{cases} 0, & \text{when } B = 0, \\ A(B\nu)^{-1}[v_y(P_-)]^2, & \text{when } B \neq 0. \end{cases}$$

From this, (8.64), and (i) or (ii) we get

$$\int_W |\nabla v|^2 \, \mathrm{d}x\mathrm{d}y \leq \nu \int_F v^2 \, \mathrm{d}x.$$

Let us put $\varphi = \exp\{\nu(y + ix)\}$. It is clear that φ is a bounded harmonic function in $\mathbb{R}^2_-$ satisfying

$$\varphi_y - \nu \varphi = 0 \quad \text{when } y = 0. \tag{8.65}$$

Also, we have $\varphi_x = i\varphi_y$, and hence,

$$\frac{\partial \varphi}{\partial n} = i \frac{\partial \varphi}{\partial s} \tag{8.66}$$

on any half-line ℓ_b emanating from $(b, 0)$ at $\pi/4$ to the vertical, where $b > a$, $\mathbf{n}$ is directed to $+\infty$, and $\mathbf{s}$ is directed to $(b, 0)$ along ℓ_b. Then (8.61), (8.65), and estimates $v(z) = O(|z|^{-1})$ and $|\nabla v| = O(|z|^{-2})$, which are true because of (8.58), allow us to write Green's formula as

$$\int_{\ell_b} \left(v \frac{\partial \varphi}{\partial n} - \varphi \frac{\partial v}{\partial n} \right) \mathrm{d}s = 0. \tag{8.67}$$

Integrating by parts in (8.67) and using (8.66), we get

$$\int_{\ell_b} \varphi \frac{\partial v}{\partial n} \, \mathrm{d}s = \int_{\ell_b} v \frac{\partial \varphi}{\partial n} \, \mathrm{d}s = i \int_{\ell_b} v \frac{\partial \varphi}{\partial s} \, \mathrm{d}s = i v(b, 0) e^{i\nu b} - i \int_{\ell_b} \varphi \frac{\partial v}{\partial s} \, \mathrm{d}s.$$

Hence,

$$v(b,0) = \int_{\ell_b} \left(\frac{\partial v}{\partial s} - i \frac{\partial v}{\partial n} \right) \exp\{\nu[y + i(x-b)]\}\, \mathrm{d}s.$$

Since ℓ_b is a half-line at $\pi/4$ to the vertical, we obtain

$$[v(b,0)]^2 \leq \left(\int_{\ell_b} |\nabla v| e^{\nu y}\, \mathrm{d}s \right)^2 = 2 \left(\int_{\ell_b} |\nabla v| e^{\nu y}\, \mathrm{d}y \right)^2.$$

By the Schwarz inequality

$$\nu[v(b,0)]^2 \leq 2\nu \left(\int_{-\infty}^{0} e^{2\nu y}\, \mathrm{d}y \right) \left(\int_{\ell_b} |\nabla v|^2\, \mathrm{d}y \right) = \int_{\ell_b} |\nabla v|^2\, \mathrm{d}y. \quad (8.68)$$

In the same way, but using half-lines inclined at $\pi/4$ to the vertical and going to $x = -\infty$, one obtains (8.68) for $b < -a$. Then integration gives

$$\int_{W \setminus W_c} |\nabla v|^2\, \mathrm{d}x\mathrm{d}y \leq 0.$$

Since $W_c \neq W$, we get that $v = 0$ in W [also, (8.61) should be taken into account]. Thus, $w = 0$ in W because w is conjugate to v and decays at infinity. The proof is complete.

Let w be a solution to the homogeneous Problem II. Then $w = 0$ in W when only a trivial solution of the homogeneous Problem II satisfies (8.58).

Since $Q = 0$ for w, it is sufficient to repeat the argument used in Subsection 7.1.4.2. Similarly to (7.30) one proves that

$$|\mathcal{D}_+|^2 + |\mathcal{D}_-|^2 = 2\,\mathrm{Im} \left\{ \int_S w \frac{\partial \bar{w}}{\partial n}\, \mathrm{d}s - \nu^{-1} \left[w(x,0) \bar{w}_x(x,0) \right]_{x=-a}^{x=+a} \right\}. \quad (8.69)$$

The homogeneous Neumann condition and (8.59) allow us to write (8.69) as follows:

$$|\mathcal{D}_+|^2 + |\mathcal{D}_-|^2 = 2\nu^{-1}\,\mathrm{Im}\{\overline{w_x(P_-)}[w(P_-) - w(P_+)]\}. \quad (8.70)$$

If $B = 0$, then (8.10), where $C = 0$, gives $w_x(P_-) = 0$. If $B \neq 0$, then the expression in braces is equal to $AB^{-1}|w_x(P_-)|^2$, and so it is real. Thus, (8.70) yields $\mathcal{D}_\pm = 0$ in both cases. Now, the assertion follows from the fact that only a trivial solution of the homogeneous Problem II satisfies (8.58).

Now we can prove the following corollary.

Let the geometric assumptions imposed in the first or second lemma in Subsection 8.1.6.2 *hold. If condition (i) or (ii) for A and B from those lemmas also holds, then any solution of the homogeneous Problem II vanishes.*

Since the first two lemmas in Subsection 8.1.6.2 guarantee that any solution of the homogeneous Problem II satisfying (8.58) vanishes, the result follows from the previous theorem.

8.1.6.3. Reduction of Problem II to an Integroalgebraic System

Here we apply the method involving representation of a solution as a sum of the logarithmic Green's potential over D along with the single-layer potential over S and two sources placed at the bow and stern points. This method allows us to avoid the so-called irregular values of ν in the same way as it was done in Subsection 3.1.1 for the water-wave problem.

First, we recall the necessary properties of potentials (see also Subsections 2.1.1 and 3.1.1). For

$$(\mathcal{V}\mu)(z) = \pi^{-1} \int_S \mu(\zeta) E(z, \zeta)\, \mathrm{d}s, \quad z \in \mathbb{R}^2_-,$$

where $\mu \in C_\kappa(S) \cap C^{0,\alpha}(\operatorname{int} S)$, $0 < \alpha, \kappa < 1$, is a complex-valued density, the properties are similar to those of $\mathcal{U}\mu$ (see Subsection 8.1.4), because Green's functions E and G differ by a smooth term. In particular, there exists the normal derivative

$$\frac{\partial(\mathcal{V}\mu)}{\partial n_z} = -\mu(z) + (\mathcal{T}\mu)(z), \quad z \in \operatorname{int} S,$$

$$(\mathcal{T}\mu)(z) = \frac{1}{\pi} \int_S \mu(\zeta) \frac{\partial E}{\partial n_z}(z, \zeta)\, \mathrm{d}s. \tag{8.71}$$

Here $\mathcal{T}$ has the same properties as T in Subsection 8.1.4, of which the most important is that $\|\mathcal{T}\| < 1$ in the space $C_\kappa(S)$ when (8.39) holds. For

$$(\mathcal{W}\rho)(z) = (2\pi)^{-1} \int_D \rho(\zeta) E(z, \zeta)\, \mathrm{d}\xi \mathrm{d}\eta, \quad z \in \overline{\mathbb{R}^2_-},$$

having a complex-valued density $\rho \in C(\bar{D})$, we shall use Poisson's equation

$$\nabla^2 \mathcal{W}\rho = -\rho \quad \text{in } D, \tag{8.72}$$

and the fact that $\mathcal{W}\rho \in C^{1,\beta}(\mathbb{R}^2_-)$ (see Remark 2 in Chapter 11, §3 in Kantorovich and Akilov [130]).

If a solution to Problem II is sought in the form

$$w(z) = (\mathcal{V}\mu)(z) + (\mathcal{W}\rho)(z) + \sum_{\pm} \mu_\pm E(z, \pm a), \tag{8.73}$$

where $\mu_\pm$ are complex numbers, then for any vector $Y = (\mu, \rho, \mu_+, \mu_-)^t$, (8.73) satisfies (8.1), (8.2), and (8.6). According to (8.56), this function admits representation (8.55), where Q is fixed, if and only if the right-hand side in the following equation,

$$w_x(P_+) - w_x(P_-) = K, \tag{8.74}$$

is prescribed. Thus (8.73) solves Problem II, if there exists a vector Y such that w satisfies (8.7), (8.10), and (8.74). Since $\mathcal{W}\rho$ is continuously differentiable, we get the following from (8.7) and (8.71):

$$-\mu(z) + (\mathcal{T}\mu)(z) + (\mathcal{R}\rho)(z) + \sum_{\pm} \mu_{\pm} \frac{\partial E}{\partial n_z}(z, \pm a) = f(z), \quad z \in \operatorname{int} S,$$

$$(\mathcal{R}\rho) = (2\pi)^{-1} \int_D \rho(\zeta) \frac{\partial E}{\partial n_z}(z, \zeta)\, \mathrm{d}\xi \mathrm{d}\eta. \tag{8.75}$$

The equation

$$\mathcal{T}_+\mu - \mathcal{T}_-\mu + \mathcal{R}_+\rho - \mathcal{R}_-\rho + \sum_{\pm} \mu_{\pm}[E_x(+a, \pm a) - E_x(-a, \pm a)] = K \tag{8.76}$$

follows from (8.74) in the same way as (8.42) follows from (8.11). Here

$$\mathcal{T}_\pm\mu = \pi^{-1} \int_S \mu(\zeta) E_x(\pm a, \zeta)\, \mathrm{d}s, \quad \mathcal{R}_\pm\rho = (2\pi)^{-1} \int_D \rho(\zeta) E_x(\pm a, \zeta)\, \mathrm{d}\xi \mathrm{d}\eta.$$

Similarly, the supplementary condition (8.10) yields

$$A\left\{\mathcal{T}_-\mu + \mathcal{R}_-\rho + \sum_{\pm} \mu_{\pm} E_x(-a, \pm a)\right\}$$
$$+B\left\{\mathcal{L}\mu + \mathcal{M}\rho + \sum_{\pm} \mu_{\pm}[E(+a, \pm a) - E(-a, \pm a)]\right\} = C. \tag{8.77}$$

Here

$$\mathcal{L}\mu = \pi^{-1} \int_S \mu(\zeta)[E(+a, \zeta) - E(-a, \zeta)]\, \mathrm{d}s,$$

$$\mathcal{M}\rho = (2\pi)^{-1} \int_D \rho(\zeta)[E(+a, \zeta) - E(-a, \zeta)]\, \mathrm{d}\xi \mathrm{d}\eta.$$

Thus we have three equations, (8.75), (8.76), and (8.77), for four unknowns. We have to complement them by an appropriate fourth equation in order to obtain a uniquely solvable integroalgebraic system. Thus we require w to satisfy

$$-\nabla^2 w = i w \quad \text{in } D. \tag{8.78}$$

From (8.72), (8.78), and the fact that other terms in (8.73) are harmonic in D, we get that $\rho = iw$ in D. This yields the fourth equation:

$$-\rho(z) + i(\mathcal{V}\mu)(z) + i(\mathcal{W}\rho)(z) + i\sum_{\pm}\mu_{\pm}E(z, \pm a) = 0, \quad z \in D. \qquad (8.79)$$

Now equations (8.75)–(8.77) and (8.79) constitute an integroalgebraic system for Y. This system can be written in the following form:

$$(-\mathcal{I} + \mathcal{S})Y = V \qquad (8.80)$$

in $C_\kappa(S) \times C(\bar{D}) \times \mathbb{C}^2$. Here $V = (2f, 0, K, C)^t$, $\mathcal{I}$ is the unit matrix operator, and $\mathcal{S}$ is given by

$$\begin{bmatrix} \mathcal{T} & \mathcal{R} & \mathcal{N}^{(+)} & \mathcal{N}^{(-)} \\ i\mathcal{V} & i\mathcal{W} & i\mathcal{E}^{(+)} & i\mathcal{E}^{(-)} \\ \mathcal{T}_+ - \mathcal{T}_- & \mathcal{R}_+ - \mathcal{R}_- & b_{33} & b_{34} \\ A\mathcal{T}_- + B\mathcal{L} & A\mathcal{R}_- + B\mathcal{M} & b_{43} & b_{44} \end{bmatrix}.$$

By $\mathcal{N}^{(\pm)}$ and $\mathcal{E}^{(\pm)}$ we denote operators of multiplication by $(\partial E/\partial n_z)(z, \pm a)$ and $E(z, \pm a)$, respectively. The elements of the second-order complex-valued matrix in the lower right-hand-side corner are as follows:

$$b_{33} = 1 - \pi\nu - E_x(-a, +a), \quad b_{34} = E_x(+a, -a) + 3\pi\nu,$$
$$b_{43} = AE_x(-a, +a) + B[2(\gamma - i\pi) - E(-a, +a)],$$
$$b_{44} = 1 - 3A\pi\nu + B[E(-a, +a) - 2(\gamma - i\pi)].$$

Here we used the definition of E, and formulae (6.134) and (6.137).

8.1.6.4. Solvability of Problem II

Let us consider (8.80) in $C_\kappa(S) \times C(\bar{D}) \times \mathbb{C}^2$ supplied with the norm

$$\max\{\|\mu\|_\kappa, \|\rho\|_{C(\bar{D})}, |\mu_+|, |\mu_-|\},$$

and let us begin by proving the following theorem.

If κ satisfies (8.39), *then Fredholm's alternative holds for* (8.80) *in* $C_\kappa(S) \times C(\bar{D}) \times \mathbb{C}^2$.

As in the proof of the theorem in Subsection 8.1.4.1, it is sufficient to show that $|\mathcal{S}| < 1$. Here $|\mathcal{S}|$ is the essential norm of $\mathcal{S}$ in $C_\kappa(S) \times C(\bar{D}) \times \mathbb{C}^2$. As in Subsection 8.1.4.1 we have to consider

$$\begin{bmatrix} \mathcal{T} & \mathcal{R} & 0 & 0 \\ i\mathcal{V} & i\mathcal{W} & 0 & 0 \\ 0 & 0 & 0 & 0 \\ 0 & 0 & 0 & 0 \end{bmatrix},$$

which differs from $\mathcal{S}$ by a finite-dimensional term. Since $|\mathcal{T}| < 1$ in $C_\kappa(S)$, for completing the proof it is sufficient for us to show that $\mathcal{R}$, $i\mathcal{V}$, and $i\mathcal{W}$ are compact operators.

First, $\mathcal{R}$ is a compact operator from $C(\bar{D})$ to $C(S)$ because $(\mathcal{W}\rho) \in C^{1,\beta}(\overline{\mathbb{R}^2_-})$. It remains for us to note that $C(S)$ is continuously embedded into $C_\kappa(S)$.

Second, $C_\kappa(S)$ is continuously embedded into $L_p(S)$ for $p \in [1, (1-\kappa)^{-1}]$. According to Theorem 7 in Chapter 11, §3 in Kantorovich and Akilov [130], $i\mathcal{V}$ is a compact operator from $L_p(S)$ to $C(\bar{D})$. The same theorem guarantees that $i\mathcal{W}$ is a compact operator in $C(\bar{D})$. This completes the proof.

We turn to the results on solvability of (8.80) and Problem II.

Let κ satisfy (8.39). *If Problem II has no more than one solution (see Subsection* 8.1.6.2 *for conditions providing this fact), then* (8.80) *is uniquely solvable in $C_\kappa(S) \times C(\bar{D}) \times \mathbb{C}^2$.*

Since Fredholm's alternative holds for (8.80) (see the previous theorem), we have to verify that

$$(-\mathcal{I} + \mathcal{S})Y^{(0)} = 0$$

has only a trivial solution. Substituting $Y^{(0)}$ into (8.73), we obtain a solution $w^{(0)}$ of the homogeneous Problem II. From the assumptions made it follows that $w^{(0)} = 0$ in W, and so $w^{(0)} = 0$ in D as well. Indeed, $w^{(0)}$ has the form of (8.73) and satisfies (8.78) and

$$w^{(0)}_{xx} + \nu w^{(0)}_y = 0 \quad \text{for } |x| < a, y = 0. \tag{8.81}$$

Multiplying (8.78) by $\overline{w^{(0)}}$ and integrating over D, we get

$$i \int_D \left|w^{(0)}\right|^2 \mathrm{d}x\mathrm{d}y = -\int_D \bar{w}^{(0)} \nabla^2 w^{(0)} \, \mathrm{d}x\mathrm{d}y.$$

Now we apply Green's formula on the right-hand side, and we use (8.81) and the fact that $w^{(0)} = 0$ on S. Then

$$i \int_D \left|w^{(0)}\right|^2 \mathrm{d}x\mathrm{d}y = \int_D \left|\nabla w^{(0)}\right|^2 \mathrm{d}x\mathrm{d}y + \nu^{-1} \int_{-a}^{+a} \overline{w^{(0)}} w^{(0)}_{xx} \, \mathrm{d}x. \tag{8.82}$$

It is clear that the last integral is equal to

$$\left[\overline{w^{(0)}(x,0)}w_x^{(0)}(x,0)\right]_{x=-a}^{x=+a} - \int_{-a}^{+a} |w_x|^2 \,\mathrm{d}x.$$

Here the first term vanishes, because $w^{(0)}(\pm a, 0) = 0$. Hence (8.82) takes the following form:

$$i\int_D \left|w^{(0)}\right|^2 \mathrm{d}x\mathrm{d}y = \int_D \left|\nabla w^{(0)}\right|^2 \mathrm{d}x\mathrm{d}y - \nu^{-1}\int_{-a}^{+a} |w_x|^2 \,\mathrm{d}x.$$

Therefore, $w^{(0)} = 0$ in D, and from (8.78) and (8.72) we have that $\rho^{(0)} = 0$. Now

$$w^{(0)}(z) = \left[\mathcal{V}\mu^{(0)}\right](z) + \sum_{\pm} \mu_{\pm}^{(0)} E(z, \pm a) = 0, \quad z \in D,$$

and using the formula for the normal derivative of $\mathcal{V}\mu^{(0)}$, we get

$$\mu^{(0)}(z) + \left[\mathcal{T}\mu^{(0)}\right](z) + \sum_{\pm} \mu_{\pm}^{(0)} \frac{\partial E}{\partial n_z}(z, \pm a) = 0, \quad z \in \operatorname{int} S.$$

Comparing this with the homogeneous equation (8.75), where $\rho^{(0)} = 0$, we find that $\mu^{(0)} = 0$. Now we have

$$w^{(0)}(z) = \sum_{\pm} \mu_{\pm} E(z, \pm a) = 0, \quad z \in \overline{\mathbb{R}^2_-}.$$

Applying the boundary operator $[\partial_x^2 + \nu\partial_y]_{y=0}$ to this function and using properties of Green's function (Subsection 6.3.1), we obtain that

$$\sum_{\pm} \mu_{\pm}^{(0)} \delta(x \mp a) = 0.$$

This holds only when $\mu_{\pm}^{(0)} = 0$, which completes the proof.

The following corollary is an immediate consequence of the proven theorem.

Let the geometric assumptions imposed in one of the lemmas on uniqueness (see Subsection 8.1.6.2*) hold along with (i) or (ii) from the corresponding lemma. Then Problem II has a unique solution for any* $(f, C, Q) \in C^{0,\alpha}(S) \times \mathbb{C}^2$.

The solvability of (8.80) in $C_\kappa(S) \times C(\bar{D}) \times \mathbb{C}^2$ is established at the previous theorem. In the same way as in the proof of the corollary at the end of Subsection 8.1.4, one ensures that $\mu \in C^{0,\alpha}(\operatorname{int} S)$. Substituting the solution of (8.80) into (8.73), we solve Problem II. The uniqueness follows from the corollary in Subsection 8.1.6.2.

8.1.7. On the Unique Solvability of Problem I

8.1.7.1. Solvability of Problem I

By $\mathcal{W}$ we denote the solution of Problem II having the following.

1. In (8.55), $Q = 0$.
2. In (8.7), $f = \partial e^{\nu(y+ix)}/\partial n$ on int S.
3. In (8.10), $C = Ai\nu e^{i\nu a} + B(e^{i\nu a} - e^{i\nu a})$.

Let $\mathcal{D}_+(\mathcal{W})$ be the coefficient in (8.57) for $\mathcal{W}$. This particular solution is helpful for establishing criteria of the unique solvability of Problem I, and we begin with the following theorem.

Let Problem II be uniquely solvable (see Subsection 8.1.6.4*). If* $\mathcal{D}_+(\mathcal{W}) \neq 1$*, then Problem I has a solution for any* $(f, C, K) \in C^{0,\alpha}(S) \times \mathbb{R}^2$.

For proving the assertion we consider w solving Problem II in the case when

$$Q = \frac{1}{\pi}\left(\int_S f\,\mathrm{d}s - \frac{K}{\nu}\right)$$

in (8.55), where f and C are right-hand-side terms in (8.7) and (8.10), respectively. Such a solution exists by the assumption made. By $\mathcal{D}_+(w)$ we denote the coefficient in (8.57) for w, and we put

$$u = w - \frac{\mathcal{D}_+(w)}{\mathcal{D}_+(\mathcal{W}) - 1}\left[\mathcal{W} - e^{\nu(y+ix)}\right]. \tag{8.83}$$

Since $Q = 0$ for $\mathcal{W}$, we get from point 2 above and (8.56) that $\mathcal{W}_x(P_+) - \mathcal{W}_x(P_-) = 0$. Hence, the right-hand-side terms in (8.7), (8.10), and (8.11) coincide for u and w. According to (8.83), $\mathcal{D}_+$ vanishes in (8.57) for u, and so this function is a solution of Problem I.

8.1.7.2. Uniqueness in Problem I

First, we recall the following theorem proved in Subsection 7.1.5.1 for a totally submerged cylinder, but the proof is the same when a cylinder is surface piercing.

Let the homogeneous Problem II have only a trivial solution having a finite energy integral (see two assertions in Subsection 8.1.6.2*). Then the kernel of Problem I is at most one dimensional.*

Let us turn to the main result of the present subsection (it is analogous to the theorem proven in Subsection 7.1.5.3) establishing necessary and sufficient conditions for the unique solvability of Problem I.

Let the homogeneous Problem II have only a trivial solution satisfying (8.58)*. Then the following assertions are equivalent:*

1. *The homogeneous Problem I has only a trivial solution (up to a constant term);*
2. $\mathcal{D}_+(\mathcal{W}) \neq 1$;
3. $|\mathcal{D}_-(\mathcal{W})| \neq 1$.

$1 \Leftrightarrow 2$. Let us suppose that $\mathcal{D}_+(\mathcal{W}) = 1$. From the asymptotics of $\mathcal{W}$, [see (8.57)], it follows that $\mathcal{W} - e^{\nu(y+ix)}$ is a nontrivial solution of the homogeneous Problem I. Thus $1 \Rightarrow 2$ is established.

Now, let us suppose that u_0 is a nontrivial solution of the homogeneous Problem I, which is real by the previous theorem. The principal term of the asymptotics of u_0 has the following form [see (8.12)]:

$$H(-x)\left[\frac{\mathcal{B} - i\mathcal{A}}{2} e^{\nu(y+ix)} + \frac{\mathcal{B} + i\mathcal{A}}{2} e^{\nu(y-ix)}\right], \quad \mathcal{A}, \mathcal{B} \in \mathbb{R}.$$

Here $\mathcal{A}$ and $\mathcal{B}$ do not vanish simultaneously, since otherwise $u_0 = 0$ in W by assumptions of the theorem. Let us consider

$$w = e^{\nu(y+ix)} - 2u_0(\mathcal{B} - i\mathcal{A})^{-1}.$$

It is clear that $\partial[w - e^{\nu(y+ix)}]/\partial n = 0$ on int S, and

$$Aw_x(P_-) + B[w(P_+) - w(P_-)] = Ai\nu e^{i\nu a} + B(e^{i\nu a} - e^{-i\nu a}).$$

From the asymptotics of w, it follows that w satisfies the radiation condition. From the third assertion in Subsection 8.1.6.2, it follows that w is the unique solution of Problem II. Therefore, $w = \mathcal{W}$ and $\mathcal{D}_+(\mathcal{W}) = 1$, and so $2 \Rightarrow 1$ is established.

$2 \Leftrightarrow 3$. Let $B = 0$, and so $A \neq 0$. Then according to the definition of $\mathcal{W}$ (see assertions 2 and 3), we can write

$$\mathcal{D}_+(\mathcal{W}) = -i\int_S \mathcal{W}\frac{\partial}{\partial n}e^{\nu(y-ix)}\,\mathrm{d}s + i\int_S e^{\nu(y-ix)}\frac{\partial}{\partial n}e^{\nu(y+ix)}\,\mathrm{d}s$$
$$- i[e^{-i\nu x}\{ie^{i\nu x} + i\mathcal{W}(x,0)\}]_{x=-a}^{x=+a}. \tag{8.84}$$

Now we note that

$$\int_S e^{\nu(y-ix)}\frac{\partial}{\partial n}e^{\nu(y+ix)}\,\mathrm{d}s - [e^{-i\nu x}\{ie^{i\nu x} + i\mathcal{W}(x,0)\}]_{x=-a}^{x=+a}$$
$$= \int_D |\nabla e^{\nu(y+ix)}|^2\,\mathrm{d}x\mathrm{d}y - \nu\int_{-a}^{+a} e^{-i\nu x}e^{i\nu x}\,\mathrm{d}x - i[e^{-i\nu x}\mathcal{W}(x,0)]_{x=-a}^{x=+a}$$
$$= 2\nu\left(\nu\int_D e^{2\nu y}\,\mathrm{d}x\mathrm{d}y - a\right) - i[e^{-i\nu x}\mathcal{W}(x,0)]_{x=-a}^{x=+a}. \tag{8.85}$$

From (8.84) and (8.85), it follows that

$$\operatorname{Re}\mathcal{D}_+(\mathcal{W}) = \operatorname{Im}\left\{\int_S \mathcal{W}\frac{\partial}{\partial n}e^{\nu(y-ix)}\,\mathrm{d}s\right\} + \operatorname{Re}\left\{[e^{-i\nu x}\mathcal{W}(x,0)]_{x=-a}^{x=+a}\right\}.$$

Comparing the last formula with (8.69), we find

$$|\mathcal{D}_+(\mathcal{W})|^2 + |\mathcal{D}_-(\mathcal{W})|^2 = 2\operatorname{Re}\mathcal{D}_+(\mathcal{W}). \tag{8.86}$$

Let us turn to the case $B \neq 0$. Then we have

$$\mathcal{W}(P_+) - \mathcal{W}(P_-) = e^{i\nu a} - e^{-i\nu a} + AB^{-1}[i\nu e^{-i\nu a} - \mathcal{W}_x(P_-)]. \tag{8.87}$$

Furthermore, it follows from (8.56) that

$$\mathcal{W}_x(P_+) - \mathcal{W}_x(P_-) = \nu\int_S \frac{\partial e^{\nu(y+ix)}}{\partial n}\,\mathrm{d}s.$$

Since $e^{\nu(y+ix)}$ is a harmonic function, we get

$$[\mathcal{W}_x(x,0) - (e^{i\nu x})_x]_{x=-a}^{x=+a} = 0. \tag{8.88}$$

Noting that A and B are real, we deduce from (8.87) and (8.88) that

$$\operatorname{Im}[\{\mathcal{W}(x,0) - e^{i\nu x}\}\{\overline{\mathcal{W}_x(x,0)} - \overline{(e^{i\nu x})}\}]_{x=-a}^{x=+a} = 0.$$

Hence,

$$\begin{aligned}&\operatorname{Im}[\mathcal{W}(x,0)\overline{\mathcal{W}_x(x,0)} - \{\overline{e^{-i\nu x}\mathcal{W}_x(x,0)} - i\nu e^{-i\nu x}\mathcal{W}(x,0) + i\nu\}]_{x=-a}^{x=+a}\\ &\quad = \operatorname{Im}[\mathcal{W}(x,0)\overline{\mathcal{W}_x(x,0)} + e^{-i\nu x}\{\mathcal{W}_x(x,0) + i\nu\mathcal{W}(x,0)\}]_{x=-a}^{x=+a} = 0.\end{aligned}$$

Thus we have

$$\begin{aligned}&\operatorname{Im}[e^{-i\nu x}\{\nu^{-1}\mathcal{W}_x(x,0) + i\mathcal{W}(x,0)\}]_{x=-a}^{x=+a}\\ &\quad = -\nu^{-1}\operatorname{Im}[\mathcal{W}(x,0)\overline{\mathcal{W}_x(x,0)}]_{x=-a}^{x=+a}.\end{aligned}$$

From this equality and (8.85), it follows that

$$\begin{aligned}&\operatorname{Re}\mathcal{D}_+(\mathcal{W})\\ &= \operatorname{Im}\left\{\int_S \mathcal{W}\frac{\partial}{\partial n}e^{\nu(y-ix)}\,\mathrm{d}s + [e^{-i\nu x}\{\nu^{-1}\mathcal{W}_x(x,0) + i\mathcal{W}(x,0)\}]_{x=-a}^{x=+a}\right\}\\ &= \operatorname{Im}\left\{\int_S \mathcal{W}\frac{\partial\bar{\mathcal{W}}}{\partial n}\,\mathrm{d}s - \nu^{-1}[\mathcal{W}(x,0)\overline{\mathcal{W}_x(x,0)}]_{x=-a}^{x=+a}\right\}.\end{aligned}$$

Comparing the last formula with (8.69), we arrive at (8.86), and so (8.86) is valid whether either B vanishes or not.

From (8.86), it follows that $\mathcal{D}_+(\mathcal{W}) = 1$ if and only if $|\mathcal{D}_-(\mathcal{W})| = 1$.

The following corollary is a consequence of the two proven theorems.

Let assumptions of either of the first two lemmas in Subsection 8.1.6.2 *hold; then Problem I is uniquely solvable for any* $(f, C, K) \in C^{0,\alpha}(S) \times \mathbb{R}^2$ *if and only if* $\mathcal{D}_+(\mathcal{W}) \neq 1$ *or equivalently* $|\mathcal{D}_-(\mathcal{W})| \neq 1$.

8.2. Total Resistance to the Forward Motion

Before turning to other supplementary conditions possible for surface-piercing cylinders, we derive formulae for the total resistance to the forward motion of such cylinders. We have to consider these formulae first, because supplementary conditions are closely related to the determination of resistance.

Let us recall some formulae from Section 7.3 concerning the resistance that is the horizontal component $\mathcal{R}$ of the force acting on the body. First, the general formula (7.71) is as follows:

$$\mathcal{R} = -\int_S p \cos(n, x)\, \mathrm{d}s,$$

where p is the pressure expressed by Bernoulli's integral (7.72):

$$p = \text{const} + \rho g y - \rho v^2/2.$$

Here ρ is the density of water, g is the acceleration due to gravity, and according to (7.73) we have $v^2 = (u_x - U)^2 + u_y^2$, where U is the body's forward speed and (u_x, u_y) is the velocity field caused by the potential u corresponding to the right-hand-side term $U \cos(n, x)$ in (8.3). Since

$$\int_S \cos(n, x)\, \mathrm{d}s = \int_S \frac{\partial y}{\partial s}\, \mathrm{d}s = 0, \quad \int_S y \cos(n, x)\, \mathrm{d}s = \int_S y \frac{\partial y}{\partial s}\, \mathrm{d}s = 0 \tag{8.89}$$

for any surface-piercing contour in $\mathbb{R}^2$ because $y = 0$ at the end points $P_\pm$, we get that

$$\begin{aligned} \mathcal{R} &= \frac{\rho}{2} \int_S \left[(u_x - U)^2 + u_y^2\right] \cos(n, x)\, \mathrm{d}s \\ &= \frac{\rho}{2} \int_S (|\nabla u|^2 - 2U u_x) \cos(n, x)\, \mathrm{d}s, \end{aligned} \tag{8.90}$$

which is convenient for use in what follows.

8.2.1. Cylinder in Deep Water

The aim of this subsection is to prove the following assertion.

Let u satisfy the boundary value problem (8.1)–(8.6). *If* $\rho = 1$, *then we have*

$$\mathcal{R} = -\left\{ \frac{\nu}{4}(\mathcal{A}^2 + \mathcal{B}^2) + \frac{1}{2\nu} \left[u_x^2(x, 0) \right]_{x=-a}^{x=+a} \right\}, \tag{8.91}$$

where $\mathcal{A}$ *and* $\mathcal{B}$ *are the coefficients in the asymptotic formula obtained in Subsection* 8.1.2.

In the same way as in Subsection 7.3.1, one obtains the following from (8.90):

$$\mathcal{R} = \int_{\partial W_\alpha \setminus S} \left[u_x \frac{\partial u}{\partial n} - \frac{1}{2} |\nabla u|^2 \cos(n, x) \right] \mathrm{d}s,$$

where $W_\alpha = W \cap \{|x| < \alpha\}$ and α is such that $\alpha > \max_S |x|$. The last integral is a sum of two integrals over the vertical half-lines $W \cap \{x = \pm\alpha\}$ and the integral over $F \cap \{|x| < \alpha\}$. The first two integrals were evaluated in Subsection 7.3.1 as $\alpha \to \infty$. However, considering the integral over $F \cap \{|x| < \alpha\}$, one obtains

$$(2\nu)^{-1} \{ \left[u_x^2(x, 0) \right]_{x=-\alpha}^{x=+\alpha} - \left[u_x^2(x, 0) \right]_{x=-a}^{x=+a} \},$$

whereas there was only the first term in braces in the case of a totally submerged cylinder. Thus the second term arising in the last expression must be added into (7.76), which was shown to express the resistance of the totally submerged cylinder, and so we arrive at (8.91).

From (8.3) and (8.89), it follows that (8.13) takes the following form:

$$\pi Q + \nu^{-1} \left[u_x(x, 0) \right]_{x=-a}^{x=+a} = 0,$$

and so (8.91) can be written as follows:

$$\mathcal{R} = -\frac{\nu}{4}(\mathcal{A}^2 + \mathcal{B}^2) + \frac{\pi Q}{2} [u_x(P_+) + u_x(P_-)]. \tag{8.92}$$

Here the first term can be naturally identified with the wave-making resistance. Since the second term contains Q, which is proportional to the extra rate of flow at infinity caused by the body's presence, this term must be associated with the so-called spray resistance.

Let us consider (8.92) when the supplementary conditions (8.10) and (8.11) have the following particular values of coefficients and right-hand-side terms:

$$A = 2\nu^{-1} \sin \nu a, \quad B = -\cos \nu a, \quad C = 0, \quad K = 0. \tag{8.93}$$

The last equality implies that the spray resistance vanishes, and so

$$\mathcal{R} = -\nu(\mathcal{A}^2 + \mathcal{B}^2)/4,$$

which coincides with (7.76) obtained in Subsection 7.3.1 for the resistance of a totally submerged cylinder. Furthermore, it follows from (8.93) that the coefficients $\mathcal{A}$ and $\mathcal{B}$ in (8.12) – see formulae (8.14) and (8.15) – do simplify and take the following form:

$$\mathcal{A} = -2 \left\{ \int_S \left[u \frac{\partial}{\partial n}(e^{\nu y} \cos \nu x) - \frac{\partial u}{\partial n} e^{\nu y} \cos \nu x \right] ds \right.$$
$$\left. + \sin \nu a [u(P_+) + u(P_-)] \right\},$$

$$\mathcal{B} = 2 \int_S \left[u \frac{\partial}{\partial n}(e^{\nu y} \sin \nu x) - \frac{\partial u}{\partial n} e^{\nu y} \sin \nu x \right] ds,$$

which is actually the same as for a totally submerged cylinder. It is clear that the term outside of the integral in the expression for $\mathcal{A}$ can be eliminated by adding an appropriate constant to the velocity potential that does not change the value of $\mathcal{A}$.

Thus, the supplementary conditions (8.10) and (8.11), where formulae (8.93) give A, B, C, and K, seem to be natural when the effects near the bow and stern points are negligible.

8.2.2. Cylinder in Shallow Water

As in the case of a cylinder totally submerged in water of finite depth d, a no-flow condition

$$u_y = 0 \quad \text{when } y = -d$$

must be added to (8.1)–(8.6) to constitute the Neumann–Kelvin problem together with (8.10) and (8.11), which can be taken as the set of supplementary conditions. Again there are two regimes of flow about S, depending on whether νd is smaller or greater than one. If $\nu d > 1$ (the corresponding value of the forward speed U is referred to as subcritical), then there is a wave pattern behind the body because the velocity potential has the asymptotics (7.53) as $|x| \to \infty$ (see Subsection 7.2.1). Of course, the formulae for coefficients must be modified and we have the following equations for them:

$$(c_+ - c_-)(1 - \nu d) = \nu \int_S \left[x \frac{\partial u}{\partial n} - u \cos(n, x) \right] ds$$
$$- [u(x, 0) + x u_x(x, 0)]_{x=-a}^{x=+a},$$

$$Q(1 - \nu d) + u_x(P_+) - u_x(P_-) = \nu \int_S \frac{\partial u}{\partial n} ds,$$

$$\mathcal{A}\lambda_0(\nu d - \cosh \lambda_0 d)/2$$
$$= \nu \int_S \left\{ u \frac{\partial}{\partial n} [\cosh \lambda_0 (y + d) \cos \lambda_0 x] - \frac{\partial u}{\partial n} \cosh \lambda_0 (y + d) \cos \lambda_0 x \right\} ds,$$
$$+ \cosh \lambda_0 d \, [u_x(x, 0) \cos \lambda_0 x + \lambda_0 u(x, 0) \sin \lambda_0 x]_{x=-a}^{x=+a},$$

$$\mathcal{B}\lambda_0(\nu d - \cosh \lambda_0 d)/2$$
$$= -\nu \int_S \left\{ u \frac{\partial}{\partial n} [\cosh \lambda_0 (y + d) \sin \lambda_0 x] - \frac{\partial u}{\partial n} \cosh \lambda_0 (y + d) \sin \lambda_0 x \right\} ds$$
$$- \cosh \lambda_0 d [u_x(x, 0) \sin \lambda_0 x - \lambda_0 u(x, 0) \cos \lambda_0 x]_{x=-a}^{x=+a},$$

where λ_0 is the positive root of $\lambda \coth \lambda d = \nu$ existing when $\nu d > 1$.

As in Subsection 8.2.1, we can use (8.90) as the starting point for expressing the resistance $\mathcal{R}$ in terms of $\mathcal{A}$, $\mathcal{B}$, and Q (again, we put $\rho = 1$). Then combining the argument applied in the previous subsection with considerations from Subsection 7.3.2, one obtains that

$$\mathcal{R} = -\frac{\lambda_0}{4} (\sinh \lambda_0 d \cosh \lambda_0 d - \lambda_0 d)(\mathcal{A}^2 + \mathcal{B}^2)$$
$$- \frac{1}{2\nu} \left\{ \left[u_x^2(x, 0) \right]_{x=-a}^{x=+a} - (\nu d - 1) Q^2 \right\}.$$

The first term on the right-hand side corresponds to the wave-making resistance, and the second one can be identified with the spray resistance.

8.3. Other Supplementary Conditions

Here we consider several sets of supplementary conditions having different hydrodynamic meaning. Some of them are of local type, that is, they are related to the behavior of the velocity potential near the bow and stern point. Other supplementary conditions are non-local, for example, conditions prescribing values of certain functionals expressed as integrals involving the unknown velocity potential.

8.3.1. Least Singular Solution and Its Generalizations

The notion of the least singular solution was introduced by Ursell in his paper [335] dealing with the Neumann–Kelvin problem for a semi-immersed circular cylinder. It follows from the asymptotic formula (8.8) that the velocity field in this case is singular unless $C_{\pm}^{(2)} = 0$, and these equalities were proposed by Ursell as supplementary conditions. A direct generalization of Ursell's definition is considered in Subsection 8.3.1.1, and the case when the velocity field is singular only at one point (the bow or stern point) is investigated in Subsection 8.3.1.2.

8.3.1.1. The Case When the Cylinder's Contour Forms Nonacute Angles With the Free Surface at the Bow and Stern Points

We will use notations introduced in Subsection 8.1.1.

Assuming that $\beta_{\pm} \geq \pi/2$, we say that u is the *solution with prescribed singularities* for the Neumann–Kelvin problem if u satisfies (8.1), (8.2), and (8.4)–(8.7), and also has given values of $C_{\pm}^{(2)}$ in (8.8). For the sake of brevity this problem will be referred to as Problem S.

Using the theorem proven in Subsection 8.1.7.1 for Problem I, we find it easy to obtain conditions for the unique solvability of Problem S. Let us suppose that the assumptions of this theorem are satisfied; then there exist solutions u_0 and u_1 of Problem I corresponding to the triples $(0, 0, 1)$ and $(0, 1, 0)$, respectively. For any given $f \in C^{0,\alpha}(S)$ there also exists u_f solving Problem I for $(f, 0, 0)$. By $C_{\pm}^{(2)}(u_i)$, $i = 0, 1$, and by $C_{\pm}^{(2)}(u_f)$ we denote the values of the constant $C_{\pm}^{(2)}$ in the asymptotics (8.8) for u_i and u_f near the point $P_{\pm}$.

Now, if a solution of Problem S is sought in the form of

$$u = u_f + p_0 u_0 + p_1 u_1, \tag{8.94}$$

where p_0 and p_1 are unknown real numbers, then u satisfies all of the conditions of Problem S except for supplementary conditions. In order to satisfy them one obtains the linear algebraic system for p_i $(i = 1, 2)$:

$$p_0 C_{+}^{(2)}(u_0) + p_1 C_{+}^{(2)}(u_1) = C_{+}^{(2)} - C_{+}^{(2)}(u_f),$$

$$p_0 C_{-}^{(2)}(u_0) + p_1 C_{-}^{(2)}(u_1) = C_{-}^{(2)} - C_{-}^{(2)}(u_f).$$

Hence, a sufficient condition for unique solvability of Problem S (provided that the homogeneous Problem I has only a trivial solution) is as follows:

$$\begin{vmatrix} C_{+}^{(2)}(u_0) & C_{+}^{(2)}(u_1) \\ C_{-}^{(2)}(u_0) & C_{-}^{(2)}(u_1) \end{vmatrix} \neq 0.$$

One has to take into account that the supplementary condition (8.10) in Problem I depends on real parameters A and B, and therefore, $C_{\pm}^{(2)}(u_i)$ ($i = 1, 2$) also depend on A and B. Thus, for the unique solvability of Problem S it is sufficient that the last determinant does not vanish for a certain choice of A and B for which the homogeneous Problem I has only a trivial solution.

8.3.1.2. The Case When Only One of the Angles at $P_{\pm}$ is Acute

To be specific we assume that $\beta_+ \geq \pi/2$ and $\beta_- < \pi/2$ (the other case can be considered in the same way). We say that u is the *solution for the Neumann–Kelvin problem with the prescribed singularity at P_+ and rate of flow at infinity*, if u satisfies (8.1), (8.2), (8.4)–(8.7), and (8.11), and it also has a given value of $C_+^{(2)}$ in (8.8). The rate of flow at infinity is determined [this follows from (8.13)] by the right-hand side in (8.11) and f in (8.7). For the sake of brevity this problem will be referred to as Problem S$^+$.

Seeking a solution to Problem S$^+$ in the form (8.94), we obtain the following algebraic system for p_i ($i = 1, 2$):

$$p_0 C_+^{(2)}(u_0) + p_1 C_+^{(2)}(u_1) = C_+^{(2)} - C_+^{(2)}(u_f),$$
$$p_0 = K.$$

It is clear that for finding p_1 it necessary and sufficient that $C_+^{(2)}(u_1) \neq 0$, in which case

$$p_0 = K, \quad p_1 = \left[C_+^{(2)}(u_1)\right]^{-1}\left[C_+^{(2)} - C_+^{(2)}(u_f) - K\right].$$

Hence, a sufficient condition for the unique solvability of Problem S+ (provided that the homogeneous Problem I has only a trivial solution) is $C_+^{(2)}(u_1) \neq 0$ and u is given by (8.94), where p_i ($i = 1, 2$) are given by the last formulae.

Another possible set of supplementary conditions in the case when $\beta_+ \geq \pi/2$ and $\beta_- < \pi/2$ consists in prescribing the value of $C_+^{(2)}$ in (8.8) and using (8.10) instead of (8.11). In particular, this means prescribing the value of circulation along the body contour S if $A = 0$ and $B = 1$.

8.3.2. Waveless Potential

Here we consider a surface-piercing cylinder in deep water and use the notations introduced in Subsection 8.1.1. We define a *waveless potential* u (or a solution of Problem L) as a function satisfying (8.1), (8.2), (8.4)–(8.7), and such that

$$\mathcal{A} = \mathcal{B} = 0, \tag{8.95}$$

where $\mathcal{A}$ and $\mathcal{B}$ are the coefficients in (8.12) and so are the linear functionals of u given by (8.14) and (8.15), respectively. Of course, one can prescribe for $\mathcal{A}$ and $\mathcal{B}$ values other than zero, but in what follows we restrict ourselves to the homogeneous conditions (8.95).

We will outline the results concerning the existence and uniqueness of the waveless potential (see Subsection 8.3.2.1) and some properties of the spray resistance calculated on the basis of this potential (it is clear that the wave resistance vanishes in the present case). The corresponding material is presented in Subsection 8.3.2.2.

8.3.2.1. On the Unique Solvability of Problem L

The solvability theorem for this problem can be proved by using the invertibility theorem in the same way as in Subsection 8.1.4. A solution is sought in the form of (8.40). Then the integral equation (8.41) follows from (8.7), and using (8.95) we can complement this equation by an algebraic system for $\mu_\pm$ containing integral functionals of μ. In fact, the wave term in the asymptotics of (8.40) as $x \to -\infty$ has the following form:

$$-4\pi e^{\nu y}\left[\int_S \mu(\zeta) e^{\nu\eta} \sin \nu(x-\xi)\, ds + \sum_{\pm} \mu_\pm \sin \nu(x \mp a)\right].$$

Comparing this with (8.12) and taking into account (8.95), we obtain

$$(\mu_+ + \mu_-) \cos \nu a + \int_S \mu(\zeta) e^{\nu\eta} \cos \nu\xi \, ds = 0, \tag{8.96}$$

$$(\mu_+ - \mu_-) \sin \nu a + \int_S \mu(\zeta) e^{\nu\eta} \sin \nu\xi \, ds = 0. \tag{8.97}$$

These equations together with (8.41) constitute an integroalgebraic system for the unknown vector $X = (\mu, \mu_+, \mu_-)^t$. Moreover, the properties of the integral operator T in (8.41) investigated in Subsection 8.1.4 imply that Fredholm's alternative holds for the integroalgebraic system (8.41), (8.96), and (8.97) in the space $C_\kappa(S) \times \mathbb{R}^2$. Since the determinant of (8.96) and (8.97) is equal to $-\sin 2\nu a$, the following assertion is true.

For any $a > 0$ and sufficiently small values $\nu > 0$, (8.96) *and* (8.97) *are solvable with respect to $\mu_\pm$.*

Substituting $\mu_\pm$ into (8.41), we get an integral equation similar to (8.48). The equation obtained is uniquely solvable for sufficiently small $\nu > 0$, which can be demonstrated in the same way as in Subsection 8.1.4.2. Of course, this fact is also true for the integroalgebraic system, and then the invertibility theorem yields that this system is uniquely solvable for all $\nu > 0$ with a possible exception of a discrete sequence tending to infinity. This leads to the solvability theorem:

For all values $\nu > 0$, except possibly for a sequence tending to infinity, Problem L has a solution for any $f \in C^{0,\alpha}(S)$.

Let us turn to the question of uniqueness, which we are going to consider by using the approach developed in Subsection 8.1.5. However, the waveless potential is not so well suited for defining a uniquely solvable adjoint problem as the solution of Problem I. In Subsection 8.1.5, the supplementary conditions of Problems I and I′ eliminated all but the integral terms in Green's identity, and this was the crucial point in the scheme of proving the uniqueness theorem. Since the supplementary conditions (8.95) are poorly adapted to vanishing the out-of-integral terms, we impose a geometric restriction when applying the same method to Problem L. Namely, we assume S to be symmetric about the y axis. This allows us to introduce symmetric and antisymmetric solutions and to consider their uniqueness separately. In either case it is possible to define an appropriate adjoint Problem L_1/L_2, whose solution is coupled with the symmetric/antisymmetric solution by Green's formula containing only integral.

Let u be a solution to the homogeneous Problem L in a symmetric domain W. Then it is possible to represent u as a sum of the even and odd functions with respect to x:

$$u(x, y) = u^{(s)}(x, y) + u^{(a)}(x, y),$$
$$2u^{(s)}(x, y) = u(x, y) + u(-x, y), \quad 2u^{(a)}(x, y) = u(x, y) - u(-x, y).$$

In order to show that $u^{(s)}$ and $u^{(a)}$ satisfy the homogeneous Problem L, let us prove an auxiliary result, which seems to have its own interest.

Let $\tilde{W} = \{(x, y) : (-x, y) \in W\}$ be the domain symmetric to W with respect to the y axis; then by $\tilde{S}$ we denote the arc symmetric to S. Let $\tilde{u}(x, y) = -u(-x, y)$, and so this function is defined in $\tilde{W}$. If u is the waveless potential in W, then $\lim_{x\to+\infty} |\nabla\tilde{u}| = 0$, and from $\lim_{x\to+\infty} |\nabla u| = 0$ it follows that there are no wave terms in the asymptotics (8.12) for $\tilde{u}$. It is easy to verify by direct calculation that the Neumann condition (8.3) is equivalent to

$$\partial\tilde{u}/\partial n = U \cos(n, x) \quad \text{on int } \tilde{S}.$$

Since all other relations in the definition of waveless potential are obviously true for $\tilde{u}$, we get the following result.

Let u be a solution of (8.1)–(8.6) *and* (8.95) *in W. Then $\tilde{u}$ is a waveless potential in $\tilde{W}$.*

This assertion ensures that $u^{(s)}$ and $u^{(a)}$ satisfy the homogeneous Problem L, and we also have

$$u_x^{(s)}(P_+) + u_x^{(s)}(P_-) = 0, \quad u_x^{(a)}(P_+) - u_x^{(a)}(P_-) = 0. \tag{8.98}$$

At last, (8.12) and (8.95) give that

$$u^{(s)}(x, y) = C + Q \log(\nu|z|) + \psi^{(s)}(x, y), \tag{8.99}$$

$$u^{(a)}(x, y) = \psi^{(a)}(x, y), \tag{8.100}$$

as $|z| \to \infty$. Here $\psi^{(s)}$, $\psi^{(a)} = O(|z|^{-1})$ and $|\nabla\psi^{(s)}|$, $|\nabla\psi^{(a)}| = O(|z|^{-2})$; C is an arbitrary constant and $\pi\nu Q = 2u^{(s)}(P_-)$ on account of (8.13) and (8.98).

Equations (8.98) show that it is reasonable to define solutions $u^{(i)}$ $(i = 1, 2)$ of Problem L_i as functions satisfying (8.1), (8.2), (8.4)–(8.6), (8.7), and the following supplementary conditions:

$$u_x^{(i)}(P_+) + (-1)^i u_x^{(i)}(P_-) = 0, \quad u^{(i)}(P_+) - (-1)^i u^{(i)}(P_-) = 0. \tag{8.101}$$

Using the method applied in Subsection 8.1.4, we find that there is no difficulty in proving the following result.

For all values $\nu > 0$, except possibly for a sequence tending to infinity, Problem L_i $(i = 1, 2)$ has a solution for any $f \in C^{0,\alpha}(S)$.

Now we are in a position to prove a theorem on the unique solvability of Problem L.

Let D be a body symmetric about the y axis; then for any $f \in C^{0,\alpha}(S)$ Problem L has one and only one solution (up to a constant term) for all $\nu > 0$ except possibly for a discrete sequence of values.

Let ν be a positive number such that Problems L, L_i $(i = 1, 2)$ are solvable for this particular ν (those values of ν include all positive numbers except possibly for a discrete sequence). In order to prove the uniqueness for Problem L we have to demonstrate that $u^{(s)}$ and $u^{(a)}$ introduced above are constants – it is obvious that $u^{(a)} = 0$ when it is a constant. Using (8.98)–(8.100) and (8.101) in the same way as in Subsection 8.1.5, one proves that

$$\int_S u^{(s)} \frac{\partial u^{(1)}}{\partial n}\, ds = 0, \quad \int_S u^{(a)} \frac{\partial u^{(2)}}{\partial n}\, ds = 0,$$

if the mean value of $\partial u^{(i)}/\partial n$ over S is zero. Since the second factors in the last two integrals are arbitrary functions orthogonal to a constant, we get that $u^{(s)}, u^{(a)} = \text{const}$ on S. For completing the proof, it remains for us to apply the uniqueness theorem for the Cauchy problem for the Laplace equation (see Subsection 8.1.5).

8.3.2.2. On the Spray Resistance of a Cylinder in the Waveless Forward Motion

For a cylinder symmetric about its midsection, the following property is true.

Let S be symmetric about the y axis, and let u be the unique (up to constant term) waveless potential in W; then the total resistance $\mathcal{R}$ vanishes.

Since W is symmetric about the y axis, we have that $\tilde{u}$ defined in Subsection 8.3.2.1 satisfies Problem L along with u. The assumption about uniqueness of the waveless potential yields that

$$u(x, y) + u(-x, y) = \text{const.}$$

This immediately implies that $u_x(a, 0) = u_x(-a, 0)$. Substituting the last equality and (8.95) into (8.91) proves the theorem.

When the cylinder's geometry is asymmetric and u satisfies the waveless statement of the problem, the behavior of spray resistance can be investigated only numerically. As an example we take a family of Pascal's snails given by the following parametric equation:

$$x(t) = b\cos^2 t + a\cos t - b, \quad y(t) = -\sin t(b\cos t + a), \quad t \in [0, \pi].$$

In Fig. 8.2(a) two patterns of this curve are given. We see in Figs. 8.2(b) and 8.2(c) that the spray resistance is not a monotonic function of b/a and there exist geometries with the towing force instead of the resistance.

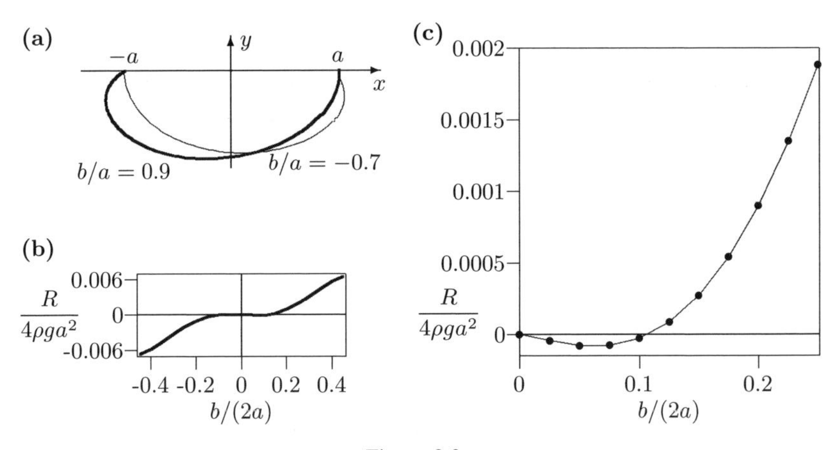

Figure 8.2.

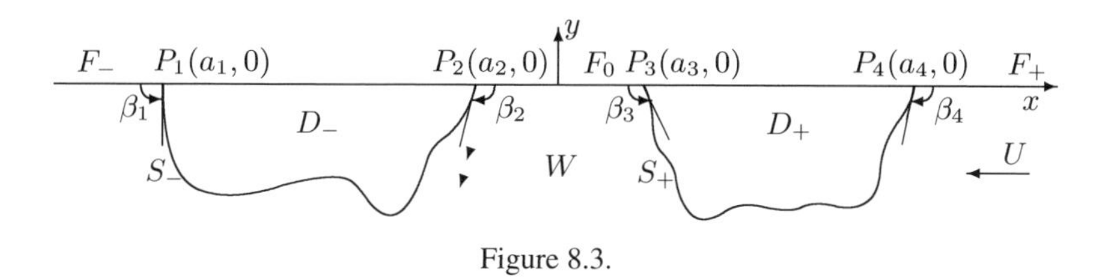

Figure 8.3.

Figure 8.2(b) demonstrates that the horizontal force takes opposite values for two snails that are symmetric to each other with respect to the y axis. The proof of this assertion for an arbitrary pair of symmetric contours follows from the assertion in Subsection 8.3.2.1 concerning $\tilde{u}$.

8.3.3. Resistanceless Potential for a Tandem

Here we consider the Neumann–Kelvin problem for a surface-piercing tandem when conditions (8.1)–(8.6) must be augmented by four supplementary conditions. In this case the waveless conditions (8.95) for a single body can be extended to the "resistanceless" set of supplementary conditions.

8.3.3.1. Statement of the Problem and Some Auxiliary Results

By D_+ and D_- we denote cross sections of surface-piercing cylinders moving forward in the direction of the x axis, and $W = \mathbb{R}^2_- \setminus (\overline{D_+} \cup \overline{D_-})$ is the cross section of the water domain (see Fig. 8.3). Three components of the free surface are F_+, F_0, and F_-, and S_+ (S_-) denotes the wetted contour of the front (back) cylinder. We assume $S = S_+ \cup S_-$ to be a C^2 curve, and the unilateral tangents to S to form angles $\beta_i \neq 0, \pi$ with $F = F_+ \cup F_0 \cup F_-$ at the points P_i, $i = 1, 2, 3, 4$. We choose the coordinate system so that $-a_2 = a_3 = a$.

Minor amendments must be done in the main set of conditions given in Subsection 8.1.1.1: E in (8.6) must be an arbitrary compact set in $\overline{\mathbb{R}^2_-}$ such that $\overline{D_+} \cup \overline{D_-} \subset E$ and $F_\pm \cap E \neq \emptyset$, $F_0 \cap E \neq \emptyset$, and the Neumann condition (8.7) must hold on $\operatorname{int} S = S \setminus \{P_1, P_2, P_3, P_4\}$.

As in Subsection 8.1.1, condition (8.6) implies that the local asymptotics (8.8) holds in a vicinity of each corner point, and so ∇u could be singular at $P_1, \ldots, P_4$, but the finite limits

$$u_x(P_i) = \lim_{x \to a_i - 0} u_x(x, 0) \quad \text{for } i = 1, 3,$$

$$u_x(P_i) = \lim_{x \to a_i + 0} u_x(x, 0) \quad \text{for } i = 2, 4$$

do exist.

Again, (8.12) gives the asymptotic behavior of any solution to the main set of conditions, but instead of (8.13), (8.14), and (8.15) we have the following equalities:

$$\pi \nu Q + \sum_{i=1}^{4} (-1)^i u_x(P_i) = \nu \int_S \frac{\partial u}{\partial n} \, ds, \tag{8.102}$$

$$\mathcal{A} = -2 \left\{ \int_S \left[u \frac{\partial}{\partial n}(e^{\nu y} \cos \nu x) - \frac{\partial u}{\partial n} e^{\nu y} \cos \nu x \right] ds \right.$$
$$\left. + \sum_{i=1}^{4} (-1)^i [\nu^{-1} u_x(P_i) \cos \nu a_i + u(P_i) \sin \nu a_i] \right\}, \tag{8.103}$$

$$\mathcal{B} = 2 \left\{ \int_S \left[u \frac{\partial}{\partial n}(e^{\nu y} \sin \nu x) - \frac{\partial u}{\partial n} e^{\nu y} \sin \nu x \right] ds \right.$$
$$\left. + \sum_{i=1}^{4} (-1)^i [\nu^{-1} u_x(P_i) \sin \nu a_i - u(P_i) \cos \nu a_i] \right\}. \tag{8.104}$$

We define a *resistanceless potential* u (or a solution of Problem R), as a function satisfying (8.1), (8.2), and (8.4)–(8.7), and such that the following supplementary conditions,

$$u_x(P_1) - u_x(P_2) = 0, \qquad u_x(P_3) - u_x(P_4) = 0, \tag{8.105}$$

$$\mathcal{A} = 0, \qquad \mathcal{B} = 0, \tag{8.106}$$

hold. Here $\mathcal{A}$ and $\mathcal{B}$ are given by (8.103) and (8.104), respectively.

The consideration leading to formula (8.91) can be applied to the case of a surface-piercing tandem and the resulting formula for the total resistance to the forward motion of a tandem is as follows:

$$\mathcal{R} = -\frac{\nu}{4}(\mathcal{A}^2 + \mathcal{B}^2) - \frac{1}{2\nu} \sum_{i=1}^{4} (-1)^i \, [u_x(P_i)]^2,$$

where $\mathcal{A}$ and $\mathcal{B}$ are again given by (8.103) and (8.104), respectively. Hence, (8.105) and (8.106) guarantee that $\mathcal{R} = 0$.

8.3.3.2. On the Unique Solvability of Problem R

Seeking a solution in the form

$$u(z) = (\mathcal{U}\mu)(z) + \sum_{i=1}^{4} \mu_i G(z, a_i),$$

one reduces Problem R to an integroalgebraic system in the same way as

Problems I and L were reduced in Subsections 8.1.4.1 and 8.3.2.1, respectively, and the same scheme is applicable for analyzing the solvability of the system obtained. The details can be found in the paper [172] by Kuznetsov and Motygin, and we simply formulate the main result.

For all values $\nu > 0$, except possibly for a sequence tending to infinity, Problem R has a solution for any $f \in C^{0,\alpha}(S)$.

As in Subsection 8.3.2.1, the uniqueness theorem can be proved under the assumption that W is symmetric about the y axis. Again, we split u satisfying the homogeneous Problem R into a sum of even and odd functions with respect to x [$u^{(s)}$ and $u^{(a)}$, respectively, which also satisfy the homogeneous Problem R]. Since we now have

$$u_x^{(s)}(P_3) + u_x^{(s)}(P_2) = 0, \quad u^{(s)}(P_4) - u^{(s)}(P_3) + u^{(s)}(P_2) - u^{(s)}(P_1) = 0,$$

$$u_x^{(a)}(P_3) - u_x^{(a)}(P_2) = 0, \quad u^{(a)}(P_4) - u^{(a)}(P_3) - u^{(a)}(P_2) + u^{(a)}(P_1) = 0$$

instead of (8.98), the supplementary conditions (8.101) used in the auxiliary Problems L_1 and L_2 must be replaced by

$$u_x^{(i)}(P_3) + (-1)^i u_x^{(i)}(P_2) = 0,$$
$$u^{(i)}(P_4) - u^{(i)}(P_3) + (-1)^i \left[u^{(i)}(P_2) - u^{(i)}(P_1)\right] = 0$$

for $u^{(i)}$ ($i = 1, 2$) satisfying auxiliary Problems R_1 and R_2. For these problems the same theorem on the unique solvability holds as for Problems L_1 and L_2. Using $u^{(i)}$ in the same way as in Subsection 8.3.2.1, one arrives at the following result.

Problem R for a tandem symmetric about the y axis has at most one solution (up to a constant term) for all $\nu > 0$ except possibly a discrete sequence of values.

8.4. Trapped Modes

The examples to be considered here present a new type of non-uniqueness comparing the non-uniqueness described in the introductory remarks to this chapter. The latter one depends on the fact that the usual conditions of the Neumann–Kelvin problem are underdefined for surface-piercing bodies (see Section 8.1). Moreover, this type of non-uniqueness occurs for all such bodies and all values of ν. The new type of non-uniqueness takes place only for special values of ν depending on the geometry. These values are point eigenvalues embedded in the continuous spectrum of the problem known to be $(0, +\infty)$. The corresponding modes have finite energy and so must be referred to as *trapped modes*.

In Section 4.1, the so-called inverse procedure was developed for constructing trapped modes satisfying the homogeneous water-wave problem. In the present section we use the same technique as in Subsection 4.1.1 and obtain trapped modes satisfying not only the main set of conditions in the Neumann–Kelvin problem, but also supplementary conditions of several types. We recall that the inverse procedure replaces finding a solution to a given problem by determining a physically reasonable water region for a given solution, and it is convenient to construct the latter by using singularities placed on the x axis at particular values of spacing.

8.4.1. Trapped Modes in Problem R

The uniqueness theorem formulated in Subsection 8.3.3.2 says that if a tandem is symmetric about a vertical axis, then the resistanceless potential is unique for all $\nu > 0$ except possibly for a certain sequence. Here we demonstrate that exceptional values of ν, admitting a nontrivial solution to the homogeneous Problem R, do exist at least for some special geometries.

In accordance with the inverse procedure we put

$$u(z) = (2\nu)^{-1}[G_x(z, \pi/\nu) - G_x(z, -\pi/\nu)], \tag{8.107}$$

and we obtain a solution to the homogeneous problem (8.1)–(8.6) for a surface-piercing tandem if at least one of the streamlines of the flow corresponding to (8.107) connects the x axis on either side of one singular point and another streamline similarly surrounds the other singular point (we interpret these streamlines as rigid contours, where the homogeneous Neumann condition holds). The streamlines are level lines of the stream function v, which is a harmonic conjugate to u and can be written as follows:

$$\begin{aligned} v(z) &= \int_0^\infty \frac{\cos k(x - \pi/\nu) - \cos k(x + \pi/\nu)}{k - \nu} e^{ky}\, \mathrm{d}k \\ &= \mathrm{Re}\{e^{-i\nu z}[\mathrm{Ei}(i\nu(z - \pi/\nu)) - \mathrm{Ei}(i\nu(z + \pi/\nu))]\}. \end{aligned} \tag{8.108}$$

The second expression in terms of the exponential integral is a consequence of 8.212.5 in Gradshteyn and Ryzhik [96]. The particular combination of singularities (8.107) is chosen so as to cancel wave terms in the asymptotics of u, and this is an immediate consequence of the first assertion in Subsection 6.3.1.2.

It follows from (8.108) that

$$v_y(x, y) - \nu v(x, y) = \frac{y}{y^2 + (x - \pi/\nu)^2} - \frac{y}{y^2 + (x + \pi/\nu)^2}, \tag{8.109}$$

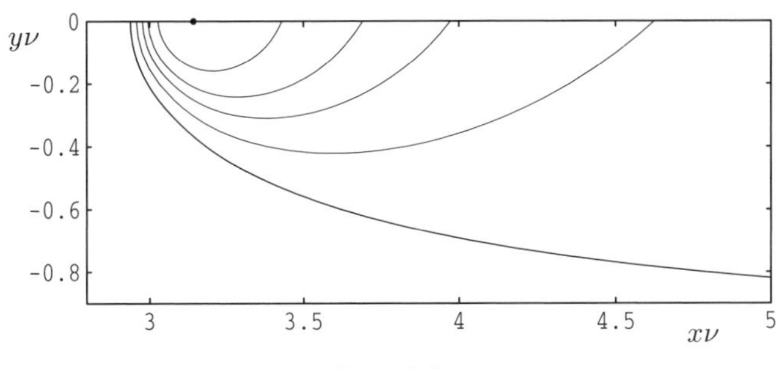

Figure 8.4.

and so

$$v_y(x, 0) - \nu v(x, 0) = 0, \quad \text{when } x \neq \pm\pi/\nu. \tag{8.110}$$

Moreover, the derivative $v_y = u_x$ has the same value at two points, where a streamline enclosing one of the singular points intersects the x axis. Taking such a contour as S_+ and a similar contour as S_-, we see that u delivers a solution to Problem R in the corresponding water domain W. The asymptotic behavior of Ei implies that $v(z) \sim \pm \log|z \pm \pi/\nu|$ as $z \to \mp\pi/\nu$, and so u given as a combination of horizontal dipoles behaves like a vortex near singular points (this has been pointed out by Ursell [335]). Hence, the streamlines enclosing the singular points do exist for sufficiently large values of v. These lines are close to semicircles that are the level lines of $\log|z \pm \pi/\nu|$.

Now, let us investigate the family of streamlines in more detail, and, in particular, prove that any streamline of positive level encloses one of the singular points as shown in Fig. 8.4, where the streamlines $v = 0$ (bold line) and 0.2, 0.4, 0.6, and 1.0 are plotted. Since $v(x, y)$ is an odd function of x, we restrict ourselves to considering only positive values of x. Properties of harmonic functions yield that streamlines can have end points either on the free surface or at infinity (see Subsection 4.1.1.2). It follows from the asymptotics of (8.107) at infinity that only nodal lines of v (the loci, where $v = 0$) are the streamlines going to infinity, and so the nodal lines divide the fluid domain into subdomains, each of which contains a family of contours with the same properties. Thus we have to investigate the behavior of the function $v(x, 0)$ and of the nodal lines of $v(x, y)$ in the quadrant $\{x > 0, y < 0\}$, and we begin with the following lemma.

There is only one zero $\xi_0 \in (2\pi/3\nu, \pi/\nu)$ of $v(x, 0)$ on the half-axis $x > 0$.

First, let us demonstrate that $v(x, 0) < 0$ for $x \in (0, 2\pi/3\nu]$. From 3.354.2 and 3.722.7 in Gradshteyn and Ryzhik [96], it follows that

$$\int_0^{+\infty} \frac{\cos ak}{k - 1}\, \mathrm{d}k = \int_0^{+\infty} \frac{ke^{-ak}}{1 + k^2}\, \mathrm{d}k - \pi \sin a, \quad a > 0. \tag{8.111}$$

Then $v(x, 0) = I(x) - 2\pi \sin \nu x$ for $x \in (0, \pi/\nu)$, and

$$I(x) = \int_0^{+\infty} \frac{k\left[e^{(x\nu-\pi)k} - e^{-(x\nu+\pi)k}\right]}{1+k^2}\,\mathrm{d}k.$$

It is obvious that $I'(x) > 0$, and so $I(x)$ is an increasing nonnegative function. Therefore,

$$I(x) \leq I(2\pi/3\nu) \leq e^{-1}\int_0^{+\infty} \frac{e^{(1-\pi/3)k}}{1+k^2}\,\mathrm{d}k < e^{-1}\int_0^{+\infty} \frac{\mathrm{d}k}{1+k^2} = \frac{\pi}{2e},$$

where it is taken into account that $e^{k-1} \geq k$ for $k \geq 0$.

Let us put

$$x^* = \nu^{-1} \arcsin(1/4e), \tag{8.112}$$

and so $2\pi \sin x\nu \geq \pi/2e$ for $x \in [x^*, 2\pi/3\nu]$, which implies that $v(x, 0) \leq 0$ on this interval. Now let us prove that $v(x, 0) < 0$ for $x \in (0, x^*]$. Assuming the contrary and taking into account that $v(0, 0) = 0$, we see that $v_x(\xi, 0)$ must have a zero at a certain $\xi \in (0, x^*)$, but this is impossible because

$$v_x(x, 0) < 2\pi\nu\{[\pi^2 - (x^*\nu)^2]^{-1} - \cos x^*\nu\} < 0,$$

and these inequalities follow from

$$v_x(x, 0) = 2\pi\nu\{[\pi^2 - (x\nu)^2]^{-1} - \cos x\nu\} - \nu\int_0^{+\infty} \frac{e^{(x\nu-\pi)k} + e^{-(x\nu+\pi)k}}{1+k^2}\,\mathrm{d}k. \tag{8.113}$$

Thus, we have proved that $v(x, 0) < 0$ for $x \in (0, 2\pi/3\nu]$. This and the obvious fact that $v(x, 0) \to +\infty$ as $x \to \pi/\nu$ imply that $v(x, 0)$ vanishes at some point $\xi_0 \in (2\pi/3\nu, \pi/\nu)$.

Let us prove that the zero of $v(x, 0)$ is unique. From (8.111) we have that

$$v(x, 0) = \int_0^{+\infty} \frac{k\left[e^{(\pi-x\nu)k} - e^{-(\pi+x\nu)k}\right]}{1+k^2}\,\mathrm{d}k > 0 \quad \text{for } x > \pi/\nu.$$

Moreover, (8.113) and the inequalities $-\cos x\nu \geq 1/2$,

$$\int_0^{+\infty} \frac{e^{(x\nu-\pi)k} + e^{-(x\nu+\pi)k}}{1+k^2}\,\mathrm{d}k \leq \pi,$$

imply that

$$v_x(x, 0) \geq 2\pi\nu[\pi^2 - (x\nu)^2]^{-1} > 0 \quad \text{for } x\nu \in [2\pi/3, \pi).$$

This completes the proof of the lemma.

Further properties of $v(x, 0)$ will follow from the second lemma.

Let x^* be defined by (8.112)*; then there is only one zero $\xi_1 \in (x^*, 2\pi/3\nu)$ of $v_x(x, 0)$ on the half-axis $x > 0$.*

In the proof of the previous lemma it was shown that $v_x(x, 0) > 0$ for $x \in [2\pi/3\nu, \pi/\nu)$ and $v_x(x, 0) < 0$ for $x \in (0, x^*] \cup (\pi/\nu, +\infty)$. Therefore, there is at least one zero $\xi_1 \in (x^*, 2\pi/3\nu)$ of $v_x(x, 0)$. In order to prove that the zero is unique it is sufficient to show that $v_{xx} \neq 0$ in this interval. Differentiating (4.59) and comparing the result with $v(x, 0)$, we get that

$$v_{xx} = -\nu^2 v(x, 0) + \frac{4\pi x \nu^3}{[\pi^2 - (x\nu)^2]^2}.$$

Since $v(x, 0) < 0$ in $(x^*, 2\pi/3\nu)$, we have that

$$v_{xx} > \frac{4\pi x \nu^3}{[\pi^2 - (x\nu)^2]^2} > 0$$

in this interval, which completes the proof of the second lemma.

The following corollary is an immediate consequence of the two proven lemmas.

The inequality $\xi_1 < \xi_0$ holds; $v(x, 0)$ is negative for $0 < x < \xi_0$ and positive for $x > \xi_0$ and $x \neq \pi/\nu$. Also, $v(x, 0)$ decreases for $0 < x < \xi_1$ and $x > \pi/\nu$ and increases between ξ_1 and π/ν.

In Fig. 8.5(a), the graph of $v(x, 0)$ is shown by solid lines (there are two of them asymptoting to $\nu x = \pi$), and the graph of $\nu^{-1} v_x(x, 0)$ is shown by dashed lines.

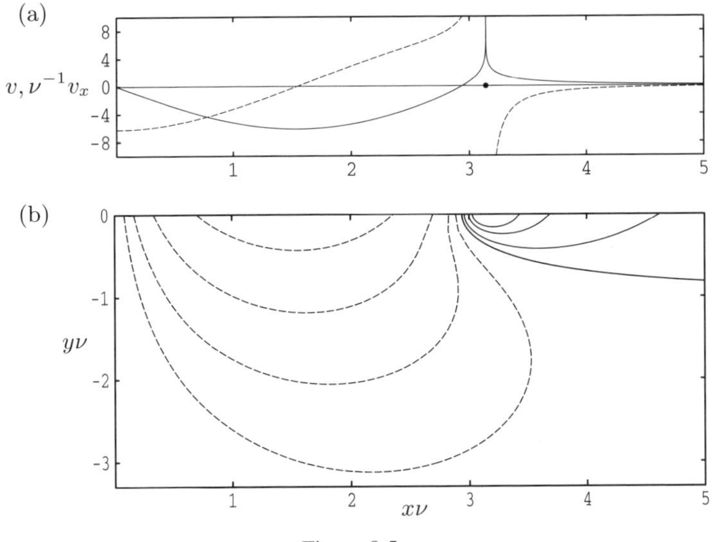

Figure 8.5.

According to the first lemma, there is only one nodal streamline in the quadrant $\{x > 0, y < 0\}$ and we proceed with describing its behavior. The nodal streamline emanates from $(\xi_0, 0)$ and its equation can be found as follows. From (8.109), we get

$$v(x, y) = e^{\nu y} \left\{ v(x, 0) - \int_y^0 \left[\frac{t}{t^2 + (x + \pi/\nu)^2} - \frac{t}{t^2 + (x - \pi/\nu)^2} \right] e^{-t} \, dt \right\},$$

and so

$$v(x, 0) = \int_y^0 \left[\frac{t}{t^2 + (x + \pi/\nu)^2} - \frac{t}{t^2 + (x - \pi/\nu)^2} \right] e^{-t} \, dt$$

is the equation we seek. Since the last integral is positive in the quadrant and $v(x, 0) > 0$ for $x > \xi_0$, the nodal line emanating from $(\xi_0, 0)$ lies below the interval $(\xi_0, +\infty)$ on the x axis. It divides the quadrant into two regions, each covered by a family of streamlines. The last corollary shows that the streamlines belonging to one of the families, say Γ, correspond to positive values of v and have the left end point in $(\xi_0, \pi/\nu)$ and the right end point in $(\pi/\nu, +\infty)$; they are shown by solid lines in Fig. 8.5(b) and correspond to $v = 0.2,\ 0.6,$ and 1.0. The streamlines from the other family (they correspond to negative values of v) have the right and left end points in $(0, \xi_1)$ and (ξ_1, ξ_0), respectively; they are shown by dashed lines in Fig. 8.5(b) and correspond to $v = -0.5, -1.0, -2.0$, and -4.0. The nodal streamline is plotted by the bold line in Fig. 8.5(b).

A streamline from Γ and a line from the family that is the symmetric image of Γ in the y axis form a tandem such that (8.107) delivers the velocity potential for a mode trapped in the water domain in $\mathbb{R}^2$ outside the streamlines chosen (this is an example of non-uniqueness in Problem R).

8.4.2. Trapped Modes in Problem S

Let us consider an arbitrary streamline $v(x, y) = \text{const} \neq 0$ making angles β_3 and β_4 with the free surface at the end points P_3 and P_4, which are on the left and on the right from the streamline, respectively (see Fig. 8.3). Then

$$\tan \beta_i = (-1)^i v_x(P_i)/v_y(P_i) = (-1)^i v_x(P_i)/\left[\nu v(P_i)\right], \quad i = 3, 4,$$

where the second equality is a consequence of (8.110). The corollary proven in Subsection 8.4.1 and illustrated in Fig. 8.5(a) implies that the last expression is negative, and so $\beta_i > \pi/2$; see Fig. 8.5(b). Of course, the same is true for streamlines surrounding the left singular point.

Since the streamlines defined by (8.108) make obtuse angles with the free surface (we recall that the velocity field can be singular in this case), and

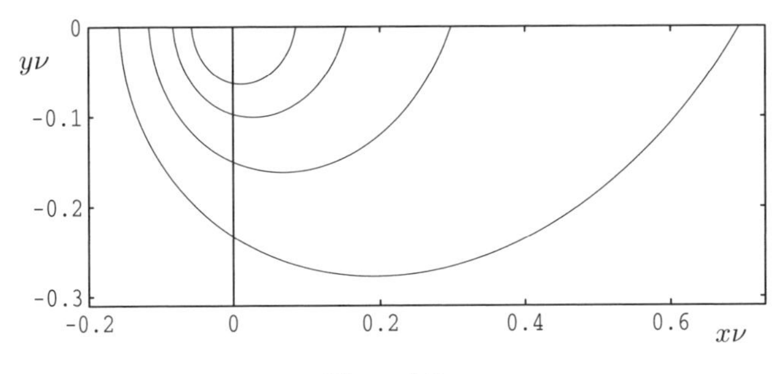

Figure 8.6.

the velocity potential (8.107) defines a nonsingular velocity field everywhere outside the singular points, we conclude that the coefficients $C_i^{(2)}$ $(i = 1, 2, 3, 4)$ in the local asymptotics (8.8) for (8.107) at P_i are equal to zero. Therefore, (8.107) is a nontrivial least singular potential in the water domain outside a pair of streamlines under consideration (or a nontrivial solution to the homogeneous Problem S, which admits the obvious generalization to the case when there are more than one surface-piercing body).

Moreover, (8.107) delivers examples of non-uniqueness for the least singular statement when three or four bodies are presented by the streamlines defined by (8.108).

An example of non-uniqueness for the least singular statement does exist when there is a single surface-piercing body. It is given by the following potential:

$$u^{(1)}(z) = (2\nu)^{-1} G_x(z; 0).$$

The corresponding streamlines $v^{(1)} = 0.5, 1.0, 1.5,$ and 2.0 are plotted in Fig. 8.6. The last potential satisfies (8.11), and so the spray resistance vanishes for the cylinder defined by $v^{(1)}$. However, in contrast to (8.107) the velocity potential $u^{(1)}$ is not waveless.

8.5. Cylinder in the Supercritical Stream

In this section we consider the two-dimensional Neumann–Kelvin problem in the case when a cylinder is partly immersed in a supercritical stream of finite depth (without loss of generality we can assume the latter to be equal to one). For this problem augmented with the supplementary conditions (8.10) and (8.11) (they were used in Section 8.1 for the problem in deep water) we obtain the same results concerning the unique solvability as those in Subsection 7.2.2 for a totally submerged cylinder. Since the methods developed earlier are applicable here, we restrict ourselves to a brief description of results.

8.5.1. Two Equivalent Statements of the Problem

Let the longitudinal section of the undisturbed stream be the strip

$$L = \{-\infty < x < +\infty, -1 < y < 0\},$$

and let the immersed cylinder's cross section be a bounded simply connected domain $D \subset L$ such that its boundary ∂D consists of a segment $\{|x| \leq a, y=0\}$, where $a > 0$, and of a simple closed C^2 arc S that lies in L with an exception for the end points $P_\pm = (\pm a, 0)$. We suppose that the x axis is not tangent to S at $P_\pm$ (see Fig. 8.1), and we set $W = L \setminus \bar{D}$, $F_\pm = \{\pm x > a, y=0\}$.

The velocity potential u describing the induced water motion must satisfy the boundary value problem consisting of (8.1), (8.2), (8.4)–(8.7), and of the no-flow condition on the horizontal bottom:

$$u_y = 0 \quad \text{when } y = -1. \tag{8.114}$$

Of course, E in (8.6) must be an arbitrary compact set in $\bar{L}$ such that $\bar{D} \subset E$ and $F_\pm \cap E \neq \emptyset$.

The local asymptotic formula (8.8) remains valid in the present case, and so formula (8.9) for the x derivative of u along the free surface is also true. This allows us to use (8.10) and (8.11) as the set of supplementary conditions, and as in Section 8.1 the corresponding problem will be referred to as Problem I.

As in the case of a totally submerged cylinder, the behavior of u as $|x| \to \infty$ is described by formula (7.62), but the relations for coefficients must be rewritten as follows:

$$\begin{aligned}(c_+ - c_-)(1-\nu) &= \nu \int_S \left(x\frac{\partial u}{\partial n} - u\frac{\partial x}{\partial n}\right) \mathrm{d}s \\ &\quad - a[u_x(P_+) + u_x(P_-)] + u(P_+) - u(P_-), \\ Q(1-\nu) + u_x(P_+) - u_x(P_-) &= \nu \int_S \frac{\partial u}{\partial n}\, \mathrm{d}s. \end{aligned} \tag{8.115}$$

For Problem I, the equivalent statement in terms of the stream function v is almost the same as in the case of deep water (see Subsection 8.1.3), but two amendments must be done: (i) conditions (8.23)–(8.29) must be complemented by

$$v = 0 \quad \text{when } y = -1; \tag{8.116}$$

(ii) conditions (8.30) and (8.31) must be rewritten as follows:

$$C_0 = \nu^{-1}[u_x(P_-) + Q(\nu - 1)], \tag{8.117}$$

$$C_- = (1-\nu)Q. \tag{8.118}$$

The last assertion in Subsection 8.1.3 providing necessary and sufficient conditions for boundedness of u at infinity remains true for the supercritical stream [of course, in (8.36) the integration must be over $(-1, 0)$ instead of $(-\infty, 0)$].

8.5.2. On the Uniqueness of the Solution

Let u satisfy the homogeneous Problem I; then the no-flow condition on S, the supplementary condition (8.11), and (8.115) imply that $Q = 0$. Therefore, we have that

$$\int_W |\nabla u|^2 \, dx dy < \infty, \tag{8.119}$$

which is similar to (8.58) in Subsection 8.1.6.2. The last inequality allows us to derive the integral identity (8.60), where the integral over the bottom vanishes because of (8.114). Then using the same considerations as in Subsection 8.1.6.2, we obtain that $u(x, y) = u(x)$. Now (8.2) and (7.62) imply that $u = c_+$ and $u_y = 0$ on F_+, and the following theorem is a consequence of the uniqueness theorem for the Cauchy problem for the Laplace equation.

Let u be a solution to the homogeneous Problem I and $\mathbf{x} \cdot \mathbf{n} \geq 0$ on S. If one of the following two assumptions holds, (i) $B = 0$, (ii) $2a + AB^{-1} \geq 0$, then $u = \text{const}$ in W.

Let us generalize the second uniqueness theorem in Subsection 8.1.6.2, which does not require the inequality $\mathbf{x} \cdot \mathbf{n} \geq 0$ to hold on S.

Let u be a solution to the homogeneous Problem I and one of the following two assumptions holds: (i) $B = 0$, (ii) $A/B \geq 0$. Then $u = \text{const}$ in W.

Let us consider the equivalent homogeneous boundary value problem for the stream function v. From (8.117) and (8.118), it follows that

$$v_y - \nu v = 0 \quad \text{on } F = F_+ \cup F_-, \tag{8.120}$$

$$v = \nu^{-1} v_y(P_-) \quad \text{on } S, \tag{8.121}$$

$$A v_y(P_-) = B \int_S \frac{\partial v}{\partial n} \, ds, \tag{8.122}$$

and (8.119) holds for v. The latter condition allows us to apply Green's identity to W, and according to (8.120)–(8.122) this identity takes the following form:

$$\int_W |\nabla v|^2 \, dx dy - \nu \int_F v^2 \, dx = -\frac{v_y(P_-)}{\nu} \int_S \frac{\partial v}{\partial n} \, ds. \tag{8.123}$$

If $B = 0$, then (8.122) yields $v_y(P_-) = 0$ and the right-hand-side term in

(8.123) vanishes. Otherwise, (8.122) and (ii) imply

$$-\frac{v_y(P_-)}{\nu}\int_S \frac{\partial v}{\partial n}\,\mathrm{d}s = -\frac{A}{B\nu}[v_y(P_-)]^2 \le 0,$$

and so we get from (8.123) that

$$\int_W |\nabla v|^2\,\mathrm{d}x\mathrm{d}y \le \nu\int_F v^2\,\mathrm{d}x. \tag{8.124}$$

Since (8.121) holds on S, v can be extended into D as a constant function. Then we have

$$v(x,0) = \int_{-1}^{0} v_y(x,y)\,\mathrm{d}y \quad \text{when } |x| > a,$$

where (8.116) is taken into account. Then the Schwarz inequality gives

$$|v(x,0)|^2 \le \int_{-1}^{0} |v_y(x,y)|^2\,\mathrm{d}y \quad \text{when } |x| > a,$$

and integrating over F we get

$$\int_F v^2\,\mathrm{d}x \le \int_{W\setminus W_a} |v_y(x,y)|^2\,\mathrm{d}x\mathrm{d}y \le \int_W |\nabla v|^2\,\mathrm{d}x\mathrm{d}y,$$

because ∇v vanishes outside W (here $W_a = W \cap \{|x| < a\}$). Comparing this with (8.124), we find that

$$(1-\nu)\int_F v^2\,\mathrm{d}x \le 0.$$

By virtue of $0 < \nu < 1$ we obtain that $v = 0$ on F. Then (8.120) allows us to apply the uniqueness theorem for the Cauchy problem for the Laplace equation, and so $v = 0$ in W. Therefore, $u = \text{const}$ in W, which completes the proof.

8.5.3. On the Existence of the Solution

Despite the fact that the problem of a supercritical stream about a surface-piercing cylinder is a real one, it is convenient to demonstrate the solvability of this problem by using the method applied in Subsection 8.1.6.4 for proving the solvability theorem for Problem II. The method is based on the following representation [cf. (8.73)]:

$$u(z) = (\mathcal{V}\mu)(z) + (\mathcal{W}\rho)(z) + \sum_{\pm} \mu_{\pm} G(z, \pm a),$$

where $\mu_{\pm}$ are unknown complex numbers and potentials $\mathcal{V}\mu$ and $\mathcal{W}\rho$ have unknown complex-valued densities $\mu \in C_\kappa(S)$ and $\rho \in C(\bar{D})$ (we recall that

$\mathcal{V}\mu$ is the single-layer potential and $\mathcal{W}\rho$ is the area potential, and both of them have Green's function G as the kernel). Substituting the given representation of u into the Neumann boundary condition on S, the supplementary conditions and Poisson's equation,

$$\nabla^2 u = -iu \quad \text{in } D,$$

one arrives at the same integroalgebraic system as in Subsection 8.1.6.4 [see equations (8.75)–(8.77) and (8.79)], and so Fredholm's alternative holds for this system in $C_\kappa(S) \times C(\bar{D}) \times \mathbb{C}^2$ when κ is chosen appropriately.

Now the same consideration as in Subsection 8.1.6.4 shows that the homogeneous integroalgebraic system has only a trivial solution when the uniqueness theorem is true for the boundary value problem (see the conditions of uniqueness in Subsection 8.5.2). Therefore, the nonhomogeneous integroalgebraic system is solvable for arbitrary right-hand-side terms, and the representation given above solves the boundary value problem. Thus the following theorem is proved.

Let the assumptions of one of the theorems proved in Subsection 8.5.2 *hold, and let* ν *be an arbitrary number in* $(0, 1)$. *Then Problem I has a unique solution for any triple* $(f, C, K) \in C^{0,\alpha}(S) \times \mathbb{R}^2$.

8.6. Bibliographical Notes

8.1. The results of this section were obtained by Kuznetsov and Maz'ya [166].

8.2. The formula for resistance of a cylinder in deep water was given without derivation in the paper [171] by Kuznetsov and Motygin, and the corresponding result for shallow water was not published earlier.

8.3.1. Results presented in this subsection are borrowed from the work [165] by Kuznetsov and Maz'ya.

8.3.2. The waveless potential investigated in this subsection was considered by Kuznetsov and Motygin [171]. Another approach allowing us to construct bodies that do not generate waves was outlined in the abstract [321] by Tuck and Tulin.

8.3.3. Resistanceless potential for a tandem investigated in this subsection was considered by Kuznetsov and Motygin [172].

8.4. Results presented in this section are borrowed from the work [172] by Kuznetsov and Motygin.

8.5. Here we present the results obtained by Kuznetsov [161].

A brief historical survey. In his work [335] published in 1981, Ursell investigated the Neumann–Kelvin problem for a surface-piercing semicircle and found that this problem has a two-parameter set of solutions. Earlier, this fact was discovered numerically (see Suzuki [316] and references cited therein). In particular, Suzuki pointed out that as early as 1963, Bessho and Mizuno had solved numerically the Neumann–Kelvin problem for a semisubmerged circular cylinder and found that there are infinitely many solutions and hence the wave resistance calculated from the velocity potential can have any value. In his G. Weinblum Memorial Lecture [25], Bessho presented results on this problem (for the most part numerical) obtained during the past several decades in Japan. In 1976, Eggers, in his discussion to Bessho [24], had suggested that the homogeneous problem possesses nontrivial solutions. However, the first mathematical proof of this assertion was given only by Ursell [335]. In the same paper, he proposed a statement in which the problem is complemented in a way leading to the so-called least singular solution (see Section 8.3).

In their work [165] published in 1988 (the original paper in Russian was published earlier than the English translation given in the Bibliography), Kuznetsov and Maz'ya proposed several versions of supplementary conditions. Later on in [166], they considered the general linear supplementary conditions (the corresponding problem is considered in Section 8.1), which include two sets of conditions considered in [165] as particular cases in which the coefficients are chosen in a special way. The problem of a supercritical stream about a two-dimensional body was considered by Kuznetsov [161] who augmented the equation, boundary conditions, and conditions at infinity by the same general linear supplementary conditions as in [166] (see Section 8.5).

There are a number of other statements of the Neumann–Kelvin problem for surface-piercing bodies (a survey of these statements is given in Section 8.3). In 1982, Lenoir [187] proposed a statement in terms of the stream function with a condition of Kutta–Zhoukovsky type at the stern point. The version of supplementary conditions proposed by Kuznetsov [158] (see also the abstract by Motygin and Kuznetsov [255]) provides the resistance to be purely wave making, that is, there are no sprays at bow and stern points, and to be expressed by just the same formula as for a totally submerged cylinder (see Section 7.3). Recently, two papers by Kuznetsov and Motygin were published, and in the first of them [171] the authors consider supplementary conditions leading to a wave-free solution and, hence, losing the wave resistance, but the second component of resistance, the so-called spray resistance, is not zero under these conditions, generally speaking. Nevertheless, the total resistance vanishes if the body contour is symmetric about its midsection. Another paper [172] by the same authors is concerned with a similar waveless statement of the Neumann–Kelvin problem for a tandem of surface-piercing cylinders. In the latter case there exists a set of four supplementary conditions canceling both the wave resistance and the spray resistance and providing a well-posed statement of the problem. This means that a unique solution exists for all values of the forward speed U except for a sequence tending to zero.

At the same time, for the exceptional values of U, examples of non-uniqueness are constructed, that is, a couple of surface-piercing contours for which the homogeneous problem augmented with the same four waveless supplementary conditions has a nontrivial solution (see Section 8.4). The construction of these examples is based on the inverse procedure considered in Section 4.1 for a non-uniqueness example in the water-wave problem. Thus, there are two different kinds of non-uniqueness in the Neumann–Kelvin problem. The non-uniqueness of the first kind can be removed by imposing proper supplementary conditions. The non-uniqueness of the second kind is intrinsic to the problem with some supplementary conditions, but it is absent in the problem with other ones.

Recently, several papers by Pagani and Pierotti [275, 278, 279, 280] were published. In the second of them, the authors treat the Neumann–Kelvin problem for a beam placed on the free surface, and the results of this paper form a base for considering the nonlinear problem of the uniform forward motion of a slender two-dimensional body. The latter problem is investigated in [279] and [280] for different regimes of flow.

An interesting approach to ship waves based on the ray theory was developed by Keller [133] in 1979, but, to our knowledge, no further development of this approach has been published.

Much attention has been paid to the numerical treatment of the three-dimensional ship-wave problem. Among numerous works on this topic we mention the paper [259] by Nakos and Sclavounos (above all, it contains a substantial list of references) in which the authors apply discretization to an integral equation arising from Green's representation of the velocity potential when the so-called double-body model is applied. This model may be considered as a kind of supplementary condition in the three-dimensional case.

In conclusion of these notes, it should be mentioned that a rigorous derivation of formulae for the resistance to the forward motion of three-dimensional surface-piercing bodies in deep and shallow water is still an open question despite the fact that Michell's pioneering paper [245] on the resistance of a knife-like ship goes back to 1898. During the first half of the 20th century, Havelock worked extensively on the theory of wave resistance (see his

Collected Papers [111] edited by C. Wigley). Results obtained by these and other authors can be found in two survey papers by Wehausen and Laitone [354] and Wehausen [353], and in the books by Kostyukov [147] and Timman et al. [318]. The questions arising in the theory of wave resistance are outlined by Weinblum [355] and Bessho [24] in their talks at International Seminars on Wave Resistance in 1963 and 1976, respectively (proceedings of these seminars present the state of the art in the field at those times). In particular, Bessho dwells on the question of uniqueness and the role of the so-called line integral (another heuristic approach to this integral was developed by Brard [32]). The answers to these questions were obtained in the two-dimensional case (they constitute the contents of this chapter), but there are no rigorous results concerning the three-dimensional case.

Part 3

Unsteady Waves

9

Submerged Obstacles: Existence and Properties of Velocity Potentials

Steady-state problems usually arise as limiting cases of time-dependent problems, and it is widely believed that the latter are more natural from the hydrodynamic point of view (see, for example, Stoker [312], pp. 175–176 for a discussion of relationships between problems describing unsteady and time-harmonic waves). Moreover, the problems describing unsteady waves are not only more attractive physically as providing a natural approach to wave phenomena, but also have another appealing feature as a mathematical challenge posed to a researcher, and up to the present only the problem involving submerged obstacles obtained a satisfactory mathematical treatment presented here.

9.1. The Initial-Boundary Value Problem and an Auxiliary Steady-State Problem

The plan of this section is as follows. In Subsection 9.1.1, we describe the water domain and formulate the initial-boundary value problem. Several function spaces that we use in what follows are also introduced in this subsection. An auxiliary steady-state problem is considered in Subsection 9.1.2.

9.1.1. Definitions

9.1.1.1. Statement of the Initial-Boundary Value Problem

Let W denote a water domain bounded above by the free surface $F = \partial\mathbb{R}^3_-$. The boundary ∂W can also contain the following disjoint parts: a bottom B and a surface S of a totally immersed body D (for the sake of simplicity, we assume that there is only one body in water, but all results remain true for a finite number of bodies). We suppose that B is a smooth surface dividing $\mathbb{R}^3_-$ into two unbounded regions and coinciding with the plane $y = -d$ $(d > 0)$ at infinity. The surface S is assumed to be smooth, connected, and closed. It is possible that either B or S (even $B \cup S$) is an empty set. We also assume that W contains a layer of constant depth $d_0 \leq d$ adjacent to the free surface; that is, $y < -d_0$ on $B \cup S$.

Let us recall the general linear initial-boundary value problem formulated in the Boundary Condition on an Immersed Rigid Surface section in the Introduction for the velocity potential $\phi(x, y, z, t)$ describing unsteady waves:

$$\nabla^2 \phi = 0 \quad \text{in } W, \tag{9.1}$$

$$\left.\begin{aligned} \phi_{tt} + \phi_y = f \quad \text{for } t > 0, \\ \phi = f_0, \ \phi_t = f_1 \quad \text{for } t = 0 \end{aligned}\right\} \quad \text{on } F, \tag{9.2}$$

$$\partial \phi / \partial n = v \quad \text{on } B \cup S, \tag{9.3}$$

where v is a continuous function vanishing at infinity when $B \neq \emptyset$. Here the units are chosen so that the acceleration due to gravity is equal to one.

9.1.1.2. Function Spaces

A solution to (9.1)–(9.3) is assumed to belong to a Hilbert space $\mathcal{H}(W)$ supplied with the inner product

$$(\phi, \psi)_{\mathcal{H}(W)} = \int_W \nabla \phi \cdot \nabla \psi \, \mathrm{d}x\mathrm{d}y\mathrm{d}z + \int_F \phi \psi \, \mathrm{d}x\mathrm{d}z \tag{9.4}$$

and obtained by completing $C_0^1(\bar{W})$ (it consists of continuously differentiable in $\bar{W}$ functions vanishing at infinity) with respect to

$$\|\phi\|_{\mathcal{H}(W)} = (\phi, \phi)^{1/2}_{\mathcal{H}(W)}.$$

A weak solution from $\mathcal{H}(W)$ solving (9.1)–(9.3) will be defined in Section 9.2.

Now we introduce some function spaces to be used in what follows. By $\psi^{(c)}$ we denote the trace of ψ on a plane $y = c$. For every $\psi \in \mathcal{H}(W)$ there exists $\psi^{(0)} \in L^2(F)$ and

$$\left\| \psi^{(0)} \right\|_{L^2(F)} \leq \|\psi\|_{\mathcal{H}(W)}. \tag{9.5}$$

Let $L_a = \{(x, y, z) \in \mathbb{R}^3 : -\infty < x, z < +\infty, \ -a < y < 0\}$, $0 \leq a \leq d_0$; then

$$\int_{\mathbb{R}^2} \left[\psi^{(-a)} - \psi^{(0)}\right]^2 \mathrm{d}x\mathrm{d}z \leq a \int_{L_a} \psi_y^2 \, \mathrm{d}x\mathrm{d}y\mathrm{d}z$$

and (9.5) yield that $\psi^{(-a)}$ belongs to $L^2(\mathbb{R}^2)$ for any $\psi \in \mathcal{H}(W)$ and

$$\left\| \psi^{(-a)} \right\|_{L^2(\mathbb{R}^2)} \leq (1 + \sqrt{a}) \|\psi\|_{\mathcal{H}(W)}. \tag{9.6}$$

Integrating this inequality with respect to a, we get

$$\|\psi\|_{L^2(L_{d_0})} \leq \left[d_0 + 2d_0^{3/2}/3\right]\|\psi\|_{\mathcal{H}(W)}. \tag{9.7}$$

As usual, $H^1(L_{d_0})$ denotes the Sobolev space supplied with the inner product

$$(\phi, \psi)_{H^1(L_{d_0})} = \int_{L_{d_0}} (\phi\psi + \nabla\phi \cdot \nabla\psi)\, \mathrm{d}x\mathrm{d}y\mathrm{d}z.$$

From (9.4) and (9.7), it follows that restricting $\psi \in \mathcal{H}(W)$ to L_{d_0} we obtain a function belonging to $H^1(L_{d_0})$ and

$$\|\psi\|_{H^1(L_{d_0})} \leq \left[1 + d_0 + 2d_0^{3/2}/3\right]\|\psi\|_{\mathcal{H}(W)}. \tag{9.8}$$

Another space to be used in what follows is the Sobolev space $H^s = H^s(F)$, which consists of functions $h \in L^2(F)$ having finite the following norm:

$$\|h\|_s = \left[\int_{\mathbb{R}^2} (1 + |\sigma|^2)^s |\hat{h}(\sigma)|^2\, \mathrm{d}\sigma\right]^{1/2},$$

where $\sigma = (\xi, \zeta)$, $|\sigma|^2 = \xi^2 + \zeta^2$, and $\hat{h}(\sigma)$ is the Fourier transform of $h(x, z)$.

By Parseval's equality we have that $H^0 = L^2(F)$, and the well-known Sobolev's embedding theorem (see, for example, Gilbarg and Trudinger [94]) guarantees that

$$\left\|\psi^{(0)}\right\|_{1/2} \leq C(d_0)\|\psi\|_{H^1(L_{d_0})}.$$

This and (9.8) imply that for $\psi \in \mathcal{H}(W)$ we have that $\psi^{(0)} \in H^{1/2}$ and the following inequality,

$$\left\|\psi^{(0)}\right\|_{1/2} \leq C(d_0)\|\psi\|_{\mathcal{H}(W)}, \tag{9.9}$$

holds.

9.1.2. Auxiliary Steady-State Problem

For studying problem (9.1)–(9.3) we need an auxiliary problem:

$$\begin{aligned} \nabla^2\phi &= 0 \quad \text{in } W, \quad \phi = \varphi \text{ on } F, \\ \partial\phi/\partial n &= v \quad \text{on } B \cup S, \end{aligned} \tag{9.10}$$

where $\varphi \in H^{1/2}$. By $\mathcal{H}_0(W)$ we denote the subspace of $\mathcal{H}(W)$ consisting of functions that are equal to zero on F.

If $\phi \in \mathcal{H}(W)$ satisfies

$$\int_W \nabla\phi \cdot \nabla\psi \, \mathrm{d}x\mathrm{d}y\mathrm{d}z = \int_{B\cup S} v\psi \, \mathrm{d}S \tag{9.11}$$

for all $\psi \in \mathcal{H}_0(W)$ and $\phi^{(0)} = \varphi$ holds, then we call ϕ a weak solution of (9.10).

It is well known (see, for example, Gilbarg and Trudinger [94]) that a weak solution ϕ is a harmonic function in W, and ϕ is smooth up to $B \cup S$ and satisfies the boundary condition on $B \cup S$. It follows from (9.9) that the trace on F is defined for any $\phi \in \mathcal{H}(W)$, and so the boundary condition $\phi^{(0)} = \varphi$ is meaningful. Besides, if φ is smooth enough, then ϕ is smooth up to F and the boundary condition on F holds in the classic sense.

By ℓ we denote the linear functional

$$\mathcal{H}_0(W) \ni \psi \mapsto \ell(\psi) = \int_{B\cup S} v\psi \, \mathrm{d}S, \tag{9.12}$$

and we let $\|\ell\|$ be its norm. By virtue of the Sobolev embedding theorem mentioned in Subsection 9.1.1 we get that

$$\|\psi\|_{L^2(B\cup S)} \leq C\|\psi\|_{\mathcal{H}(W)},$$

and so ℓ is a bounded functional because v is continuous function having a compact support.

The following theorem solves the question about the unique solvability of the auxiliary problem (9.10).

For every $\varphi \in H^{1/2}$ *there exists a unique weak solution* $\phi \in \mathcal{H}(W)$ *of* (9.10), *and the following inequality,*

$$\|\phi\|_{\mathcal{H}(W)} \leq C\|\varphi\|_{1/2} + \|\ell\|, \tag{9.13}$$

holds for this solution.

For proving the uniqueness, we substitute $\psi = \phi$ into (9.11), where $v = 0$. This immediately gives that $\nabla\phi = 0$ in W, and so $\phi = \text{const}$. Since $\phi^{(0)} = \varphi = 0$, we get that $\phi = 0$, which completes the proof of uniqueness.

Let us turn to the existence of the solution. First, we note that results concerning extension of functions (see, for example, Gilbarg and Trudinger [94]) imply that for any $\varphi \in H^{1/2}$ there exists $\phi_1 \in \mathcal{H}(W)$ such that

$$\phi_1^{(0)} = \varphi, \quad \|\phi_1\|_{\mathcal{H}(W)} \leq C\|\varphi\|_{1/2}. \tag{9.14}$$

Boundedness of ℓ implies that the following functional,

$$\ell_1 : \mathcal{H}_0(W) \ni \psi \mapsto \ell(\psi) - (\phi_1, \psi)_{\mathcal{H}(W)},$$

is bounded. Then the Riesz theorem gives that $\phi_2 \in \mathcal{H}_0(W)$ exists such that

$$\|\phi_2\|_{\mathcal{H}(W)} = \|\ell_1\| \leq \|\ell\| + \|\phi_1\|_{\mathcal{H}(W)}, \tag{9.15}$$

and $\ell_1(\psi) = (\phi_2, \psi)_{\mathcal{H}(W)}$ for every $\psi \in \mathcal{H}(W)$. The latter equality implies that (9.11) holds for $\phi = \phi_1 + \phi_2$. Then ϕ is a solution of (9.10) because $\phi \in \mathcal{H}(W)$ and $\phi^{(0)} = \varphi$. At last, combining (9.14) and (9.15), one obtains (9.13), which completes the proof.

It is clear that the solution proved to exist (see the previous theorem) is as smooth in L_{d_0} as the smoothness of φ allows, which is made precise in the following assertion.

Let $\varphi \in H^s$, $s \geq 1/2$, *and let a be an arbitrary number from* $[0, d_0 - \delta]$, *where* $0 < \delta < d_0$. *Then the weak solution of* (9.10) *satisfies the following estimate,*

$$\left\| \left(\partial_y^j \phi\right)^{(-a)} \right\|_{s-j} \leq C(s, j, \delta) \left[\|\varphi\|_s + \|\ell\|\right], \tag{9.16}$$

for every integer $j \geq 0$.

The weak solution of (9.10) is harmonic function in L_{d_0}, and (9.6) and (9.13) imply that

$$\left\|\phi^{(-d_0)}\right\|_{L^2(\mathbb{R}^2)} \leq C\left[\,\|\varphi\|_{1/2} + \|\ell\|\,\right]. \tag{9.17}$$

Applying the Fourier transform in x and z – we will denote it $F(\cdot)$ when necessary to simplify the notation – to ϕ, we use its harmonicity and the Fourier transforms of its boundary values on sides of L_{d_0}, and we obtain that

$$\hat{\phi} = \left[\hat{\varphi} \sinh|\sigma|(y + d_0) - F\left(\phi^{(-d_0)}\right) \sinh|\sigma| y\right] / \sinh|\sigma| d_0. \tag{9.18}$$

Combining two obvious inequalities,

$$\left[\partial_y^j \sinh|\sigma|(y + d_0)\right]^{(-a)} / \sinh|\sigma| d_0 \leq C(j)(1 + |\sigma|^2)^{j/2},$$
$$\left[\partial_y^j \sinh|\sigma| y\right]^{(-a)} / \sinh|\sigma| d_0 \leq C(s, j, \delta)(1 + |\sigma|^2)^{(s-j)/2},$$

with (9.15) and (9.17) proves (9.16).

9.2. Operator Equation for the Unsteady Problem

First, we recall two facts: first, if $\phi \in \mathcal{H}(W)$, then (9.9) implies that $\phi^{(0)} \in H^{1/2}$, second, if $\phi \in \mathcal{H}(W)$ satisfies (9.11) for every $\psi \in \mathcal{H}(W)$, then the last assertion in Subsection 9.1.2 implies that $\phi_y^{(0)} \in H^{-1/2}$. According to these remarks the following definition is meaningful.

By $C^k([0, T], H^s)$ we denote the set of functions depending on t and having values in H^s that are assumed to be continuous together with their

t derivatives up to the order k. Let f, f_0, and f_1 be three functions belonging to $C^0([0,T], H^0)$, $H^{1/2}$, and H^0, respectively. Then ϕ is said to be a weak solution of (9.1)–(9.3), if the following conditions are fulfilled: (i) $\phi \in \mathcal{H}(W)$ for $t \in [0,T]$; (ii) ϕ satisfies (9.11) for every $\psi \in \mathcal{H}_0(W)$; (iii) the trace on the free surface $\phi^{(0)}$ belongs to $C^k([0,T], H^{(1-k)/2})$ for $k = 0, 1, 2$; and (iv) conditions (9.2) hold and the first of them means that for every $t \in [0,T]$ both sides coincide as the elements of $H^{-1/2}$.

Let us introduce an operator required for reducing the unsteady problem to an operator equation.

By K we denote the Dirichlet–Neumann operator mapping $H^{1/2}$ into $H^{-1/2}$ and defined as follows:

$$K\varphi = \phi_y^{(0)}, \tag{9.19}$$

where $\phi \in \mathcal{H}(W)$ solves (9.10) for zero Neumann data on $B \cup S$. The following lemma gives the properties of K to be used below.

For any $s \geq 1/2$, K is a bounded linear operator mapping H^s into H^{s-1} and such that $K = \eta + T$. Here $\eta = F^{-1}\,\hat{\eta}\,F(\cdot)$, $\hat{\eta}$ is the operator of multiplication by $|\sigma| \coth |\sigma| d_0$, and the properties of η and T are as follows.

Let $\lambda_0 > 0$ and I be the identity operator; then $(\lambda_0 I + \eta)^{-1}$ is a bounded operator mapping H^{s-1} into H^s for any s, and T is such that the estimate

$$\|T\varphi\|_\alpha \leq C_\alpha \|\varphi\|_{1/2}$$

holds for any α.

Since (9.10) is a linear problem, K is a linear operator, and from (9.16), where $\|\ell\| = 0$ because $v = 0$, it follows that K is bounded. The required representation is a consequence of (9.18), which gives that

$$T\varphi = -F^{-1}\left(|\sigma| F\left(\phi^{(-d_0)}\right) / \sinh |\sigma| d_0\right).$$

The estimate for T immediately follows from the definition of norm in H^s, Parseval's equality, and (9.17), where $\|\ell\| = 0$. From the following inequality,

$$(\lambda_0 + |\sigma| \coth |\sigma| d_0)^{-1} \leq C(\lambda_0)(1 + |\sigma|^2)^{-1/2},$$

one immediately obtains that $(\lambda_0 I + \eta)^{-1}$ is bounded in the indicated spaces, which completes the proof.

Let us turn to the unsteady problem (9.1)–(9.3) and begin with getting rid of the inhomogeneity in (9.3). For this purpose a solution of (9.1)–(9.3) is sought in the form $\phi = \phi_1 + \phi_2$, where ϕ_1 is the weak solution of (9.10v) [we refer to the auxiliary problem (9.10) as (9.10v) and (9.10φ), if $\varphi = 0$ and

$v = 0$, respectively]. Then ϕ_2 must satisfy the following problem:

$$\nabla^2 \phi = 0 \quad \text{in } W, \tag{9.20}$$

$$\left.\begin{aligned} \phi_{tt} + \phi_y = f_* \quad \text{for } t > 0, \\ \phi = f_0, \phi_t = f_1 \quad \text{for } t = 0 \end{aligned}\right\} \quad \text{on } F, \tag{9.21}$$

$$\partial\phi/\partial n = 0 \quad \text{on } B \cup S, \tag{9.22}$$

where $f_* = f - (\partial_y \phi_1)^{(0)}$. According to the last assertion in Subsection 9.1.2, we have that $f_* \in C^0([0, T], H^0)$ when f is from this class.

Now, let ϕ be weak solution of (9.20)–(9.22) and $\varphi = \phi^{(0)}$; then

$$\varphi \in C^k\big([0, T], H^{(1-k)/2}\big), \quad k = 0, 1, 2, \tag{9.23}$$

and (9.21) can be written in the form

$$\left.\begin{aligned} \varphi_{tt} + K\varphi = f_* \quad \text{for } t \geq 0, \\ \varphi = f_0, \ \varphi_t = f_1 \quad \text{for } t = 0 \end{aligned}\right\}. \tag{9.24}$$

Hence, any weak solution of (9.20)–(9.22) is a solution of the auxiliary problem (9.10φ), where φ satisfies (9.23) and (9.24). It is obvious that the converse is also true, that is, if ϕ is weak solution of (9.10φ) and φ satisfies (9.23) and (9.24), then ϕ is weak solution of (9.20)–(9.22), and so the following theorem is proved.

Let $f \in C^0([0, T], H^0)$, $f_0 \in H^{1/2}$, and $f_1 \in H^0$; then the following two assertions are equivalent: (i) ϕ is weak solution of (9.1)–(9.3); *(ii) $\phi = \phi_1 + \phi_2$, where ϕ_1 is weak solution of* (9.10v) *and ϕ_2 is weak solution of* (9.10φ)*; here φ satisfies* (9.23) *and solves* (9.24)*, where*

$$f_* = f - (\partial_y \phi_1)^{(0)} \in C^0([0, T], H^0). \tag{9.25}$$

This result, together with the first theorem in Subsection 9.1.2, yields the existence and uniqueness of weak solution of (9.1)–(9.3) provided the unique solvability of (9.24) is proven in classes (9.23), and so in the next section, we investigate in detail properties of solutions to (9.24), but we have to consider further properties of K first.

9.3. Main Results

9.3.1. Further Properties of the Operator K

We preserve the notation K for an unbounded operator in H^0 having the domain H^1 and defined by (9.19). For K understood in this way, the properties used below are given in the following lemma.

The operator K is self-adjoint and nonnegative in H^0, and for any $s \geq 0$ there exists a constant a_s such that

$$a_s^{-1}\|\varphi\|_s \leq \|(I+K)^s\varphi\|_0 \leq a_s\|\varphi\|_s \tag{9.26}$$

for every $\varphi \in H^s$.

First, let us show that if ϕ is weak solution of (9.10φ), where $\varphi \in H^1$, then

$$\int_W \nabla\phi \cdot \nabla\psi \,\mathrm{d}x\mathrm{d}y\mathrm{d}z - \int_F \phi_y \psi \,\mathrm{d}x\mathrm{d}z = 0 \tag{9.27}$$

holds for every $\psi \in \mathcal{H}(W)$. Let $\{\varphi_k\}$ be a sequence of infinitely differentiable functions on F such that φ_k tends to φ in H^1 as $k \to \infty$, and let ϕ_k be solutions of (9.10φ_k). The last theorem in Subsection 9.1.2 implies that ϕ_k is infinitely differentiable in $\overline{L_{d_0}}$, and so ϕ_k is harmonic in L_{d_0} and the following relations hold:

$$\|\nabla(\phi_k - \phi)\|_{L^2(L_{d_0})} \to 0, \qquad \left\|(\phi_k - \phi)_y^{(0)}\right\|_{L^2(F)} \to 0 \quad \text{as } k \to \infty. \tag{9.28}$$

Let $\psi \in C^\infty(\bar{W})$ be equal to zero at infinity. We split ψ into a sum $\psi_1 + \psi_2$, where $\psi_i \in C^\infty(\bar{W})$ vanishes at infinity, and also $\psi_1 = 0$ when $y = 0$ and $\psi_2 = 0$ when $y = -d_0$. Letting k to infinity in Green's formula

$$\int_{L_{d_0}} \nabla\phi_k \cdot \nabla\psi_2 \,\mathrm{d}x\mathrm{d}y\mathrm{d}z - \int_F \partial_y\phi_k\, \psi_2 \,\mathrm{d}x\mathrm{d}z = 0,$$

we get (9.27) for $\psi = \psi_2$. For $\psi = \psi_1$, the same follows by the definition of weak solution of (9.10φ), and so (9.27) holds for any $\psi \in C^\infty(\bar{W})$ vanishing at infinity. Since such functions are dense in $\mathcal{H}(W)$, (9.27) holds for all $\psi \in \mathcal{H}(W)$.

It is obvious that (9.27) can be written in the form

$$\int_W \nabla\phi \cdot \nabla\psi \,\mathrm{d}x\mathrm{d}y\mathrm{d}z = \int_F \psi K\varphi \,\mathrm{d}x\mathrm{d}z,$$

where $\varphi \in H^1$, $\psi \in \mathcal{H}(W)$, and ϕ is weak solution of (9.10φ). Therefore, K is a symmetric, nonnegative operator having the domain H^1. In order to show that K is self-adjoint it is sufficient to verify that its deficiency indices are equal to zero. For this purpose, in its turn, it is sufficient to demonstrate that $\lambda_0 I + K$, where λ_0 is a positive number, maps H^1 onto $L^2(F)$; that is, the equation

$$(\lambda_0 I + K)\varphi = h \tag{9.29}$$

is solvable in H^1 for any $h \in H^0$. Let us write (9.29) in the form

$$(\lambda_0 I + \eta + T)\varphi = h,$$

which is possible by virtue of the first assertion in Section 9.2, and let us seek a solution of the latter equation in the form $\varphi = (\lambda_0 I + \eta)^{-1}\psi$. Applying the first assertion in Section 9.2 again, we see that $\varphi \in H^1$ if $\psi \in H^0$. Now, the unknown function ψ must satisfy

$$\psi + T(\lambda_0 I + \eta)^{-1}\psi = h.$$

Since the relation

$$(1 + |\sigma|^2)^{1/4}(\lambda_0 + |\sigma| \coth |\sigma| d_0)^{-1} \to 0 \quad \text{as } \lambda_0 \to \infty$$

holds uniformly in $|\sigma| \in \mathbb{R}$, the norm of $(\lambda_0 I + \eta)^{-1}$ acting from H^0 into $H^{1/2}$ tends to zero as $\lambda_0 \to \infty$. This and the fact that T is a bounded operator from $H^{1/2}$ into H^0 (see the first assertion in Section 9.2) prove that (9.29) is solvable when $\lambda_0 > 0$ is sufficiently large. This completes the proof that K is a self-adjoint operator.

It remains for us to prove (9.26) and this requires the following two facts. First, $I + K$ is a bounded operator mapping H^n into H^{n-1}, where n is a positive integer. Second, $I + K$ considered in fact 1 has a bounded inverse operator. Since the first assertion in Section 9.2 implies fact 1, let us turn to proving fact 2. In order to show that the image of $I + K$ is the whole space H^{n-1}, we have to demonstrate that

$$(I + K)\varphi = f \tag{9.30}$$

is solvable in H^n for every $f \in H^{n-1}$. This equation has a solution $\varphi \in H^1$ because K is nonnegative. Now let us write (9.30) in the form

$$(I + \eta)\varphi = f - T\varphi. \tag{9.31}$$

Then the estimate for T obtained in the first assertion in Section 9.2 gives that $f - T\varphi \in H^{n-1}$, and so the part of the same assertion concerning $(I + \eta)^{-1}$ can be applied to (9.31) providing that $\varphi \in H^n$. Thus (9.30) has a solution in H^n for any $f \in H^{n-1}$; that is, $I + K$ and its inverse are bounded, and so (9.26) holds when s is an arbitrary positive integer. Since it also holds for $s = 0$, the well-known results on interpolation (see, for example, Gilbarg and Trudinger [94]) imply that (9.26) is true for every $s \geq 0$. The proof is complete.

9.3.2. Existence of a Unique Weak Solution

The aim of this subsection is to prove the following theorem.

Let $f \in C^0([0, T], H^0)$, $f_0 \in H^{1/2}$, and $f_1 \in H^0$; then there exists a unique weak solution ϕ of (9.1)–(9.3), and we have that $\phi = \phi_1 + \phi_2$, where ϕ_1

and ϕ_2 *are weak solutions of* (9.10v) *and* (9.10φ), *respectively, where*

$$\varphi = \int_0^t K^{-1/2} \sin\left[(t-\tau)K^{1/2}\right] f_*(\tau)\,\mathrm{d}\tau + \cos\left(tK^{1/2}\right) f_0 + K^{-1/2} \sin\left(tK^{1/2}\right) f_1 \tag{9.32}$$

and

$$f_* = f - (\partial_y \phi_1)^{(0)}. \tag{9.33}$$

For proving the uniqueness of the weak solution to (9.1)–(9.3), let us consider $\varphi = \phi^{(0)}$, where ϕ solves the corresponding homogeneous problem. According to the definition of the weak solution we have that $\varphi \in C^k([0,T], H^{(1-k)/2})$, $k = 0, 1, 2$. Thus if $\psi = \int_0^t \varphi(\tau)\,\mathrm{d}\tau$, then

$$\psi \in C^{k+1}\left([0,T], H^{(1-k)/2}\right), \quad k = 0, 1, 2. \tag{9.34}$$

The first theorem in Subsection 9.1.2 and the last theorem in Section 9.2 show that $\varphi_{tt} + K\varphi = 0$. Integrating this equation with respect to t and using the initial condition $\varphi_t(x, z, 0) = 0$, we obtain that $\psi_{tt} + K\psi = 0$. This and (9.34) imply that

$$(\psi_{tt}, \psi_t)_{H^0} + (K\psi, \psi_t)_{H^0} = 0. \tag{9.35}$$

Since K is a positive operator, the second term on the right-hand side can be written as follows:

$$\left(K^{1/2}\psi,\ K^{1/2}\psi_t\right)_{H^0} = \left(K^{1/2}\psi,\ \left(K^{1/2}\psi\right)_t\right)_{H^0},$$

and so (9.35) takes the form

$$\frac{\mathrm{d}}{\mathrm{d}t}\left(\|\psi_t\|_0^2 + \|K^{1/2}\psi\|_0^2\right) = 0,$$

which means that

$$\|\psi_t\|_0^2 + \|K^{1/2}\psi\|_0^2 = \text{const}. \tag{9.36}$$

Since $\psi = \psi_t = 0$ when $t = 0$, both terms on the right-hand side in (9.36) do vanish identically. Therefore, $\psi_t(x, z, t) = \varphi(x, z, t) = 0$, and the uniqueness of the weak solution to (9.1)–(9.3) follows from the first theorem in Subsection 9.1.2 and the last theorem in Section 9.2.

Let us turn to the existence of the solution. In Subsection 9.3.3, it will be demonstrated that

$$\left\| \partial_t^n \varphi \right\|_{s-n/2} \le C \Bigg\{ \|f_0\|_s + \|f_1\|_{s-1/2} + \sum_{j=0}^{n-2} \left\| \partial_t^j f_* \right\|_{s-1-j/2} + \int_0^t \|f_*(\tau)\|_{s-1/2} \, \mathrm{d}\tau \Bigg\} \tag{9.37}$$

holds for φ defined by (9.32) and $0 \le n \le 2s + 1$. Moreover, considerations leading to (9.37) show that usual rules are applicable for differentiating (9.32), and so this function satisfies (9.23) [to demonstrate this it is sufficient to set $s = 1/2$ in (9.37) and to note that ϕ_1 does not depend on t] and solves (9.24). Then the last theorem in Section 9.2 guarantees that ϕ defined in theorem's formulation solves (9.1)–(9.3). Thus we need (9.37) for completing the proof.

9.3.3. Regularity of Solution

Here we complete the investigation of the unsteady problem (9.1)–(9.3) by proving the regularity theorem for solutions of (9.24).

Let $s \ge 1/2$ and let $f \in C^k([0, T], H^{s-(k+1)/2})$ for all k such that $0 \le k \le 2s - 1$, $f_0 \in H^s$, and $f_1 \in H^{s-1/2}$. If ϕ solves (9.1)–(9.3), *then $\varphi(t) = \phi^{(0)}$ belongs to $C^n([0, T], H^{s-n/2})$ for all n such that $0 \le n \le 2s + 1$, and the following estimate,*

$$\left\| \partial_t^n \varphi \right\|_{s-n/2} \le C \Bigg\{ \|f_0\|_s + \|f_1\|_{s-1/2} + \|\ell\| + \sum_{j=0}^{n-2} \left\| \partial_t^j f \right\|_{s-1-j/2} + \int_0^t \big(\|f(\tau)\|_{s-1/2} + \|\ell\| \big) \, \mathrm{d}\tau \Bigg\}, \tag{9.38}$$

holds, where $\|\ell\|$ is the norm of the linear functional defined by (9.12).

Before proving this theorem, we note that it implies two facts: first, the proof of the unique solvability theorem in Subsection 9.3.2 is complete because (9.38) justifies (9.37); second, estimate (9.16) holds for all $a \in [0, d_0]$.

First, it is convenient to assume that

$$f_0 = f_1 = 0, \qquad f_* \in C^k\big([0, T], H^{s-(1+k)/2}\big),$$

where f_* is defined by (9.33). Let us consider

$$h(t) = \left[(I + K)^{s-1/2} f_* \right](t), \tag{9.39}$$

which allows us to write (9.32) in the form

$$\varphi = (I + K)^{-s+n/2} \int_0^t (I + K)^{(1-n)/2} K^{-1/2} \sin\left[(t - \tau)K^{1/2}\right] h(\tau)\, d\tau. \tag{9.40}$$

Here we also used the fact that functions of K do commute and we took into account the assumption that $f_0 = f_1 = 0$. Since

$$(1 + \lambda)^{1/2} \lambda^{-1/2} \sin(t - \tau)\lambda^{1/2}$$

is a bounded function for $\lambda > 0$, the operator in the integrand in (9.40) is a bounded operator in H^0 when $n = 0$. Therefore, combining (9.40), where $n = 0$, with (9.26), we obtain that

$$\|\varphi\|_s \leq C \int_0^t \|h(\tau)\|_0\, d\tau.$$

Using (9.39) and applying (9.26) once more, we get (9.37) for $n = 0$ under the assumption that $f_0 = f_1 = 0$. If $0 < n \leq 2s + 1$, then one can differentiate (9.40) n times, which gives

$$\partial_t^n \varphi = (I + K)^{-s+n/2} \int_0^t (I + K)^{(1-n)/2} K^{(n-1)/2} s_n\left[(t - \tau)K^{1/2}\right] h(\tau)\, d\tau$$
$$+ \sum_{j=0}^{n-2} C_j \partial_t^j \left\{\left[(I + K)^{-s+1/2} K^{(n-j-2)/2} h\right](t)\right\},$$

where $s_n(\lambda) = \sin(\lambda + \pi n/2)$ and $C_j,\ j = 0, 1, \ldots, n-2$, are constants. The integral in the last equality can be estimated in the same way as that in (9.40). Let us substitute h expressed through f_* [see (9.39)] into the last sum and apply (9.26). This proves (9.37) for $0 < n \leq 2s + 1$, provided $f_0 = f_1 = 0$. If $f_* = 0$, but f_0 and $f_1 = 0$ are nonzero functions, then (9.37) is an immediate consequence of (9.26), which completes the proof.

9.4. Bibliographical Notes

The results presented in this chapter were obtained by Garipov [92]. A slightly different approach leading to similar results was developed by Friedman and Shinbrot [90]. Lau [180] extended the approach applied in [90], allowing the bottom to intersect the free surface of water, but there are restrictions on the bottom in a neighborhood of the shoreline. The bottom must intersect the free surface vertically in the three-dimensional case, but in the two-dimensional problem a less restrictive assumption is imposed. The bottom must be presented by a straight segment in a neighborhood of the intersection point, and this segment must be inclined at an angle less than $\pi/2$ to the free surface.

Let us list some other works treating unsteady water waves. Much attention was paid to the investigation of various statements of the Cauchy–Poisson problem in water of constant depth when an explicit solution could be obtained by virtue of integral transforms. In particular, such formulae were use by Isakova [122, 123], who justified the limiting amplitude principle for the three-dimensional Cauchy–Poisson problem with the homogeneous initial data, and by Clarisse, Newman, and Ursell [39], who considered the classical problem of two- and tree-dimensional waves generated by an impulsive pressure. They concentrated on the asymptotic behavior at large distances and large times in a zone near the wave front and developed a systematic procedure for successive terms in an asymptotic expansion.

A more complicated case of the three-dimensional Cauchy–Poisson problem was investigated by Garipov [93]. In his work, the bottom is assumed to be flat but having a ridge-like protrusion. Assuming the protrusion's height to be sufficiently small, Garipov investigated the asymptotic behavior of waves for large times. The same question of asymptotic behavior was treated by Buslayeva [34] for the two-dimensional problem under the assumption that local intrusions of the bottom were present as well as local protrusions.

An interesting contribution to the theory of unsteady waves was made as early as in the 1910 by Hadamard [102]. In three brief notes, he derives what is now known as various versions of Hadamard's equation. In the case in which water occupies a basin having vertical walls, this integrodifferential equation becomes a differential equation, which was demonstrated by Bouligand in his doctoral thesis (see the book [31], where Bouligand summarized his studies). A brief account of these works can be found in Section 14.4 of Hadamard's biography [228] by Maz'ya and Shaposhnikova.

Maslov's method was applied to various statements of the Cauchy–Poisson problem under the assumption that the bottom is slowly varying in the horizontal direction. The corresponding results can be found in the following papers: Dobrokhotov [48, 49], Dobrokhotov and Zhevandrov [51, 52], Dobrokhotov, Zhevandrov, and Kuz'mina [53, 370], and Dobrokhotov et al. [54].

The coupled problem describing unsteady water waves over an elastic bottom was considered by Dobrokhotov, Tolstova, and Chudinovich [50].

In a number of papers, the linearized initial-boundary value problem describing waves in the presence of a surface-piercing body is considered, and we list some of them. The limiting amplitude principle applied to the motion of floating bodies was justified by Vullierme-Ledard [349]. Ursell [327] and

Maskell and Ursell [221] considered unsteady waves in the presence of a freely floating body and investigated the decay of a body's motion at large time. A kind of wave equation was obtained by Athanassoulis and Makrakis [7], who applied a time-dependent complex potential. Another paper [8] by these authors treats the problem with the help of a series expansion of that potential. Jami [124] combines a theoretical approach to a problem of unsteady waves with numerical modeling. A brief survey of other works treating unsteady waves was given by Euvrard [65].

10

Waves Caused by Rapidly Stabilizing and High-Frequency Disturbances

Results presented in Chapter 9 provide no details of the transient behavior of flows and do not yield direct hydrodynamic corollaries. However, there are situations in which information about developing waves in time can be extracted so that it leads to specific properties of hydrodynamic characteristics. In particular, an asymptotic analysis allows us to do this at least for two classes of disturbances. One of these classes constitutes rapidly stabilizing disturbances (this class includes brief disturbances as an important subclass), and the second class is formed by high-frequency disturbances. Both of these classes can be treated by using the same technique of two-scale asymptotic expansions for velocity potentials. The latter allows us to derive principal terms in asymptotics of some hydrodynamic characteristics.

10.1. Rapidly Stabilizing Surface Disturbances

In this section we are concerned with the effect of rapidly stabilizing disturbances on magnitudes characterizing unsteady water waves. For this purpose we consider several initial-boundary value problems describing waves caused by surface and underwater disturbances. The main example of the first kind is given by a pressure system applied to the free surface at the initial moment and rapidly stabilizing to a given distribution (a particular case is an impulsive pressure system). Underwater disturbances are presented by a source having a strength rapidly stabilizing in time to a constant value, and a rapidly stabilizing bottom movement. Complete asymptotic expansions in powers of a nondimensional small duration of disturbance are constructed for velocity potentials. These expansions allow us to obtain asymptotic formulae for various characteristics of waves and to interpret their principal terms in hydrodynamic terms. For simple geometries, there are explicit formulae for terms in asymptotic expansions.

10.1.1. Objectives for the Application of Asymptotic Methods

Let us give a brief account of results obtained in this section for the case of an impulsive surface pressure. We assume that water (or, generally speaking, an inviscid, incompressible fluid of density ρ) occupies an infinite domain W that is a part of a layer outside a finite number of bounded domains occupied by totally submerged rigid bodies. Let Cartesian coordinates (x, y, z) be chosen so that

$$F = \{-\infty < x, z < +\infty, y = 0\}$$

corresponds to the free surface at rest bounding W from above, and the y axis is directed vertically upward. Two other parts of the boundary ∂W are sufficiently smooth surfaces B and S placed at a certain finite distance from F, and so $\partial W = F \cup B \cup S$, but either B or S may be empty. The seabed B is unbounded and separates the water layer of variable depth from the lower rigid part of the half-space $\{y < 0\}$, whereas S is the wetted boundary of an immersed body (or of a finite number of bodies).

The unsteady motion of water is formulated (see the Linearized Unsteady Problem section in the Introduction) in terms of a velocity potential $\phi(P;t)$, $P = (x, y, z)$, and we assume it to belong to the class of functions having a finite kinetic and potential energy:

$$\int_W |\nabla\phi|^2 \,\mathrm{d}x\mathrm{d}y\mathrm{d}z + \int_F \eta^2 \,\mathrm{d}x\mathrm{d}z < \infty. \tag{10.1}$$

Here η is the free surface elevation linked to ϕ by a linearized Bernoulli's equation:

$$\eta(x, z; t) = -\left[\partial_t\phi(x, 0, z; t) + p(x, z; t)\right], \tag{10.2}$$

where p stands for pressure. Throughout this chapter we use nondimensional magnitudes (they are specified in each section), and so there are no ρ and g in (10.2). Another relation between η and ϕ is the linearized kinematic condition on the free surface:

$$-\partial_y\phi(x, 0, z; t) + \partial_t\eta(x, z; t) = 0.$$

This and (10.2) combine to give

$$\partial_t^2\phi + \partial_y\phi = -\partial_t p \quad \text{on } F \text{ for } t \geq 0. \tag{10.3}$$

The continuity equation for the velocity field implies that ϕ satisfies the Laplace equation

$$\nabla^2\phi = 0 \quad \text{in } W \text{ for } t \geq 0. \tag{10.4}$$

There is no flow through any rigid surface and so

$$\partial_n \phi = 0 \quad \text{on } B \cup S \text{ for } t \geq 0, \tag{10.5}$$

where ∂_n indicates differentiation with respect to a unit normal on the surface. For motions starting from the rest, equations (10.3)–(10.5) must be complemented by two initial conditions:

$$\phi(x, 0, z; 0) = 0, \tag{10.6}$$

$$\partial_t \phi(x, 0, z; 0) = 0. \tag{10.7}$$

Our aim is to apply the technique of the singular perturbations theory to (10.3)–(10.7) in order to rigorously justify the widely known heuristic consideration deriving the nonhomogeneous initial condition (10.6) from (10.2). A typical pattern of the hydrodynamic approach to this condition is given by Stoker [312], pp. 149–150 (see also Lamb [179], Section 11):

> In water wave problems it is of particular interest to consider cases in which the motion of the water is generated by applying an impulsive pressure to the surface when the water is initially at rest. To obtain the condition appropriate for an initial impulse we start from (10.2) and integrate it over the small interval $0 \leq t \leq \epsilon$. The result is
>
> $$\int_0^\epsilon p \, \mathrm{d}t = -\phi(x, 0, z, \epsilon) - \int_0^\epsilon \eta \, \mathrm{d}t, \tag{10.8}$$
>
> since $\phi(x, y, z, 0)$ can be assumed to vanish. One now imagines that $\epsilon \to +0$ while $p \to \infty$ in such a way that the integral on the left tends to a finite value – the impulse I per unit area. Since it is natural to assume that η is finite it follows that the integral on the right vanishes as $\epsilon \to +0$, and we have the formula
>
> $$I = -\phi(x, 0, z, +0) \tag{10.9}$$
>
> for the initial impulse per unit area at the free surface in terms of the value of ϕ there. If I is prescribed on the free surface (together with appropriate conditions at other boundaries), it follows that $\phi(x, y, z, +0)$ can be determined, or, in other words, the initial velocity of particles is known.

Despite the fact that this argument looks very convincing, it gives rise to some questions:

- How should the limit be understood in (10.8) as $\epsilon \to 0$?
- Why do we get the finite value of the force

  $$\mathbf{F}(t) = \int_S \phi_t \, \mathbf{n} \, \mathrm{d}S \tag{10.10}$$

  acting on a submerged body, when $\mathbf{F}(t)$ is calculated by virtue of ϕ determined with the help of (10.9), in spite of the fact that the pressure p is infinite at the initial moment?

These and other questions are answered here by using two-scale asymptotic expansions. To give an idea of the results obtained below, we present a simple consequence concerning (10.3)–(10.7). Let p in (10.3) be prescribed as follows:

$$p(x, z; t, \epsilon) = \epsilon^{-1} Q(t/\epsilon) I(x, z), \tag{10.11}$$

where $I(x, z)$ is a smooth function decaying at infinity, and $Q(\tau)$ is a continuous function of $\tau \geq 0$, vanishing at $\tau = 0$, and tending to zero as $\tau \to \infty$ so that

$$\int_0^\infty Q(\mu)\, \mathrm{d}\mu = 1. \tag{10.12}$$

Under these assumptions, an infinite two-scale asymptotic expansion holds for the potential $\phi(P; t, \epsilon)$ satisfying this problem, but we restrict ourselves to the following initial terms,

$$\phi(P; t, \epsilon) = \phi(P; t, 0) - v(P) \int_{t/\epsilon}^\infty Q(\tau)\, \mathrm{d}\tau + O(\epsilon), \tag{10.13}$$

which are of importance for hydrodynamic corollaries. Here the velocity potential $\phi(P; t, 0)$ satisfies (10.9) and (10.7), but the right-hand-side term in (10.3) is zero, and v is a unique solution to the following time-independent problem:

$$\nabla^2 v = 0 \text{ in } W, \qquad v = -I \quad \text{on } F, \qquad \partial_n v = 0 \quad \text{on } S \cup B.$$

Now, we get the asymptotic formula for the force:

$$\mathbf{F}(t, \epsilon) = \mathbf{F}(t, 0) + \epsilon^{-1} Q(t/\epsilon) \int_S v\, \mathbf{n}\, \mathrm{d}S + O(\epsilon),$$

where $\mathbf{F}(t, 0)$ must be determined from (10.10). Thus, for a small initial interval, (10.10) gives only a negligible part of the force, and the principal term in the force asymptotics, tending to infinity as $\epsilon \to 0$, requires the knowledge of v. However, the second term rapidly tends to zero with time.

For other characteristics of waves, asymptotics can be also obtained from (10.13). In particular, we have that the energy of waves caused by an impulsive pressure evolves as follows:

$$E(t, \epsilon) = E(0, 0) \left[\int_0^{t/\epsilon} Q(\mu)\, \mathrm{d}\mu \right]^2 + O(\epsilon), \tag{10.14}$$

where

$$E(0,0) = 2^{-1}\left[\int_W |\nabla\phi(P;t,0)|^2\, dxdydz + \int_F |\partial_t\phi(x,0,z;t,0)|^2\, dxdz\right]$$
$$= 2^{-1}\int_W |\nabla v|^2\, dxdydz$$

is the constant energy of waves described by the limit potential $\phi(P;t,0)$.

In conclusion of this subsection we note that evolution of the principal term in the asymptotics of energy is similar to (10.14) in the case of a bottom movement. For a surface pressure system in the rapidly accelerating forward motion, we consider asymptotics for the wave-making resistance instead of that for energy.

The plan of the present section is as follows. A rapidly stabilizing surface pressure is considered in Subsection 10.1.2, and some other types of such disturbances (in particular, a bottom movement and a submerged source) are treated in Subsection 10.1.3. Each subsection begins with a derivation of a formal two-scale asymptotic expansion. Then hydrodynamic corollaries and justification of asymptotics (the remainder term is estimated in certain function spaces including the space of functions having a finite Dirichlet integral) are given.

10.1.2. Rapidly Stabilizing Surface Pressure

In this subsection, we first assume that the pressure in (10.3) is given in the following form:

$$p(x,z;t,\epsilon) = q(t/\epsilon)\mathcal{P}(x,z), \tag{10.15}$$

where $\mathcal{P}$ is a function decaying at infinity, and q is a differentiable function such that $q(0) = 0$ and $q(\tau) \to q(\infty) = \text{const}$ as $\tau \to \infty$ so that

$$\tau^m q'(\tau) \to 0 \quad \text{as } \tau \to \infty \quad \text{for } m = 1, 2, \ldots. \tag{10.16}$$

Nondimensional variables applied here are defined as follows. Let d be the diameter of a region in the (x, z) plane outside of which $\mathcal{P}$ is negligibly small. As characteristic values we take d for length, $(d/g)^{1/2}$ for the time interval, and $\rho d g$ for the pressure magnitude (we recall that ρ is water's density, and g is the acceleration gravity). Characteristic values for all other variables follow from those given here.

10.1.2.1. Formal Asymptotic Expansion

Let us consider the velocity potential $\phi(P; t, \epsilon)$ satisfying (10.4)–(10.7), and (10.3), where the right-hand-side term is defined by (10.15), and $\mathcal{P}$ is infinitely smooth. Since the pressure in the latter condition depends on the so-called rapid time $\tau = t/\epsilon$, we seek the potential in the form of a two-time-scaled asymptotic series:

$$\phi(P; t, \epsilon) = \sum_{m=0}^{\infty} \epsilon^m \left[\varphi_m(P; \tau) + \psi_m(P; t)\right], \tag{10.17}$$

where $\varphi_m(P; \tau)$ tends to zero as $\tau \to \infty$, $m = 0, 1, \ldots$, and these functions decay as $|P| = (x^2 + y^2 + z^2)^{1/2} \to \infty$, as well as $\psi_m(P; t)$.

Boundary value problems for φ_m and ψ_m must be obtained as follows. After substituting (10.17) into (10.3)–(10.7), we equate coefficients at each power of ϵ. Moreover, coefficients depending on τ and t are equated separately. Thus, we get the following equations for φ_m holding when $\tau \geq 0$:

$$\nabla^2 \varphi_m = 0 \quad \text{in } W, \qquad \partial_n \varphi_m = 0 \quad \text{on } B \cup S, m = 0, 1, \ldots; \tag{10.18}$$

$$\partial_\tau^2 \varphi_0 = 0, \qquad \partial_\tau^2 \varphi_0 = -q'(\tau)\mathcal{P}(x, z) \quad \text{on } F; \tag{10.19}$$

$$\partial_\tau^2 \varphi_m + \partial_y \varphi_{m-2} = 0 \quad \text{on } F, m = 2, 3, \ldots. \tag{10.20}$$

For $\psi_m, m = 0, 1, \ldots$, we arrive at the following equations valid for $t \geq 0$:

$$\nabla^2 \psi_m = 0 \quad \text{in } W, \qquad \partial_n \psi_m = 0 \quad \text{on } B \cup S, \tag{10.21}$$

$$\partial_t^2 \psi_m + \partial_y \psi_m = 0 \quad \text{on } F. \tag{10.22}$$

Besides, the initial relations must hold:

$$\psi_m(x, 0, z; 0) = -\varphi_m(x, 0, z; 0), \quad m = 0, 1, \ldots; \tag{10.23}$$

$$\partial_\tau \varphi_0(x, 0, z; 0) = 0; \tag{10.24}$$

$$\partial_t \psi_m(x, 0, z; 0) = -\partial_\tau \varphi_{m+1}(x, 0, z; 0), \quad m = 0, 1, \ldots. \tag{10.25}$$

A unique solution of the first equation (10.19) that satisfies (10.24) and tends to zero as $\tau \to \infty$ is $\varphi_0(x, 0, z; \tau) = 0$ for $\tau \geq 0$. From this and (10.18) we get

$$\varphi_0(P; \tau) = 0 \quad \text{in } W \text{ for } \tau \geq 0$$

as the only solution decaying as $|P| \to \infty$. Furthermore, (10.20) and the same argument provides that

$$\varphi_m(P; \tau) = 0 \quad \text{in } W \text{ for } \tau \geq 0 \text{ when } m = 0, 2, \ldots, 2k, \ldots. \tag{10.26}$$

Now integration of the second equation of (10.19) gives the boundary condition for φ_1 on F, which together with (10.18) gives

$$\varphi_1(P;\tau) = \beta_1(\tau)v_1(P), \tag{10.27}$$

where

$$\nabla^2 v_1 = 0 \quad \text{in } W, \qquad \partial_n v_1 = 0 \quad \text{on } B \cup S, \qquad v_1 = -\mathcal{P} \quad \text{on } F, \tag{10.28}$$

and here and below we use the following definition:

$$\beta_m(\tau) = (m!)^{-1} \int_\tau^\infty (\mu - \tau)^m \, q'(\mu)\, \mathrm{d}\mu, \tag{10.29}$$

which is consistent under assumption (10.16). Then seeking remaining functions φ_m in the form

$$\varphi_m(P;\tau) = \beta_m(\tau)v_m(P), \tag{10.30}$$

we see that (10.20) and (10.29) lead to a recurrent sequence of boundary value problems,

$$\begin{aligned} \nabla^2 v_m &= 0 \quad \text{in } W, \qquad \partial_n v_m = 0 \quad \text{on } B \cup S, \\ v_m &= -\partial_y v_{m-2} \quad \text{on } F,\, m = 3, 5, \ldots, 2k+1, \ldots, \end{aligned} \tag{10.31}$$

complementing (10.28).

Since (10.26), (10.27), and (10.29)–(10.31) provide the right-hand-side terms in the initial conditions (10.23) and (10.25), we are able to solve the initial-boundary value problem (10.21)–(10.23) and (10.25) for determining ψ_m, $m = 0, 1, \ldots$ [see Chapter 9, where the solvability and uniqueness theorems are established for this problem in the class defined by (10.1), and the same is true for (10.28) and (10.31)].

Let us summarize the results of the present section. We developed the following algorithm for finding terms in the asymptotic expansion (10.17).

First, solutions to the recurrent sequence of time-independent boundary value problems (10.28) *and* (10.31) *must be found, and* (10.29) *and* (10.30) *determine* φ_m, $m = 1, 3, \ldots$, *whereas for* $m = 2k$, φ_m *vanish identically. Since all* φ_m *are defined, they provide the initial data in the sequence of the initial-boundary value problems* (10.21)–(10.23) *and* (10.25) *for* ψ_m. *Then, solving the latter problems, one completes construction of* (10.17).

10.1.2.2. Hydrodynamic Corollaries

We begin with a list of characteristics whose asymptotics are considered in this subsection:

- the free surface elevation given by (10.2);
- the force acting on a submerged body expressed by (10.10), and the moment of this force about a certain point P_0

$$\mathbf{M}(t, \epsilon) = \int_S \phi_t \, \mathbf{r}_0 \times \mathbf{n} \, \mathrm{d}S, \tag{10.32}$$

  where $\mathbf{r}_0$ is directed from P_0 to a point on S;
- the impulse of $\mathbf{F}$ during the time interval $(0, t)$,

$$\mathbf{S}(t, \epsilon) = \int_0^t \mathbf{F}(\mu) \, \mathrm{d}\mu = \int_S \phi \, \mathbf{n} \, \mathrm{d}S, \tag{10.33}$$

  where the last expression is obtained by using (10.10) and the initial condition (10.6); and
- the energy of wave motion equal to the half of the expression in (10.1) that can be also written in the form of

$$E(t, \epsilon) = 2^{-1} \int_F (\phi \partial_y \phi + \eta^2) \, \mathrm{d}x \mathrm{d}z, \tag{10.34}$$

  where integration by parts is applied as well as (10.5).

Let us turn to asymptotics of the characteristics listed above, but first, consider the leading terms in (10.17) in more detail. We have

$$\phi(P; t, \epsilon) = \psi_0(P; t) + \epsilon \, [\beta_1(\tau) v_1(P) + \psi_1(P; t)] + O(\epsilon^2). \tag{10.35}$$

Here the coefficient at ϵ^2 in the discarded part of (10.17) depends on t only, and ψ_0, ψ_1 satisfy (10.21), (10.22), and the initial conditions

$$\begin{aligned} &\psi_0(x, 0, z; 0) = 0, & &\partial_t \psi_0(x, 0, z; 0) = -\mathcal{P}(x, z); \\ &\psi_1(x, 0, z; 0) = \beta_1(0) \mathcal{P}(x, z), & &\partial_t \psi_1(x, 0, z; 0) = 0. \end{aligned} \tag{10.36}$$

The limit in (10.15) as $\epsilon \to 0$ gives the pressure jumping from zero to one at $t = 0$, and the limit velocity potential in (10.35) is $\psi_0(P; t)$, but $\phi(P; t, \epsilon)$ does not converge to it uniformly in t. As a consequence, the boundary condition (10.3) is nonhomogeneous in the original problem whereas in the problem for ψ_0 such a condition is the second initial condition (10.36) prescribing the initial elevation of the free surface in the absence of pressure. Thus we arrived at the following conclusion.

A jump of the surface pressure is equivalent, up to $O(\epsilon)$, to the initial free surface elevation.

Differentiating (10.35) with respect to t [β_1 is given by (10.29)] and taking into account (10.10), one obtains asymptotics of the force acting on the submerged body:

$$\mathbf{F}(t,\epsilon) = \left[q\left(\frac{t}{\epsilon}\right) - 1\right]\int_S v_1 \, \mathbf{n} \, \mathrm{d}S + \int_S \partial_t \psi_0 \, \mathbf{n} \, \mathrm{d}S + O(\epsilon).$$

Here the second integral is the contribution of the limit velocity potential, and the first term on the right-hand side gives an additional contribution when one takes into account the duration of the pressure variation, but the latter decays rapidly [during the time interval $O(\epsilon)$]. For the moment of force (10.32) the asymptotic formula has the same form, but one has to substitute $\mathbf{r}_0 \times \mathbf{n}$ instead of $\mathbf{n}$. Now we get from (10.33) that

$$\mathbf{S}(t,\epsilon) = \int_S \psi_0 \, \mathbf{n} \, \mathrm{d}S + O(\epsilon),$$

where only the limit potential ψ_0 is used.

From (10.2) and (10.35), it is obvious that the asymptotics of the free surface elevation has the following form:

$$\eta(x, z; t, \epsilon) = -\left[\mathcal{P}(x, z) + \partial_t \psi_0(x, 0, z; t)\right] + O(\epsilon).$$

This expression is distinguished by the first term in brackets from that resulting from the limit problem only. Substituting the last formula into (10.1), we get

$$E(t,\epsilon) = 2^{-1}\left[\int_W |\nabla \psi_0|^2 \, \mathrm{d}x\mathrm{d}y\mathrm{d}z + \int_F (\mathcal{P} + \partial_t \psi_0)^2 \, \mathrm{d}x\mathrm{d}z\right] + O(\epsilon).$$

Since

$$\int_W |\nabla \psi_0|^2 \, \mathrm{d}x\mathrm{d}y\mathrm{d}z + \int_F (\partial_t \psi_0)^2 \, \mathrm{d}x\mathrm{d}z$$

does not depend on t (see Stoker [312], Section 6.9 for the proof of the energy conservation law), we have

$$\begin{aligned} E(t,\epsilon) &= \int_F (\mathcal{P}^2 + \mathcal{P} \, \partial_t \psi_0) \, \mathrm{d}x\mathrm{d}z + O(\epsilon) \\ &= -\int_F \mathcal{P}(x, z) \, \eta(x, z; t, 0) \, \mathrm{d}x\mathrm{d}z + O(\epsilon). \end{aligned}$$

The last expression means that the principal part of the energy is equal to the work of pressure $\mathcal{P}(x, z)$ resulting in variation of the free surface level from $y = 0$ to $y = \eta(x, z; t, 0)$.

The results of the present subsection can be summarized as follows.

Only for $\mathbf{F}(t,\epsilon)$ *does the principal term in asymptotics depend explicitly through* $q(t/\epsilon)$ *and* $v_1(P)$ *on the duration of the pressure variation. Formulae for* $\eta(x, z; t, \epsilon)$ *and* $E(t,\epsilon)$ *contain only the limit velocity potential, but in these cases the duration manifests itself implicitly. This means that the formulae obtained here are distinguished from those arising when only the limit problem is involved.*

10.1.2.3. Justification of Asymptotics

In order to justify (10.17), we have to estimate a remainder term:

$$r_N(P;t,\epsilon) = \phi(P;t,\epsilon) - \sum_{m=0}^{N} \epsilon^m \left[\varphi_m(P;\tau) + \psi_m(P;\tau)\right].$$

Since

$$r_{2N}(P;t,\epsilon) = r_{2N+1}(P;t,\epsilon) + \epsilon^{2N+1}\left[\varphi_{2N+1}(P;\tau) + \psi_{2N+1}(P;t)\right], \tag{10.37}$$

it is sufficient to estimate only r_{2N+1} (it is more convenient to estimate the remainder with an odd number because φ_m vanishes identically when m is even), and each term in (10.17). From (10.18) to (10.25) we obtain that r_{2N+1} must satisfy the following initial-boundary value problem:

$$\nabla^2 r_{2N+1} = 0 \quad \text{in } W, \qquad \partial_n r_{2N+1} = 0 \quad \text{on } B \cup S \text{ for } t \geq 0, \tag{10.38}$$

$$\partial_t^2 r_{2N+1} + \partial_y r_{2N+1} = f_{2N+1}(x,z;t,\epsilon) \quad \text{on } F \text{ for } t \geq 0, \tag{10.39}$$

$$r_{2N+1}(x,0,z;0,\epsilon) = 0, \qquad \partial_t r_{2N+1}(x,0,z;0,\epsilon) = 0. \tag{10.40}$$

Here (10.38) and the first initial condition in (10.40) are immediate consequences of (10.18) and (10.21) and of (10.23) and (10.25), respectively. Let us find f_{2N+1} in (10.39). According to (10.22) this function cannot depend on t, and (10.20) gives that

$$f_{2N+1}(x,z;t,\epsilon) = -\epsilon^{2N+1}\beta_{2N+1}(\tau)\partial_y v_{2N+1}(x,0,z). \tag{10.41}$$

In order to estimate r_{2N+1}, we note that there is an explicit formula for the trace on F of a solution to (10.38)–(10.40):

$$r_{2N+1}(x,0,z;t,\epsilon) = \int_0^t K^{-1/2} \sin\left[(t-\mu)K^{1/2}\right] f_{2N+1}(x,z;\mu,\epsilon)\,\mathrm{d}\mu. \tag{10.42}$$

This formula is derived in Subsection 9.3.2, and K is a linear operator defined as follows (see Section 9.2). For $\varphi(x,z)$ from a certain space of functions given on F (function spaces are specified below), we solve the boundary value problem,

$$\nabla^2 u = 0 \quad \text{in } W, \qquad \partial_n u = 0 \quad \text{on } B \cup S, \qquad u = \varphi \quad \text{on } F,$$

and put $(K\varphi)(x,z) = u_y(x,0,z)$. It is established in the first assertion in Section 9.2 that K satisfies the estimate

$$\|K\varphi\|_{\ell-1} \le C_\ell \,\|\varphi\|_\ell, \quad -\infty < \ell < +\infty. \tag{10.43}$$

We recall that $\|\cdot\|_\ell$ denotes the norm in the Sobolev space $H^\ell(F)$ introduced in Subsection 9.1.1.2.

The recurrent definition of v_{2n+1} in (10.28) and (10.31) gives

$$v_{2n-1} = (-1)^n K^{n-1}\mathcal{P} \quad \text{on } F, \quad n = 1, 2, \dots \tag{10.44}$$

Combining this with (10.41) and (10.42), we get

$$\begin{aligned} &r_{2N+1}(x,0,z;t,\epsilon) \\ &= (-1)^N \epsilon^{2N+1} \int_0^t \beta_{2N+1}\left(\frac{\mu}{\epsilon}\right) \sin\left[(t-\mu)K^{1/2}\right]\left(K^{N+1/2}\mathcal{P}\right)(x,z)\,\mathrm{d}\mu. \end{aligned}$$

From here

$$\begin{aligned} \|r_{2N+1}\|_{1/2} &\le \epsilon^{2N+1}\left\|K^{N+1/2}\mathcal{P}\right\|_{1/2}\int_0^t \left|\beta_{2N+1}\left(\frac{\mu}{\epsilon}\right)\right|\mathrm{d}\mu \\ &\le C(N)\epsilon^{2N+2}\|\mathcal{P}\|_{N+1}\int_0^\infty |\beta_{2N+1}(\mu)|\,\mathrm{d}\mu, \end{aligned}$$

where the second inequality is a consequence of (10.43), and by $C(N)$ we denote (here and below) various constants depending on N only. In view of (10.16) we have for the last integral

$$\begin{aligned} \int_0^\infty |\beta_m(\mu)|\,\mathrm{d}\mu &\le (m!)^{-1}\int_0^\infty \mathrm{d}\mu \int_\mu^\infty (\lambda-\mu)^m |q'(\lambda)|\,\mathrm{d}\lambda \\ &= [(m+1)!]^{-1}\int_0^\infty \mu^{m+1}|q'(\mu)|\,\mathrm{d}\mu, \end{aligned}$$

where the equality follows by integration by parts $m+1$ times. Hence, we have

$$\|r_{2N+1}\|_{1/2} \le C(N)\epsilon^{2N+2}\|\mathcal{P}\|_{N+1}\int_0^\infty \mu^{2N+2}|q'(\mu)|\,\mathrm{d}\mu. \tag{10.45}$$

Now, (10.29) and (10.30), and (10.43) and (10.44), produce

$$\|\varphi_{2N+1}\|_{1/2} \le C(N)\|\mathcal{P}\|_{N+1/2}\int_0^\infty \mu^{2N+1}|q'(\mu)|\,\mathrm{d}\mu. \tag{10.46}$$

In order to estimate $\|\psi_{2N+1}\|_{1/2}$ we note that (10.23) and (10.25), and (10.30) and (10.31) give the following initial conditions:

$$\begin{aligned} &\psi_{2N+1}(x,0,z;0) = -\beta_{2N+1}(0)v_{2N+1}(x,0,z), \\ &\partial_t\psi_{2N+1}(x,0,z;0) = 0, \end{aligned}$$

complementing (10.21) and (10.22). Then we find that

$$\psi_{2N+1}(x,0,z;t) = -\beta_{2N+1}(0)\cos\left(tK^{1/2}\right) v_{2N+1}(x,0,z), \quad (10.47)$$

and again (10.29), (10.43), and (10.44) lead to

$$\|\psi_{2N+1}\|_{1/2} \le C(N)\|\mathcal{P}\|_{N+1/2} \int_0^\infty \mu^{2N+1}|q'(\mu)|\,\mathrm{d}\mu. \quad (10.48)$$

In view of (10.37), (10.45), (10.46), and (10.48) combine to produce

$$\|r_{2N}\|_{1/2} \le C(N)\epsilon^{2N+1}\|\mathcal{P}\|_{N+1} \max_{k=1,2} \int_0^\infty \mu^{2N+k}|q'(\mu)|\,\mathrm{d}\mu. \quad (10.49)$$

Finally, the amalgamated form of (10.45) and (10.49) is as follows:

$$\|r_N\|_{1/2} \le C(N)\epsilon^{N+1}\|\mathcal{P}\|_{[N/2]+1} \max_{k=1,2} \int_0^\infty \mu^{N+k}|q'(\mu)|\,\mathrm{d}\mu, \quad (10.50)$$

where $[s]$ denotes the integer part of $s \in \mathbb{R}$.

Applying the same considerations to (10.42), (10.30), and (10.47) differentiated with respect to t, one arrives at

$$\|\partial_t r_{2N+1}\|_{1/2} \le C(N)\epsilon^{2N+2}\|\mathcal{P}\|_{N+3/2} \int_0^\infty \mu^{2N+2}|q'(\mu)|\,\mathrm{d}\mu,$$

$$\|\partial_t \varphi_{2N+1}\|_{1/2} \le C(N)\|\mathcal{P}\|_{N+1/2} \int_0^\infty \mu^{2N}|q'(\mu)|\,\mathrm{d}\mu,$$

$$\|\partial_t \psi_{2N+1}\|_{1/2} \le C(N)\|\mathcal{P}\|_{N+1} \int_0^\infty \mu^{2N+1}|q'(\mu)|\,\mathrm{d}\mu,$$

respectively, which combine to give

$$\|\partial_t r_N\|_{1/2} \le C(N)\epsilon^{N+1}\|\mathcal{P}\|_{N/2+1} \max_{k=0,1} \int_0^\infty \mu^{N+k}|q'(\mu)|\,\mathrm{d}\mu. \quad (10.51)$$

Let us recall that $\mathcal{H}(W)$ (see Subsection 9.1.1.2) denotes the Hilbert space, where the norm is defined as follows,

$$|\phi|^2 = \int_W |\nabla\phi|^2\,\mathrm{d}x\mathrm{d}y\mathrm{d}z + \int_F |\phi|^2\,\mathrm{d}x\mathrm{d}z,$$

and is related to the energy of waves (10.1). It is established in Subsection 9.1.1 that

$$|\phi| \le C\|\phi\|_{1/2}. \quad (10.52)$$

We conclude the present subsection by summarizing the obtained results.

Let $\mathcal{P} \in H^{N+1}$ and $q'(\tau)$ decays as $\tau \to \infty$ so that

$$\int_0^\infty \tau^{N+2} \, |q'(\tau)| \, d\tau < \infty.$$

Then (10.50) *and* (10.51) *hold for the remainder term r_N, and* (10.52) *provides that*

$$|r_N| \leq C(N)\epsilon^{N+1} \|\mathcal{P}\|_{[N/2]+1} \max_{k=1,2} \int_0^\infty \mu^{N+k} |q'(\mu)| \, d\mu$$

holds as well.

10.1.3. Impulsive Surface Pressure

10.1.3.1. Asymptotics of the Velocity Potential and Hydrodynamic Characteristics

When p is given by (10.11), then results obtained in the previous subsection are applicable. Since we assume here that $q(t/\epsilon) = \epsilon^{-1} Q(t/\epsilon)$, and

$$\tau^m Q(\tau) \to 0 \quad \text{as } \tau \to \infty \text{ for any } m = 1, 2, \ldots, \tag{10.53}$$

the asymptotic expansion can be simplified. For this purpose we replace q by Q in (10.29) and integrate by parts, which gives

$$\beta_0(\tau) = -Q(\tau);$$

$$\beta_m(\tau) = -\,[(m-1)!]^{-1} \int_\tau^\infty (\mu - \tau)^{m-1} Q(\mu) \, d\mu, \quad m = 1, 2, \ldots. \tag{10.54}$$

Now (10.17) takes the form of

$$\phi(P; t, \epsilon) = \sum_{m=0}^{\infty} \epsilon^{m-1} \left[\varphi_m(P; \tau) + \psi_m(P; t) \right],$$

where φ_m and ψ_m are defined in the same way as in Subsection 10.1.2.1, but using β_m in the form (10.54) and I instead of $\mathcal{P}$ in (10.28).

From (10.27) and the second equation of (10.54) we get that ψ_0 satisfies the homogeneous initial-boundary value problem because $Q(0) = 0$. Then ψ_0 vanishes identically in W for $t \geq 0$. Since the same is true for φ_0, we obtain the following asymptotic series:

$$\phi(P; t, \epsilon) = \sum_{m=0}^{\infty} \epsilon^{m} \left[\varphi_{m+1}(P; \tau) + \psi_{m+1}(P; t) \right],$$

and its justification is already given in Subsection 10.1.2.3.

The leading term in the case of the impulsive pressure is as follows [cf. (10.13)]:

$$\phi(P;t,\epsilon) = -v_1(P)\int_{t/\epsilon}^{\infty} Q(\mu)\,\mathrm{d}\mu + \psi_1(P;t) + O(\epsilon),$$

where the coefficient at ϵ depends on t only. We have

$$\nabla^2 v_1 = 0 \quad \text{in } W, \qquad \partial_n v_1 = 0 \quad \text{on } B \cup S, \qquad v_1 = -I \quad \text{on } F,$$

and

$$\nabla^2 \psi_1 = 0 \quad \text{in } W, \qquad \partial_n \psi_1 = 0 \quad \text{on } B \cup S \text{ for } t > 0,$$
$$\partial_t^2 \psi_1 + \partial_y \psi_1 = 0 \quad \text{on } F \text{ for } t > 0,$$
$$\psi_1(x,0,z;0) = -I(x,z), \qquad \partial_t \psi_1(x,0,z;0) = 0,$$

for determining v_1 and ψ_1, respectively.

Now,

$$\mathbf{F}(t,\epsilon) = \epsilon^{-1} Q\left(\frac{t}{\epsilon}\right)\int_S v_1\,\mathbf{n}\,\mathrm{d}S + \int_S \partial_t \psi_1(P;t)\,\mathbf{n}\,\mathrm{d}S + O(\epsilon)$$

is the asymptotic expression for the force acting on the body bounded by S. For the impulse of $\mathbf{F}(t,\epsilon)$ during the time interval $(0,t)$ we get, according to (10.33),

$$\begin{aligned}\mathbf{S}(t,\epsilon) &= \left[\int_0^{t/\epsilon} Q(\mu)\,\mathrm{d}\mu\right]\int_S v_1\,\mathbf{n}\,\mathrm{d}S \\ &\quad + \int_S [\psi_1(P;t) - \psi_1(P;0)]\,\mathbf{n}\,\mathrm{d}S + O(\epsilon) \\ &= \left[\int_{t/\epsilon}^{\infty} Q(\mu)\,\mathrm{d}\mu\right]\int_S v_1\,\mathbf{n}\,\mathrm{d}S + \int_S \psi_1(P;t)\,\mathbf{n}\,\mathrm{d}S + O(\epsilon),\end{aligned}$$

and the last equality follows from these two facts:

$$\psi_1(P;0) = v_1(P) \quad \text{in } W, \qquad \int_0^{\infty} Q(\mu)\,\mathrm{d}\mu = 1. \tag{10.55}$$

The second expression for $\mathbf{S}(t,\epsilon)$ demonstrates how the contribution arising from the duration of the pressure impulse decays as $t \to \infty$.

From (10.2) and the boundary condition for v_1 on F, one immediately obtains

$$\eta(x,z;t,\epsilon) = -\partial_t \psi_1(x,0,z;t) + O(\epsilon). \tag{10.56}$$

Unlike $\mathbf{F}(t,\epsilon)$ and $\mathbf{S}(t,\epsilon)$, the leading term of this asymptotics depends on ψ_1 only.

Substituting (10.56) and the principal term in the expansion for ϕ into (10.34), we get

$$E(t,\epsilon) = 2^{-1}\int_F [\beta_1(\tau)v_1 + \psi_1]\,\partial_y\,[\beta_1(\tau)v_1 + \psi_1]\,\mathrm{d}x\mathrm{d}z$$
$$+2^{-1}\int_F (\partial_t\psi_1)^2\,\mathrm{d}x\mathrm{d}z + O(\epsilon).$$

Since $\beta_1(\infty) = 0$, we have

$$E(t,0) = 2^{-1}\left[\int_W |\nabla\psi_1|^2\,\mathrm{d}x\mathrm{d}y\mathrm{d}z + \int_F (\partial_t\psi_1)^2\,\mathrm{d}x\mathrm{d}z\right]$$
$$= 2^{-1}\int_W |\nabla v_1|^2\,\mathrm{d}x\mathrm{d}y\mathrm{d}z.$$

Here the first expression follows from Green's formula, and the second one is a consequence of the conservation energy law,

$$E(t,0) = E(0,0) \quad \text{for } t \geq 0$$

(see, for example, Stoker [312], Section 6.9), and the first equation (10.55). Thus

$$E(t,\epsilon) = E(0,0)\{[\beta_1(\tau)]^2 + 1\}$$
$$+2^{-1}\beta_1(\tau)\int_F (\psi_1\partial_y v_1 + v_1\partial_y\psi_1)\,\mathrm{d}x\mathrm{d}z + O(\epsilon)$$
$$= E(0,0)\{[\beta_1(\tau)]^2 + 1\} + \beta_1(\tau)\int_F \psi_1(x,0,z;t)\partial_y v_1\,\mathrm{d}x\mathrm{d}z + O(\epsilon),$$

where Green's formula is applied again as well as the homogeneous Neumann condition on $B \cup S$. By the first equation of (10.55),

$$\psi_1(x,0,z;t) = v_1(x,0,z) + O(t).$$

Then (10.53) and (10.54) give

$$\beta_1(\tau)\int_F \psi_1(x,0,z;t)\partial_y v_1\,\mathrm{d}x\mathrm{d}z = 2\beta_1(\tau)E(0,0) + O(\epsilon),$$

and substituting this into the last expression for $E(t,\epsilon)$, we arrive at

$$E(t,\epsilon) = E(0,0)\{[\beta_1(\tau)]^2 + 2\beta_1(\tau) + 1\} + O(\epsilon)$$
$$= E(0,0)\left[\int_0^{t/\epsilon} Q(\mu)\,\mathrm{d}\mu\right]^2 + O(\epsilon),$$

where the second equality of (10.55) is used.

10.1.3.2. Waves in a Layer of Constant Depth

Let the water domain be a layer of constant depth; that is, $S = \emptyset$ and

$$W = \{-\infty < x, z < +\infty, -d < y < 0\}.$$

Then the problem formulated in Subsection 10.1.1 with the pressure given by (10.15) can be solved by means of the Fourier transform

$$\hat{\mathcal{P}}(\sigma) = \int_{\mathbb{R}^2} \mathcal{P}(x, z) e^{-i(x\xi + z\zeta)} \, \mathrm{d}x \mathrm{d}z, \quad \text{where } \sigma = (\xi, \zeta),$$

and so one obtains

$$\phi(P; t, \epsilon) = \frac{1}{(2\pi)^2 \epsilon} \int_{\mathbb{R}^2} \hat{\mathcal{P}}(\sigma) \frac{\cosh |\sigma|(y + d)}{|\sigma|^{1/2} \cosh |\sigma| d} e^{i(x\xi + z\zeta)} \, \mathrm{d}\xi \mathrm{d}\zeta$$
$$\times \int_0^t q'\left(\frac{\mu}{\epsilon}\right) \sin\left[(\mu - t)(|\sigma| \tanh |\sigma| d)^{1/2}\right] \mathrm{d}\mu.$$

Since $q(0) = 0$, we can integrate by parts, which gives that

$$\phi(P; t, \epsilon) = \frac{1}{(2\pi)^2} \int_{\mathbb{R}^2} \hat{\mathcal{P}}(\sigma) \frac{\tanh^{1/2} |\sigma| d \cosh |\sigma|(y + d)}{\cosh |\sigma| d} e^{i(x\xi + z\zeta)} \, \mathrm{d}\xi \mathrm{d}\zeta$$
$$\times \int_0^t q\left(\frac{\mu}{\epsilon}\right) \cos\left[(\mu - t)(|\sigma| \tanh |\sigma| d)^{1/2}\right] \mathrm{d}\mu. \qquad (10.57)$$

In the case of deep water, $W = \mathbb{R}^3_-$ and the last formula must be changed as follows: first, omit $\tanh |\sigma| d$; second, replace $[\cosh |\sigma|(y + d)]/(\cosh |\sigma| d)$ by $e^{|\sigma| y}$.

The behavior of (10.57) is not obvious as $\varepsilon \to 0$, and so the asymptotic expansion (10.17) is preferable because its terms are simpler. Thus

$$v_m(P) = \frac{-\sin(m\pi/2)}{(2\pi)^2} \int_{\mathbb{R}^2} \hat{\mathcal{P}}(\sigma)(|\sigma| \tanh |\sigma| d)^{(m-1)/2}$$
$$\times \frac{\cosh |\sigma|(y + d)}{\cosh |\sigma| d} e^{i(x\xi + z\zeta)} \, \mathrm{d}\xi \mathrm{d}\zeta, \qquad (10.58)$$

where $m = 1, 2, \ldots$, and writing $\psi_m(P; t) = \beta_m(0) w_m(P; t)$, we get that

$$w_m(P; t) = \frac{1}{(2\pi)^2} \int_{\mathbb{R}^2} \hat{\mathcal{P}}(\sigma) (|\sigma| \tanh |\sigma| d)^{(m-1)/2}$$
$$\times \frac{\cosh |\sigma|(y + d)}{\cosh |\sigma| d} \sin\left[\frac{m\pi}{2} - t(|\sigma| \tanh |\sigma| d)^{1/2}\right] e^{i(x\xi + z\zeta)} \mathrm{d}\xi \mathrm{d}\zeta,$$
$$(10.59)$$

where $m = 0, 1, \ldots$, and $v_m(P)$, $w_m(P; t)$ for deep water can be obtained in the same way as above.

Example

Let us consider the two-dimensional problem of impulsive pressure applied to the free surface of deep water. The pressure is assumed to have the form of (10.11); that is,

$$p(x;t,\epsilon) = \epsilon^{-1} Q(t/\epsilon)\, I(x),$$

where we set the time-depending factor to be as follows:

$$Q(\tau) = 4\pi^{-1} b\tau (b^2 + \tau^4)^{-1}, \quad \tau = t/\epsilon.$$

The x-depending factor $I(x) = (a^2 + x^2)^{-1}$ must replace $\mathcal{P}(x)$ in the previous formulae, and a and b are assumed to be positive parameters. According to formula 3.767.2 in Gradshteyn and Ryzhik [96], we have $\hat{I}(\xi) = \pi a^{-1} e^{-a|\xi|}$, and so (10.57) simplifies to produce

$$\phi(P;t,\epsilon) = \frac{4b\epsilon^2}{\pi a} \int_0^\infty e^{(y-a)\xi} \cos x\xi \,\mathrm{d}\xi \int_0^t \frac{\mu \cos \xi^{1/2}(\mu - t)}{b^2\epsilon^4 + \mu^4} \,\mathrm{d}\mu. \quad (10.60)$$

On the other hand, after substitution of (10.58) and (10.59) into the leading term of asymptotics for ϕ (see Subsection 10.1.3.1), one obtains

$$\phi(P;t,\epsilon) = -\frac{4b}{\pi a} \int_0^\infty e^{(y-a)\xi} \cos x\xi \,\mathrm{d}\xi \int_{t/\epsilon}^\infty \frac{\mu \,\mathrm{d}\mu}{b^2 + \mu^4}$$
$$+ a^{-1} \int_0^\infty e^{(y-a)\xi} \cos x\xi \cos t\xi^{1/2} \,\mathrm{d}\xi + O(\epsilon), \quad (10.61)$$

where all of the integrals have explicit expressions involving well-known transcendental functions. In fact, the first term on the right-hand side is equal to

$$\frac{2(y-a)}{\pi a[x^2 + (y-a)^2]} \left(\frac{\pi}{2} - \arctan \frac{t^2}{b\epsilon^2} \right).$$

In order to find the second term explicitly, we change the variable and represent the first cosine as the real part of the exponential function, which gives

$$\frac{2}{a} \operatorname{Re} \int_0^\infty \xi \exp\{-[(a-y) + ix]\xi^2\} \cos t\xi \,\mathrm{d}\xi$$
$$= \frac{a-y}{a[(a-y)^2 + x^2]} - \operatorname{Im}\left(\frac{t\sqrt{\pi}}{[(a-y)+ix]^{3/2}} \exp\left\{ \frac{-t^2}{4[(a-y)+ix]} \right\} \right.$$
$$\left. \times \operatorname{erf}\left(\frac{it}{2\sqrt{a-y+ix}} \right) \right),$$

where the equality is a consequence of formula 2.5.36.7 in Prudnikov et al. [293] and erf$(\cdot)$ denotes the error function. These expressions for integrals in (10.61) are used for a numerical evaluation of the principal term in the

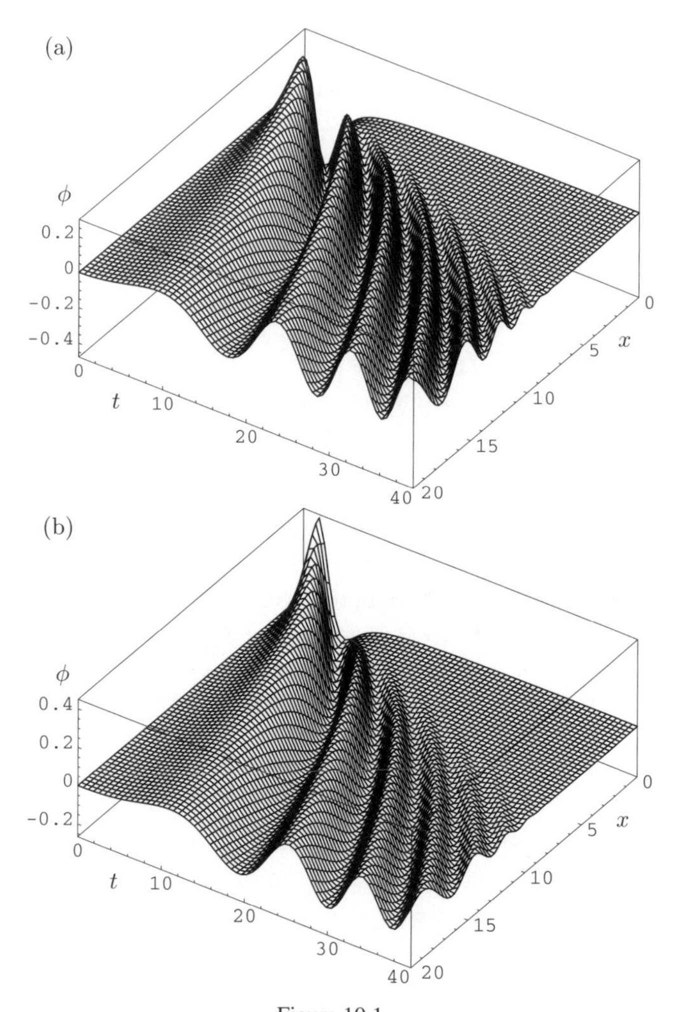

Figure 10.1.

asymptotics for $\phi(P; t, \epsilon)$ when $\epsilon = 3/2$, $a = 1$, $b = 1$, and $y = 0$, and the result is shown in Fig. 10.1(a).

The exact expression (10.60) for $\phi(P; t, \epsilon)$ can be also simplified by using 2.5.36.7 in [293], and we get

$$\phi(P; t, \epsilon) = \frac{2b(a-y)}{\pi a[(a-y)^2 + x^2]} \arctan\left(\frac{t}{\epsilon}\right)^2$$
$$- \frac{4b\sqrt{\pi}\epsilon^2}{\pi a} \operatorname{Im} \int_0^t \exp\left\{\frac{-(\mu - t)^2}{4[(a-y) + ix]}\right\} \operatorname{erf}\left(\frac{i(\mu - t)}{2\sqrt{a - y + ix}}\right)$$
$$\times \frac{\mu(\mu - t)\,\mathrm{d}\mu}{(b^2\epsilon^4 + \mu^4)[(a-y) + ix]^{3/2}}.$$

To our knowledge, there is no explicit formula for the last integral. Numerical calculations based on the last formula are shown in Fig. 10.1(b) for $\epsilon = 3/2$, $a = 1, b = 1$, and $y = 0$. Comparing this figure with Fig. 10.1(a), where the same values of parameters were used, demonstrates that the leading term in the asymptotic formula provides a good appriximation for the exact solution even for a sufficiently large value of ϵ.

10.2. Rapidly Stabilizing Underwater Disturbances

This section is concerned with two problems. The first one describes a brief movement of underwater rigid surfaces B and S. The second problem deals with a source having a strength rapidly stabilizing in time to a constant value.

10.2.1. Brief Movement of Rigid Surfaces

10.2.1.1. Statement of the Problem and a Formal Asymptotic Expansion

Waves arising in the initially resting water caused by a brief movement of B or S (which are assumed to be sufficiently smooth) are described by a velocity potential $\phi(P; t, \epsilon)$, which must satisfy the following boundary value problem:

$$\nabla^2 \phi = 0 \quad \text{in } W, \qquad \partial_t^2 \phi + \partial_y \phi = 0 \quad \text{on } F \text{ for } t \geq 0; \tag{10.62}$$

$$\phi(x, 0, z; 0, \epsilon) = \partial_t \phi(x, 0, z; 0, \epsilon) = 0; \tag{10.63}$$

$$\partial_n \phi = f \quad \text{on } B \cup S \text{ for } t \geq 0, \tag{10.64}$$

where $f(P; t) = Q(t/\epsilon) w(P)$ and $\epsilon \ll 1$. We assume that $Q(0) = 0$ and $Q(\tau)$ satisfies (10.12) and (10.53), and w is a sufficiently smooth function given on $B \cup S$.

As in Section 10.1, we seek the potential in the form of a two-time-scaled asymptotic series:

$$\phi(P; t, \epsilon) = \sum_{m=0}^{\infty} \epsilon^m \left[\varphi_m(P; \tau) + \psi_m(P; t) \right], \tag{10.65}$$

where $\varphi_m(P; \tau)$ tends to zero as $\tau \to \infty$, $m = 0, 1, \ldots$, and these functions decay as $|P| \to \infty$, as well as $\psi_m(P; t)$.

The procedure applied in Subsection 10.1.2.1 gives equations for determining φ_m and ψ_m, and so we have the following for $\tau \geq 0$:

$$\nabla^2 \varphi_m = 0 \quad \text{in } W, m = 0, 1, \ldots; \tag{10.66}$$

$$\partial_\tau^2 \varphi_m = 0 \quad \text{on } F, m = 0, 1; \tag{10.67}$$

$$\partial_\tau^2 \varphi_m + \partial_y \varphi_{m-2} = 0 \quad \text{on } F, m = 2, 3, \ldots; \tag{10.68}$$

$$\partial_n \varphi_0 = Q(\tau) w(P) \quad \text{on } B \cup S; \tag{10.69}$$

$$\partial_n \varphi_m = 0 \quad \text{on } B \cup S, m = 1, 2, \dots . \tag{10.70}$$

Equations (10.21) and (10.22) obtained in Subsection 10.1.2.1 for ψ_m remain valid in the present situation, as well as the initial conditions (10.23)–(10.25).

Again, it is convenient to put [cf. (10.30)]

$$\varphi_m(P;\tau) = \beta_m(\tau) v_m(P), \quad m = 0, 1, \dots, \tag{10.71}$$

where β_m is given by (10.54). From (10.67) we get that

$$\varphi_0 = \varphi_1 = 0 \quad \text{on } F. \tag{10.72}$$

This, (10.66), and (10.70) give that

$$\varphi_1(P;\tau) = 0 \quad \text{in } W \text{ for } \tau \geq 0$$

is the only solution decaying as $|P| \to \infty$. In the same way as in Subsection 10.1.2.1, we obtain from (10.68)

$$\varphi_m(P;\tau) = 0 \quad \text{in } W \text{ for } \tau \geq 0 \text{ when } m = 1, 3, \dots, 2k-1, \dots, \tag{10.73}$$

but now it follows from (10.69) that v_0 does not vanish identically and satisfies the following boundary value problem [cf. (10.28)]:

$$\nabla^2 v_0 = 0 \quad \text{in } W, \qquad v_0 = 0 \quad \text{on } F, \qquad \partial_n v_0 = -w \quad \text{on } B \cup S. \tag{10.74}$$

Then (10.68) produces a recurrent sequence of boundary value problems:

$$\begin{aligned} \nabla^2 v_m &= 0 \quad \text{in } W, \qquad \partial_n v_m = 0 \quad \text{on } B \cup S, \\ v_m &= -\partial_y v_{m-2} \quad \text{on } F, \quad m = 2, 4, \dots, 2k, \dots, \end{aligned} \tag{10.75}$$

which is similar to (10.31). Now, the initial-boundary value problem (10.21)–(10.23) and (10.25) serves for determining ψ_m, $m = 0, 1, \dots$. Thus we arrive at the following conclusion.

The algorithm described at the end of Subsection 10.1.2.1 *works for* (10.62)–(10.64), *and the only amendment to be made is that the recurrent sequence* (10.74) *and* (10.75) *of time-independent boundary value problems must be used instead of* (10.28) *and* (10.31).

It is important to note that (10.72) requires both initial conditions for ψ_0 to vanish. Hence, ψ_0 is equal to zero identically in W, and this distinguishes the present expansion from that in Subsection 10.1.2.

10.2.1.2. Hydrodynamic Corollaries

Let us begin with some details about the leading terms in (10.65):

$$\phi(P;t,\epsilon) = -Q(\tau)v_0(P) + \epsilon\psi_1(P;t) + \epsilon^2\left[\beta_2(\tau)v_2(P) + \psi_2(P;t)\right] + O(\epsilon^3). \tag{10.76}$$

Here we give more terms than in (10.35) for the rapidly stabilizing pressure because there are cases in which the coefficient $\beta_2(\tau)v_2(P)$ is involved in the principal term of asymptotics. Furthermore, the coefficient at ϵ^3 in the discarded part of (10.65) depends on t only, and ψ_1, ψ_2 satisfy (10.21), (10.22), and the initial conditions

$$\psi_1(x,0,z;0) = 0, \qquad \partial_t\psi_1(x,0,z;0) = \partial_y v_0(x,0,z);$$

$$\psi_2(x,0,z;0) = \partial_y v_0(x,0,z)\int_0^\infty \mu Q(\mu)\,\mathrm{d}\mu, \qquad \partial_t\psi_2(x,0,z;0) = 0,$$

where (10.12), (10.54), and the boundary condition on F in (10.75) are applied.

From (10.76) we see that $\phi(P;t,\epsilon)$ describing waves from a brief movement of underwater rigid surfaces tends to zero for every $t > 0$ as $\epsilon \to 0$, but the convergence is not uniform. For instance, if $Q(\tau) = 0$ for $\tau > 1$, then $\epsilon\psi_1(P;t)$ becomes the leading term after $t = \epsilon$, which allows interpretation of waves from a brief bottom movement as arising from the initial elevation of the free surface equal to $\epsilon\partial_y v_0(x,0,z)$. Hence, the long-term consequences of a finite bottom movement are similar to those arising from a small initial free surface elevation.

Let us assume that there is a brief movement of bottom taking place while S is at rest, and let us consider the force acting on the body bounded by S. Differentiation of (10.76) with respect to t then gives the asymptotics of force in the following form:

$$\mathbf{F}(t,\epsilon) = -\epsilon^{-1}Q'\left(\frac{t}{\epsilon}\right)\int_S v_0\,\mathbf{n}\,\mathrm{d}S + \epsilon\left[\left(\int_\tau^\infty Q(\mu)\,\mathrm{d}\mu\right)\int_S v_2\,\mathbf{n}\,\mathrm{d}S + \int_S \partial_t\psi_1(P;t)\,\mathbf{n}\,\mathrm{d}S\right] + O(\epsilon^2), \tag{10.77}$$

and so $\mathbf{F}(t,\epsilon)$ tends to infinity as $\epsilon \to 0$, but as $Q'(\tau)$ decays rapidly (and might vanish for $\tau > 1$), and the coefficient at ϵ^0 is zero, the term of the first order occurs to represent the principal part of $\mathbf{F}(t,\epsilon)$ after an initial time interval. Furthermore, the large force corresponding to the first term on the right-hand side in (10.77) produces a finite impulse during the time interval $(0, t)$, as we see from

$$\mathbf{S}(t, \epsilon) = -Q\left(\frac{t}{\epsilon}\right)\int_S v_0\,\mathbf{n}\,\mathrm{d}S + \epsilon\int_S \psi_1(P;t)\,\mathbf{n}\,\mathrm{d}S + O(\epsilon^2). \quad (10.78)$$

Also, the contribution of the first term in brackets in (10.77) reduces to $O(\epsilon^2)$ here.

The free surface elevation is expressed by

$$\eta(x, z; t, \epsilon) = \epsilon\left[\partial_t\psi_1(x, 0, z; t) - \partial_y v_0(x, 0, z)\int_\tau^\infty Q(\mu)\,\mathrm{d}\mu\right] + O(\epsilon^2), \quad (10.79)$$

and it is small from the very beginning because of the homogeneous Dirichlet condition on F in (10.74); also, the free surface condition in (10.75) is taken into account here.

The last formula shows that for the potential energy [see the second integral in (10.1)], asymptotics has the first nontrivial coefficient at ϵ^2. Thus substituting expansions into the formula for energy (10.1), we truncate them, keeping coefficients at ϵ^2:

$$2E(t, \epsilon) = \int_W |\nabla\{-Q(\tau)v_0 + \epsilon\psi_1 + \epsilon^2[\beta_2(\tau)v_2 + \psi_2]\}|^2\,\mathrm{d}x\mathrm{d}y\mathrm{d}z$$

$$+\epsilon^2\int_F [\partial_t\psi_1 + \beta_1(\tau)\partial_y v_0]^2\,\mathrm{d}x\mathrm{d}z + O(\epsilon^3). \quad (10.80)$$

Let us consider coefficients at ϵ^0, ϵ^1, and ϵ^2 on the right-hand side:

$$\epsilon^0:\ [Q(\tau)]^2\int_W |\nabla v_0|^2\,\mathrm{d}x\mathrm{d}y\mathrm{d}z;$$

$$\epsilon^1:\ -2Q(\tau)\int_W \nabla v_0\cdot\nabla\psi_1\,\mathrm{d}x\mathrm{d}y\mathrm{d}z = 0;$$

$$\epsilon^2:\ \int_W |\nabla\psi_1|^2\,\mathrm{d}x\mathrm{d}y\mathrm{d}z + \int_F |\partial_t\psi_1|^2\,\mathrm{d}x\mathrm{d}z$$

$$+ [\beta_1(\tau)]^2\int_F |\partial_y v_0|^2\,\mathrm{d}x\mathrm{d}z + 2\beta_1(\tau)\int_F \partial_t\psi_1\,\partial_y v_0\,\mathrm{d}x\mathrm{d}z$$

$$- 2Q(\tau)\int_W \nabla v_0\cdot\nabla(\beta_2(\tau)v_2 + \psi_2)\,\mathrm{d}x\mathrm{d}y\mathrm{d}z.$$

The coefficient at ϵ^1 vanishes because Green's formula transforms the corresponding integral into

$$\int_F v_0\,\partial_y\psi_1\,\mathrm{d}x\mathrm{d}z - \int_{B\cup S} v_0\,\partial_n\psi_1\,\mathrm{d}S,$$

which is equal to zero because v_0 and $\partial_n \psi_1$ do vanish on F and $B \cup S$, respectively. The same is true for the last integral in the coefficient at ϵ^2.

Remaining terms in the coefficient at ϵ^2 can be simplified as follows. The sum

$$\int_W |\nabla \psi_1|^2 \, \mathrm{d}x\mathrm{d}y\mathrm{d}z + \int_F |\partial_t \psi_1|^2 \, \mathrm{d}x\mathrm{d}z$$

is twice the conserved energy corresponding to ψ_1, and so it is equal to

$$\int_F |\partial_t \psi_1(x, 0, z; 0)|^2 \, \mathrm{d}x\mathrm{d}z = \int_F |\partial_y v_0|^2 \, \mathrm{d}x\mathrm{d}z,$$

where the initial conditions are taken into account. By the initial condition we have

$$\psi_1(x, 0, z; t) = \partial_y v_0(x, 0, z) + O(t).$$

Then (10.53) and (10.54) give

$$2\beta_1(\tau) \int_F \partial_t \psi_1(x, 0, z; t) \partial_y v_0 \, \mathrm{d}x\mathrm{d}z = 2\beta_1(\tau) \int_F |\partial_y v_0|^2 \, \mathrm{d}x\mathrm{d}z + O(\epsilon).$$

Hence, the coefficient at ϵ^2 takes the following form:

$$\{[\beta_1(\tau)]^2 + 2\beta(\tau) + 1\} \int_F |\partial_y v_0|^2 \, \mathrm{d}x\mathrm{d}z.$$

Substituting this and the coefficients at ϵ^0 and ϵ^1 into (10.80), we arrive at

$$2E(t, \epsilon) = [Q(t/\epsilon)]^2 \int_W |\nabla v_0|^2 \, \mathrm{d}x\mathrm{d}y\mathrm{d}z + \epsilon^2 \left[\int_0^{t/\epsilon} Q(\mu) \, \mathrm{d}\mu \right]^2 \int_F |\partial_y v_0|^2 \, \mathrm{d}x\mathrm{d}z + O(\epsilon^3),$$

where the second equality of (10.55) is used. The second term on the right-hand side describes the behavior of energy, in particular when the first term vanishes because of vanishing $Q(\tau)$.

10.2.1.3. Justification of Asymptotics

Since the functions φ_{2k-1}, $k = 1, 2, \ldots$, vanish identically in (10.65), it is convenient to begin with an estimation of r_{2N}, $N \geq 1$ (in contrast to Subsection 10.1.2.3). In the same way as equations (10.38)–(10.40) were obtained, we get

$$\nabla^2 r_{2N} = 0 \quad \text{in } W, \qquad \partial_n r_{2N} = 0 \quad \text{on } B \cup S \text{ for } t \geq 0,$$
$$\partial_t^2 r_{2N} + \partial_y r_{2N} = f_{2N}(x, z; t, \epsilon) \quad \text{on } F \text{ for } t \geq 0,$$
$$r_{2N}(x, 0, z; 0, \epsilon) = 0, \qquad \partial_t r_{2N}(x, 0, z; 0, \epsilon) = 0,$$

where

$$f_{2N}(x, z; t, \epsilon) = \epsilon^{2N} \beta_{2N}(\tau) \partial_y v_{2N}(x, 0, z).$$

Then similarly to (10.42) we have

$$r_{2N}(x, 0, z; t, \epsilon) = \int_0^t K^{-1/2} \sin\left[(t-\mu)K^{1/2}\right] f_{2N}(x, z; \mu, \epsilon)\, d\mu,$$

and K is defined in Section 9.2. The recurrent definition of v_{2N} in (10.75) gives

$$v_{2N} = (-1)^N K^N v_0 \quad \text{on } F,$$

and we arrive at

$$\begin{aligned} &r_{2N}(x, 0, z; t, \epsilon) \\ &\quad = (-1)^N \epsilon^{2N} \int_0^t \beta_{2N}\left(\frac{\mu}{\epsilon}\right) \sin\left[(t-\mu)K^{1/2}\right] \left(K^{N+1/2} v_0\right)(x, 0, z)\, d\mu. \end{aligned}$$

Hence,

$$\begin{aligned} \|r_{2N}\|_{1/2} &\le \epsilon^{2N} \left\|K^{N+1/2} v_0\right\|_{1/2} \int_0^t \left|\beta_{2N}\left(\frac{\mu}{\epsilon}\right)\right| d\mu \\ &\le C(N)\epsilon^{2N+1} \|v_0\|_{N+1} \int_0^\infty |\beta_{2N}(\mu)|\, d\mu, \end{aligned}$$

where the second inequality follows from (10.43), and $C(N)$ is a constant depending on N only. Also,

$$\begin{aligned} \int_0^\infty |\beta_m(\mu)|\, d\mu &\le [(m-1)!]^{-1} \int_0^\infty d\mu \int_\mu^\infty (\lambda - \mu)^{m-1} |Q(\lambda)|\, d\lambda \\ &= [m!]^{-1} \int_0^\infty \mu^m |Q(\mu)|\, d\mu \end{aligned}$$

is a consequence of (10.53) and (10.54). Therefore,

$$\|r_{2N}\|_{1/2} \le C(N)\epsilon^{2N+1} \|v_0\|_{N+1} \int_0^\infty \mu^{2N} |Q(\mu)|\, d\mu. \qquad (10.81)$$

Using

$$r_{2N-1}(P; t, \epsilon) = r_{2N}(P; t, \epsilon) + \epsilon^{2N} \left[\varphi_{2N}(P; \tau) + \psi_{2N}(P; t)\right],$$

and estimating $\|\varphi_{2N}\|_{1/2}$ and $\|\psi_{2N}\|_{1/2}$ similarly to (10.46) and (10.48), one obtains

$$\|r_{2N-1}\|_{1/2} \le C(N)\epsilon^{2N} \|v_0\|_{N+1} \max_{k=0,1} \int_0^\infty \mu^{2N-k} |Q(\mu)|\, d\mu,$$

which combines with (10.81) to give

$$\|r_N\|_{1/2} \leq C(N)\epsilon^{N+1}\|v_0\|_{[N/2]+1} \max_{k=0,1} \int_0^\infty \mu^{N-k}|Q(\mu)|\,\mathrm{d}\mu. \quad (10.82)$$

Similar estimates hold for $\|\partial_t r_N\|_{1/2}$ and $|r_N|$.

10.2.2. Source Having a Rapidly Stabilizing Strength

Let a source be placed at $P_s = (x_s, y_s, z_s) \in W$, and let its strength be equal to $Q(t/\epsilon)$, which has the same properties as in Subsection 10.2.1. The arising wave motion is described by a velocity potential $G(P, P_s; t, \epsilon)$, which must satisfy the following initial-boundary value problem:

$$\nabla^2 G = Q(t/\epsilon)\delta(|P - P_s|) \quad \text{in } W, \text{ for } t \geq 0;$$

$$\partial_t^2 G + \partial_y G = 0 \quad \text{on } F, \qquad \partial_n G = 0 \quad \text{on } B \cup S \text{ for } t \geq 0;$$

$$\phi(x, 0, z; 0, \epsilon) = \partial_t \phi(x, 0, z; 0, \epsilon) = 0,$$

where $\delta(|P - P_s|)$ denotes Dirac's measure at P_s.

Seeking $G(P, P_s; t, \epsilon)$ in the same form of (10.65) as in Subsection 10.2.1, we see that equations (10.67) and (10.68) for φ_m remain be true, but instead of (10.66), (10.69), and (10.70) we obtain the following for $\tau > 0$:

$$\nabla^2 \varphi_0 = Q(\tau)\delta(|P - P_s|) \quad \text{in } W;$$

$$\nabla^2 \varphi_m = 0 \quad \text{in } W, m = 1, 2, \ldots;$$

$$\partial_n \varphi_m = 0 \quad \text{on } B \cup S, m = 0, 1, \ldots$$

Also, the initial-boundary value problem (10.21)–(10.23) and (10.25) determines ψ_m.

Again, representing φ_m in the form of (10.71) where β_m is given by (10.54), we obtain that (10.73) holds. Besides, we get

$$\nabla^2 v_0 = -\delta(|P - P_s|) \quad \text{in } W, \qquad v_0 = 0 \quad \text{on } F, \qquad \partial_n v_0 = 0 \quad \text{on } B \cup S,$$

and the rest of v_m, $m = 2, 4, \ldots, 2k, \ldots$, must be found from the sequence of recurrent boundary value problems (10.75). After that we obtain ψ_m, $m = 0, 1, \ldots$, by solving (10.21)–(10.23) and (10.25).

Concerning hydrodynamic corollaries, we note that they repeat literally those given in the case of a brief movement of rigid surfaces with one exception. To be exact, (10.76) presents the leading terms of asymptotics, and (10.77), (10.78), and (10.79) remain true, giving the asymptotics of the force, its impulse during $(0, t)$, and the free surface elevation, respectively. Since the Dirichlet integral of v_0 diverges in the case of a source, it is meaningless to consider the asymptotics of energy.

In the present case, estimate (10.82) for the remainder term r_N can be obtained in the same way as in Subsection 10.1.2.3, as well as estimates for $\|\partial_t r_N\|$ and $|r_N|$.

Concerning explicit formulae in the case in which W is a layer of constant depth d, they can be derived in the same way as those in Subsection 10.1.3.2.

10.3. High-Frequency Surface Pressure

In the present section we study waves caused by the pressure

$$p(x, z; t, \varepsilon) = \kappa(t/\varepsilon)\mathcal{P}(x, z) \tag{10.83}$$

applied to the horizontal free surface resting at the initial moment $t = 0$. Here κ is a 1-periodic function and the frequency ε^{-1} is assumed to be high in comparison with the inverse of the characteristic time $(g/d)^{1/2}$, where g is the gravity acceleration and d is a characteristic length. Our aim is to show that the asymptotic expansion

$$\phi(P; t, \varepsilon) \sim \sum_{m=0}^{\infty} \varepsilon^m \alpha_m w_m(P, t) + \sum_{m=1}^{\infty} \varepsilon^{2m-1} \beta_m \left(\frac{t}{\varepsilon}\right) v_m(P) \tag{10.84}$$

holds for the velocity potential. Here β_m $(m = 1, 2, 3, \dots)$ are certain 1-periodic functions, α_m $(m = 0, 1, 2, \dots)$ are constants expressed in terms of β_m, and the harmonic functions v_m, w_m do not depend on ε. Moreover, v_m satisfies the Dirichlet condition on the free surface, and the Neumann condition on rigid surfaces, whereas w_m are solutions of the Cauchy–Poisson problem. Both sequences of functions are defined recurrently.

An analysis of the principal term in (10.84) demonstrates that up to $O(\varepsilon)$ the waves are the same as those resulting from the initial elevation of the free surface equal to

$$[\langle\kappa\rangle - \kappa(0)]\mathcal{P}(x, z),$$

where $\langle\kappa\rangle$ is the mean value of κ. Furthermore, if $\langle\kappa\rangle = \kappa(0)$, then the wave pattern is stationary up to $O(\varepsilon)$, but along with this slow wave motion there exists a high-frequency wave of amplitude $O(\varepsilon)$ having zero average and giving a finite contribution to the force acting on a submerged body [see the first term in the second sum in (10.84)].

As in Sections 10.1 and 10.2, we begin with the formal derivation of (10.84) and then discuss asymptotic formulae for hydrodynamic characteristics, and the concluding subsection is concerned with proving estimates for the remainder term. We present our results for the three-dimensional case, but the same argument and similar asymptotic formulae are true for the two-dimensional problem.

10.3.1. Statement of the Problem and Formal Asymptotics

10.3.1.1. Statement of the Problem

Let the geometric assumptions made in Section 10.1 hold; then the initial-boundary value problem takes the following form in the nondimensional variables introduced in Subsection 10.1.2:

$$\nabla^2\phi = 0 \quad \text{in } W \text{ for } t \geq 0, \tag{10.85}$$

$$\phi_{tt} + \phi_y = -p_t \quad \text{on } F \text{ for } t \geq 0, \tag{10.86}$$

$$\partial\phi/\partial n = 0 \quad \text{on } B \cup S \text{ for } t \geq 0, \tag{10.87}$$

$$\phi(x, 0, z; 0) = \phi_t(x, 0, z; 0) = 0. \tag{10.88}$$

We assume that the surface pressure p is given by (10.83), where $\mathcal{P}$ is a sufficiently smooth function decaying at infinity. Our aim is to construct an asymptotic expansion for ϕ valid for $\varepsilon \ll 1$.

10.3.1.2. Formal Asymptotic Expansion

The velocity potential is sought in the form of a two-scale asymptotic series:

$$\phi(P; t, \varepsilon) \sim \sum_{m=0}^{\infty} \varepsilon^m \left[\varphi_m\left(P; \frac{t}{\varepsilon}\right) + \psi_m(P; t)\right], \tag{10.89}$$

where $\varphi_m(P; \tau)$ $(m = 1, 2, \ldots)$ are 1-periodic functions of the second argument and both $\varphi_m(P; \tau)$ and $\psi_m(P; t)$ decay as $|P| \to \infty$. Substituting (10.89) into (10.85)–(10.88) and equating the coefficients at each power of ε, one obtains, for $m = 0, 1, \ldots$, the following:

$$\nabla^2(\varphi_m + \psi_m) = 0 \quad \text{in } W, \tag{10.90}$$

$$\partial_\tau^2\varphi_{m+2} + \partial_y\varphi_m + \partial_t^2\psi_m + \partial_y\psi_m = 0 \quad \text{on } F, \tag{10.91}$$

$$\partial(\varphi_m + \psi_m)/\partial n = 0 \quad \text{on } B \cup S \tag{10.92}$$

for $t, \tau \geq 0$ and

$$\varphi_m + \psi_m = 0, \tag{10.93}$$

$$\partial_\tau\varphi_{m+1} + \partial_t\psi_m = 0 \tag{10.94}$$

on F when $t = \tau = 0$. Furthermore, one gets the following on F:

$$\partial_\tau^2 \varphi_0 = 0 \quad \text{for } \tau \geq 0, \qquad \partial_\tau \varphi_0 = 0 \quad \text{when } \tau = 0, \tag{10.95}$$

$$\partial_\tau^2 \varphi_1 = -\kappa'(\tau)\mathcal{P}(x, z) \quad \text{for } \tau \geq 0. \tag{10.96}$$

Moreover, functions of τ and t must be equated to zero separately in (10.90)–(10.92). Let us analyze the arising equations.

From (10.95) we find that

$$\varphi_0 = C_0 = \text{const} \quad \text{on } F \text{ for } \tau \geq 0,$$

and $C_0 = 0$ because φ_0 vanishes at infinity. Also, φ_0 satisfies the Laplace equation in W and the homogeneous Neumann condition on $B \cup S$, and so

$$\varphi_0 = 0 \quad \text{in } W \text{ for } \tau \geq 0. \tag{10.97}$$

From (10.91) it follows that

$$\partial_\tau^2 \varphi_m + \partial_y \varphi_{m-2} = 0 \quad \text{on } F \text{ for } \tau \geq 0. \tag{10.98}$$

Since φ_m is a periodic function of τ and decays as $|P| \to \infty$, (10.97) and (10.98) yield that

$$\varphi_{2k} = 0 \quad \text{in } W \text{ for } \tau \geq 0, \quad k = 1, 2, \ldots, \tag{10.99}$$

and so it remains to determine $\varphi_{2k-1}(P; \tau)$, which is sought in the form of $\beta_k(\tau) v_k(P)$. According to (10.96),

$$\beta_1'' = \kappa' \quad \text{for } \tau \geq 0, \qquad v_1 = -\mathcal{P} \quad \text{on } F. \tag{10.100}$$

Taking into account (10.90), (10.92), and the last condition, we arrive at the boundary value problem:

$$\nabla^2 v_1 = 0 \quad \text{in } W, \qquad v_1 = -\mathcal{P} \quad \text{on } F, \qquad \partial v_1 / \partial n = 0 \quad \text{on } B \cup S. \tag{10.101}$$

The first equation (10.100) has a periodic solution

$$\beta_1(\tau) = \int_0^\tau [\kappa(\mu) - \langle \kappa \rangle] \, \mathrm{d}\mu + c, \tag{10.102}$$

where $\langle \kappa \rangle = \int_0^1 \kappa(\mu) \, \mathrm{d}\mu$ and c is an arbitrary constant. Another form of (10.102) is as follows:

$$\beta_1(\tau) = \frac{-i}{2\pi} \sum_{n \neq 0} \frac{\kappa_n}{n} e^{2\pi i n \tau} + c_1, \tag{10.103}$$

where κ_n ($n = 0, \pm 1, \pm 2, \ldots$) are the Fourier coefficients of κ and c_1 is a constant.

From (10.98) we get the following for odd $m \geq 3$:

$$\beta_k'' = \beta_{k-1} \quad \text{for } \tau \geq 0, \qquad v_k = -\partial v_{k-1}/\partial y \quad \text{on } F, \quad k = 2, 3, \ldots \tag{10.104}$$

Combining the second of these relations with (10.90) and (10.92), we get a sequence of boundary value problems:

$$\nabla^2 v_k = 0 \ \text{ in } W, \quad \partial v_k/\partial n = 0 \ \text{ on } B \cup S, \quad v_k = -\partial_y v_{k-1} \ \text{ on } F, \tag{10.105}$$

where $k = 2, 3, \ldots$.

In order to obtain a periodic solution of the first equation (10.104) for $k = 2$, one has to put $c_1 = 0$ in (10.103). Then

$$\beta_2(\tau) = \frac{i}{(2\pi)^3} \sum_{n \neq 0} \frac{\kappa_n}{n^3} e^{2\pi i n \tau} + c_2,$$

where c_2 is an arbitrary constant that must be taken equal to zero on the next step. Proceeding in the same manner, we find that

$$\beta_k(\tau) = \frac{i(-1)^k}{(2\pi)^{2k-1}} \sum_{n \neq 0} \frac{\kappa_n}{n^{2k-1}} e^{2\pi i n \tau}, \quad k = 1, 2, \ldots. \tag{10.106}$$

Thus, (10.97), (10.99), (10.101), (10.105), and (10.106) give a complete description of the first term in brackets in (10.89) for all values of m, and so, the second series in (10.84) is obtained.

Putting

$$\alpha_m = \frac{-i^m}{(2\pi)^m} \sum_{n \neq 0} \frac{\kappa_n}{n^m}, \quad m = 0, 1, \ldots,$$

we see that

$$\alpha_m = \begin{cases} -\beta_{k+1}'(0) & \text{for } m = 2k \quad (k = 0, 1, \ldots) \\ -\beta_k(0) & \text{for } m = 2k - 1 \quad (k = 1, 2, \ldots) \end{cases}.$$

Seeking ψ_m in the form $\alpha_m w_m$, we get from (10.90)–(10.94) a sequence of initial-boundary value problems:

$$\nabla^2 w_m = 0 \quad \text{in } W, \tag{10.107}$$

$$\partial_t^2 w_m + \partial_y w_m = 0 \quad \text{on } F, \tag{10.108}$$

$$\partial w_m/\partial n = 0 \quad \text{on } B \cup S, \tag{10.109}$$

where $t \geq 0$ and $m = 0, 1, \ldots$; the initial conditions on F at $t = 0$ have the form:

$$w_m = \begin{cases} 0 & \text{for } m = 2k \quad (k = 0, 1, \ldots), \\ v_k & \text{for } m = 2k - 1 \quad (k = 1, 2, \ldots), \end{cases} \tag{10.110}$$

$$\partial_t w_m = \begin{cases} v_{k+1} & \text{for } m = 2k \quad (k = 0, 1, \ldots), \\ 0 & \text{for } m = 2k - 1 \quad (k = 1, 2, \ldots). \end{cases} \tag{10.111}$$

Here we take into account the formulae for φ_m obtained above. Thus the formal derivation of the asymptotic expansion (10.84) is complete.

10.3.2. Hydrodynamic Corollaries

Let us consider the terms in (10.84) containing ε^0 and ε in detail. By virtue of (10.102) and the definition of α_m we have

$$\phi(P; t, \varepsilon) = [\langle\kappa\rangle - \kappa(0)]\, w_0(P; t) + \varepsilon \left\{ \int_0^{t/\varepsilon} [\kappa(\mu) - \langle\kappa\rangle]\, \mathrm{d}\mu + \int_0^1 \mu[\kappa(\mu) - \langle\kappa\rangle]\, \mathrm{d}\mu \right\} v_1(P) + \cdots. \tag{10.112}$$

Here v_1 satisfies (10.101) and w_0 satisfies

$$\nabla^2 w_0 = 0 \quad \text{in } W \text{ for } t \geq 0,$$
$$\partial_t^2 w_0 + \partial_y w_0 = 0 \quad \text{on } F \text{ for } t \geq 0,$$
$$\partial w_0 / \partial n = 0 \quad \text{on } B \cup S \text{ for } t \geq 0,$$
$$w_0(x, 0, z; 0) = 0, \qquad \partial_t w_0(x, 0, z; 0) = -\mathcal{P}(x, z),$$

and so we can interpret w_0 as the velocity potential describing waves caused by the initial elevation of the free surface equal to $\mathcal{P}(x, z)$.

Turning to hydrodynamic characteristics defined in Subsection 10.1.2.2, we begin with the asymptotics of the hydrodynamic pressure $p = -\phi_t$ in the interior of water. Differentiating (10.84) and using (10.102), we get

$$p(P; t, \varepsilon) = -[\langle\kappa\rangle - \kappa(0)]\partial_t w_0(P; t) - [\kappa(t/\varepsilon) - \langle\kappa\rangle]\, v_1(P) + O(\varepsilon). \tag{10.113}$$

Substituting this into (10.10), we arrive at the asymptotic formula for the force acting on a submerged body:

$$\mathbf{F}(t, \varepsilon) = [\langle\kappa\rangle - \kappa(0)] \int_S \partial_t w_0\, \mathbf{n}\, \mathrm{d}S + [\kappa(t/\varepsilon) - \langle\kappa\rangle] \int_S v_1\, \mathbf{n}\, \mathrm{d}S + O(\varepsilon).$$

Replacing $\mathbf{n}$ by $\mathbf{r}_0 \times \mathbf{n}$, we get the asymptotics for the moment about P_0. Thus the principal parts of the force and moment include slowly as well as rapidly oscillating terms. We need v_1 as well as w_0 for calculating the principal term of force, but the analogous term for the force impulse during $(0, t)$ requires only w_0:

$$\mathbf{S}(t, \varepsilon) = [\langle \kappa \rangle - \kappa(0)] \int_S w_0 \, \mathbf{n} \, \mathrm{d}S + O(\varepsilon).$$

This is a consequence of the fact that v_1 is involved only in the rapidly oscillating part of the force having the zero mean value.

For the elevation of the free surface given by (10.2), we obtain the following asymptotics:

$$\eta(x, z; t, \varepsilon) = [\kappa(0) - \langle \kappa \rangle] \partial_t w_0(x, 0, z; t) - \langle \kappa \rangle \mathcal{P}(x, z) + O(\varepsilon). \quad (10.114)$$

Using it in (10.34), we get the asymptotic representation of the wave energy in the following form:

$$E(t, \varepsilon) = \frac{[\langle \kappa \rangle - \kappa(0)]^2}{2} \int_W |\nabla w_0|^2 \, \mathrm{d}x\mathrm{d}y\mathrm{d}z$$

$$+ \frac{1}{2} \int_F \{[\langle \kappa \rangle - \kappa(0)] \partial_t w_0 + \langle \kappa \rangle \mathcal{P}\}^2 \, \mathrm{d}x\mathrm{d}z + O(\varepsilon).$$

The conservation law for energy (see Stoker [312], Section 6.9) gives that

$$\int_W |\nabla w_0|^2 \, \mathrm{d}x\mathrm{d}y\mathrm{d}z + \int_F (\partial_t w_0)^2 \, \mathrm{d}x\mathrm{d}z = \int_F \mathcal{P}^2 \, \mathrm{d}x\mathrm{d}z,$$

and so

$$E(t, \varepsilon) = \frac{1}{2} \{[\langle \kappa \rangle - \kappa(0)]^2 + \langle \kappa \rangle^2\} \int_F \mathcal{P}^2 \, \mathrm{d}x\mathrm{d}z$$

$$+ \langle \kappa \rangle [\langle \kappa \rangle - \kappa(0)] \int_F \mathcal{P} \, \partial_t w_0 \, \mathrm{d}x\mathrm{d}z + O(\varepsilon) \quad (10.115)$$

is the final asymptotics for energy.

10.3.3. Waves in a Layer of Constant Depth

Let the water domain be a layer of constant depth; that is, $S = \emptyset$ and

$$W = \{-\infty < x, z < +\infty, -d < y < 0\}.$$

Then the Fourier transform

$$\hat{\mathcal{P}}(\sigma) = \int_{\mathbb{R}^2} \mathcal{P}(x, z) e^{-i(x\xi + z\zeta)} \, \mathrm{d}x\mathrm{d}z, \quad \text{where } \sigma = (\xi, \zeta),$$

allows us to solve the initial-boundary value problem (10.85)–(10.88) in the following form (cf. Subsection 10.1.3.2):

$$\phi(P;t,\varepsilon) = \frac{1}{(2\pi)^2\varepsilon}\int_{\mathbb{R}^2} \hat{\mathcal{P}}(\sigma)\frac{\cosh|\sigma|(y+d)}{|\sigma|^{1/2}\cosh|\sigma|d} e^{i(x\xi+z\zeta)}\,\mathrm{d}\xi\,\mathrm{d}\zeta$$
$$\times \int_0^t \kappa'\left(\frac{\mu}{\varepsilon}\right)\sin\left[(\mu-t)\left(|\sigma|\tanh|\sigma|d\right)^{1/2}\right]\mathrm{d}\mu.$$

In the case of deep water, $W=\mathbb{R}^3_-$ and the last formula must be changed as follows: first, omit $\tanh|\sigma|d$; second, replace $[\cosh|\sigma|(y+d)]/(\cosh|\sigma|d)$ by $e^{|\sigma|y}$.

The behavior of the explicit expression for ϕ is not obvious as $\varepsilon\to 0$, and so the asymptotic expansion (10.84) is preferable because its terms are simpler:

$$w_m(P;t) = \frac{-1}{(2\pi)^2}\int_{\mathbb{R}^2} \hat{\mathcal{P}}(\sigma)\left(|\sigma|\tanh|\sigma|d\right)^{(m-1)/2}$$
$$\times\frac{\cosh|\sigma|(y+d)}{\cosh|\sigma|d}\sin\left[\frac{m\pi}{2}+t(|\sigma|\tanh|\sigma|d)^{1/2}\right]e^{i(x\xi+z\zeta)}\,\mathrm{d}\xi\,\mathrm{d}\zeta,$$

where $m=0,1,\dots,$ and

$$v_m(P) = \frac{(-1)^m}{(2\pi)^2}\int_{\mathbb{R}^2}\hat{\mathcal{P}}(\sigma)(|\sigma|\tanh|\sigma|d)^{m-1}\frac{\cosh|\sigma|(y+d)}{\cosh|\sigma|d}e^{i(x\xi+z\zeta)}\,\mathrm{d}\xi\,\mathrm{d}\zeta,$$

where $m=1,2,\dots,$ and formulae for deep water can be obtained in the same way as above.

Furthermore, considering the energy of waves in a layer, we note that Parseval's equality gives an explicit formula,

$$\int_F \mathcal{P}\,\partial_t w_0\,\mathrm{d}x\mathrm{d}z = -(2\pi)^{-2}\int_{\mathbb{R}^2}|\tilde{\mathcal{P}}(\sigma)|^2\cos\left[t(|\sigma|\tanh|\sigma|d)^{1/2}\right]\mathrm{d}\xi\,\mathrm{d}\zeta,$$

for the time-dependent integral in the principal term for the energy.

Example

Let $W=\mathbb{R}^2_-$ and $\mathcal{P}(x)=(a^2+x^2)^{-1}$, where $a>0$. Then formula 3.767.2 in Gradshteyn and Ryzhik [96] gives that $\hat{\mathcal{P}}(\xi)=\pi a^{-1}e^{-a|\xi|}$, and combining this with the equation preceding example we get

$$\int_F \mathcal{P}\partial_t w_0\,\mathrm{d}x = -(2\pi)^{-1}\int_{-\infty}^{+\infty}|\mathcal{P}(\xi)|^2\cos\left(t|\xi|^{1/2}\right)\mathrm{d}\xi$$
$$= \left(\frac{2\pi}{a}\right)^2\int_0^{\infty}\mu e^{-2a\mu^2}\cos t\mu\,\mathrm{d}\mu.$$

Applying formula 3.953.4 in [96], we obtain

$$\int_F \mathcal{P} \partial_t w_0 \, dx = -\frac{\pi}{2a} \left\{ 1 + \frac{i\pi^{1/2}}{\sqrt{8a}} \exp\left[-\frac{t^2}{(8a)}\right] \operatorname{erf}\left(\frac{it}{\sqrt{8a}}\right) \right\},$$

where erf $(\cdot)$ denotes the error function (see 7.1.1 in Abramowitz and Stegun [1]). On the other hand,

$$\int_F \mathcal{P}^2 \, dx = 2 \int_0^\infty \frac{dx}{(a^2 + x^2)^2} = \frac{\pi}{2a}.$$

Substituting these expressions into (10.115), we get

$$E(t, \varepsilon) = \frac{\pi \kappa^2(0)}{4a} + [\kappa(0) - \langle \kappa \rangle] \langle \kappa \rangle \frac{it}{4\sqrt{2}} \left(\frac{\pi}{a}\right)^{3/2} \exp\left\{-\frac{t^2}{8a}\right\} \operatorname{erf}\left(\frac{it}{\sqrt{8a}}\right) + O(\varepsilon).$$

Here the first term corresponds to the initial energy at $t = 0$ and the second term describes how energy's principal part evolves in time.

Using (10.113), we can express the principal part of the average pressure in water as follows:

$$[\kappa(0) - \langle \kappa \rangle] \left[\frac{y - a}{a[x^2 + (y - a)^2]^2} - \frac{t\pi^{1/2}}{2a} \times \operatorname{Im} \left(\frac{\exp\{(-t^2)/[4(a - y - ix)]\}}{(a - y - ix)^{3/2}} \operatorname{erf}\left[\frac{it}{2(a - y - ix)^{1/2}}\right] \right) \right]. \tag{10.116}$$

We can also write the rapidly oscillating part of pressure explicitly:

$$\left[\langle \kappa \rangle - \kappa\left(\frac{t}{\varepsilon}\right) \right] \frac{y - a}{a[x^2 + (y - a)^2]^2},$$

which combines with (10.116) to produce the asymptotics of pressure:

$$p(P; t, \varepsilon) = \left[\kappa(0) - \kappa\left(\frac{t}{\varepsilon}\right) \right] \frac{y - a}{a[x^2 + (y - a)^2]} + [\langle \kappa \rangle - \kappa(0)] \frac{t\pi^{1/2}}{2a} \times \operatorname{Im} \left(\frac{\exp\{(-t^2)/[4(a - y - ix)]\}}{(a - y - ix)^{3/2}} \operatorname{erf}\left[\frac{it}{2(a - y - ix)^{1/2}}\right] \right) + O(\varepsilon). \tag{10.117}$$

Starting from (10.114) and proceeding in the same way, we obtain the asymptotics of the free surface elevation:

$$\eta(x,t,\varepsilon) = \frac{-\kappa(0)}{a^2+x^2} + [\langle\kappa\rangle - \kappa(0)]\frac{t\pi^{1/2}}{2a} \times \operatorname{Im}\frac{\exp\{(-t^2)/[4(a-ix)]\}}{(a-ix)^{3/2}} \operatorname{erf}\left[\frac{it}{2(a-ix)^{1/2}}\right] + O(\varepsilon). \tag{10.118}$$

Here the second term is the time-dependent perturbation having a zero order with respect to ε and corresponding to the initial elevation.

Now we note that

$$\frac{1}{(a^2+x^2)^{N+1}} = N!\left(\frac{-1}{2a}\frac{\partial}{\partial a}\right)^N \frac{1}{a^2+x^2},$$

and so applying the operator $N![-(2a)^{-1}\partial_a]^N$ to the right-hand side in (10.118), we get a formula for the free surface elevation corresponding to $\mathcal{P}(x) = (a^2+x^2)^{-N-1}$.

Let us consider the evolution of the free surface profile and pressure in the case in which the pressure applied to the free surface is given as follows:

$$p(x;t,\varepsilon) = \frac{\lambda\sin^2(2\pi t/\varepsilon) - \alpha}{a^2+x^2}.$$

For $a = 6.0$, $\alpha = 0.2$, $\lambda = 0.21/\pi^{1/2}$, and $\varepsilon = 10.0$, the evolution is shown in Fig. 10.2. One sees that even for surprisingly large ε, the principal term – see (10.118) and Fig. 10.2(a) – provides a good approximation for the exact free surface shown in Fig. 10.2(b). The evolution of the hydrodynamic pressure at a depth of 0.5 is shown in Fig. 10.3 for the same p applied to the free surface. Figure 10.3(a) presents the principal term (10.117), which combines the mean value (10.117) shown in Fig. 10.3(c) with oscillations. One sees that the asymptotic behavior in Fig. 10.3(a) is in good agreement with the exact pressure shown in Fig. 10.3(b).

10.3.4. Justification of the Asymptotic Expansion

In the present subsection we justify (10.84) [see also (10.89)] and begin with estimating every term in (10.89). As in Subsection 10.1.2.3 we use the operator K mapping φ defined on F into $\phi_y(x,0,z)$, where ϕ solves the following boundary value problem:

$$\nabla^2\phi = 0 \quad \text{in } W, \qquad \partial\phi/\partial n = 0 \quad \text{on } B\cup S, \qquad \phi = \varphi \quad \text{on } F.$$

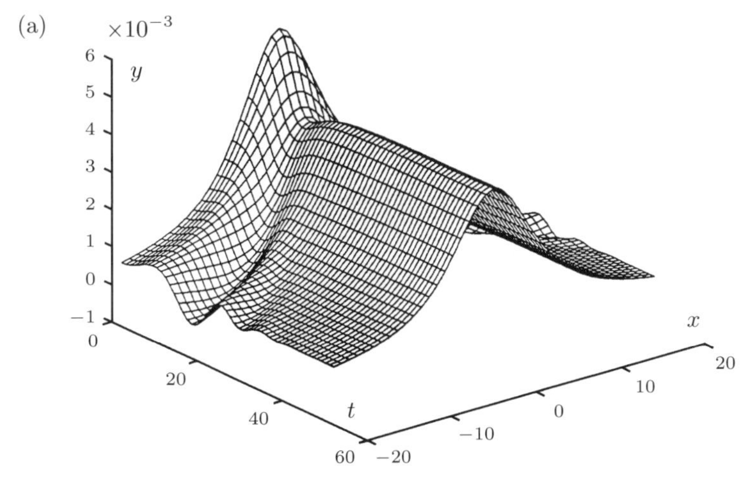

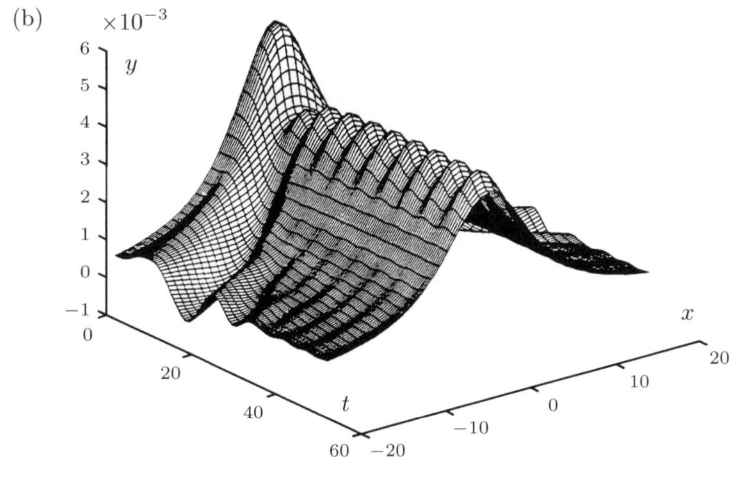

Figure 10.2.

It was demonstrated that for any $s \geq 1/2$ we have

$$\|K\varphi\|_{s-1} \leq M_s \|\varphi\|_s, \tag{10.119}$$

where $\|\cdot\|_s$ is the norm in the scale of Sobolev spaces $H^s(F)$ and this norm is involved in the following assertion.

Let $\kappa \in L^2(0, 1)$, and $\mathcal{P}$ belongs to $H^{m/2}(F)$ when m is odd and to $H^{(m+1)/2}(F)$ when m is even. Then we have two inequalities for functions in (10.89)*:*

$$\|\varphi_{2\ell-1}\|_{1/2} \leq C(2\pi)^{-2\ell+1} \|\kappa\|_{L^2(0,1)} \|K^{\ell-1}\mathcal{P}\|_{1/2}, \tag{10.120}$$

$$\|\psi_m\|_{1/2} \leq C(2\pi)^{-m} \|\kappa\|_{L^2(0,1)} \|w_m\|_{1/2}, \tag{10.121}$$

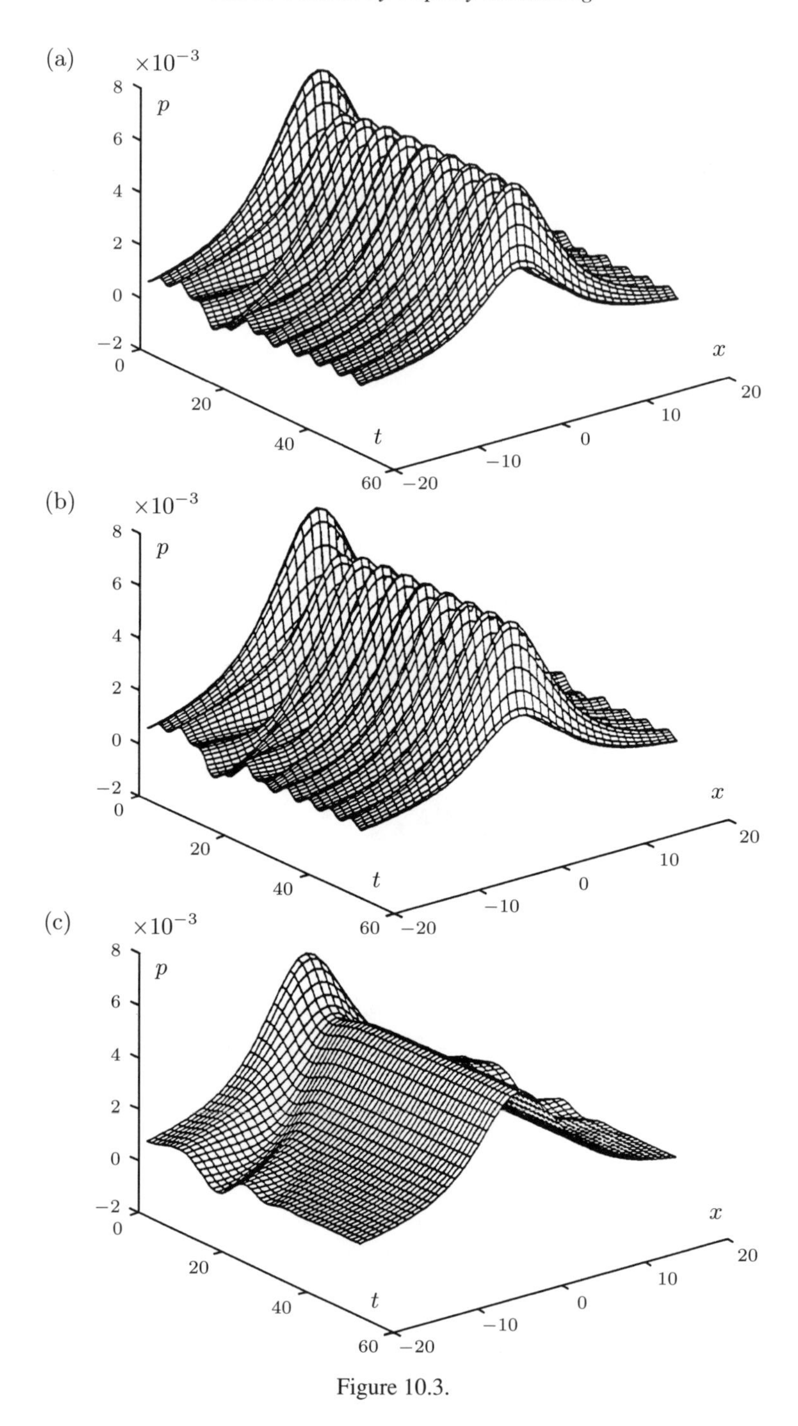
(a)
×10−3
8
6
4
2
0
−2
p
0
20
40
60
t
x
−20
−10
0
10
20
(b)
×10−3
8
6
4
2
0
−2
p
0
20
40
60
t
x
−20
−10
0
10
20
(c)
×10−3
8
6
4
2
0
−2
p
0
20
40
60
t
x
−20
−10
0
10
20

Figure 10.3.

where

$$\|w_m\|_{1/2} \leq \begin{cases} Ct\|K^\ell \mathcal{P}\|_{1/2} & \text{when } m = 2\ell, \\ C\|K^{\ell-1}\mathcal{P}\|_{1/2} & \text{when } m = 2\ell - 1. \end{cases} \tag{10.122}$$

Since $\varphi_{2\ell} = 0$, we have to estimate $\varphi_{2\ell-1}(P;\tau) = \beta_\ell(\tau)v_\ell(P)$. Using the Cauchy inequality, we get from (10.106) that

$$|\beta_\ell| \leq \frac{C}{(2\pi)^{2\ell-1}}\|\kappa\|_{L^2(0,1)}. \tag{10.123}$$

Taking into account that the functions v_m are defined recurrently, see (10.101) and (10.105), we get that

$$v_\ell = K^{\ell-1}\mathcal{P} \quad \text{on } F, \tag{10.124}$$

and so (10.123) and (10.124) imply (10.120).

Turning to $\psi_m(P;t) = \alpha_m w_m(P;t)$, we note that the definition of α_m yields (10.121) in the same way as (10.120). Since w_m solves (10.107)–(10.111), the trace of w_m on F (we shall use the same notation w_m for the trace) can be found as follows:

$$w_m(\cdot, t) = \begin{cases} K^{-1/2}\sin\left(K^{1/2}t\right)v_{\ell+1} & \text{for } m = 2\ell \\ \cos\left(K^{1/2}t\right)v_\ell & \text{for } m = 2\ell - 1 \end{cases} \quad \ell = 1, 2, \ldots.$$

This yields (10.122), if (10.124) is taken into account.

Let us estimate the remainder term,

$$\begin{aligned} R_{2N+1}(P;t,\varepsilon) &= \phi(P;t,\varepsilon) \\ &\quad - \sum_{m=0}^{2N+1}\varepsilon^m \alpha_m w_m(P,t) - \sum_{m=1}^{N+1}\varepsilon^{2m-1}\beta_m\left(\frac{t}{\varepsilon}\right)v_m(P), \end{aligned} \tag{10.125}$$

which will complete justification of (10.84). From (10.85), (10.88), (10.101), and (10.105), it follows that for $t \geq 0$,

$$\nabla^2 R_{2N+1} = 0 \quad \text{in } W, \tag{10.126}$$

$$\partial R_{2N+1}/\partial n = 0 \quad \text{on } B \cup S. \tag{10.127}$$

According to (10.86), (10.108), and (10.125), we have the following on F for $t \geq 0$:

$$\begin{aligned} \partial_t^2 R_{2N+1} + \partial_y R_{2N+1} &= -\varepsilon^{-1}\kappa'(\tau)\mathcal{P}(x,z) \\ &\quad - \sum_{m=1}^{N+1}\varepsilon^{2m-1}[\varepsilon^{-2}\beta_m''(\tau)v_m(x,0,z) \\ &\quad + \beta_m(\tau)\partial_y v_m(x,0,z)], \end{aligned}$$

and (10.100) and (10.104) reduce this condition to

$$\partial_t^2 R_{2N+1} + \partial_y R_{2N+1} = -\varepsilon^{2N+1} \beta_{N+1}(t/\varepsilon)\partial_y v_{N+1} \quad \text{on } F \text{ for } t \geq 0. \tag{10.128}$$

Substituting $t = 0$ into (10.125) and using the first condition (10.88) and (10.110), we get

$$R_{2N+1} = 0 \quad \text{on } F \text{ for } t = 0. \tag{10.129}$$

Differentiating (10.125) with respect to t and taking into account the second condition (10.88), (10.111), and the definition of α_m, we obtain that

$$\partial_t R_{2N+1} = \sum_{m=0}^{N} \varepsilon^m \beta'_{m+1}(0) v_{m+1} - \sum_{m=1}^{N+1} \varepsilon^{2m-2} \beta'_m(0) v_m$$

holds on F for $t = 0$, and so

$$\partial_t R_{2N+1} = 0 \quad \text{on } F \text{ for } t = 0. \tag{10.130}$$

Thus R_{2N+1} satisfies the initial-boundary value problem (10.126)–(10.130), and we are in a position to prove the following proposition.

Let $\mathcal{P} \in H^{(2N+3)/2}(F)$ and $\kappa \in L^2(0, 1)$. Then

$$\|R_{2N+1}\|_{1/2} \leq \left(\frac{\varepsilon}{2\pi}\right)^{2N+2} Ct\|K^{N+1}\mathcal{P}\|_{1/2}, \tag{10.131}$$

where C does not depend on t and $\mathcal{P}$.

Since R_{2N+1} satisfies (10.126)–(10.130), the trace of R_{2N+1} on F has the following explicit form:

$$R_{2N+1}(\cdot; t, \varepsilon) = -\varepsilon^{2N+1} \left\{ \int_0^t \beta_{N+1}\left(\frac{\mu}{\varepsilon}\right) \sin\left[(t-\mu)K^{1/2}\right] \mathrm{d}\mu \right\} K^{N+1/2}\mathcal{P}, \tag{10.132}$$

where (10.124) is taken into account. Integrating by parts in the last integral, we get

$$\begin{aligned} &R_{2N+1}(\cdot; t, \varepsilon) \\ &\quad = -\varepsilon^{2N+2} \left\{ \int_0^t \left[\int_0^{\mu/\varepsilon} \beta_{N+1}(\gamma)\,\mathrm{d}\gamma \right] \cos\left[(t-\mu)K^{1/2}\right] \mathrm{d}\mu \right\} K^{N+1}\mathcal{P}. \end{aligned} \tag{10.133}$$

Since β_{N+1} is a 1-periodic function having a zero mean value, we have that

$$\left| \int_0^{\mu/\varepsilon} \beta_{N+1}(\gamma)\,\mathrm{d}\gamma \right| \leq \max_\gamma |\beta_{N+1}(\gamma)|.$$

Combining this inequality with (10.123) and (10.133), we arrive at (10.131), which completes the proof.

In the following theorem justifying the asymptotic expansion (10.84), the expression

$$\phi(P;t,\varepsilon) - \sum_{m=0}^{2N} \varepsilon^m \alpha_m w_m(P;t) - \sum_{m=1}^{N} \varepsilon^{2m-1} \beta_m \left(\frac{t}{\varepsilon}\right) v_m(P)$$

is denoted by R_{2N}.

Let $\mathcal{P} \in H^{(n+2)/2}(F)$ and $\kappa \in L^2(0,1)$; then

$$|R_n| + \|R_n\|_{1/2} \leq \varepsilon^{n+1} C_n t \|\mathcal{P}\|_{(n+2)/2}, \tag{10.134}$$

where $|\cdot|$ is the energy norm defined at the end of Subsection 10.1.2.3, *and C_n does not depend on t and $\mathcal{P}$.*

Let $n = 2N+1$ first; then (10.134) follows from the second proposition in the present subsection combined with (10.119) and the known inequality [see (10.52) at the end of Subsection 10.1.2.3]:

$$|R_n| \leq C \|R_n\|_{1/2}. \tag{10.135}$$

For $n = 2N$ we write

$$R_{2N} = R_{2N+1} + \varepsilon^{2N+1} (\varphi_{2N+1} + \psi_{2N+1}),$$

and the required estimate for the sum in parentheses follows from the first proposition in the present subsection combined with (10.119). Applying (10.119) and (10.123) to (10.132), we get

$$\|R_{2N+1}\|_{1/2} \leq \varepsilon^{2N+1} t C_N \|\mathcal{P}\|_{N+1},$$

which together with (10.135) completes the proof.

When W is a layer, the following assertion provides the convergence of the asymptotic series (10.89).

Let W be a layer having the constant depth d, or the lower half-space $\{y < 0\}$. If $\kappa \in L^2(0,1)$ and the support of the Fourier transform $\hat{\mathcal{P}}$ is contained in $\{|\sigma| < (2\pi/\varepsilon)^2\}$, then (10.89) *converges in the $H^{1/2}(F)$ norm.*

It follows from the first proposition in the present subsection that the following inequality

$$\left(\frac{\varepsilon}{2\pi}\right)^n \left\|K^{n/2}\mathcal{P}\right\|_{1/2} \leq C q^n \tag{10.136}$$

with C not depending on n and $q < 1$ guarantees that (10.89) converges in the $H^{1/2}(F)$ norm. Using the Fourier transform, one obtains that

$$(K\mathcal{P})(x, z) = \frac{1}{(2\pi)^2} \int_{\mathbb{R}^2} \hat{\mathcal{P}}(\sigma)|\sigma| \tanh(|\sigma|d)\, e^{i(x\xi + z\zeta)}\, \mathrm{d}\xi\, \mathrm{d}\zeta$$

when W is the layer, and if $W = \{y < 0\}$, then $\tanh(|\sigma|d)$ must be omitted. For both cases, the Parseval theorem gives that

$$\left\| K^{n/2}\mathcal{P} \right\|_{1/2}^2 \le C \int_{\mathbb{R}^2} (1 + |\sigma|^2)^{1/2} |\sigma|^n |\hat{\mathcal{P}}(\sigma)|^2\, \mathrm{d}\xi\, \mathrm{d}\zeta.$$

By the assertion's assumption there exists $q \in (0, 1)$ such that $\hat{\mathcal{P}}(\sigma) = 0$ for $|\sigma| \ge (q2\pi/\varepsilon)^2$. Therefore,

$$\begin{aligned} \left\| K^{n/2}\mathcal{P} \right\|_{1/2}^2 &\le C \left[\int_0^{(q2\pi/\varepsilon)^2} |\sigma|^{n+1}\, \mathrm{d}|\sigma| + \int^{(q2\pi/\varepsilon)^2} |\sigma|^{n+2}\, \mathrm{d}|\sigma| \right] \\ &\le C \left(q\frac{2\pi}{\varepsilon} \right)^{2n}, \end{aligned}$$

and so (10.136) is proved, which completes the assertion's proof.

10.4. High-Frequency Underwater Disturbances

In this section we consider a source pulsating in water (in Subsection 10.4.1) and oscillations of rigid surfaces (in Subsection 10.4.2). In the case of a source, the water motion proves to be a superposition of two motions up to $O(\varepsilon)$. The principal term in the asymptotics for the velocity potential is as follows:

$$[\kappa(t/\varepsilon) - \langle\kappa\rangle] v_0(P, P_0) + \langle\kappa\rangle w_0(P, P_0, t),$$

where $v_0(P, P_0)$ is the waveless potential describing the stationary unit source at P_0, and $\langle\kappa\rangle w_0$ is the velocity potential describing the wave motion that is due to the source having the average strength starting at $t = 0$. Hence, the force acting on a submerged body has the form of a sum of two components. The first one is the large high-frequency force

$$\varepsilon^{-1}\kappa'\left(\frac{t}{\varepsilon}\right) \int_S v_0\, \mathbf{n}\, \mathrm{d}S.$$

Here S is the body surface, and $\mathbf{n}$ is the normal directed into water. The second component is the slowly varying force equal to

$$\langle\kappa\rangle \int_S \partial_t w_0\, \mathbf{n}\, \mathrm{d}S.$$

However, the impulse of force during the time interval $(0, t)$ has the following asymptotics:

$$\mathbf{S}(t, \varepsilon) = \left[\kappa\left(\frac{t}{\varepsilon}\right) - \langle\kappa\rangle\right]\iint_S v_0 \mathbf{n}\, dS + \langle\kappa\rangle \int_S w_0 \mathbf{n}\, dS + O(\varepsilon).$$

We see that both terms in the formula for force having different orders produce contributions of the same order into $\mathbf{S}(t, \varepsilon)$.

Similar results are obtained for waves caused by high-frequency oscillations of rigid surfaces. For both problems, we present our results for the three-dimensional case, but the same argument and similar asymptotic formulae are true for the two-dimensional problems.

10.4.1. Waves Caused by a Submerged Source

In this subsection we consider the problem of waves arising when a high-frequency submerged source starts making pulsations in resting water. As in the previous sections of this chapter, we seek an asymptotic expansion for the velocity potential assuming that $\varepsilon \ll 1$, where ε is the nondimensional period of oscillation. We begin with the formal derivation of (10.84), then discuss estimates for the remainder term, and conclude the subsection with asymptotic formulae for hydrodynamic characteristics.

10.4.1.1. Statement of the Problem

We assume the water domain W to be the same as in Section 10.3. Let the source's strength be equal to $-\kappa(t/\varepsilon)$, where κ is a 1-periodic function. If the source is placed at $P_0 \in W$, then the corresponding velocity potential $G(P, P_0; t, \varepsilon)$ must satisfy the initial-boundary value problem:

$$\nabla^2 G = -\kappa(t/\varepsilon)\delta_{P_0}(P) \quad \text{in } W \text{ for } t \geq 0, \tag{10.137}$$

$$G_{tt} + G_y = 0 \quad \text{on } F \text{ for } t \geq 0, \tag{10.138}$$

$$\partial G/\partial n = 0 \quad \text{on } B \cup S \text{ for } t \geq 0, \tag{10.139}$$

$$G = 0, \qquad G_t = 0 \quad \text{on } F \text{ for } t = 0. \tag{10.140}$$

Here δ_{P_0} is Dirac's measure located at P_0. Our aim is to construct an asymptotic series for G.

10.4.1.2. Asymptotic Expansion

It is natural to suppose that waves caused by the high-frequency source are the same, up to a term $O(\varepsilon)$, as the waves produced by the source having the average strength. Let $\langle\kappa\rangle w_0$ denote the velocity potential corresponding to the latter waves. Then w_0 satisfies

$$\nabla^2 w_0 = -\delta_{P_0}(P) \quad \text{in } W, \qquad \partial_t^2 w_0 + \partial_y w_0 = 0 \quad \text{on } F, \tag{10.141}$$

$$\partial w_0/\partial n = 0 \quad \text{on } B \cup S \text{ for } t \geq 0, \tag{10.142}$$

$$w_0 = 0, \qquad \partial_t w_0 = 0 \quad \text{on } F \text{ for } t = 0. \tag{10.143}$$

Let us seek the asymptotic series for $G - \langle\kappa\rangle w_0$ in the form of (10.89). Then we arrive at (10.90)–(10.96) for $m = 0, 1, \ldots$, but there are two exceptions. First, we have

$$\nabla^2 \varphi_0 = [\langle\kappa\rangle - \kappa(\tau)]\delta_{P_0}(P) \quad \text{in } W \text{ for } \tau \geq 0 \tag{10.144}$$

instead of the Laplace equation, and second,

$$\partial_\tau^2 \varphi_1 = 0 \quad \text{on } F \text{ for } \tau \geq 0 \tag{10.145}$$

instead of (10.96). Since (10.95) implies that

$$\varphi_0 = 0 \quad \text{on } F \text{ for } \tau \geq 0,$$

we put $\varphi_0(P;\tau) = b_0(\tau)v_0(P)$, where $b_0(\tau) = \kappa(\tau) - \langle\kappa\rangle$, and we get the boundary value problem for v_0:

$$\nabla^2 v_0 = -\,\delta_{P_0}(P) \quad \text{in } W, \qquad v_0 = 0 \quad \text{on } F, \qquad \partial v_0/\partial n = 0 \quad \text{on } B \cup S. \tag{10.146}$$

Here (10.144) and (10.92) are also taken into account. Then (10.145), (10.92), and (10.90) give that

$$\varphi_1 = 0 \quad \text{in } W \text{ for } \tau \geq 0$$

because φ_1 is a periodic function of τ decaying as $|P| \to \infty$. Using (10.98) in the same manner as in Subsection 10.3.2, we obtain

$$\varphi_{2k+1} = 0, \qquad \varphi_{2k}(P;\tau) = b_k(\tau)v_k(P) \quad \text{in } W \text{ for } \tau \geq 0, k = 1, 2, \ldots,$$

and so v_k satisfies (10.105), and

$$b_k(\tau) = \frac{(-1)^k}{(2\pi)^{2k}} \sum_{n\neq 0} \frac{b_n^{(0)}}{n^{2k}}\, e^{2\pi i n\tau}, \tag{10.147}$$

where by $b_n^{(0)}$ we denote the Fourier coefficients of $b_0 = \kappa - \langle\kappa\rangle$.

Let us consider

$$a_m = -\left(\frac{i}{2\pi}\right)^{m+1+(-1)^{m+1}} \sum_{n\neq 0} \frac{b_n^{(0)}}{n^{m+1+(-1)^{m+1}}},$$

which gives

$$a_m = \begin{cases} -b_k(0) & \text{for } m = 2k, \\ -b'_{k+1}(0) & \text{for } m = 2k-1, \end{cases} \quad \text{for } k = 1, 2, \ldots.$$

Seeking ψ_m ($m = 1, 2, \ldots$) in the form $a_m w_m$, we get from (10.90)–(10.94) the Cauchy–Poisson problem [see (10.107)–(10.109)] with the following initial conditions on F for $t = 0$ [cf. (10.110) and (10.111)]:

$$w_m = \begin{cases} 0 & \text{for } m = 2k-1, \\ v_k & \text{for } m = 2k, \end{cases} \tag{10.148}$$

$$\partial_t w_m = \begin{cases} v_{k+1} & \text{for } m = 2k-1, \\ 0 & \text{for } m = 2k, \end{cases} \tag{10.149}$$

where $k = 1, 2, \ldots$.

In order to keep the notation uniform we put $a_0 = \langle \kappa \rangle$. Then the asymptotic representation for the source potential can be written as follows:

$$G(P, P_0; t, \varepsilon) \sim \sum_{m=0}^{\infty} \varepsilon^m a_m w_m(P; t) + \sum_{m=0}^{\infty} \varepsilon^{2m} b_m \left(\frac{t}{\varepsilon}\right) v_m(P), \tag{10.150}$$

where a_m and $b_m(\tau)$ are defined above; see, for example, (10.147). Moreover, w_0 and v_0 satisfy (10.141)–(10.143) and (10.146), respectively, and other functions must be found recurrently by using the boundary value problem (10.105) for v_m, and the Cauchy–Poisson problem (10.107)–(10.109), (10.148), and (10.149) for w_m.

For justifying the asymptotic expansion (10.150), we consider the initial-boundary value problem for a remainder term:

$$R_{2N}(P; t, \varepsilon) = G(P, P_0; t, \varepsilon) - \sum_{m=0}^{2N} \varepsilon^m a_m w_m(P; t) - \sum_{m=0}^{N} \varepsilon^{2m} b_m \left(\frac{t}{\varepsilon}\right) v_m(P).$$

In the same way as in Subsection 10.3.4, one obtains the problem for R_{2N}:

$$\nabla^2 R_{2N} = 0 \quad \text{in } W \text{ for } t \geq 0,$$

$$\partial_t^2 R_{2N} + \partial_y R_{2N} = -\varepsilon^{2N} b_n(t/\varepsilon) \partial_y v_N \quad \text{on } F \text{ for } t \geq 0,$$

$$\partial R_{2N}/\partial n = 0 \quad \text{on } B \cup S \text{ for } t \geq 0,$$

$$R_{2N} = \partial R_{2N}/\partial t = 0 \quad \text{on } F \text{ for } t = 0.$$

We see that this problem is the same as (10.126)–(10.130), and so the asymptotic expansion (10.150) can be justified in the same way as that in Section 10.3, and the only new point is that the trace of $\partial_y v_0$ on F must be used instead of $\mathcal{P}$.

10.4.1.3. Source in a Layer

Explicit solutions can be obtained for v_m and w_m when W has a constant depth d. For this purpose one has to use a well-known identity

$$(a^2 + b^2)^{-1/2} = \int_0^\infty e^{-\mu b} J_0(\mu a)\, d\mu$$

in the same way as in Section 6.9 of Stoker's book [312]. This gives

$$v_0 = \frac{1}{4\pi}\left[\frac{1}{R} - \frac{1}{R_0} + 2\int_0^\infty e^{-\mu d} \frac{\sinh \mu y \sinh \mu y_0}{\cosh \mu d} J_0(\mu r)\, d\mu\right], \quad (10.151)$$

$$w_0 = \frac{1}{4\pi}\left\{\frac{1}{R} - \frac{1}{R_0} + 2\int_0^\infty e^{-\mu d} \frac{\sinh \mu y \sinh \mu y_0}{\cosh \mu d} J_0(\mu r)\, d\mu \right.$$
$$+ 4\int_0^\infty \frac{1 - \cos\left[(\mu \tanh \mu d)^{1/2} t\right]}{\sinh 2\mu d} \cosh \mu(y + d)$$
$$\left. \times \cosh \mu(y_0 + d) J_0(\mu r)\, d\mu \right\}, \quad (10.152)$$

where $r = [(x - x_0)^2 + (z - z_0)^2]^{1/2}$, $R = [r^2 + (y - y_0)^2]^{1/2}$, and $R_0 = [r^2 + (y + y_0)^2]^{1/2}$. For $m \geq 1$ we have

$$v_m = \frac{(-1)^{m+1}}{2\pi}$$
$$\times \int_0^\infty \mu(\mu \tanh \mu d)^{m-1} \frac{\cosh \mu(y + d) \cosh \mu(y_0 + d)}{\cosh^2 \mu d} J_0(\mu r)\, d\mu,$$

$$w_m = \frac{1}{2\pi}\int_0^\infty \mu(\mu \tanh \mu d)^{(m-2)/2} \frac{\cosh \mu(y + d) \cosh \mu(y_0 + d)}{\cosh^2 \mu d}$$
$$\times J_0(\mu r) \cos\left[\frac{m\pi}{2} - t(\mu \tanh \mu h)^{1/2}\right] d\mu.$$

The explicit solution of (10.137)–(10.140) in the layer of depth d is as follows:

$$G(P, P_0; t, \varepsilon)$$
$$= \frac{\kappa(t/\varepsilon)}{4\pi}\left[\frac{1}{R} - \frac{1}{R_0} + 2\int_0^\infty e^{-\mu d} \frac{\sinh \mu y \sinh \mu y_0}{\cosh \mu d} J_0(\mu r)\, d\mu\right]$$
$$+ \frac{1}{2\pi}\int_0^\infty \left(\frac{\mu}{\tanh \mu d}\right)^{1/2} \frac{\cosh \mu(y + d) \cosh \mu(y_0 + d)}{\cosh^2 \mu d} J_0(\mu r)\, d\mu$$
$$\times \int_0^t \kappa\left(\frac{\gamma}{\varepsilon}\right) \sin\left[(\mu \tanh \mu d)^{1/2}(t - \gamma)\right] d\gamma, \quad (10.153)$$

and so (10.150) gives an expansion of the time-dependent integral in (10.153).

If the depth is infinite, then the first integrals in (10.152) and (10.153) must be omitted as well as the integral in (10.151). Also, $\tanh \mu d$ should be replaced by one, and $\exp\{\mu(y + y_0)\}$ must replace

$$\frac{\cosh \mu(y+h) \cosh \mu(y_0+d)}{\cosh^2 \mu d}.$$

Moreover, $v_m = -(2\pi)^{-1}\partial_y^m R_0^{-1}$ for $m = 1, 2, \ldots$ in this case.

10.4.1.4. Asymptotics of Wave Characteristics

Let us begin with the principal term in the asymptotics for a source. From (10.150) we get that

$$G(P, P_0; t, \varepsilon) = [\kappa(t/\varepsilon) - \langle\kappa\rangle] v_0(P, P_0) + \langle\kappa\rangle w_0(P, P_0; t) + O(\varepsilon).$$

According to (10.146), the first term on the right-hand side describes the steady-state ε-periodic source having an average strength equal to zero. It produces no waves because of the homogeneous Dirichlet condition on F in (10.146). The second term is proportional to $\langle\kappa\rangle$ and w_0 satisfies (10.143). Therefore, it describes a source starting at $t = 0$, and its strength is equal to the mean value of κ. There are waves corresponding only to $\langle\kappa\rangle w_0$, and the asymptotics of the free surface elevation has the following form:

$$\eta(x, z; t, \varepsilon) = -\langle\kappa\rangle \partial_t w_0(x, 0, z; t) + O(\varepsilon).$$

The force and moment about P_* acting on a submerged body depend on v_0 and w_0:

$$\mathbf{F}(t, \varepsilon) = \varepsilon^{-1}\kappa'\left(\frac{t}{\varepsilon}\right)\int_S v_0\, \mathbf{n}\, dS + \langle\kappa\rangle \int_S \partial_t w_0\, \mathbf{n}\, dS + O(\varepsilon),$$

$$\mathbf{M}(t, \varepsilon) = \varepsilon^{-1}\kappa'\left(\frac{t}{\varepsilon}\right)\int_S v_0\, \mathbf{r}_* \times \mathbf{n}\, dS + \langle\kappa\rangle \int_S \partial_t w_0\, \mathbf{r}_* \times \mathbf{n}\, dS + O(\varepsilon).$$

We see that the principal terms of $\mathbf{F}$ and $\mathbf{M}$ tend to infinity as $\varepsilon \to 0$, while the contributions that are due to w_0 remain finite. The impulse of force during the time interval $(0, t)$ has the following asymptotics:

$$\mathbf{S}(t, \varepsilon) = \left[\kappa\left(\frac{t}{\varepsilon}\right) - \langle\kappa\rangle\right]\int_S v_0\, \mathbf{n}\, dS + \langle\kappa\rangle \int_S w_0\, \mathbf{n}\, dS + O(\varepsilon),$$

and so two terms having different orders in the formula for force produce contributions of the same order into $\mathbf{S}(t, \varepsilon)$. This is due to the high-frequency oscillations in the term containing ε^{-1}.

10.4.2. Oscillations of Underwater Surfaces

10.4.2.1. Main Results

Let B, S, or both execute high-frequency oscillations of small amplitude. Then the velocity potential $\phi(P; t, \varepsilon)$ must satisfy the following initial-boundary value problem:

$$\nabla^2 \phi = 0 \quad \text{in } W \text{ for } t \geq 0, \tag{10.154}$$

$$\phi_{tt} + \phi_y = 0 \quad \text{on } F \text{ for } t \geq 0, \tag{10.155}$$

$$\partial\phi/\partial n = \kappa(t/\varepsilon) f(P) \quad \text{on } B \cup S \text{ for } t \geq 0, \tag{10.156}$$

$$\phi = 0, \qquad \phi_t = 0 \quad \text{on } F \text{ for } t = 0. \tag{10.157}$$

It is assumed that $f(P)$ decays as $|P| \to \infty$ and $P \in B$.

As $\varepsilon \to 0$, one obtains the same asymptotic expansion (10.150) for ϕ solving (10.154)–(10.157), and so there is no difference with the case of a submerged source. Moreover, if $m \geq 1$, then v_m and w_m satisfy the same problems (10.105) and (10.107)–(10.109), (10.148), and (10.149), respectively. However, v_0 must be determined from the steady-state boundary value problem:

$$\nabla^2 v_0 = 0 \quad \text{in } W, \qquad v_0 = 0 \quad \text{on } F, \qquad \partial v_0/\partial n = f \quad \text{on } B \cup S. \tag{10.158}$$

Similarly, w_0 solves the initial-boundary value problem:

$$\begin{aligned} \nabla^2 w_0 = 0 \text{ in } W, \qquad \partial_t^2 w_0 + \partial_y w_0 = 0 \quad \text{on } F, \\ \partial w_0/\partial n = f \quad \text{on } B \cup S \text{ for } t \geq 0, \\ w_0 = \partial_t w_0 = 0 \quad \text{on } F \text{ for } t = 0. \end{aligned} \tag{10.159}$$

If $\langle\kappa\rangle = 0$, then (10.150) implies that the water motion is determined by v_0 up to $O(\varepsilon)$. Therefore, the amplitude of surface waves is proportional to ε because v_0 satisfies the homogeneous Dirichlet condition on F.

As in Subsections 10.3.3 and 10.4.1.3, the explicit expressions for v_m and w_m can be obtained when W is a layer of constant depth d. Using the Fourier transform, we get

$$v_0(P) = \frac{1}{(2\pi)^2} \int_{\mathbb{R}^2} \hat{f}(\sigma) \frac{\sinh|\sigma|y}{|\sigma|\cosh|\sigma|d} e^{i(x\xi + z\zeta)}\, d\sigma,$$

and for $m \geq 1$,

$$v_m(P) = \frac{(-1)^m}{(2\pi)^2} \int_{\mathbb{R}^2} \hat{f}(\sigma)(|\sigma|\tanh|\sigma|d)^{m-1} \frac{\cosh|\sigma|(y+d)}{\cosh^2|\sigma|d} e^{i(x\xi + z\zeta)}\, d\sigma.$$

The time-dependent functions w_m are given by

$$w_m(P;t) = \frac{1}{(2\pi)^2} \int_{\mathbb{R}^2} \hat{f}(\sigma)(|\sigma| \tanh |\sigma| d)^{(m-1)/2} e^{i(x\xi + z\zeta)} \times \frac{\cosh |\sigma|(y+d)}{\cosh^2 |\sigma| d} \sin\left[\frac{m\pi}{2} - t(|\sigma| \tanh |\sigma| d)^{1/2}\right] \mathrm{d}\sigma$$

for $m = 0, 1, \ldots$.

10.4.2.2. Asymptotics for Energy, Force, and Moment

Since the same asymptotic expansion (10.150) is valid for both submerged sources and surfaces, we do not repeat formulae from Subsection 10.4.1.4 but instead turn to the energy defined by (10.1) and immediately obtain

$$E(t, \varepsilon) = \frac{1}{2} \left\{ \int_W |[\kappa(t/\varepsilon) - \langle\kappa\rangle]\nabla v_0 + \langle\kappa\rangle \nabla w_0|^2 \,\mathrm{d}x\mathrm{d}y\mathrm{d}z + \langle\kappa\rangle^2 \int_F (\partial_t w_0)^2 \,\mathrm{d}x\mathrm{d}z \right\} + O(\varepsilon).$$

Using Green's formula and the boundary value problems (10.158) and (10.159), we obtain that

$$2\langle\kappa\rangle\, (\kappa - \langle\kappa\rangle) \int_W \nabla w_0 \cdot \nabla v_0 \,\mathrm{d}x\mathrm{d}y\mathrm{d}z + (\kappa - \langle\kappa\rangle)^2 \int_W |\nabla v_0|^2 \,\mathrm{d}x\mathrm{d}y\mathrm{d}z = (\kappa^2 - \langle\kappa\rangle^2) \int_W |\nabla v_0|^2 \,\mathrm{d}x\mathrm{d}y\mathrm{d}z,$$

and so the asymptotic formula for energy takes the following form:

$$E(t, \varepsilon) = \frac{\langle\kappa\rangle^2}{2} \left[\int_W |\nabla w_0|^2 \,\mathrm{d}x\mathrm{d}y\mathrm{d}z + \int_F (\partial_t w_0)^2 \,\mathrm{d}x\mathrm{d}z \right] + \frac{\kappa^2(t/\varepsilon) - \langle\kappa\rangle^2}{2} \int_W |\nabla v_0|^2 \,\mathrm{d}x\mathrm{d}y\mathrm{d}z + O(\varepsilon). \quad (10.160)$$

Averaging the energy over the period ε, we get

$$\langle E\rangle = \frac{\langle\kappa\rangle^2}{2} \left[\int_W |\nabla w_0|^2 \,\mathrm{d}x\mathrm{d}y\mathrm{d}z + \int_F (\partial_t w_0)^2 \,\mathrm{d}x\mathrm{d}z \right] + \frac{\langle\kappa^2\rangle - \langle\kappa\rangle^2}{2} \int_W |\nabla v_0|^2 \,\mathrm{d}x\mathrm{d}y\mathrm{d}z + O(\varepsilon). \quad (10.161)$$

In particular, if $\langle\kappa\rangle = 0$, then

$$\langle E\rangle = \frac{\langle\kappa^2\rangle}{2} \int_W |\nabla v_0|^2 \,\mathrm{d}x\mathrm{d}y\mathrm{d}z + O(\varepsilon). \quad (10.162)$$

We note that formulae (10.160)–(10.162) are also true in the two-dimensional case, in which the Dirichlet integral in (10.162) can be calculated explicitly for particular geometries.

Example

Let $W = \mathbb{R}^2_- \setminus \{|z - i| \leq a\}$, where $z = x + iy$ and $a < 1$. For solving the boundary value problem

$$\nabla^2 v_0 = 0 \quad \text{in } W, \qquad v_0 = 0 \quad \text{when } y = 0,$$
$$\partial v_0/\partial n = f(e^{i\varphi}) \quad \text{when } z + i = ae^{i\varphi},$$

we apply the conformal mapping

$$\zeta = \frac{z + i(1 - a^2)^{1/2}}{z - i(1 - a^2)^{1/2}},$$

which maps W onto $\{r < |\zeta| < 1\}$ on the ζ plane, where

$$r = \frac{a}{1 + (1 - a^2)^{1/2}}. \tag{10.163}$$

We put $\zeta = \rho e^{i\theta}$, and so the boundary value problem takes the following form:

$$\nabla^2 u = 0 \quad \text{in } \{r < \rho < 1\}, \qquad u|_{\rho=1} = 0, \qquad u_\rho|_{\rho=r} = 2(1 - a^2)^{1/2} g(e^{i\theta}).$$

Here $u(\rho, \theta) = v_0[z(\zeta)]$ and

$$g(e^{i\theta}) = \frac{f[z(re^{i\theta})]}{1 - 2r\cos\theta + r^2}.$$

Seeking a solution in the form

$$u(\rho, \theta) = 2(1 - a^2)^{1/2}\left[a_0 \log \rho + \sum_{n=1}^{\infty}(\rho^n - \rho^{-n})(a_n \cos n\theta + b_n \sin n\theta)\right],$$

we see that a_n and b_n must be determined from the Neumann condition, and so

$$a_0 = \frac{r\alpha_0}{2}, \qquad a_n + ib_n = \frac{r^{n+1}}{n(1 + r^{2n})}(\alpha_n + i\beta_n) \quad \text{for } n \geq 1,$$

where α_n and β_n are the Fourier coefficients of g.

Using the obvious equalities

$$\int_W |\nabla v_0|^2 \, \mathrm{d}x\mathrm{d}y = \int_{r<\rho<1} |\nabla u|^2 \, \mathrm{d}\xi \mathrm{d}\eta = -r \int_0^{2\pi} [u_\rho u]_{\rho=r} \, \mathrm{d}\theta$$

and the series obtained for u, we get

$$\int_W |\nabla v_0|^2 \,\mathrm{d}x\mathrm{d}y = 4\pi r^2(1-a^2)\left[-\frac{\alpha_0^2}{2}\log r + \sum_{n=1}^{\infty} n^{-1}\frac{1-r^{2n}}{1+r^{2n}}(\alpha_n^2+\beta_n^2)\right].$$

Therefore, the Dirichlet integral in (10.162) is expressed in terms of the Fourier coefficients of g.

Now, let us calculate principal terms in the asymptotic formulae for $\mathbf{F}(t,\varepsilon)$ and $M(t,\varepsilon)$. We have

$$\int_{|z+i|=a} v_0\,\mathbf{n}\,\mathrm{d}S = 2(1-a^2)^{1/2} r \int_0^{2\pi} \frac{u(r,\theta)\,\nu}{1-2r\cos\theta + r^2}\,\mathrm{d}\theta,$$

where $\nu = -(\sin\theta, \cos\theta)$. Substituting the series for u and calculating the integrals, we get the explicit asymptotic formula for the force acting per unit length of the circular cylinder:

$$\mathbf{F}(t,\varepsilon) = \frac{4\pi(1-a^2)^{1/2} r}{\varepsilon}\kappa'\left(\frac{t}{\varepsilon}\right)\sum_{n=1}^{\infty}\frac{r^n(1-r^{2n})}{(1+r^{2n})}\begin{pmatrix}\beta_n\\ \alpha_n\end{pmatrix} + O(1). \quad (10.164)$$

As above, α_n and β_n are the Fourier coefficients of g. Similarly, the moment about $P_0 = (x_0, y_0)$ is equal to

$$M(t,\varepsilon) = \frac{4\pi(1-a^2)^{1/2} r}{\varepsilon}\kappa'\left(\frac{t}{\varepsilon}\right)$$
$$\times \sum_{n=1}^{\infty}\frac{r^n(1-r^{2n})}{n(1+r^{2n})}[\beta_n(y_0+1) - \alpha_n x_0] + O(1). \quad (10.165)$$

In the case of a circular cylinder oscillating so that $\langle\kappa\rangle = 0$ and $f(e^{i\varphi}) = A = \text{const}$, formulae (10.162), (10.164), and (10.165) have very simple forms because

$$\alpha_n = \frac{2r^n A}{1-r^2}, \qquad \beta_n = 0,$$

and so

$$\langle E\rangle = 2\langle\kappa^2\rangle\pi a^2 A^2[-2-1\log r + \varphi(r)] + O(\varepsilon),$$

$$\mathbf{F}(t,\varepsilon) = \frac{4\pi Aa}{\varepsilon}\kappa'\left(\frac{t}{\varepsilon}\right)\varphi(r)\begin{pmatrix}0\\1\end{pmatrix} + O(\varepsilon),$$

$$M(t,\varepsilon) = -\frac{4\pi Aax_0}{\varepsilon}\kappa'\left(\frac{t}{\varepsilon}\right)\varphi(r) + O(\varepsilon),$$

where

$$\varphi(r) = \sum_{n=1}^{\infty} \frac{r^{2n}(1 - r^{2n})}{n(1 + r^{2n})} = \log \left[\frac{1 - r^2}{2r^{1/2}} \theta_2(0, r^2)\theta_3(0, r^2) \right].$$

Here r is defined by (10.163) and θ_2, θ_3 are theta functions (see Section 22.5 in Whittaker and Watson [360]).

10.5. Bibliographical Notes

10.1 and 10.2. The material in these sections was obtained by Kuznetsov and Maz'ya [164]. Another approach to waves generated by brief motions of underwater surfaces was developed by Sabatier [301].

10.3 and 10.4. The material in these sections is borrowed from Kuznetsov and Maz'ya's paper [167]. The same method was applied by Kuznetsov [160] for obtaining formal asymptotics in the problem describing the forward motion of a submerged body at a speed having high-frequency oscillations about a certain mean value. Waves in shallow channels of rapidly varying depth were considered by Nachbin and Papanicolaou [258]. Various singularly perturbed elliptic boundary value problems are considered in the book [226] by Maz'ya, Nazarov, and Plamenevsky.

Bibliography

[1] **Abramowitz, M., Stegun, I. A.,** *Handbook of Mathematical Functions.* National Bureau of Standards, 1964.

[2] **Aranha, J. A. P.,** Existence and some properties of waves trapped by submerged cylinder. *J. Fluid Mech.* **192** (1988) 421–433.

[3] **Aranha, J. A. P.,** Excitation of waves trapped by submerged slender structures, and nonlinear resonance. *J. Fluid Mech.* **192** (1988) 435–453.

[4] **Athanassoulis, G. A.,** An expansion theorem for water-wave potentials. *J. Eng. Math.* **18** (1984) 181–194.

[5] **Athanassoulis, G. A.,** On the solvability of a two-dimensional water-wave radiation problem. *Quart. Appl. Math.* **44** (1987) 601–620.

[6] **Athanassoulis, G. A., Kakilis, P. D., Politis, C. G.,** Low-frequency oscillations of a partially submerged cylinder of arbitrary shape. *J. Ship Res.* **39** (1995) 123–138.

[7] **Athanassoulis, G. A., Makrakis, G. N.,** An unusual wave equation arising in water-wave theory. *Differential Equations (Proc. of the EQUADIFF '87 Conf.)* New York: Marcel Dekker, 1987, pp. 49–55.

[8] **Athanassoulis, G. A., Makrakis, G. N.,** A function-theoretic approach to a two-dimensional transient wave-body interaction problem. *Applicable Anal.* **54** (1994) 283–303.

[9] **Athanassoulis, G. A., Politis, C. G.,** On the solvability of a two-dimensional wave-body interaction problem. *Quart. Appl. Math.* **48** (1990) 1–30.

[10] **Aubin, J.-P.,** *Approximation of Elliptic Boundary Value Problems.* New York: Wiley-Interscience, 1972.

[11] **Angell, T. S., Hsiao, G. C., Kleinman, R. E.,** An integral equation for the floating-body problem. *J. Fluid Mech.* **166** (1986) 161–171.

[12] **Angell, T. S., Hsiao, G. C., Kleinman, R. E.,** An optimal design problem for submerged bodies. *Math. Meth. Appl. Sci.* **8** (1986) 50–76.

[13] **Angell, T. S., Hsiao, G. C., Kleinman, R. E.,** Recent developments in floating body problems. *Mathematical Approaches in Hydrodynamics.* Philadelphia: SIAM, 1991.

[14] **Angell, T. S., Kleinman, R. E.,** A Galerkin procedure for optimization in radiation problems. *SIAM J. Appl. Math.* **44** (1984) 1246–1257.

[15] **Angell, T. S., Kleinman, R. E.,** On a domain optimization problem in hydrodynamics. *Optimal Control of Partial Differential Equations. II.* Basel: Birkhäuser, 1987.

[16] **Angell, T. S., Kleinman, R. E.,** A constructive method for shape optimization: a problem in hydrodynamics. *IMA J. Appl. Math.* **47** (1991) 265–281.

[17] **Baar, J. J. M., Price, W. G.**, Evaluation of the wavelike disturbance in the Kelvin wave source potential. *J. Ship Res.* **32** (1988) 44–53.

[18] **Bai, K. J., Yeung, R. W.,** Numerical solutions to free surface flow problems. *Proceedings of the Tenth Symposium on Naval Hydrodynamics.* Cambridge: MIT Press, 1974, pp. 609–647.

[19] **Barber, N. F., Ursell, F.,** The generation and propagation of ocean waves and swell. I. Wave periods and velocities. *Phil. Trans. Roy. Soc. Lond. A* **240** (1948) 527–560.

[20] **Bauer, M. G.,** Le problème de Neumann–Kelvin. I, II. *Annali di Mat.* **74** (1980) 233–256, 257–280.

[21] **Beale, J. T.,** Eigenfunction expansions for objects floating in an open sea. *Comm. Pure Appl. Math.* **30** (1977) 283–313.

[22] **Beale, J. T., Hou, T. Y., Lowengrub, J. S.,** Growth rates for the linearized motion of fluid interfaces away from equilibrium. *Comm. Pure Appl. Math.* **46** (1993) 1269–1301.

[23] **Bessho, M.,** On the fundamental function in the theory of the wave-making resistance of ships. *Mem. Def. Acad. Japan* **4** (1964) 99–119.

[24] **Bessho, M.,** Line integral, uniqueness and diffraction of waves in the linearized theory. *International Seminar on Wave Resistance.* Tokyo: Society of Naval Architects of Japan, 1976, pp. 45–55.

[25] **Bessho, M.,** On a consistent linearized theory of the wave-making resistance of ships. *J. Ship Res.* **38** (1994) 83–95.

[26] **Bhattacharya, R. N.,** *Ship Waves and Wave Resistance.* Calcutta: Jadavpur University Press, 1967.

[27] **Birman, M. S., Solomyak, M. Z.,** *Spectral Theory of Self-Adjoint Operators in Hilbert Spaces.* Dordrecht: Reidel, 1987.

[28] **Bochner, S.,** *Lectures on Fourier Integrals.* Princeton: Princeton University Press, 1959.

[29] **Bolton, W. E., Ursell, F.,** The wave force on an infinitely long circular cylinder in an oblique sea. *J. Fluid Mech.* **57** (1973) 241–256.

[30] **Bonnet–Ben Dhia, A.-S., Joly, P.,** Mathematical analysis of guided water waves. *SIAM J. Appl. Math.* **53** (1993) 1507–1550.

[31] **Bouligand, G.,** *Sur Divers Problèmes de la Dynamique des Liquides.* Paris: Gauthier-Villars, 1930.

[32] **Brard, R.,** The representation of a given ship form by singularity distributions when the boundary condition on the free surface is linearized. *J. Ship Res.* **16** (1972) 79–92.

[33] **Burago, Yu. D., Maz'ya, V. G.,** *Potential Theory and Function Theory for Irregular Regions. Steklov Mathematical Institute Seminars in Mathematics* (Leningrad) **5**. New York: Consultant Bureau, 1969.

[34] **Buslayeva, M. V.,** An application of scattering theory to one hydrodynamic problem. *Steklov Mathematical Institute Seminars in Mathematics* (Leningrad) **77**. Leningrad: Nauka, 1978, pp. 57–75. (In Russian).

[35] **Callan, M., Linton, C. M., Evans, D. V.,** Trapped modes in two-dimensional wave guides. *J. Fluid Mech.* **229** (1991) 51–64.

[36] **Carleman, T.,** *Über das Neumann–Poincarésche Problem für ein Gebiet mit Ecken.* Uppsala: Almqvist & Wiksells, 1916.

[37] **Chakrabarti, A.,** A survey on two mathematical methods used in scattering of surface waves. *Mathematical Techniques for Water Waves.* Southampton: Computational Mechanics Publications, 1997, pp. 231–253.

[38] **Clarisse, J.-M., Newman, J. N.,** Evaluation of the wave-resistance Green function. Part 3. The single integral near the singular axis. *J. Ship Res.* **38** (1994) 1–8.

[39] **Clarisse, J.-M., Newman, J. N., Ursell, F.,** Integrals with a large parameter: water waves on finite depth due to an impulse. *Proc. Roy. Soc. Lond. A* **450** (1995) 67–87.

[40] **Colton, D., Kress, R.,** *Integral Equation Methods in Scattering Theory.* New York: Wiley-Interscience, 1983.

[41] **Courant, R., Hilbert, D.,** *Methods of Mathematical Physics.* Vol. 2. New York: Interscience, 1961.

[42] **Crapper, G. D.,** *Introduction to Water Waves.* Chichester: Ellis Horwood, 1984.

[43] **Davies, E. B., Parnovski, L.,** Trapped modes in acoustic waveguides. *Quart. J. Mech. Appl. Math.* **51** (1998) 477–492.

[44] **Davis, A. M. J.,** On the short surface waves due to an oscillating, partially immersed body. *J. Fluid Mech.* **75** (1976) 791–807.

[45] **Dean, R. G., Ursell, F., Yu, Y. S.,** Forced small-amplitude water waves: comparison of theory and experiment. *J. Fluid Mech.* **7** (1959) 33–52.

[46] **Debnath, L.,** *Nonlinear Water Waves.* London: Academic, 1994.

[47] **Dern, J. C.,** Existence, uniqueness and regularity of Neumann–Kelvin problem for two or three-dimensional submerged bodies. *Proceedings of the Second International Conference on Numerical Ship Hydrodynamics.* Berkeley: University of California Press, 1977, pp. 57–77.

[48] **Dobrokhotov, S. Yu.,** Maslov's methods in the linearized theory of gravity waves on fluid's surface. *Dokl. Acad. Nauk USSR* **269** (1983) 76–80. (In Russian).

[49] **Dobrokhotov, S. Yu.,** Asymptotics for surface waves captured by by beaches and rough bottom. *Dokl. Acad. Nauk USSR* **289** (1986) 575–579. (In Russian).

[50] **Dobrokhotov, S. Yu., Tolstova, O. L., Chudinovich, I. Yu.,** Waves in a fluid on an elastic base. An existence theorem and exact solutions. *Math. Notes Russian Acad. Sci.* **54**, No. 6 (1993) 33–55.

[51] **Dobrokhotov, S. Yu., Zhevandrov, P. N.,** Maslov's operator method for the problem of water waves generated by a pressure point travelling above an uneven bottom. *Izv. Acad. Nauk USSR, Atmosph. Ocean Phys.* **21** (1985) 744–751. (In Russian).

[52] **Dobrokhotov, S. Yu., Zhevandrov, P. N.,** Nonstandard characteristics and the Maslov operator method in linear problems of time-dependent waves in water. *Functional Anal. Appl.* **19** (1985) 285–295.

[53] **Dobrokhotov, S. Yu., Zhevandrov, P. N., Kuz'mina, V. M.,** Asymptotic behaviour of the solution to the Cauchy–Poisson problem in a layer of non-constant thickness. *Math. Notes Russian Acad. Sci.* **53** (1993) 657–660.

[54] **Dobrokhotov, S. Yu., Zhevandrov, P. N., Korobkin, A. A., Sturova, I. V.,** Asymptotic theory of propagation of nonstationary surface and internal waves over

uneven bottom. *International Series of Numerical Mathematics* **106**. Basel: Birkhäuser, 1992, pp. 105–111.

[55] **Doppel, K.,** On the weakly formulated floating body problem. *Tech. Rep.* **87-4** (1987). University of Delaware, Newark.

[56] **Doppel, K., Hochmuth, R.,** An application of the limiting absorption principle to a mixed boundary value problem in an infinite strip. *Math. Meth. Appl. Sci.* **18** (1995) 529–548.

[57] **Doppel, K., Hsiao, G. C.,** On weak solutions of the floating body problem. *Tech. Rep.* **87-5** (1987). University of Delaware, Newark.

[58] **Doppel, K., Schomburg, B.,** On a weakly formulated exterior problem from linear hydrodynamics. *Math. Meth. Appl. Sci.* **10** (1988) 595–608.

[59] **Doppel, K., Schomburg, B.,** Regularity of solutions of the weak floating beam problem. *Zeit. Anal. Anwend.* **10** (1991) 461–477.

[60] **Doppel, K., Schomburg, B.,** The floating beam problem: weighted regularity of weak solutions. *Applic. Anal.* **45** (1992) 69–93.

[61] **Ehrenmark, U. T.,** Oblique wave incidence on a plane beach: the classical problem revisited. *J. Fluid Mech.* **368** (1998) 291–319.

[62] **Ehrenmark, U. T.,** A note on tuning in Roseau's alternative edge waves. *J. Fluid Mech.* **382** (1999) 245–262.

[63] **Eggers, K. W. H., Sharma, S. D., Ward, L. W.,** An assessment of some experimental methods for determining the wavemaking characteristics of a ship form. *Trans. SNAME* **75** (1967) 112–157.

[64] **Euvrard, D.,** Les mille et une facéties de la fonction de Green de la résistance de vagues. *ENSTA Rapp. Recherche* **144** (1981).

[65] **Euvrard, D.,** A short review of mathematical and numerical methods in transient ship hydrodynamics. *Proceedings of the Fifteenth Symposium on Naval Hydrodynamics*. Washington, DC: National Academy of Sciences Press, 1985, pp. 151–162.

[66] **Evans, D. V.,** Diffraction of water waves by a submerged vertical plate. *J. Fluid Mech.* **40** (1970) 433–451.

[67] **Evans, D. V.,** The application of a new source potential to the problem of the transmission of waves over a shelf of arbitrary profile. *Proc. Camb. Phil. Soc.* **71** (1972) 391–410.

[68] **Evans, D. V.,** The solution of a class of boundary-value problems with smoothly varying boundary conditions. *Quart. J. Mech. Appl. Math.* **38** (1985) 521–536.

[69] **Evans, D. V.,** Mechanisms for the generation of edge waves over a sloping beach. *J. Fluid Mech.* **186** (1988) 379–391.

[70] **Evans, D. V.,** Edge waves over a sloping beach. *Quart. J. Mech. Appl. Math.* **42** (1989) 131–142.

[71] **Evans, D. V.,** Trapped acoustic modes. *IMA J. Appl. Math.* **49** (1992) 45–60.

[72] **Evans, D. V.,** Vertical barriers, sloping beaches and submerged bodies. *Wave Asymptotics*. Cambridge: Cambridge University Press, 1992, pp. 202–219.

[73] **Evans, D. V., Fernyhough, M.,** Edge waves along periodic coastlines. Part 2. *J. Fluid Mech.* **297** (1995) 307–325.

[74] **Evans, D. V., Levitin, M., Vassiliev, D.,** Existence theorems for trapped modes. *J. Fluid Mech.* **261** (1994) 21–31.

[75] **Evans, D. V., Kuznetsov, N.,** Trapped modes. *Gravity Waves in Water of Finite Depth.* Southampton: Computational Mechanics, 1997, pp. 127–168.

[76] **Evans, D. V., Linton, C. M.,** Trapped modes in open channels. *J. Fluid Mech.* **225** (1991) 153–175.

[77] **Evans, D. V., Linton, C. M.,** Edge waves along periodic coastlines. *Quart. J. Mech. Appl. Math.* **46** (1993) 644–656.

[78] **Evans, D. V., Linton, C. M., Ursell, F.,** Trapped mode frequencies embedded in the continuous spectrum. *Quart. J. Mech. Appl. Math.* **46** (1993) 253–274.

[79] **Evans, D. V., McIver, P.,** Edge waves over a shelf: full linear theory. *J. Fluid Mech.* **142** (1984) 79–95.

[80] **Evans, D. V., Porter, R.,** Trapped modes about multiple cylinders in a channel. *J. Fluid Mech.* **339** (1997) 331–356.

[81] **Evans, D. V., Porter, R.,** Complementary methods for scattering by thin barriers. *Mathematical Techniques for Water Waves.* Southampton: Computational Mechanics, 1997, pp. 1–44.

[82] **Evans, D. V., Porter, R.,** Trapped modes embedded in the continuous spectrum. *Quart. J. Mech. Appl. Math.* **52** (1998) 263–274.

[83] **Evans, D. V., Porter, R.,** An example of non-uniqueness in the two-dimensional linear water-wave problem involving a submerged body. *Proc. Roy. Soc. Lond. A* **454** (1998) 3145–3165.

[84] **Fedoryuk, M. V.,** *Asymptotics, Integrals and Series.* Moscow: Nauka, 1987. (In Russian).

[85] **Fernyhough, M.,** *Application of the residue and modified residue calculus method in linear acoustic and water wave theory.* PhD thesis, University of Bristol, 1994.

[86] **Feynman, R. P., Leighton, R. B., Sands, M.,** *The Feynman Lectures on Physics.* Vol. 1. Reading, MA: Addison-Wesley, 1963.

[87] **Fitz-Gerald, G. F.,** The reflection of plane gravity waves travelling in water of variable depth. *Phil. Trans. Roy. Soc. Lond. A* **284** (1976) 49–89.

[88] **Fitz-Gerald, G. F., Grimshaw, R. H.,** A note on the uniqueness of small-amplitude water waves travelling in a region of varying depth. *Proc. Camb. Phil. Soc.* **86** (1979) 511–519.

[89] **Fox, D. D., Kuttler, J. R.,** Sloshing frequencies. *Z. Angew. Math. Phys.* **34** (1983) 668–696.

[90] **Friedman, A., Shinbrot, M.,** The initial value problem for the linearized equation of water waves. I, II. *J. Math. Mech.* **17** (1967) 107–180, **18** (1969) 1177–1194.

[91] **Friis, A., Grue, J., Palm, E.,** Application of Fourier transform to the second order 2D wave diffraction problem. *Mathematical Approaches in Hydrodynamics.* Philadelphia: SIAM, 1991.

[92] **Garipov, R. M.,** On the linear theory of gravity waves: the theorem of existence and uniqueness. *Arch. Rat. Mech. Anal.* **24** (1967) 352–362.

[93] **Garipov, R. M.,** Asymptotics of the Cauchy–Poisson waves. *Some Problems in Mathematics and Mechanics.* Moscow: Nauka, 1970, pp. 135–145. (In Russian).

[94] **Gilbarg, D., Trudinger, N. S.,** *Elliptic Partial Differential Equations of Second Order.* Berlin: Springer-Verlag, 1983.

[95] **Gohberg, I., Krein, M. G.,** *Introduction to the Theory of Linear Nonselfadjoint Operators in Hilbert Space.* Translations of Mathematical Monographs **18**. Providence, RI: American Mathematical Society, 1969.

[96] **Gradshteyn, I. S., Ryzhik, I. M.,** *Table of Integrals, Series and Products*. New York: Academic, 1980.

[97] **Greenspan, H. P.,** A note on edge waves in a stratified fluid. *Stud. Appl. Math.* **49** (1970) 381–388.

[98] **Grimshaw, R. H.,** Edge waves: a long-wave theory for oceans of finite depth. *J. Fluid Mech.* **62** (1974) 775–791.

[99] **Groves, M. D.,** On the existence of trapped modes in channels of arbitrary cross-section. *Math. Meth. Appl. Sci.* **20** (1997) 521–545.

[100] **Groves, M. D., Lesky, P. H.,** Resonance phenomena for water waves in channels of arbitrary cross-section. *Math. Meth. Appl. Sci.* **22** (1999) 837–865.

[101] **Gutmann, C.,** Resultates theoriques et numeriques sur la résistance de vagues d'un corps tridimensionnel immerse. *ENSTA Rapp. Recherche* **177** (1983).

[102] **Hadamard, J.,** Sur les ondes liquides. *C. R. Acad. Sci. Paris* **150** (1910) 609–611, 772–774, *Rend. Accad. Lincei* (5) **25** (1916) 716–719.

[103] **Hamdache, K.,** Forward speed motions of a submerged body. The Cauchy problem. *Math. Meth. Appl. Sci.* **6** (1984) 371–392.

[104] **Haskind, M. D. (Khaskind),** On wave motions of a heavy fluid. *Prikl. Mat. Mekh.* **18** (1954) 15–26. (In Russian).

[105] **Haskind, M. D. (Khaskind),** Translation of bodies under the free surface of a heavy fluid of finite depth. *Prikl. Mat. Mekh.* **9** (1945) 67–78. (In Russian; English transl. in *NACA Tech. Memo.* **1345** (1952) 20 pp.).

[106] **Haskind, M. D. (Khaskind),** *The Hydrodynamic Theory of Ship Oscillations*. Moscow: Nauka, 1973. (In Russian).

[107] **Havelock, T. H.,** Forced surface waves. *Phil. Mag.* (7) **8** (1929) 569–576.

[108] **Havelock, T. H.,** Ship waves: the calculation of wave profiles. *Proc. Roy. Soc. Lond. A* **135** (1932) 1–13.

[109] **Havelock, T. H.,** The damping of the heaving and pitching motion of a ship. *Phil. Mag.* (7) **33** (1942) 666–673.

[110] **Havelock, T. H.,** Waves due to a floating sphere making periodic heaving oscillations. *Proc. Roy. Soc. Lond. A* **231** (1955) 1–7.

[111] **Havelock, T. H.,** *Collected Papers*. Washington, DC: U.S. Government Printing Office, 1963.

[112] **Hazard, C.,** The singularity expansion method: an application in hydrodynamics. *The 5th Conference on Mathematical and Numerical Aspects of Wave Propagation*. Philadelphia: SIAM, 2000, pp. 494–498.

[113] **Hazard, C., Lenoir, M.,** Stabilité et résonances pour le problème du mouvement sur la houle. *ENSTA Rapp. Recherche* **310** (1998).

[114] **Hille, E.,** *Lectures on Ordinary Differential Equations*. Reading, MA: Addison-Wesley, 1969.

[115] **Hogner, E.,** A contribution to the theory of ship waves. *Arkiv Mat. Astr. och Fysik* **17** (1923) 1–50.

[116] **Hogner, E.,** Notes on some new contributions to the theory of ship waves. *Arkiv Mat. Astr. och Fysik* **18** (1924) 1–9.

[117] **Hsiao, G. C., Kleinman, R. E., Roach, G. F.,** Weak solutions of fluid-solid interaction problem. *Preprint* **1917** (1997). Technische Hochschule Darmstadt, Fachbereich Mathematik.

[118] **Hulme, A.,** The potential of a horizontal ring of wave sources in a fluid with a free surface. *Proc. Roy. Soc. Lond. A* **375** (1981) 295–305.

[119] **Hulme, A.,** A ring-source/integral-equation method for the calculation of hydrodynamic forces exerted on floating bodies of revolution. *J. Fluid Mech.* **128** (1983) 387–412.

[120] **Hulme, A.,** Some applications of Maz'ja's uniqueness theorem to a class of linear water wave problems. *Math. Proc. Camb. Phil. Soc.* **95** (1984) 511–519.

[121] **Hurd, R. A.,** Propagation of an electromagnetic wave along an infinite corrugated surface. *Can. J. Phys.* **32** (1954) 727–734.

[122] **Isakova, E. K.,** The asymptotic behaviour of solution to the Cauchy–Poisson problem. *Proc. Steklov Math. Inst.* **103** (1968) 73–95. (In Russian).

[123] **Isakova, E. K.,** The limiting amplitude principle for the Cauchy–Poisson problem. I, II, III. *Diff. Equations.* **6** (1970) 56–71, 721–730, 1289–1297. (In Russian).

[124] **Jami, A.,** Étude théorique et numérique de phénomènes transitoires en hydrodynamique navale. *ENSTA Rapp. Recherche* **154** (1982).

[125] **John, F.,** On the motion of floating bodies. I. *Comm. Pure Appl. Math.* **2** (1949) 13–57.

[126] **John, F.,** On the motion of floating bodies. II. *Comm. Pure Appl. Math.* **3** (1950) 45–101.

[127] **Jones, D. S.,** The eigenvalues of $\nabla^2 u - \lambda u = 0$ when the boundary conditions are given on semi-infinite domains. *Proc. Camb. Phil. Soc.* **49** (1953) 668–684.

[128] **Jones, D. S.,** Integral equations for the exterior acoustic problem. *Quart. J. Mech. Appl. Math.* **27** (1974) 129–142.

[129] **Kamotskii, I. V., Nazarov, S. A.,** Wood's anomalies and surface waves in the problem of scattering by a periodic boundary. I, II. *Sbornik: Math.* **190** (1999) 111–141, 205–231.

[130] **Kantorovich, L. V., Akilov, G. P.,** *Functional Analysis.* Oxford: Pergamon, 1982.

[131] **Keldysh, M. V.,** Remarks on certain motions of a heavy fluid. *Technical Notes of TsAGI* **52** (1935) 5–9. (In Russian).

[132] **Keldysh, M. V., Lavrentiev, M. A.,** On the motion of a wing under the surface of a heavy fluid. *Proceedings of the Conference on the Wave Resistance Theory.* Moscow: TsAGI, 1937, pp. 31–62. (In Russian).

[133] **Keller, J. B.,** The ray theory of ship waves and the class of streamlined ships. *J. Fluid Mech.* **91** (1979) 465–488.

[134] **Kelvin, Lord (Sir W. Thomson),** On ship waves. *Proc. Inst. Mech. Eng.* 3 August 1887.

[135] **Kelvin, Lord (Sir W. Thomson),** Deep-sea ship waves. *Phil. Mag.* (6) **11** (1906) 1–25.

[136] **Kellogg, O. D.,** *Foundations of Potential Theory.* Berlin: Springer-Verlag, 1929.

[137] **Kenig, C. E.,** *Harmonic Analysis Techniques for Second Order Elliptic Boundary Value Problems.* Providence, RI: American Mathematical Society, 1994.

[138] **Kirchgässner, K.,** Nonlinearly resonant surface waves and homoclinic bifurcation. *Adv. Appl. Mech.* **26** (1988) 135–181.

[139] **Kleinman, R. E.,** On the mathematical theory of motion of floating bodies – an update. *D.W. Taylor Naval Ship Res. & Devel. Center Report* **82/074** (1982).

[140] **Kleinman, R. E., Roach, G. F.,** On modified Green's functions in exterior problems for the Helmholtz equation. *Proc. Roy. Soc. Lond. A* **383** (1982) 313–332.

[141] **Kochin, N. E.,** On the wave resistance and lift of bodies submerged in a fluid. *Proceedings of the Conference on the Wave Resistance.* Moscow: TsAGI, 1937, pp. 65–134. (In Russian; English transl. in *SNAME Tech. & Res. Bull.* No. **1–8** (1951)).

[142] **Kochin, N. E.,** The two-dimensional problem of steady oscillations of bodies under the free surface of a heavy incompressible fluid. *Acad. Sci. USSR, Izvestia OTN,* No. **4** (1939) 37–62. (In Russian; English transl. in *SNAME Tech. & Res. Bull.* No. **1–10** (1952)).

[143] **Kochin, N. E.,** The theory of waves generated by oscillations of a body under the free surface of a heavy incompressible fluid. *Trans. Moscow Univ.* **46** (1940) 85–106. (In Russian; English transl. in *SNAME Tech. & Res. Bull.* No. **1–10** (1952)).

[144] **Komech, A. I., Merzon, A. E., Zhevandrov, P. N.,** On the completeness of Ursell's trapping modes. *Russian J. Math. Phys.* **4** (1997) 457–486.

[145] **Kondratyev, V. A.,** Boundary value problems for elliptic equations in regions with conical or angular points. *Trans. Moscow Math. Soc.* **16** (1967) 227–313.

[146] **Kopachevskiy, N. D., Krein, S. G., Ngo Zuy Can,** *Operator Methods in Hydrodynamics. Evolution and Spectral Problems.* Moscow: Nauka, 1989. (In Russian).

[147] **Kostyukov, A. A.,** *The Theory of Ship Waves and Wave Resistance.* Iowa City: Effective Communication, 1968.

[148] **Kozlov, V., Maz'ya, V.,** *Differential Equations with Operator Coefficients.* Berlin: Springer-Verlag, 1999.

[149] **Kozlov, V. A., Maz'ya, V. G., Rossmann, J.,** *Elliptic Boundary Value Problems in Domains with Point Singularities.* Providence, RI: American Mathematical Society, 1997.

[150] **Kozlov, V. A., Maz'ya, V. G., Rossmann, J.,** *Spectral Problems Associated with Corner Singularities of Solutions to Elliptic Equations.* Providence, RI: American Mathematical Society, 1999.

[151] **Král, J.,** *Integral Operators in Potential Theory.* Lect. Notes in Math. **823**. Berlin: Springer-Verlag, 1980.

[152] **Kreisel, G.,** Surface waves. *Quart. Appl. Math.* **7** (1949) 21–44.

[153] **Kuznetsov, N. G.,** Plane problem of the steady-state oscillations of a fluid in the presence of two semiimmersed cylinders. *Math. Notes Acad. Sci. USSR* **44** (1988) 685–690.

[154] **Kuznetsov, N. G.,** Steady waves on the surface of a fluid of variable depth in the presence of floating bodies. *Regular Asymptotic Algorithms in Mechanics.* Novosibirsk: Nauka, 1989. (In Russian).

[155] **Kuznetsov, N. G.,** Uniqueness of a solution of a linear problem for stationary oscillations of a liquid. *Diff. Equat.* **27** (1991) 187–194.

[156] **Kuznetsov, N. G.,** Integral equations for the problem of stationary waves produced by a floating body. *Math. Notes Acad. Sci. USSR* **50** (1991) 1036–1042.

[157] **Kuznetsov, N. G.,** The lower bound of the eigenfrequencies of plane oscillations of a fluid in a channel. *J. Appl. Math. Mech.* **56** (1992) 293–297.

[158] **Kuznetsov, N. G.,** Waves due to the forward motion of a surface-piercing cylinder in a fluid of infinite depth. *Modelling in Mech.* **6**, No. 4 (1992) 70–84. (In Russian).

[159] **Kuznetsov, N. G.,** The Maz'ya identity and lower estimates of eigenfrequencies of steady-state oscillations of a liquid in a channel. *Russian Math Surveys* **48** (1993) 222.

[160] **Kuznetsov, N. G.,** Asymptotic analysis of wave resistance of a submerged body moving with an oscillating velocity. *J. Ship Res.* **37** (1993) 119–125.

[161] **Kuznetsov, N. G.,** On uniqueness and solvability in the linearized two-dimensional problem of a supercritical stream about a surface-piercing body. *Proc. Roy. Soc. Lond. A* **450** (1995) 233–253.

[162] **Kuznetsov, N. G.,** Trapped modes of surface and internal waves in a channel occupied by a two-layer fluid. *Third International Conference on Mathematical and Numerical Aspects of Wave Propagation* (eds. G. Cohen et al.). Philadelphia: SIAM, 1995, pp. 624–633.

[163] **Kuznetsov, N. G., Maz'ya, V. G.,** Problem concerning steady-state oscillations of a layer of fluid in the presence of an obstacle. *Sov. Phys. Dokl.* **19** (1974) 341–343.

[164] **Kuznetsov, N. G., Maz'ya, V. G.,** Asymptotic expansions for the surface waves due to brief disturbances. *Asymptotic Methods. Problems of Mechanics.* Novosibirsk: Nauka, 1986.

[165] **Kuznetsov, N. G., Maz'ya, V. G.,** On unique solvability of the plane Neumann-Kelvin problem. *Math. USSR Sbornik.* **63** (1989) 425–446.

[166] **Kuznetsov, N. G., Maz'ya, V. G.,** On a well-posed formulation of the two-dimensional Neumann–Kelvin problem for a surface-piercing body. *Problemi Attuali dell'Analisi e della Fisica Matematica.* Roma: Aracne, 2000, pp. 77–109. See also: *Preprint LiTH-MAT-R-92-42.* Department of Mathematics, Linköping University, 1992.

[167] **Kuznetsov, N. G., Maz'ya, V. G.,** Asymptotic analysis of surface waves due to high-frequency disturbances. *Rend. Mat. Accad. Lincei, Ser. 9,* **8** (1997) 5–29.

[168] **Kuznetsov, N. G., Maz'ya, V. G.,** Water-wave problem for a vertical shell. *Math. Bohemica,* **126** (2001) 411–420.

[169] **Kuznetsov, N., McIver, P.,** On uniqueness and trapped modes in the water-wave problem for a surface-piercing axisymmetric structure. *Quart. J. Mech. Appl. Math.* **50** (1997) 565–580.

[170] **Kuznetsov, N., McIver, P., Linton, C. M.,** On uniqueness and trapped modes in the water-wave problem for vertical barriers. *Wave Motion* **33** (2001) 283–307.

[171] **Kuznetsov, N., Motygin, O.,** On waveless statement of the two-dimensional Neumann–Kelvin problem for a surface-piercing body. *IMA J. Appl. Math.* **59** (1997) 25–42.

[172] **Kuznetsov, N., Motygin, O.,** On the resistanceless statement of the two-dimensional Neumann–Kelvin problem for a surface-piercing tandem. *IMA J. Appl. Math.* **62** (1999) 81–99.

[173] **Kuznetsov, N. G., Porter, R., Evans, D. V., Simon, M. J.,** Uniqueness and trapped modes for surface-piercing cylinders in oblique waves. *J. Fluid Mech.* **365** (1999) 351–368.

[174] **Kuznetsov, N. G., Simon, M. J.,** On uniqueness in linearized two-dimensional water-wave problems for two surface-piercing bodies. *Quart. J. Mech. Appl. Math.* **48** (1995) 507–515.

[175] **Kuznetsov, N. G., Simon, M. J.,** On uniqueness in the two-dimensional water-wave problem for surface-piercing bodies in fluid of finite depth. *Appl. Math. Rep.* **95/4** (1995). University of Manchester.

[176] **Kuznetsov, N. G., Simon, M. J.,** A note on uniqueness in the linearized water-wave problem. *J. Fluid Mech.* **386** (1999) 5–14.

[177] **Kuznetsov, N. G., Vainberg, B. R.,** Maz'ya's works in the linear theory of water waves. *The Maz'ya Anniversary Collection,* **1**. Basel: Birkhäuser, 1999, pp. 17–34.

[178] **Lahalle, D.,** Calcul des efforts sur un profil portant d'hydroptere par couplage elements finis–representation integrale. *ENSTA Rapp. Recherche* **187** (1984).

[179] **Lamb, H.,** *Hydrodynamics.* Cambridge: Cambridge University Press, 1932.

[180] **Lau, B. R.,** The linearized equations of water waves. *Indiana Univ. Math. J.* **22** (1972) 233–266.

[181] **Lau, S. M., Hearn, G. E.,** Suppression of irregular frequency effects in fluid-structure interaction problems using a combined boundary integral equation method. *Int. J. Numer. Meth. Fluids* **9** (1989) 763–782.

[182] **Lavrentiev, M. A., Chabat, B. V.,** *Effets Hydrodynamiques et Modeles Mathematique.* Moscow: Mir, 1973.

[183] **Le Blond, P. H., Mysak, L. A.,** *Waves in the Ocean.* Amsterdam: Elsevier, 1978.

[184] **Lee, C.-H., Sclavounos, P. D.,** Removing the irregular frequencies from integral equations in wave-body interactions. *J. Fluid Mech.* **207** (1989) 393–418.

[185] **Lehman, R. S., Lewy, H.,** Uniqueness of water waves on a sloping beach. *Comm. Pure. Appl. Math.* **14** (1961) 521–546.

[186] **Le Méhauté, B.,** *An Introduction to Hydrodynamics and Water Waves.* New York: Springer-Verlag, 1976.

[187] **Lenoir, M.,** Méthodes de couplage en hydrodynamique navale et application à la résistance de vagues bidimensionnelle. *ENSTA Rapp. Recherche* **164** (1982).

[188] **Lenoir, M., Jami, A.,** A variational formulation for exterior problems in linear hydrodynamics. *Comput. Meth. Appl. Mech. Eng.* **16** (1978) 341–359.

[189] **Lenoir, M., Martin, D.,** An application of the principle of limiting absorption to the motions of floating bodies. *J. Math. Anal. Appl.* **79** (1981) 370–383.

[190] **Lenoir, M., Tounsi, A.,** The localized finite element method and its application to the two-dimensional sea-keeping problem. *SIAM J. Numer. Anal.* **25** (1988) 729–752.

[191] **Lenoir, M., Vullierme-Ledard, M., Hazard, C.,** Variational formulations for the determination of resonant states in scattering problems. *SIAM J. Math. Anal.* **23** (1992) 579–608.

[192] **Leppington, F. G.,** On the scattering of short surface waves by a finite dock. *Proc Camb. Phil. Soc.* **64** (1968) 1109–1129.

[193] **Leppington, F. G.,** On the radiation of short surface waves by a finite dock. *J. Inst. Math. Appl.* **6** (1970) 319–340.

[194] **Levi-Civita, T.,** Détermination rigoureuse des ondes permanentes d'ampleur finie. *Math. Ann.* **93** (1925) 264–314.

[195] **Levine, H., Rodemich, E.,** Scattering of surface waves on an ideal fluid. *Stanford Tech. Rep.* **78** (1958).

[196] **Liapis, S.,** A method for suppressing the irregular frequencies from integral equations in water wave-structure interaction problems. *Comp. Mech.* **12** (1993) 59–68.

[197] **Licht, C.,** Évolution d'un système fluide-flotteur. *J. Méc. Théor. Appl.* **1** (1982) 211–235.

[198] **Licht, C.,** Etude de quelques modèles décrivant les vibrations d'une structure élastique dans la mer. *ENSTA Rapp. Recherche* **163** (1982).

[199] **Licht, C.,** Trois modèles décrivant les vibrations d'une structure élastique dans la mer. *C. R. Acad. Sci. Paris, Sér.* I, **296** (1983) 341–344.

[200] **Licht, C.,** Time-dependent behaviour of floating bodies. *Proceedings of the Fifteenth Symposium on Naval Hydrodynamics*. Washington, DC: National Academy of Sciences Press, 1985, pp. 221–233.

[201] **Lighthill, M. J.,** *Waves in Fluids*. Cambridge: Cambridge University Press, 1978.

[202] **Linton, C. M.,** The use of multipoles in channel problems. *Mathematical Techniques for Water Waves*. Southampton: Computational Mechanics Publications, 1997, pp. 45–78.

[203] **Linton, C. M., Evans, D. V.,** Trapped modes above a submerged horizontal plate. *Quart. J. Mech. Appl. Math.* **44** (1991) 487–506.

[204] **Linton, C. M., Evans, D. V.,** Integral equations for a class of problems concerning obstacles in waveguides. *J. Fluid Mech.* **245** (1992) 349–365.

[205] **Linton, C. M., Evans, D. V.,** The radiation and scattering of surface waves by a vertical cylinder in a channel. *Phil. Trans. Roy. Soc. Lond. A* **338** (1992) 325–357.

[206] **Linton, C. M., Evans, D. V.,** Acoustic scattering by an array of parallel plates. *Wave Motion* **18** (1993) 51–65.

[207] **Linton, C. M., Kuznetsov, N. G.,** Non-uniqueness in two-dimensional water wave problems: numerical evidence and geometrical restrictions. *Proc. Roy. Soc. Lond. A* **453** (1997) 2437–2460.

[208] **Linton, C. M., McIver, P.,** *Handbook of Mathematical Techniques for Wave/Structure Interactions*. Boca Raton, FL: CRC Press, 2001.

[209] **Liu, Y. W.,** A boundary integral equation for the two-dimensional floating-body problem. *SIAM J. Math. Anal.* **22** (1991) 973–981.

[210] **Livshits, M. L.,** On steady-state oscillations of a sphere submerged in deep water. *Proc. Leningrad Shipbuild. Inst.* **91** (1974) 133–139.

[211] **Livshits, M., Maz'ya, V.,** Solvability of the two-dimensional Kelvin–Neumann problem for a submerged circular cylinder. *Applic. Analysis* **64** (1997) 1–5.

[212] **Ludwig, D.,** Geometrical theory for surface waves. *SIAM Rev.* **17** (1975) 1–15.

[213] **Mandal, B. N., Banerjea, S.,** On singular integral equation and its use to some barrier problems. *Mathematical Techniques for Water Waves*. Southampton: Computational Mechanics Publications, 1997, pp. 255–283.

[214] **Maniar, H. D., Newman, J. N.,** Wave diffraction by a long array of cylinders. *J. Fluid Mech.* **339** (1997) 309–330.

[215] **Martin, P. A.,** On the null-field equations for water-wave radiation problems. *J. Fluid Mech.* **113** (1981) 315–332.

[216] **Martin, P. A.,** Acoustic scattering and radiation problems, and the null-field method. *Wave Motion* **4** (1982) 391–408.

[217] **Martin, P. A.,** On the null-field equations for water-wave scattering problems. *IMA J. Appl. Math.* **33** (1984) 55–69.

[218] **Martin, P. A.,** Multiple scattering of surface water waves and the null-field method. *Proceedings of the 15th Symposium on Naval Hydrodynamics*. Washington, DC: National Academy of Sciences Press, 1985.

[219] **Martin, P. A.,** Integral-equation methods for multiple-scattering problems. II. Water waves. *Quart. J. Mech. Appl. Math.* **38** (1985) 119–133.

[220] **Martin, P. A., Ursell, F.,** On the null-field equations for water-wave radiation problems. *Proceedings of the Third International Conference on Numerical Ship Hydrodynamics*. Paris: Bassin d'Essais des Carènes, 1981.

[221] **Maskell, S. J., Ursell, F.,** The transient motion of a floating body. *J. Fluid Mech.* **44** (1970) 303–313.

[222] **Maz'ya, V. G.,** On the steady problem of small oscillations of a fluid in the presence of a submerged body. *Proceedings of the Sobolev's Seminar* No. **2**, (1977) 57–79. Novosibirsk: Institute of Mathematics, Sibirian Branch, Acad. Sci. USSR. (In Russian).

[223] **Maz'ya, V. G.,** Solvability of the problem on the oscillations of a fluid containing a submerged body. *J. Soviet Math.* **10** (1978) 86–89.

[224] **Maz'ya, V. G.,** Boundary integral equations. *Encyclopaedia of Math. Sciences* **27**, *Analysis IV*. Berlin: Springer-Verlag, 1991, pp. 127–222.

[225] **Maz'ya, V. G., Morozov, N. F., Plamenevsky, B. A., Stupyalis, L.,** *Elliptic Boundary Value Problems*. Translations Ser. II, **123**, Providence, RI: American Mathematical Society, 1984.

[226] **Maz'ya, V., Nazarov, S., Plamenevsky, B.,** *Asymptotic Theory of Elliptic Boundary Value Problems in Singularly Perturbed Domains*. Vols. 1, 2. Basel: Birkhäuser, 2000.

[227] **Maz'ya, V. G., Rossmann, J.,** Über die Asymptotic der Lösungen elliptisher Randwertaufgaben in der Umgebung von Kanten. *Math. Nachr.* **138** (1988) 27–53.

[228] **Maz'ya, V., Shaposhnikova, T.,** *Jaques Hadamard*. Providence, RI: American Mathematical Society, 1998.

[229] **Maz'ya, V. G., Vainberg, B. R.,** On ship waves. *Wave Motion* **18** (1993) 31–50.

[230] **McIver, M.,** An example of non-uniqueness in the two-dimensional linear water wave problem. *J. Fluid Mech.* **315** (1996) 257–266.

[231] **McIver, M.,** Uniqueness below cut-off frequency for the two-dimensional linear water-wave problem. *Proc. Roy. Soc. Lond. A* **455** (1999) 1435–1441.

[232] **McIver, M.,** Trapped modes supported by submerged obstacles. *Proc. Roy. Soc. Lond. A* **456** (2000) 1851–1860.

[233] **McIver, M., Linton, C. M.,** On the non-existence of trapped modes in acoustic waveguides. *Quart. J. Mech. Appl. Math.* **48** (1995) 543–555.

[234] **McIver, M., Porter, R.,** Trapping of waves by a submerged elliptical torus. *Proceedings of the Sixteenth International Workshop on Water Waves and Floating Bodies*. Hiroshima: University of Hiroshima, 2001.

[235] **McIver, P.,** Trapping of surface water waves by fixed bodies in a channel. *Quart. J. Mech. Appl. Math.* **44** (1991) 193–208.

[236] **McIver, P.,** Low-frequency asymptotics of hydrodynamic forces on fixed and floating structures. *Ocean Waves Engineering*. Southampton: Computational Mechanics Publications, 1994, pp. 1–49.

[237] **McIver, P.,** The dispersion relation and eigenfunction expansions for water waves in a porous structure. *J. Eng. Math.* **34** (1998) 319–334.

[238] **McIver, P., Evans, D. V.,** The trapping of surface waves above a submerged horizontal cylinder. *J. Fluid Mech.* **151** (1985) 243–255.

[239] **McIver, P., Linton, C. M., McIver, M.,** Construction of trapped modes for wave guides and diffraction gratings. *Proc. Roy. Soc. Lond. A* **454** (1998) 2593–2616.

[240] **McIver, P., McIver, M.,** Trapped modes in an axisymmetric water-wave problem. *Quart. J. Mech. Appl. Math.* **50** (1997) 165–178.

[241] **McIver, P., Newman, J. N.,** Non-axisymmetric trapping structures in the three-dimensional water-wave problem. *Proceedings of the Sixteenth International Workshop on Water Waves and Floating Bodies.* Hiroshima: University of Hiroshima, 2001.

[242] **Mei, C. C.,** *The Applied Dynamics of Ocean Surface Waves.* New York: Wiley, 1983.

[243] **Merzon, A. E.,** On Ursell's problem. *Third International Conference on Mathematical and Numerical Aspects of Wave Propagation.* Philadelphia: SIAM, 1995.

[244] **Merzon, A. E., Zhevandrov, P. N.,** High-frequency asymptotics of edge waves on a beach of nonconstant slope. *SIAM J. Appl. Math.* **59** (1998) 529–546.

[245] **Michell, J. H.,** The wave resistance of a ship. *Phil. Mag.* (5) **46** (1898) 106–123.

[246] **Mihlin, S. G.,** *Vorlesungen über lineare Integralgleichungen.* Berlin: VEB Deutscher Verlag der Wissenschaften, 1962.

[247] **Miles, J. W.,** Edge waves on a gently sloping beach. *J. Fluid Mech.* **199** (1989) 125–131.

[248] **Mittra, R., Lee, S. W.,** *Analytical Techniques in the Theory of Guided Waves.* New York: Macmillan, 1971.

[249] **Moiseev, N. N.,** Introduction to the theory of oscillations of liquid-containing bodies. *Adv. Appl. Mech.* **8** (1964) 233–289.

[250] **Morris, C. A. N.,** The generation of surface waves over a sloping beach by an oscillating line-source. I. The general solution. II. The existence of wave-free source positions. *Proc. Camb. Phil. Soc.* **76** (1974) 545–554, 555–562.

[251] **Morse, M., Feshbach, H.,** *Methods of Theoretical Physics*, II. New York: McGraw-Hill, 1953.

[252] **Motygin, O. V.,** Trapped modes of oscillation of a liquid for surface-piercing bodies in oblique waves. *J. Appl. Math. Mech.* **63** (1999) 257–264.

[253] **Motygin, O. V.,** Uniqueness and solvability in the linearized two-dimensional problem of a body in finite depth subcritical stream. *Euro. J. Appl. Math.* **10** (1999) 141–155.

[254] **Motygin, O. V.,** On frequency bounds for modes trapped near a channel-spanning cylinder. *Proc. Roy Soc. Lond. A* **456** (2000) 2911–2930.

[255] **Motygin, O., Kuznetsov, N.,** The 2D Neumann–Kelvin problem for a surface-piercing tandem. *Proceedings of the Tenth International Workshop on Water Waves and Floating Bodies.* Oxford: University of Oxford, 1995, pp. 181–185.

[256] **Motygin, O., Kuznetsov, N.,** Non-uniqueness in the water-wave problem: an example violating the inside John condition. *Proceedings of the Thirteenth International Workshop on Water Waves and Floating Bodies.* Delft: Delft University of Technology, 1998.

[257] **Mysak, L. A.,** Topographically trapped waves. *Ann. Rev. Fluid Mech.* **12** (1980) 45–76.

[258] **Nachbin, A., Papanicolaou, G. C.,** Water waves in shallow channels of rapidly varying depth. *J. Fluid Mech.* **241** (1992) 311–332.

[259] **Nakos, D. E., Sclavounos, P. D.,** Ship motions by a three-dimensional Rankine panel method. *Proceedings of the Eighteenth Symposium on Naval Hydrodynamics.* Ann Arbor: University of Michigan, 1990, pp. 21–39.

[260] **Nazarov, S., Plamenevsky, B.,** *Elliptic Problems in Domains with Piecewise Smooth Boundaries.* Berlin: W. De Gruyter, 1994.

[261] **Nekrasov, A. I.,** On waves of permanent type. I, II. *Izvestia Ivanovo-Voznesensk Polytech. Inst.* **3** (1921) 52–65, **6** (1922) 155–171. (In Russian).

[262] **Newman, J. N.,** *Marine Hydrodynamics.* Cambridge: MIT Press, 1977.

[263] **Newman, J. N.,** The theory of ship motions. *Adv. Appl. Mech.* **18** (1978) 221–283.

[264] **Newman, J. N.,** Evaluation of the wave-resistance Green function. Part 1. The double integral. *J. Ship Res.* **31** (1987) 79–90.

[265] **Newman, J. N.,** Evaluation of the wave-resistance Green function. Part 2. The single integral on the centerplane. *J. Ship Res.* **31** (1987) 145–150.

[266] **Newman, J. N.,** Radiation and diffraction analysis of the McIver toroid. *J. Eng. Math.* **35** (1999) 135–147.

[267] **Newman, J. N.,** Diffraction of water waves by an air chamber. *Proceedings of the Fifteenth International Workshop on Water Waves and Floating Bodies.* Tel-Aviv: University of Tel-Aviv, 2000.

[268] **Noblesse, F.,** The fundamental solution in the theory of steady motion of a ship. *J. Ship Res.* **21** (1977) 82–88.

[269] **Noblesse, F.,** The steady wave potential of a unit source at the centerplane. *J. Ship Res.* **22** (1978) 80–88.

[270] **Noblesse, F.,** Alternative integral representation for the Green function of the theory of ship wave resistance. *J. Eng. Math.* **15** (1981) 241–265.

[271] **Olver, F. W. J.,** *Asymptotics and Special Functions.* New York: Academic Press, 1974.

[272] **Olver, P. J.,** Conversation laws of free boundary problems and the classification of conservation laws for water waves. *Trans. Amer. Math. Soc.* **277** (1983) 353–380.

[273] **Ovsyannikov, L. V. et al.,** *Nonlinear Problems in the Theory of Surface and Internal Waves.* Novosibirsk: Nauka, 1985. (In Russian).

[274] **Packham, B. A.,** A note on generalized edge wave on a sloping beach. *Quart. J. Mech. Appl. Math.* **42** (1989) 441–446.

[275] **Pagani, C. D., Pierotti, D.,** The Neumann–Kelvin problem in a bounded domain. *J. Math. Anal. Appl.* **192** (1995) 41–62.

[276] **Pagani, C. D., Pierotti, D.,** Exact solution of the wave-resistance problem for a submerged cylinder: I. Linearized theory. *Arch. Rat. Mech. Anal.* **149** (1999) 271–288.

[277] **Pagani, C. D., Pierotti, D.,** Exact solution of the wave-resistance problem for a submerged cylinder: II. The non-linear problem. *Arch. Rat. Mech. Anal.* **149** (1999) 289–327.

[278] **Pagani, C. D., Pierotti, D.,** The Neumann–Kelvin problem for a beam. *J. Math. Anal. Appl.* **240** (1999) 60–79.

[279] **Pagani, C. D., Pierotti, D.,** On solvability of the non-linear wave resistance problem for a surface-piercing symmetric cylinder. *SIAM J. Math. Anal.* **32** (2000) 214–233.

[280] **Pagani, C. D., Pierotti, D.,** The forward motion of an unsymmetric surface-piercing cylinder: the solvability of non linear problem in the supercritical case. *Quart. J. Mech. Appl. Math.* **54** (2001) 85–106.

[281] **Parker, R.,** Resonance effects in wake shedding from parallel plates: some experimental observations. *J. Sound Vibrat.* **4** (1966) 63–72.

[282] **Parker, R., Stoneman, S. A. T.,** The excitation and consequences of acoustic resonances in enclosed fluid flow around solid bodies. *Proc. Inst. Mech. Eng.* **203** (1989) 9–19.

[283] **Parsons, N., Martin, P. A.,** Trapping of water waves by submerged plates using hypersingular integral equations. *J. Fluid Mech.* **284** (1995) 359–375.

[284] **Peters, A. S.,** A new treatment of the ship wave problem. *Comm. Pure Appl. Math.* **2** (1949) 123–148.

[285] **Peters, A. S.,** Water waves over sloping beaches and the solution of a mixed boundary value problem for $\nabla^2\phi - k^2\phi = 0$ in a sector. *Comm. Pure Appl. Math.* **5** (1952) 87–108.

[286] **Peters, A. S., Stoker, J. J.,** The motion of a ship, as a floating rigid body, in a seaway. *Comm. Pure Appl. Math.* **10** (1957) 399–490.

[287] **Petrovski, I. G.,** *Ordinary Differential Equations.* Englewood Cliffs, NJ: Prentice-Hall, 1966.

[288] **Petrovskii, I. G.,** *Partial Differential Equations.* Philadelphia: Saunders, 1967.

[289] **Pinkster, J. A.,** The effect of air cushions under floating offshore structures. *Proceedings of the 8th International Conference on the Behavior of Offshore Structures* **2** (1997) 143–158.

[290] **Porter, R., Evans, D. V.,** Rayleigh–Bloch surface waves along periodic gratings and their connection with trapped modes in waveguides. *J. Fluid Mech.* **386** (1999) 233–258.

[291] **Porter, R., Evans, D. V.,** The trapping of surface waves by multiple submerged horizontal cylinders. *J. Eng. Math.* **34** (1998) 417–433.

[292] **Protter, M. H., Weinberger, H. F.,** *Maximum Principles in Differential Equations.* New York: Springer, 1984.

[293] **Prudnikov, A. P., Brychkov, Yu. A., Marichev, O. I.,** *Integrals and Series. Vol. 2. Special Functions.* New York: Gordon & Breach, 1986.

[294] **Quenez, J.-M.,** Etude des résonances pour le problèmes de tenue à la mer et de résistance de vagues. *ENSTA Rapp. Recherche* **285** (1995).

[295] **Radon, J.,** Über Randwertaufgaben beim logarithmischen Potential. *Sitzber. Akad. Wiss. Wien.* **128** (1919) 1123–1167.

[296] **Reed, M., Simon, B.,** *Methods of Modern Mathematical Physics.* Vols. 1–4. New York: Academic, 1981.

[297] **Rhodes-Robinson, P. F.,** On the short-wave asymptotic motion due to a cylinder heaving on water of finite depth. I, II, III. *Proc. Camb. Phil. Soc.* **67** (1970) 423–442, 443–468, **72** (1972) 83–94.

[298] **Roseau, M.,** Short waves parallel to the shore over a sloping beach. *Comm. Pure Appl. Math.* **11** (1958) 433–493.

[299] **Roseau, M.,** *Asymptotic Wave Theory.* Amsterdam: North-Holland, 1976.

[300] **Rosenblat, S.,** An existence theorem for waves in a channel. *SIAM J. Appl. Math.* **30** (1976) 381–390.

[301] **Sabatier, P. C.,** On water waves produced by ground motions. *J. Fluid Mech.* **126** (1983) 27–58.

[302] **Sanchez-Palencia, E.,** *Non-Homogeneous Media and Vibration Theory.* Lecture Notes in Physics **127**. New York: Springer-Verlag, 1980.

[303] **Shen, M. C.,** Ray method for surface waves in fluid of variable depth. *SIAM Rev.* **17** (1975) 38–55.

[304] **Shen, M. C., Meyer, R. E., Keller, J. B.,** Spectra of waves in channels and around islands. *Phys. Fluids* **11** (1968) 2289–2304.

[305] **Simon, M. J.,** On a bound for the frequency of surface waves trapped near a cylinder spanning a channel. *Theor. Comp. Fluid Dynam.* **4** (1992) 71–78.

[306] **Simon, M. J., Kuznetsov, N. G.,** On uniqueness of the water-wave problem for a floating toroidal body. *Appl. Math. Rep.* **96/1** (1996). University of Manchester.

[307] **Simon, M. J., Ursell, F.,** Uniqueness in linearized two-dimensional water-wave problems. *J. Fluid Mech.* **148** (1984) 137–154.

[308] **Sollitt, C. K., Cross, R. H.,** Wave transmission through permeable breakwaters. *Proceedings of the 13th Conference on the Coastal Engineering.* Vancouver: ASCE, 1972, pp. 1827–1846.

[309] **Sretensky, L. N.,** Periodic waves due to a source over a sloping bottom. *Applications of Theory of Functions to Continuum Mechanics. Proc. Int. Symp. Tbilissi.* Vol. 2. Moscow: Nauka, 1965. (In Russian).

[310] **Sretensky, L. N.,** *The Theory of Wave Motions of a Fluid.* Moscow: Nauka, 1977. (In Russian).

[311] **Staziker, D. J., Porter, D., Stirling, S. G.,** The scattering of surface waves by local bed elevations. *Appl. Ocean Res.* **18** (1996) 283–291.

[312] **Stoker, J. J.,** *Water Waves. The Mathematical Theory with Applications.* New York: Interscience, 1957.

[313] **Stokes, G. G.,** Report on recent researches in hydrodynamics. *Report to 16th Meeting British Association for Advancement of Science, Southampton.* London: Murrey, 1846, pp. 1–20.

[314] **Struik, D. J.,** Détermination rigoreuse des ondes irrotationelles periodiques dans un canal à profondeur finie. *Math. Ann.* **95** (1926) 595–634.

[315] **Sun, S. M., Shen, M. C.,** Linear water waves over a gently sloping beach. *Quart. Appl. Math.* **52** (1994) 243–259.

[316] **Suzuki, K.,** Numerical studies of the Neumann–Kelvin problem for a two-dimensional semisubmerged body. *Proceedings of the Third International Conference on Numerical Ship Hydrodynamics.* Paris: Bassin d'Essais des Carènes, 1982, pp. 83–95.

[317] **Thorne, R. C.,** Multipole expansions in the theory of surface waves. *Proc. Camb. Phil. Soc.* **49** (1953) 707–716.

[318] **Timman, R., Hermans, A. J., Hsiao, G. C.,** *Water Waves and Ship Hydrodynamics.* Dordrecht: Nijhoff, 1985.

[319] **Troesch, B. A.,** Free oscillations of a fluid in a container. *Boundary Problems in Differential Equations.* Madison: University of Wisconsin Press, 1960, pp. 279–299.

[320] **Tuck, E. O.,** On line distributions of Kelvin sources. *J. Ship Res.* **8** (1964) 45–52.

[321] **Tuck, E. O., Tulin, M. P.,** Submerged bodies which do not generate waves. *Proceedings of the Seventh International Workshop on Water Waves and Floating Bodies.* Val de Reuil: Bassin d'Essais des Carènes, 1992, pp. 275–279.

[322] **Ursell, F.,** Surface waves on deep water in the presence of a submerged circular cylinder. I, II. *Proc. Camb. Phil. Soc.* **46** (1950) 141–152, 153–158.

[323] **Ursell, F.,** Trapping modes in the theory of surface waves. *Proc. Camb. Phil. Soc.* **47** (1951), 347–358.

[324] **Ursell, F.,** Discrete and continuous spectra in the theory of gravity waves. In *Gravity Waves, NBS Circular* **521** (1952) 1–5.

[325] **Ursell, F.,** Edge waves on a sloping beach. *Proc. Roy. Soc. Lond. A* **214** (1952) 79–97.

[326] **Ursell, F.,** On Kelvin's ship-wave pattern. *J. Fluid Mech.* **8** (1960) 418–431.

[327] **Ursell, F.,** The decay of the free motion of a floating body. *J. Fluid Mech.* **19** (1964) 305–319.

[328] **Ursell, F.,** Integrals with a large parameter. The continuation of uniform asymptotic expansions. *Proc. Camb. Phil Soc.* **61** (1965) 113–128.

[329] **Ursell, F.,** Slender oscillating ships at zero forward speed. *J. Fluid Mech.* **14** (1968) 496–516.

[330] **Ursell, F.,** On head seas travelling along a horizontal cylinder. *J. Inst. Math. Appl.* **4** (1968) 414–427.

[331] **Ursell, F.,** Integrals with a large parameter. Paths of descent and conformal mapping. *Proc. Camb. Phil Soc.* **67** (1970) 371–381.

[332] **Ursell, F.,** A problem in the theory of water waves. *Numerical Solution of Integral Equations.* Oxford: Clarendon Press, 1974.

[333] **Ursell, F.,** On the exterior problem of acoustics. II. *Math. Proc. Camb. Phil. Soc.* **84** (1978) 545–548.

[334] **Ursell, F.,** Irregular frequencies and the motion of floating bodies. *J. Fluid Mech.* **105** (1981) 143–156.

[335] **Ursell, F.,** Mathematical note on the two-dimensional Kelvin–Neumann problem. *Proceedings of the Thirteenth Symposium on Naval Hydrodynamics.* Tokyo: Shipbuilding Research Association of Japan, 1981, pp. 245–251.

[336] **Ursell, F.,** Mathematical note on the fundamental solution (Kelvin source) in ship hydrodynamics. *IMA J. Appl. Math.* **32** (1984) 335–351.

[337] **Ursell, F.,** Mathematical observations on the method of multipoles. *Schiffstechnik* **33** (1986) 113–128.

[338] **Ursell, F.,** Mathematical aspects of trapping modes in the theory of surface waves. *J. Fluid Mech.* **183** (1987) 421–437.

[339] **Ursell, F.,** On the theory of the Kelvin ship-wave source: asymptotic expansion of an integral. *Proc. Roy. Soc. Lond. A* **418** (1988) 81–93.

[340] **Ursell, F.,** On the theory of the Kelvin ship-wave source: the near-field convergent expansion of an integral. *Proc. Roy. Soc. Lond. A* **428** (1990) 15–26.

[341] **Ursell, F.,** Some unsolved and unfinished problems in the theory of waves. *Wave Asymptotics.* Cambridge: Cambridge University Press, 1992, pp. 220–244.

[342] **Ursell, F.,** *Collected Papers*, Vol. 1. Singapore: World Scientific, 1994.

[343] **Ursell, F., Yu, Y. S.,** Surface waves generated by an oscillating circular cylinder on shallow water: theory and experiment. *J. Fluid Mech.* **9** (1960) 529–551.

[344] **Utsunomiya, T., Eatock Taylor, R.,** Trapped modes around a row of circular cylinders in a channel. *J. Fluid Mech.* **386** (1999) 259–279.

[345] **Vainberg, B. R.,** *Asymptotic Methods in Equations of Mathematical Physics.* New York: Gordon & Breach, 1989.

[346] **Vainberg, B. R., Maz'ya, V. G.,** On the plane problem of the motion of a body immersed in a fluid. *Trans. Moscow Math. Soc.* **28** (1973) 33–55.

[347] **Vainberg, B. R., Maz'ya, V. G.,** On the problem of the steady state oscillations of a fluid layer of variable depth. *Trans. Moscow Math. Soc.* **28** (1973) 56–73.

[348] **Vladimirov, V. S.,** *Equations of Mathematical Physics.* New York: Marcel Dekker, 1971.

[349] **Vullierme-Ledard, M.,** The limiting amplitude principle applied to the motion of floating bodies. *Math. Model. Numer. Anal.* **21** (1987) 125–170.

[350] **Watson, G. N.,** *A Treatise on the Theory of Bessel Functions.* Cambridge: Cambridge University Press, 1944.

[351] **Weck, N.,** On a boundary value problem in the theory of linear water-waves. *Math. Meth. Appl. Sci.* **12** (1990) 393–404.

[352] **Wehausen, J. V.,** The motion of floating bodies. *Ann. Rev. Fluid Mech.* **3** (1971) 237–268.

[353] **Wehausen, J. V.,** The wave resistance of ships. *Adv. Appl. Mech.* **13** (1973) 93–245.

[354] **Wehausen, J. V., Laitone, E. V.,** Surface waves. *Handbuch der Physik* **9**. Berlin: Springer-Verlag, 1960, pp. 446–778.

[355] **Weinblum, G.,** On problems of wave resistance research. *International Seminar on Wave Resistance.* Ann Arbor: University of Michigan, 1963, pp. 1–44.

[356] **Weinstein, A.,** Sur un problème aux limites dans une bande indéfinite. *C.R. Acad. Sci. Paris* **184** (1927) 497–499.

[357] **Werner, P.,** Randwertprobleme der mathematischen Akustik. *Arch. Rat. Mech. Anal.* **10** (1962) 29–66.

[358] **Whitehead, E. A. N.,** The theory of parallel-plate media for microwave lenses. *Proc. IEE (London)* **98** (1951) 133–140.

[359] **Whitham, G. B.,** *Lectures on Wave Propagation.* New York: Springer, 1979.

[360] **Whittaker, E. T., Watson, G. N.,** *A Course of Modern Analysis.* Cambridge: Cambridge University Press, 1927.

[361] **Wienert, L.,** An existence proof for a boundary-value problem with non-smooth boundary from the theory of water waves. *IMA J. Appl. Math.* **40** (1988) 95–112.

[362] **Wilcox, C. H.,** *Scattering Theory for Diffraction Gratings.* Berlin: Springer, 1984.

[363] **Williams, W. E.,** Waves on a sloping beach. *Proc. Camb. Phil. Soc.* **57** (1961) 160–165.

[364] **Wu, G. X.,** Wave radiation and diffraction by a submerged sphere in a channel. *Quart. J. Mech. Appl. Math.* **51** (1998) 647–666.

[365] **Yeung, R. W.,** A hybrid integral-equation method for time-harmonic free-surface flow. *Proceedings of the First International Conference on Numerical Ship Hydrodynamics.* Gaithersburg, MD: D. W. Taylor Naval Ship Res. & Devel. Center, 1975.

[366] **Zachmanoglou, E. C., Thoe, D. W.,** *Introduction to Partial Differential Equations with Applications.* Baltimore, MD: Williams and Wilkins, 1976.

[367] **Zargaryan, S. S., Maz'ya, V. G.,** On the asymptotics of the solutions of the integral equations of potential theory in a neighbourhood of angular points of the contour. *J. Appl. Math. Mech.* **48** (1985) 120–124.

[368] **Zhevandrov, P.,** Edge waves on a gently sloping beach: uniform asymptotics. *J. Fluid Mech.* **233** (1991) 483–493.

[369] **Zhevandrov, P., Merzon, A.,** On the Neumann problem for the Helmholtz equation in a plane angle. *Preprint* **2076** (2000). Technische Universität Darmstadt, Fachbereich Mathematik.

[370] **Zhevandrov, P., Dobrokhotov, S. Yu., Kuz'mina, V. M.,** Asymptotics of linear water waves over a gradually developing bottom. *Second International Conference on Mathematical and Numerical Aspects of Wave Propagation*. Philadelphia: SIAM, 1993.

Name Index

Abramowitz, M., 12, 145, 146, 150, 152, 159, 160, 163, 188, 310, 467
Akilov, G. P., 61, 371, 385, 388
Angell, T. S., 98, 109, 112, 140
Aranha, J. A. P., 224, 230
Athanassoulis, G. A., 141, 434
Aubin, J.-P., 105

Baar, J. J. M., 317
Bai, K. J., 113
Banerjea, S., 98, 213
Barber, N. F., 6
Bauer, 316
Beale, J. T., 6, 10
Bessho, M., 316, 416, 418
Bhattacharya, R. N., xii
Birman, M. S., 106, 225, 257
Bochner, S., 39, 314
Bolton, W. E., 218
Bonnet–Ben Dhia, A.-S., 220, 223, 229
Bouligand, G., 433
Brard, R., 16, 418
Burago, Yu. D., 96
Buslayeva, M. V., 433

Callan, M., 256
Carleman, T., 65, 96, 103, 375
Cauchy, A. L., xi
Chabat, B. V., 230, 235
Chudinovich, I. Yu., 433
Clarisse, J.-M., 317, 433
Colton, D., 53, 96
Courant, R., 22
Crapper, G. D., xi
Cross, R.H., 14

Davies, E. B., 256
Davis, A. M. J., 141
Dean, R. G., 6
Debnath, L., xii, 5
Dern, J. C., 360
Dobrokhotov, S. Yu., 433
Doppel, K., 98, 141

Eatock Taylor, R., 256
Eggers, K. W. H., 316, 416
Ehrenmark, U. T., 223
Euler, L., xi
Euvrard, D., 316, 434
Evans, D. V., xvii, 49, 98, 212–214, 216, 220–222, 229, 231–233, 255–258, 260–262

Fedoryuk, M. V., 291
Fernyhough, M., 232, 262
Feshbach, H., 193, 206
Feynman, R. P., xi
Fitz-Gerald, G. F., 97
Fox, D. D., 105
Friedman, A., 432
Friis, A., 141

Garipov, R. M., 230, 235, 432, 433
Gilbarg, D., 59, 90, 118, 140, 423, 424, 429
Gohberg, I., 55, 359
Gradshteyn, I. S., 12, 24, 25, 27, 35, 36, 40–42, 65, 131, 146, 182, 310, 314, 315, 374, 406, 407, 451, 466
Greenspan, H. P., 222
Grimshaw, R. H., 97, 220, 227, 228, 234
Groves, M. D., 218
Grue, J., 141
Gutmann, C., 359

Hadamard, J., 433
Hamdache, K., 17
Haskind, M. D. (Khaskind), xii, 48, 359
Havelock, T. H., xi, 48, 238, 239, 316, 417

Hazard, C., 14, 98
Hearn, G. E., 110
Hermans, A. J., xii, 418
Hilbert, D., 22, 55, 62, 257, 422
Hille, E., 254
Hochmuth, R., 141
Hogner, E., 316
Hou, T. Y., 6
Hsiao, G. C., xii, 98, 109, 112, 141, 418
Hulme, A., 48, 76, 97
Hurd, R. A., 262

Isakova, E. K., 433

Jami, A., 98, 434
John, F., xi, xiv, 5, 8, 10, 48, 99, 100, 116, 117, 120, 121, 126, 134, 136–140, 143, 164, 176, 177, 179, 180, 189–192, 195, 205, 208, 231, 238–242
Joly, P., 220, 223, 229
Jones, D. S., 113, 217, 229, 230

Kakilis, P. D., 141
Kamotskii, I. V., 256
Kantorovich, L. V., 61, 371, 385, 388
Keldysh, M. V., 317
Keller, J. B., 227, 417
Kellogg, O. D., 53, 56, 96, 357
Kelvin, Lord (Sir W. Thomson), xi, xiv, 265, 283, 316
Kenig, C. E., 96
Kirchgässner, K., 5
Kirchhoff, G. R., xi
Kleinman, R. E., 98, 109, 110, 112, 113, 141
Kochin, N. E., xi, 48, 51, 55, 87, 96, 316, 317, 359, 360
Komech, A. I., 219, 220, 222, 223
Kondratyev, V. A., 96, 101, 364
Kopachevskiy, N. D., 105
Korobkin, A. A., 433
Kostyukov, A. A., xii, 316, 317, 359, 360, 418
Kozlov, V. A., 55, 62, 92, 96, 101
Král, J., 96
Krein, S. G., 55, 105, 359
Kreisel, G., 97
Kress, R., 53, 96
Kuttler, J. R., 105
Kuz'mina, V. M., 433
Kuznetsov, N. G., 97, 140, 141, 192, 212–214, 216, 231, 237, 241, 359, 405, 415, 416, 484

Lagrange, J. L., xi
Lahalle, D., 359
Laitone, E. V., xi, xvi, 5, 48, 227, 315, 316, 359, 360, 418
Lamb, H., xi, 2, 3, 214, 437
Lau, S. M., 110, 432
Lavrentiev, M. A., 230, 235, 317
Le Méhauté, B., 2
Lee, S. W., 110, 262
Lehman, R. S., 220
Lenoir, M., 14, 48, 98, 141, 317, 416
Leppington, F. G., 141
Lesky, P. H., 218
Levi-Civita, T., 5
Levine, H., 171
Levitin, M., 256, 257
Lewy, H., 220
Le Blond, P. H., 214
Le Méhauté, B., 6
Liapis, S., 110
Licht, C., 10, 14
Lighthill, M. J., xi
Linton, C. M., xii, xvii, 98, 192, 213, 228, 231, 232, 255, 260–262
Liu, Y. W., 113
Livshits, M. L., 97, 359
Lowengrub, J. S., 6
Ludwig, D., 141

Makrakis, G. N., 434
Mandal, B. N., 98, 213
Maniar, H. D., 256
Martin, P. A., 114, 115, 140, 141, 232
Maskell, S. J., 434
Maz'ya, V. G., 48, 49, 54, 55, 62, 92, 96–98, 101, 165, 169, 213, 231, 235, 283, 316, 317, 359, 415, 416, 433, 484
McIver, M., xvii, 98, 116, 141, 212, 228, 256, 261
McIver, P., xii, xvii, 15, 141, 212, 213, 220, 229, 231–233, 238, 239, 255, 256
Mei, C. C., xii
Merzon, A. E., 220, 222, 223, 227
Meyer, R. E., 227
Michell, J. H., 417
Mihlin, S. G., 53, 55, 56, 96, 357
Miles, J. W., 227
Mittra, R., 262
Moiseev, N. N., 105
Morris, C. A. N., 49

Morse, M., 193, 206
Motygin, O. V., xvii, 213, 234, 238, 359, 405, 415, 416
Mysak, L. A., 214, 215

Nachbin, A., 484
Nakos, D. E., 417
Nalimov, V. I., 7
Nazarov, S. A., 96, 101, 256, 484
Nekrasov, A. I., 5
Newman, J. N., xii, xvi, 15, 17, 212, 213, 256, 311, 316, 317, 433
Ngo Zuy Can, 105
Noblesse, F., 316

Olver, F. W. J., 5, 42
Ovsyannikov, L. V., xii, 5, 7

Packham, B. A., 222
Pagani, C. D., 360, 417
Palm, E., 141
Papanicolaou, G. C., 484
Parker, R., 258
Parnovski, L., 256
Parsons, N., 232
Peters, A. S., 16, 221, 316
Petrovski, I. G., 63, 254, 321, 357
Pierotti, D., 360, 417
Pinkster, J. A., 15
Plamenevsky, B., 96, 101
Plamenevsky, B. A., 484
Poisson, S., xi
Politis, C. G., 141
Porter, D., xvii, 98, 212, 213, 231, 256
Price, W. G., 317
Protter, M. H., 228
Prudnikov, A. P., 41, 451

Quenez, J.-M., 360

Radon, J., 96
Reed, M., 225
Rhodes-Robinson, P. F., 141
Roach, G. F., 113, 141
Rodemich, E., 171
Roseau, M., 212, 221, 223
Rosenblat, S., 218
Rossmann, J., 92, 96, 101, 166, 169
Ryzhik, I. M., 12, 24, 25, 27, 35, 36, 40–42, 65, 131, 146, 182, 310, 314, 315, 374, 406, 407, 451, 466

Sabatier, P. C., 484
Sanchez-Palencia, E., 107, 140
Schomburg, B., 98, 141
Sclavounos, P. D., 110, 417
Shaposhnikova, T., 433
Sharma, S. D., 316
Shen, M. C., 98, 227
Shinbrot, M., 432
Simon, M. J., xvii, 97, 124, 140, 213, 225, 238, 241, 382
Sollitt, C. K., 14
Solomyak, M. Z., 106, 225, 257
Sretensky, L. N., xi, 49
Staziker, D. J., 98
Stegun, I. A., 12, 131, 145, 146, 150, 152, 159, 160, 163, 188, 190, 310, 467
Stoker, J. J., xi, xvi, 2, 3, 5, 16, 17, 316, 421, 437, 443, 449, 465, 478
Stokes, G. G., xi, 159, 160, 189, 214, 215, 219–221, 227, 241
Stoneman, S. A. T., 258
Struik, D. J., 5
Sun, S. M., 227
Suzuki, K., 416

Thoe, D. W., 254
Thorne, R. C., 48
Timman, R., xii, 418
Tolstova, O. L., 433
Tounsi, A., 141
Troesch, B. A., 143, 212
Trudinger, N. S., 59, 90, 118, 140, 423, 424, 429
Tuck, E. O., 316, 415
Tulin, M.P., 415

Ursell, F., xi–xiii, xvii, 6, 97, 98, 113–115, 124, 140, 141, 218, 219, 221, 226, 227, 229–234, 238, 245, 255, 283, 287, 316, 317, 382, 397, 407, 416, 433, 434
Utsunomiya, T., 256

Vainberg, B. R., 49, 96–98, 231, 235, 283, 291, 316, 317, 359
Vassiliev, D., 256, 257
Vladimirov, V. S., 53, 55, 56, 96, 357
Vullierme-Ledard, M., 10, 433

Watson, G. N., 36, 41, 484
Weck, N., 97

Wehausen, J. V., xi, xii, xvi, 5, 48, 227, 315, 316, 359, 360, 418
Weinberger, H. F., 228
Weinblum, G., 416, 418
Weinstein, A., 46, 48
Werner, P., 140
Whitehead, E. A. N., 262
Whitham, G. B., xi, 221
Whittaker, E. T., 36, 484
Wienert, L., 110, 111
Wilcox, C. H., 261
Williams, W. E., 222
Wu, G. X., 98

Yeung, R. W., 113
Yu, Y. S., 6

Zachmanoglou, E. C., 254
Zargaryan, S. S., 97
Zhevandrov, P. N., 223, 227, 433

Subject Index

Acceleration caused by gravity, 1, 2, 87, 265, 266, 349, 393, 422, 439
Airy function, 283, 286, 287, 293, 297
Artificial boundary, 224, 225
Assertion on continuity of solutions, 53
Asymptotic
  behavior, 21, 22, 27, 32, 39, 42, 117, 148, 152, 171, 188, 266, 283, 284, 287, 289–291, 295, 303, 314, 316, 319, 332, 404, 407, 433, 468, 487, 491
  formula, 12, 28, 41, 46, 66, 67, 69, 96, 181, 227, 283, 286, 309, 311, 317, 320, 324, 325, 327, 329–335, 341–343, 350–352, 364–366, 376–378, 394, 397, 412, 435, 438, 443, 460, 461, 464, 475, 481, 483
  representation, 283
Asymptotic behavior, 28
Asymptotics
  at infinity, 25, 327, 341, 343, 362, 379, 381
Auxiliary integral identity, 70, 134, 135, 231, 234
Axisymmetric
  problem, 97, 122, 130, 140, 142, 143, 158, 212, 213
  structures, 162
Azimuthal mode, 143, 163, 164, 188, 189, 212, 213
A priori estimate, 90

Banach space, 60, 62, 292, 370
Barrier, 98, 164, 165, 170–172, 213, 260, 262, 488, 489, 493, 495
Beam, 417
Bernoulli's equation, 3, 436
Bessel function, 12, 25, 26, 41, 132, 133, 159–161, 163, 186, 188, 231, 232, 259, 502
Bipolar coordinates, 201
Bottom, 1, 5, 9, 11, 42, 49, 50, 54, 58, 68–71, 78, 80, 81, 88, 92–95, 97, 98, 112, 117, 121, 124–126, 134, 137, 139, 156, 165, 169, 173, 175, 185, 192, 198, 199, 202, 203, 212, 215–217, 219, 220, 224–227, 230, 231, 234, 235, 241, 256, 261, 325, 341, 344, 350, 352, 360, 412, 413, 421, 432, 433, 455, 487, 488, 503
Bottom condition, 352
Bottom movement, 435, 439, 455
Boundary conditions, 3, 5, 43, 44, 46, 54, 69, 88–90, 94, 105, 118, 139, 171, 214, 230, 312, 344, 348, 361, 362, 416, 488, 491

Cauchy's inequality, 167, 174
Cauchy–Poisson problem, 4, 7, 433, 460, 477, 487, 491
Cauchy–Riemann equations, 338, 348, 368–370
Cauchy principal value, 23, 34, 39, 306, 312
Cauchy problem for the Laplace equation, 59, 169, 326, 348, 380, 402, 413, 414
Circular cylinder, 43, 49, 97, 229, 231, 255, 256, 258, 318, 337, 397, 416, 483, 486, 495, 500, 501
Circulation, 16, 360, 365, 398
Cliff, 156, 197, 220, 225, 226
Comparison method, 228
Complete elliptic integral, 41
Complex potential, 307, 434
Condition of the Kutta–Zhoukovsky type, 416
Continuity equation, 2, 436
Corner point, 51, 60–63, 65–67, 170, 171, 362, 403
Cutoff, 233, 236, 238–241, 244, 249, 255, 258, 496
Cutoff function, 278, 366
Cylindrical wave, 12, 37

Deep water, 11, 12, 15, 16, 21, 37, 39, 40, 42, 46, 48, 68, 69, 76, 80, 121, 124–127, 130, 143, 158, 175, 185, 192, 204, 205, 211, 213, 229, 230, 232, 239, 241, 255, 265, 305, 316–318, 351, 353, 359, 362, 398, 411, 412, 415, 450, 451, 466, 495, 500
Dipole, 177, 178, 180, 239, 407
Dirac's measure, 21, 22, 94, 281, 313, 459, 475
Direct integral equation, 54
Dirichlet–Neumann operator, 104, 106, 107, 426
Divergence theorem, 14, 23, 70, 103, 228, 234, 239, 351
Dynamic boundary condition, 3

Edge waves, 214–216, 219, 220, 222, 226, 227, 229, 233, 255, 262, 488–490, 497, 501, 502
Ellipse, 76, 77, 126
Energy, 4, 5, 13, 60, 71, 72, 101, 102, 105, 118, 142, 144, 166, 202, 208, 213, 217, 219–222, 227, 229, 232, 237, 318, 327, 390, 405, 438, 439, 442, 443, 446, 456, 457, 460, 465–467, 481
Energy conservation law, 443, 449, 457, 465
Energy norm, 446, 473
Error function, 451, 467
Essential norm, 62, 65, 103, 108, 371, 373, 387
Euler's constant, 151, 310
Exponential integral, 150, 406
Extended auxiliary integral identity, 135, 198, 241

Floating body, 6, 156, 434, 485, 488, 492, 496, 501
Flow of energy, 13
Forward speed, 265, 349, 361, 393, 395, 417, 490, 501
Fourier
  coefficients, 462, 476, 482, 483
  transform, 21, 24, 48, 141, 223, 268, 269, 274, 280–282, 313, 317, 423, 425, 450, 465, 473, 474, 480, 489
Fredholm
  operator, 62, 373
  theory, 58, 59
Fredholm's alternative, 50, 58, 61, 87, 103, 108, 358, 371, 373, 375, 387, 388, 399, 415

Freeg surface, 397
Free surface, 1, 3–5, 7, 9–11, 15, 17, 22, 39, 42, 48–50, 54, 56, 69, 70, 87, 88, 92–95, 99, 100, 105, 112, 113, 116, 117, 123–126, 130, 131, 134, 136, 137, 139, 142–145, 157, 162–165, 169–173, 177, 179, 185, 188–191, 193, 196, 198, 203, 205, 208, 209, 212, 215–217, 219, 224, 231, 232, 234, 238, 239, 241, 244, 245, 248, 249, 252, 261, 265–267, 283, 287–289, 309, 311, 316, 344, 349, 358, 363, 364, 370, 403, 407, 410, 412, 417, 421, 422, 426, 432, 433, 435–437, 442, 443, 451, 455, 456, 460, 464, 465, 468, 486, 490–492
  boundary condition, 8, 17, 31, 41, 52, 70, 72, 112, 130, 169, 177, 185, 274, 342, 356, 367
  elevation, 4, 7, 365, 370, 442, 443, 455, 456, 459, 468, 479

Galerkin approximation, 262
Gamma function, 292
Geometric criteria of uniqueness, 50
Green's
  formula, 51, 69, 92, 101, 103, 106, 122, 124, 127, 166, 167, 324–326, 330, 352, 366, 367, 379, 383, 388, 400, 428, 449, 456, 481
  function, 21–25, 28, 30, 32, 33, 36–40, 42, 43, 48–50, 52, 90, 93, 100, 105, 109–114, 139, 177, 232, 233, 245, 253, 265–268, 280, 283, 284, 305, 308, 309, 311, 312, 314–320, 324, 326, 327, 343, 349, 353, 354, 356, 359, 366, 370, 376, 378, 380, 381, 385, 389, 415
  representation, 21, 48, 109–112, 116, 319, 417

Half-axis, 56, 83, 142, 147, 149, 150, 184, 245, 270, 274, 297, 313, 407, 409
Half-plane, 38, 66, 114, 177, 194, 200, 209, 232, 237, 270, 283, 287, 306, 307
Half-space, 22, 32, 185, 206, 273, 281, 286, 287, 436, 473
H function, 359
Hankel function, 12, 13, 36, 131, 159
Harmonic function, 22, 24, 33, 43, 46, 47, 52, 53, 62, 67, 107, 124, 130, 138, 144, 148, 157, 177, 178, 253, 308, 312, 337, 338,

347, 357, 368, 380, 383, 392, 407, 424, 425, 460
Heaviside function, 147, 267, 309, 312, 313, 342, 347, 365
Helmholtz equation, 117, 255, 491, 503
Hilbert space, 446, 486, 489
Hypersingular operator, 109, 110, 112

Impulsive pressure, 433, 435–438, 448, 451
Initial-boundary value problem, 10, 14, 17, 421, 422, 433, 435, 441, 444, 447, 454, 459, 461, 463, 466, 472, 475, 477, 480
Initial conditions, 7, 9, 437, 441, 442, 445, 454, 455, 457, 464, 477
Integral
  equation, 50–61, 65, 87, 88, 96, 98–100, 102–104, 107, 109–113, 115, 116, 121, 122, 232, 233, 255, 262, 318, 321–323, 331–333, 336, 341, 345–348, 357–359, 370, 375, 380, 399, 417, 485, 487, 491, 492, 494–496, 499, 501, 502
  operator, 50–52, 54, 61, 64, 90, 100, 103, 109, 111, 112, 232, 318, 321, 333, 357, 359, 399, 492
Integroalgebraic system, 370–372, 377, 380, 385–387, 399, 404, 415
Inverse procedure, 158, 212, 361, 406, 417
Invertibility theorem, 55, 61, 62, 65, 96, 101, 103, 321, 323, 359, 373, 377, 399
Irregular frequency, 140, 494

Jacobian, 89, 200, 210
Jump formula, 311, 322, 332, 337

Kelvin's angle, 283, 285, 287, 288, 291, 295, 316
Kelvin's source, 283, 316
Kinematic boundary condition, 4
Kinetic energy, 4, 68, 105, 116, 165, 167, 173, 177, 186, 196, 362, 363
Knife-like ship, 417
Kronecker delta, 40, 46, 71, 186, 234

Laplace equation, 3, 24, 34, 41, 46, 47, 50, 70, 72, 88, 96, 118, 119, 167, 185, 216, 274, 281, 339, 356, 357, 362, 436, 462, 476
Layer of variable depth, 49, 70, 83, 85, 87, 97, 436, 502
Least singular solution, 397, 416

Linearization, 6–8
Line source, 21, 244, 305, 317
Lipschitz condition, 71, 253, 254
Local asymptotics, 170, 362, 366, 403, 411

McIver toroid, 498
Mean value theorem, 22, 47, 148
Method of stationary phase, 290, 291, 296–298
Mittag-Leffler's theorem, 36
Modified Green's function, 113, 115, 491
Moon pool, 212

Neumann–Kelvin problem, 16, 283, 284, 311, 313, 318, 319, 323, 324, 326, 337, 341, 342, 347, 349, 358–362, 395, 397, 398, 403, 405, 406, 411, 416, 417, 487, 493, 497, 498, 500, 501
Neumann condition, 8, 52, 53, 56, 58, 59, 61, 66, 69, 74, 92, 94, 95, 102, 105, 107, 112, 139, 142, 145, 160, 167, 216, 261, 330, 333, 342, 346, 347, 356, 364, 380, 384, 400, 403, 406, 449, 460, 462, 482
Nodal line, 138–140, 146, 148–156, 161, 179–181, 407, 410
Normal velocity, 4, 5, 9
No-flow condition, 3, 145, 341, 352, 363, 395, 412, 413
Null-field equations, 115, 495, 496

Obstacle, 4, 14, 21, 44, 50, 51, 65, 68, 98, 116, 121, 122, 125, 126, 142, 144, 169, 216, 231, 256, 421, 493, 495, 496

Parseval's equality, 423, 426, 466
Parseval theorem, 474
Pascal's snail, 402
Perturbation procedure, 7, 17
Piecewise smooth contour, 61, 65, 127
Point source, 21, 40, 48, 52, 265, 353
Poisson's equation, 385, 415
Potential energy, 4, 68, 97, 117, 122, 165, 166, 170, 173, 177, 196, 209, 210, 239, 241, 436, 456
Pressure, 1–3, 87, 88, 218, 316, 349, 393, 435–437, 439, 440, 442, 443, 450, 451, 455, 460, 461, 464, 467, 468, 487
Progressive wave, 12, 13, 49

Radiation condition, 13, 14, 16, 25, 28, 31, 41, 42, 44–46, 54, 89, 94, 95, 118, 159, 167, 170, 245, 343, 381, 391
Rate of flow, 320, 365, 370, 394, 398
Rayleigh–Bloch wave, 256
Rayleigh quotient, 257
Ray theory, 417
Regular normal derivative, 52, 58, 61, 102, 323
Rellich's radiation condition, 42
Residue theorem, 184, 270, 273
Resistance, 311, 351, 353, 361, 393–396, 402–404, 415–417, 498
Resistanceless potential, 404, 406, 415, 493
Riesz theorem, 425
Rigid body, 10, 15, 17, 143, 155, 157, 351, 353, 436, 499
Ring source, 21, 40
Ring Green's function, 40–42, 48
Ring source, 159, 163, 190, 191
Roseau's method, 221
Runge–Kutta scheme, 246

Schwarz inequality, 31, 106, 117, 119, 120, 123, 128, 132, 139, 168, 174, 186, 187, 194, 236, 348, 384, 414
Semisubmerged body, 99, 500
Shallow water, 265, 305, 311, 313, 317, 318, 346, 349, 359, 415, 417, 501
Shell, 164, 165, 169, 213, 493
Ship wave, 265, 316, 318, 417, 486, 490–492, 496, 498, 499
Simple wave, 46, 48, 116–120, 166, 173, 238, 240
Single-layer potential, 51, 54, 56, 66, 88, 93, 100–102, 104, 115, 233, 321, 322, 331, 349, 356, 370, 385, 415
Sloping beach, 214, 216, 219–222, 227, 488, 494, 497–502
Sloshing problem, 105, 143, 176, 212, 229
Sobolev's embedding theorem, 423
Sobolev space, 88, 118, 165, 221, 423, 445, 469
Solvability, 21, 50, 51, 55, 57, 58, 60, 65, 76, 87, 88, 91, 92, 95–97, 99, 103, 104, 110, 113, 116, 120, 121, 140, 142, 265, 318, 319, 322, 323, 326, 334, 336, 337, 341, 345, 358, 359, 362, 373, 375, 377, 380, 381, 387–390, 397–401, 404, 405, 411, 414, 424, 427, 431, 441, 485, 493, 495–498
Sommerfeld's radiation condition, 28, 45, 52
Source's track, 283, 285, 287, 288, 295, 316
Spray resistance, 361, 394–396, 399, 402, 411, 416, 417
Standing wave, 12, 13
Steady-state problem, 10, 421
Streamline, 247–253, 406, 407, 410, 411, 491
Stream function, 130, 144, 145, 148, 150, 151, 157, 159–162, 178, 189–191, 247, 338, 347, 362, 368, 406, 412, 413, 416
Subcritical stream, 345, 352, 359, 360, 497
Submerged body, 50, 51, 59, 67, 68, 71, 78, 80, 81, 87, 97–100, 102, 125, 126, 140, 169, 212, 318, 349, 359–362, 378, 437, 442, 443, 460, 464, 474, 479, 484, 485, 487–490, 493, 496, 500
Supercritical stream, 341, 359–361, 411, 413, 414, 416, 493
Supplementary conditions, 16, 318, 361, 362, 364, 368, 378, 393–398, 400, 401, 403–406, 411, 412, 415–417
Surface-piercing body, 60, 68, 81, 99, 100, 104, 109, 116, 117, 121, 125, 134–140, 143, 144, 146, 164, 165, 172, 177, 196–198, 202, 204, 212, 239–241, 318, 359, 361, 382, 405, 411, 416, 417, 433, 493, 497

Theta functions, 484
Time-harmonic waves, 265, 318, 421
Toroidal body, 143, 164, 165, 205, 208, 209, 500
Trapped mode, 142, 143, 146, 160, 162, 164, 176, 185, 189–191, 212, 214–216, 219, 220, 224–232, 234, 236–238, 240, 249, 252, 255, 256, 258–262, 405, 406, 486–490, 493, 495–497, 499, 501
Two-dimensional problem, 1, 2, 16, 21, 37, 38, 68, 71, 76, 115, 121, 122, 140, 141, 172, 191, 214, 216, 238, 242, 246, 252, 305, 318, 349, 432, 433, 451, 461, 475, 492, 493, 497
Two-scale asymptotic expansion, 435, 438, 439

Uniqueness of solution, 50, 55, 84, 113, 136, 142, 241, 330, 341, 359
Unsteady waves, 17, 421, 422, 433, 434

Vector field, 67, 71, 75, 76, 78–82, 84, 85, 89, 136, 234, 235, 237
Velocity potential, 3, 4, 7, 9, 10, 12, 15, 16, 21, 22, 37, 42, 44, 48, 66, 67, 93, 95, 109, 112, 116, 130, 138, 144, 145, 158, 160, 178, 216, 217, 222, 240, 246, 265, 267, 283, 312, 316, 330, 353, 362, 363, 395, 396, 410–412, 416, 417, 422, 435, 436, 438, 440, 442, 443, 447, 453, 459–461, 464, 474, 475, 480
Vertical cylinders, 169, 188, 215, 216, 256, 257, 260–262
Vortex, 407

Water
  depth, 6, 15, 52, 57, 101, 157, 172, 198, 261
  domain, 11, 13, 42, 49–51, 54, 68, 70, 71, 81, 87, 88, 93, 95, 116, 117, 122, 127, 130, 134, 135, 138, 140, 159, 160, 165, 170–172, 178, 179, 185, 192, 200, 202, 204, 205, 207, 209, 215, 219, 221, 224, 232, 237, 249, 319, 341, 353, 362, 403, 407, 410, 411, 421, 450, 465, 475
Water-wave problem, 4, 5, 14, 16, 21, 50–53, 55–61, 65, 68, 70–72, 74–76, 79, 81–88, 92, 93, 95–104, 107–109, 111–113, 115, 116, 118, 120, 121, 123–125, 127, 134–136, 138, 140–142, 146, 158, 166, 167, 171–173, 175–178, 185, 188, 193, 194, 199, 202–205, 208, 210, 212, 213, 231, 235, 239, 241, 311, 318, 361, 382, 385, 406, 417, 489, 493, 494, 496, 497, 500
Water domain, 11, 13
Watson's lemma, 42
Wave
  height, 6
  resistance, 318, 349, 352, 359–361, 399, 416–418, 486, 487, 491–493, 497, 498, 502
Wave-making resistance, 394, 396, 416, 486
Wavelength, 6, 218
Waveless potential, 398–400, 402, 415, 474
Wave making resistance, 439
Wiener–Hopf technique, 260
Wronskian, 133, 160, 161